Dielectric Materials and Technology

Dielectric Materials and Technology

Edited by **Doreen Rowe**

NY RESEARCH PRESS
New York

Published by NY Research Press,
23 West, 55th Street, Suite 816,
New York, NY 10019, USA
www.nyresearchpress.com

Dielectric Materials and Technology
Edited by Doreen Rowe

International Standard Book Number: 978-1-63238-508-6 (Hardback)

Printed in the United States of America.

Contents

Preface

This book was inspired by the evolution of our times; to answer the curiosity of inquisitive minds. Many developments have occurred across the globe in the recent past which has transformed the progress in the field.

This book covers in detail some existent theories and innovative concepts revolving around dielectric materials and technology. The various advancements in this area are glanced at and their applications as well as ramifications are looked at in detail, in this text. Dielectric materials are those materials which are used to hold electrostatically charged objects because they are bad conductors of electricity. These materials are generally known as electric insulators and they can be polarized by applied electric field. Some of the common dielectric materials are glass, mica, rubber, porcelain, dry air, etc. As this field of study is emerging at a rapid pace, the contents of this book will help the readers understand the modern concepts and technological developments related to the subject. Students, researchers, experts and all associated with this field will benefit alike from this book.

This book was developed from a mere concept to drafts to chapters and finally compiled together as a complete text to benefit the readers across all nations. To ensure the quality of the content we instilled two significant steps in our procedure. The first was to appoint an editorial team that would verify the data and statistics provided in the book and also select the most appropriate and valuable contributions from the plentiful contributions we received from authors worldwide. The next step was to appoint an expert of the topic as the Editor-in-Chief, who would head the project and finally make the necessary amendments and modifications to make the text reader-friendly. I was then commissioned to examine all the material to present the topics in the most comprehensible and productive format.

I would like to take this opportunity to thank all the contributing authors who were supportive enough to contribute their time and knowledge to this project. I also wish to convey my regards to my family who have been extremely supportive during the entire project.

Editor

1

Dielectric and Electric Modulus Behavior of Chlorinated Poly(Vinyl Chloride) Stabilized with Phenyl Maleimide

Taha A. Hanafy[1,2]
[1]Department of Physics, Faculty of Science, Tabuk University, Tabuk, KSA
[2]Physics Department, Faculty of Science, Fayoum University, El Fayoum, Egypt
Email: tahanafy2@yahoo.com

ABSTRACT

Dielectric constant, ε', dielectric loss factor, ε'', electric modulus, *M*, and ac conductivity, σ_{ac}, of pure CPVC and that stabilized with 10 wt% of phenyl maleimide, PM, have been carried out. The dielectric properties have been studied in the temperature and frequency ranges; 310 K - 450 K and 1 kHz- 4 MHz, respectively. The incorporation of 10 wt% of PM as stabilizer for CPVC leads to reduce its *Tg* from 405K to 378K at 10 kHz. PM molecules within CPVC structure reduce the double bond, stabilizer effect, and cause the widely spacing between CPVC main chains, plasticizer effect. Three dielectric relaxation processes namely ρ, α', and α were observed for pure CPVC. The first process was explained based on space charge formation or Maxwell-Wagner-Sillers, MWS, polarization. The second one is due to the segmental motion of the branching of CPVC. The third process occurs around the glass-rubber temperature, *Tg*, and is related to the micro-Brownian motion of the main polymer chain. Electric modulus and ac conductivity reveal that the conduction mechanism of CPVC is follow the correlated barrier hopping, CBH, while stabilized sample exhibits a quantum mechanical tunneling, QMT, type conduction.

Keywords: CPVC; Stabilizer; Plasticizer; Dielectric Relaxation; Hopping Conduction

1. Introduction

Chlorinated polyvinyl chloride, CPVC, is produced by post-chlorination of polyvinyl chloride, PVC. CPVC has high heat distortion temperature, chemical inertness, and outstanding mechanical. It has good flame retardant and smoke density properties [1-5]. Chlorination of PVC enhances the mechanical properties of the resulting polymer. Chlorination of PVC increases the glass rubber transition temperature, *Tg*, [4] from 80˚C to 90˚C and makes the materials suitable for use in high temperature applications [6]. Both PVC and CPVC can not be processed without the addition of stabilizer and plasticizer. Commercial CPVC stabilizers are usually either basic lead salt [7,8], which can trap the evolved HCl gas, thus delaying the polymer degradation or metallic soap esters [9]. The stabilizing action occurs through displacement of the labile chlorine atoms on the polymer chains by the ester from the decomposed stabilizer. Recently, it was reported that the incorporation of organic materials like N-phenyl maleimides as a stabilizers for PVC and CPVC [10,11]. Organic stabilizers have the advantage of being metal free and environmentally acceptable. N-phenyl maleimides are known have high thermal stability [9,10]. These stabilizers trap the radical species in the degradation process by blocking the newly formed radical sites on the polymeric chains and absorption of liberated HCl gas.

Dielectric relaxation spectroscopy, DRS, of polymers provides information about the orientation and translational adjustment of mobile charge present in the dielectric medium. The energy transferred to the dielectric material is a function not only of the applied electric field but also depends on the physical properties of the material. The dependence of contributions of different components of dielectric polarization, such as electric, ionic, and orientational polarization on the frequency of the applied electric field, is responsible for the change in the value of the dielectric constant. The relative permittivity, $\varepsilon^*(\omega)$, of a dielectric as a function of the frequency is given by [12]:

$$\varepsilon^* = \varepsilon'(\omega) - i\varepsilon''(\omega) \quad (1)$$

where $\varepsilon'(\omega)$ is the dielectric constant characterizes the most important property of the dielectric material. The dielectric loss index, $\varepsilon''(\omega)$, characterizes the dissipation of energy of the electric oscillation within the dielectric medium. The magnitude of $\varepsilon'(\omega)$ of the polymeric material is determined by its chemical constitution, structure, and composition. The parameters that characterize the dielectric loss index and the dielectric loss tangent, $\tan(\delta)$, depend upon the specific features of molecular main chain motion within the polymer. Real dielectric polymers are commonly described by a spectrum of relaxation times. The relaxation spectra appear due to presence of long polymeric chain and specific inter-molecular interactions. The simplest form of $\varepsilon'(\omega)$ and $\varepsilon''(\omega)$ for the relaxation process that characterized by a single relaxation time (τ) is given by[12]:

$$\varepsilon'(\omega) = \varepsilon'(\infty) + \frac{\Delta\varepsilon}{1+\omega^2\tau^2} \quad (2)$$

$$\varepsilon''(\omega) = \frac{\Delta\varepsilon\omega\tau}{1+\omega^2\tau^2} \quad (3)$$

where $\Delta\varepsilon = \varepsilon'(o) - \varepsilon'(\infty)$ is the dielectric strength, $\varepsilon'(o)$, and $\varepsilon'(\infty)$ are the relaxed and unrelaxed dielectric constant, respectively.

Different experimental techniques, including differential scanning calorimetry, DSC, dynamic mechanical thermal analysis, DMTA, and dielectric relaxation spectroscopy, DRS, are used for the determination of glass rubber transition, Tg of the polymeric materials [13]. In this work the temperature and frequency dependence of ε'' is used to determine Tg of the investigated samples, where the α-relaxation process is accompanied with the Tg of the polymeric material. Also, we investigate the effect of an additive phenyl maleimide to improve the performance of CPVC. Also, the dielectric properties, electric modulus, and ac conductivity of pure CPVC and that stabilized with 10 wt% of PM were investigated to shed light on their molecular relaxation and conduction mechanism.

2. Experimental

CPVC used in the present work was supplied by Weihai Jinhong Chemicals Ltd. with chlorine content 69% and density 1.4 g/cm^3. Phenyl maleimide was prepared by the method described by Searle [10,14]. CPVC film stabilized with 10 wt% of PM was prepared by casting as follows: CPVC was dissolved in tetrahydrofuran (THF, Aldrich) at 50˚C for 30 min with continuous stirring. A 10 wt% of PM was also dissolved in THF at 50˚C and added to CPVC solution. The mixture was left 8 h, with continuous stirring, to reach a suitable viscosity. The solution of the mixture was cast into a Petri dish placed on a leveled plate at 30˚C for 10 days until the solvent was completely evaporated. The obtained film of 0.1 mm thickness was cut into square pieces and then coated with silver paste to achieve ohmic contacts.

The dielectric measurements were carried out with a Hioko 3532 RLC bridge (Nagano, Japan) in the frequency and temperature ranges of 50 kHz to 1 MHz and 300 - 450 K, respectively. The dielectric constant, dielectric loss index, and alternating current conductivity, σ_{ac}, were calculated by [5]:

$$\varepsilon' = \frac{Cd}{\varepsilon_o l} \quad (4)$$

$$\varepsilon'' = \varepsilon' \tan(\delta) \quad (5)$$

$$\sigma_{ac} = \omega\varepsilon_o\varepsilon'' \quad (6)$$

where C is the capacitance of the sample-filled capacitor, d is the sample thickness, ε_o is the permittivity of the free space, l is the electrode area, and ω is the angular frequency. The temperature of the sample was measured with a T-type thermocouple. The measurement accuracy for the temperature was about ±1 "K".

3. Results and Discussion

3.1. Dielectric Formalism

The frequency dependence of ε' for pure CPVC and that stabilized with PM at some fixed temperatures is shown in **Figures 1(a)** and **(b)**. It is clear that the values of ε' decrease with increasing the frequency due to decreasing number of dipoles which contribute to polarization. At very low frequency ($\omega \ll 1/\tau$), where τ is the relaxation time, the dipoles can follow the field and $\varepsilon' \approx \varepsilon_o$. As the frequency increases ($\omega < 1/\tau$), the dipoles begin to lag the field and ε' slightly decreases. When the frequency reaches a characteristic value ($\omega = 1/\tau$), the dielectric constant exhibits relaxation process [15]. The dielectric constant of the stabilized CPVC sample increases slightly compared to that of pure one. This can be discussed as follows: the stabilized molecules are compatible with CPVC and are distributed inter-structurally within the free volume of the non dense globular areas of CPVC [5,16]. This will increase the segmental mobility of the non array macromolecules and passage chains [17]. In addition, with the increase of temperature the thermo-mechanical motion of PM molecules become significance and cooperative nature. Also, PM molecules can form channel cluster in the polymeric glassy state [18,19]. These clusters have many polar groups (C=O and phenyl groups) and began to move at high temperature in cooperation with C-Cl dipolar groups of CPVC. This will increase ε' of the stabilized CPVC sample.

Figures 2(a) and **(b)** represent the frequency dependence of ε'' for pure CPVC and that stabilized with 10 wt% of PM at some fixed temperatures. From **Figure 2 (a)**, ε'' of pure CPVC undergoes three relaxation processes namely, ρ-, α'-, and α-peaks. The first is located at 10 kHz. It originates from the motion of space charges that are accumulated on the polymeric sample close to electrode [20]. The nature of ρ-relaxation process for crystalline regions is due to the chain trapping of interfaces or Maxwell-Wagner-Sillers, MWS, polarization [21].

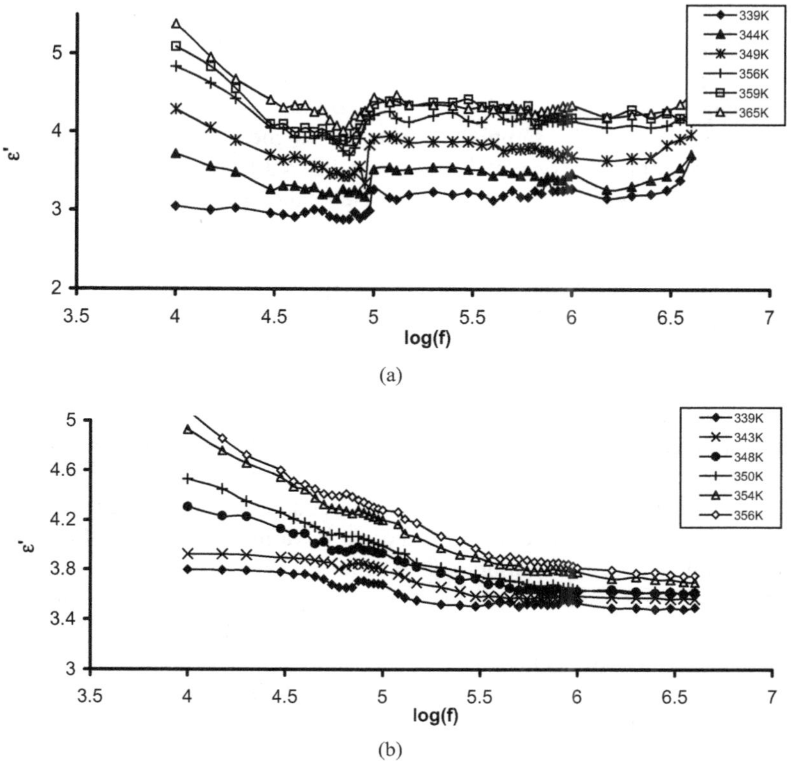

Figure 1. Frequency dependence of ε' (a) Pure CPVC and (b) CPVC stabilized with 10 wt% of PM at some constant temperatures.

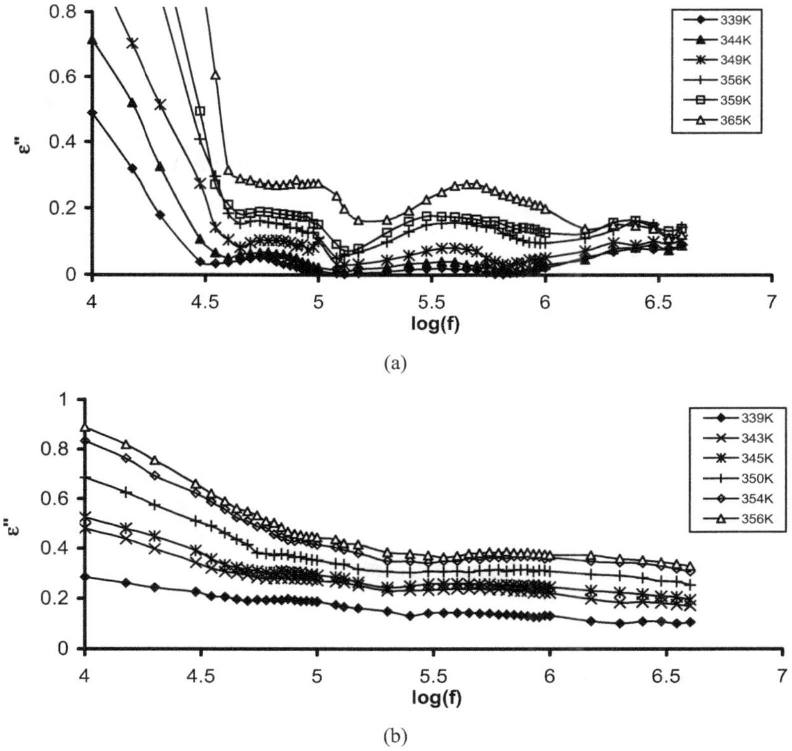

Figure 2. Frequency dependence of ε'' (a) Pure CPVC and (b) CPVC stabilized with 10 wt% of PM at some constant temperatures.

However, in the amorphous phases, ρ-process is related to conductive impurities, injected space change and electrode polarizations. In semi-crystalline polymer such as PVC and CPVC, chain trapping or MWS polarization is more acceptable [20]. For stabilized CPVC sample, ρ-relaxation can be related to the multiple phases which come from the channel cluster formation within the sample [18,19]. The second process, α'-relaxation, is observed at 80 kHz. The process has been reported for PVC [17,21] but this effect is weak and is not always observed in comparison with the main α-process. It has been found that, α'-process is present in methyl ester form polymer. It represents the cooperative dipole motion of C-O-CF_2-CF-(CF_3). For the present work, α'-relaxation can be assigned to the cooperative dipole motion of CPVC end groups [21] or the segmental motion of branching of CPVC. However, this effect in the present work is very weak for the stabilized CPVC sample. This may refer to the crooslinking formation between maleimide molecules and the π-bond of CPVC main chains. Crosslinking hinders the segmental motion of the branching of CPVC main chins. So, the α' process has little chance to appear.

The third process, α-peak, is obtained at 560 kHz. The peak position of this process shifts towards higher temperature with increasing frequency. This behavior results in the micro-Brownian motion between the polymeric main chains [5]. The peak height of α-relaxation process for stabilized CPVC sample is higher than that obtained for pure component. Also, the shape of α-peak for stabilized sample is broader than that for pure CPVC. This can be assigned to the diminution of the crystallinity of the stabilized CPVC sample, which is caused by the increase of the crosslinking density [5,20]. However, there is a competitive action of least three factors during the crosslinking:

1) Diminution of the existing physical network due to the hydrogen bonding;

2) Formation of a chemical network;

3) Crosslinking formation inside the amorphous regions.

Figures 3(a) and **(b)** show the temperature dependence of ε' for pure CPVC and that stabilized with 10 wt% of PM at some fixed frequencies. It is clear that, the values of ε' varies slowly with the increase in the temperature up to $Tg \approx 363K$. After Tg, the magnitude of ε' increases gradually with further increasing in the temperature. Then, the dipoles of CPVC have sufficient energy, via heating, to orient themselves easily in the direction of the applied filed. So, the chain segments get sufficient thermal energy to speed up its rotational motion. This leads to increase in the polarization of the investigated sample [4,5]. Moreover, the increase of ε' with the temperature can be assigned to the disentanglement of the

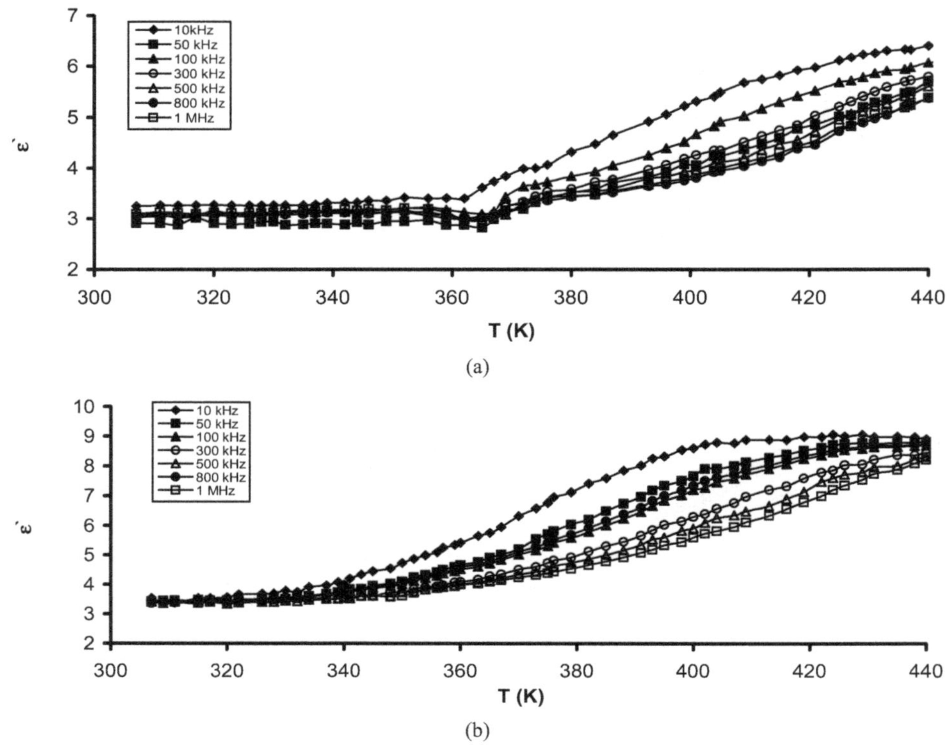

Figure 3. Temperature dependence of ε' (a) Pure CPVC and (b) CPVC stabilized with 10 wt% of PM at some constant frequencies.

molecular chains. It becomes easier due to the temperature induced molecular vibrations whereas at low temperature the chains are tightly packed and strongly held [8].

Figures 4(a) and **(b)** show the temperature dependence of ε'' for pure CPVC and that stabilized with 10 wt% of PM at some fixed frequencies. It is observed that, ε'' for both pure and stabilized CPVC sample undergoes α-relaxation process. The peak of this process for pure and stabilized CPVC sample was obtained at 405 K and 387 K, at 10 kHz, respectively. Consequentially, the addition 10 wt% of PM to CPVC structure leads to decrease Tg of CPVC by about 27 K. So, one can suggest that, PM molecules increase the free volume within the CPVC structure. In addition, the increment of the free volume lowers the chain-chain interaction within the backbone structure of CPVC. Then, PM molecules may screen the polar groups of CPVC and prevent the formation of polymer-polymer bond. Also, it indicates that PM molecules behave as a plasticizer for CPVC. The plasticization effect was observed earlier for the addition of Cadmium laurate [22], Gadolinium chloride $GdCl_3$ [20] and bisphenol A corncobs [21] as stabilizers for PVC.

The temperature dependence of the maximum loss peak frequency f_{max} is displayed in **Figure 5**. The behavior of f_{max} can be described by Arrheniuss relation[23]:

$$f_{max} = f_o \exp\left(-\frac{E_a}{kT}\right) \quad (7)$$

where f_o is constant, k is the Boltzmann's constant and E_a is the activation energy. The calculated values of E_a for pure and stabilized CPVC sample are 307 "kJ/mol" and 152 "kJ/mol" respectively. The values of E_a are disagreement with those reported in literature [4,20,24-26]. This may be assigned to the presence of the micro-inhomogeneities inside the investigated sample [14].

The temperature dependence of $\Delta\varepsilon$ for pure CPVC and that stabilized with 10 wt% of PM is displayed in **Figure 6**. It is seen that, $\Delta\varepsilon$ is proportional to the temperature. This can be attributed to the cooperation that exists between thermal energy and the electric field effects for dipole alignments. The increment of the thermal energy of the C-Cl, C=O, and phenyl groups will tend to enhance the alignment of themselves with the direction of the electric field and therefore the enhancement of $\Delta\varepsilon$ will expected.

3.2. Electric Modulus Formalism

The complex modulus, M^*, formulism is a very important and convenient tool to analyze and interpret the dielectric relaxation of polymeric material [4,5]. The main advantage of M^* formulism is that the electrode polarization effect can be suppressed [27]. As the electrode effect is very important in the system, we have analyzed the dielectric spectrum by M^* which can be evolved from the following relations [4.5]:

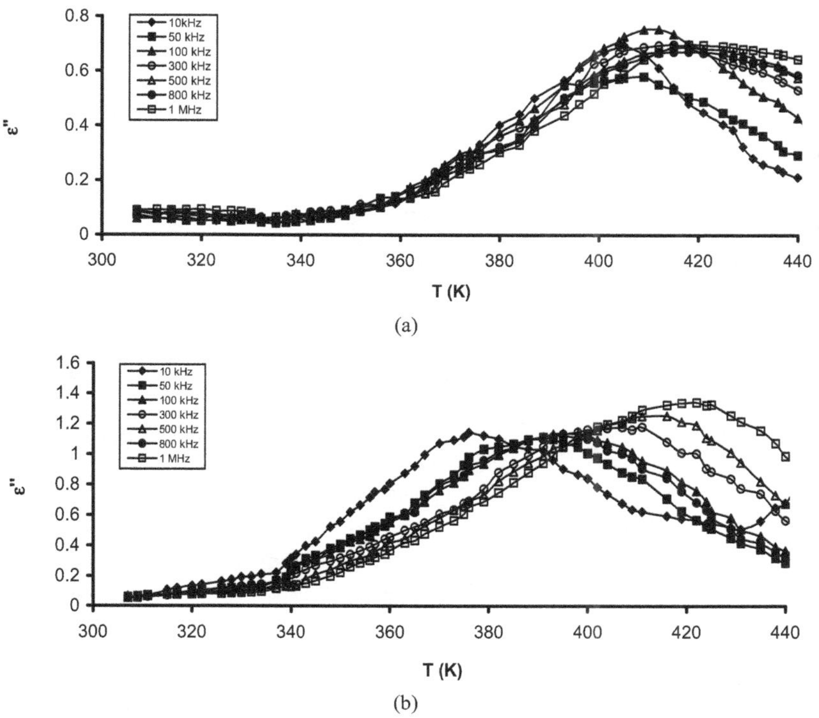

(b)

Figure 4. Temperature dependence of ε'' (a) Pure CPVC and (b) CPVC stabilized with 10 wt% of PM at some constant frequencies.

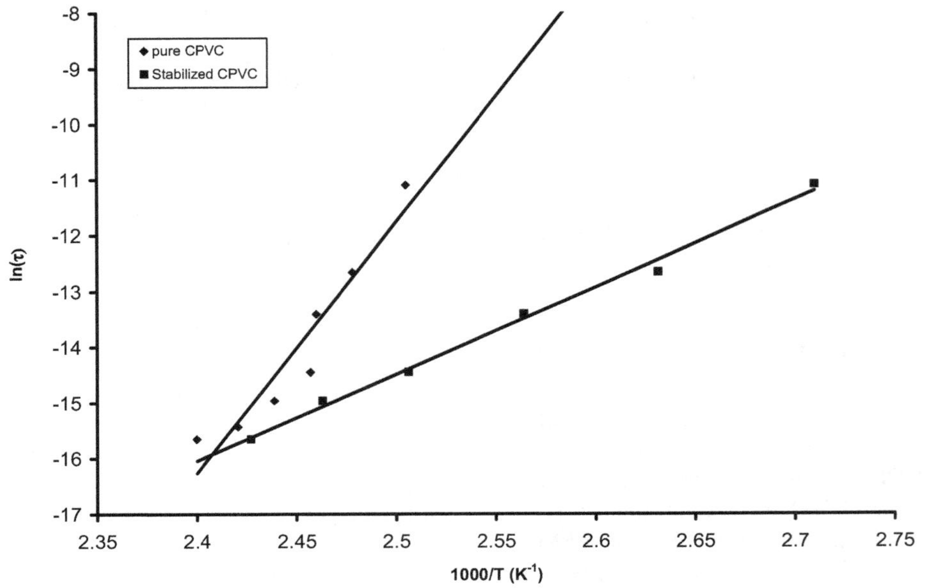

Figure 5. The relation between ln(τ_σ) and 1000/T for pure CPVC and that stabilized with 10 wt% of P.

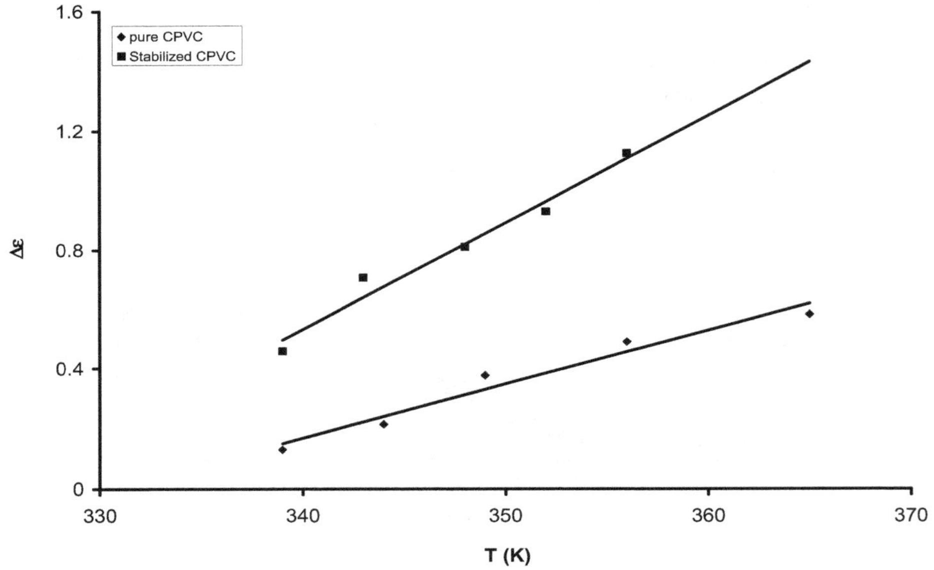

Figure 6. The temperature dependence of $\Delta\varepsilon$ for pure CPVC and that stabilized with 10 wt% of PM.

$$M' = \frac{\varepsilon'}{\left(\varepsilon'^2 + \varepsilon''^2\right)} \quad (8)$$

$$M'' = \frac{\varepsilon''}{\left(\varepsilon'^2 + \varepsilon''^2\right)} \quad (9)$$

where M' and M'' are the real and the imaginary parts of the dielectric modulus, respectively. The frequency dependence of M' for pure CPVC and that stabilized with 10 wt% of PM at some fixed temperatures is shown in **Figures 7(a)** and **(b)**. As the frequency increased M' of all CPVC samples is slightly increased and then reach to approximately constant value. The temperature dependence of M' for pure CPVC and that stabilized with 10 wt% of PM at some fixed frequencies is shown in **Figures 8(a)** and **(b)**. It is noticed that, at lower temperatures M' exhibits nearly constant value in the temperature ranges of 310 - 360 K and 310 - 330 K for pure and stabilized CPVC sample, respectively. This indicates that, the thermal stability of pure CPVC is higher than that obtained for the stabilized one. Moreover, the plasticization effect of PM molecules plays an important role for this process. At higher temperatures, M' for pure and stabilized CPVC samples tends to reach a constant value. This

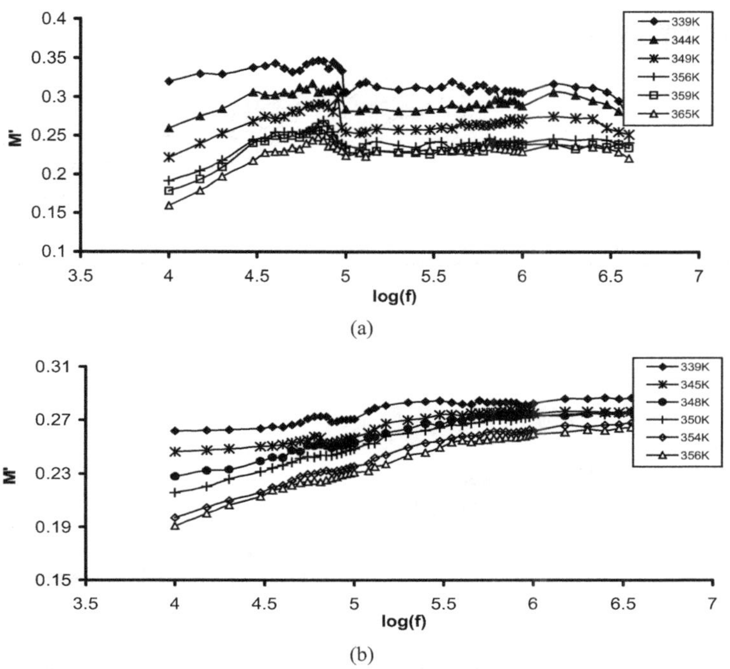

Figure 7. Frequency dependence of M' (a) Pure CPVC and (b) CPVC stabilized with 10 wt% of PM at some constant temperatures.

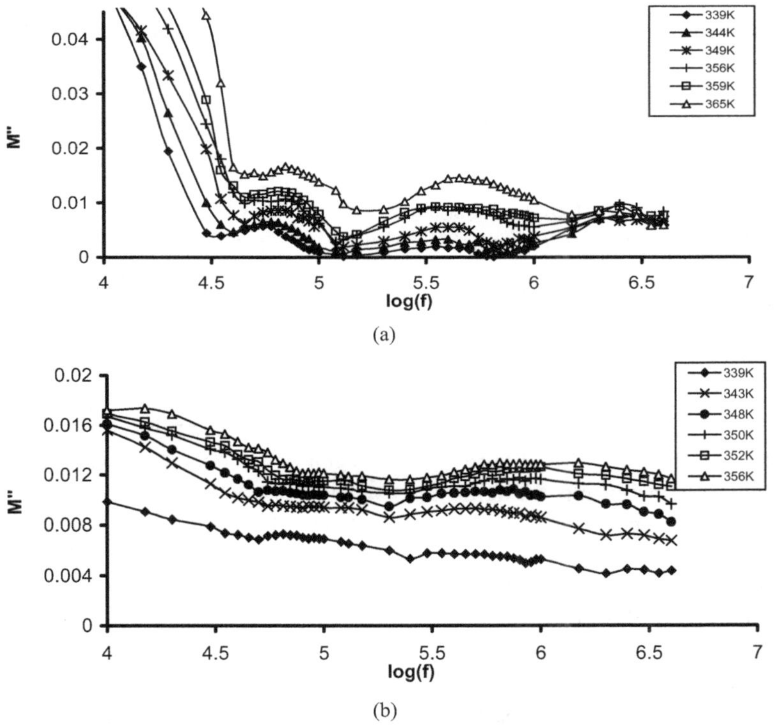

Figure 8. Frequency dependence of M'' (a) Pure CPVC and (b) CPVC stabilized with 10 wt% of PM at some constant temperatures.

indicates thermally activated nature of the dielectric constant. In addition, the behavior of M' as a function of frequency, **Figure 6**, and as a function of temperature, **Figure 7**, reveals that there is a role of electrode polarization for the dielectric relaxation.

Figures 9(a) and **(b)** illustrates the frequency dependence

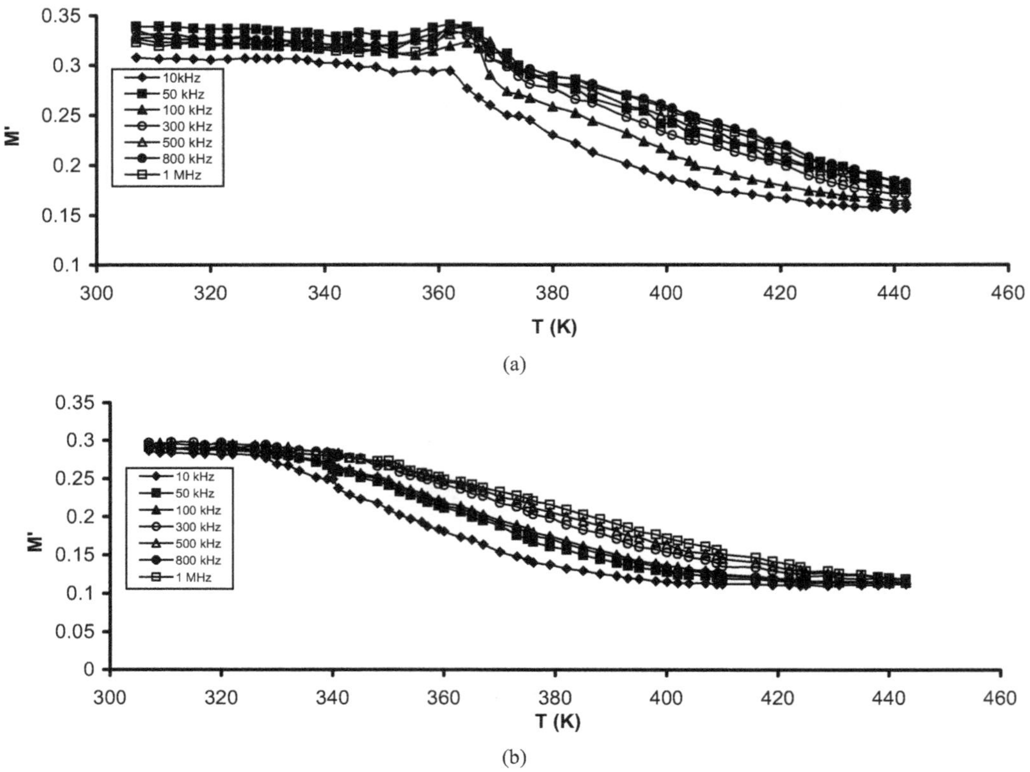

Figure 9. Temperature dependence of M' (a) Pure CPVC and (b) CPVC stabilized with 10 wt% of PM at some constant frequencies.

of M'' for pure CPVC and that stabilized with 10 wt% of PM at some fixed temperatures. The behavior of M'' for all CPVC sample exhibits same trend that obtained for the frequency dependence of ε'', **Figure 2**. Three peaks ρ-, α'- and the main α-relaxation process were observed for pure CPVC at 20 kHz, 63 kHz, and 400 kHz, respectively. Also, M'' of stabilized CPVC exhibits ρ-, α-process at 25 kHz and 1 MHz, respectively. This pattern provides information relating charge transport processes such as mechanism of electrical transport, conductivity relaxation, and ion dynamics as a function of temperature and frequency. The peak position of α-process shifts toward the higher frequency side on increasing the temperature. The low-frequency side of the peak signifies the range of frequency in which ions can perform successful hopping from one site to the neighboring site. The high-frequency side of M'' peak represents the range of frequency in which the ions are spatially confined to their potential wells and the ions can make localized motion within the well. The region where the peak occurs is indicative of the transition from long-range to short-range mobility with increase in temperature. This type of behavior of the modulus spectrum is suggestive of a temperature-dependent hopping type mechanism for electrical conduction in the system [28]. The broadening of the peak points out the spread of relaxation with different mean time constants and non-Debye type of relaxation in CPVC.

The temperature dependence of M'' for pure and that stabilized CPVC sample at some fixed frequencies is shown in **Figures 10(a)** and **(b)**. It is noticed that M'' for both pure and stabilized sample undergoes the main α-relaxation process at 395 K and 365 K for pure and stabilized CPVC sample, respectively. This is due to the release of frozen-in all of the dipolar groups and their cooperative motion with adjoining segments of the main chains [29-31]. The relaxation time, τ, has been calculating using the relaxation $\omega_{max}\tau = 1$ where ω_{max} is the angular frequency at the peak of M''. The temperature dependence of τ for pure and stabilized CPVC is shown in **Figure 11**. The variation of τ with temperature follows the Arrhenius Equation [31]:

$$\tau = \tau_0 \exp\left(\frac{E_a}{kT}\right) \quad (10)$$

where τ_0 is constant. The calculated values of E_a are 300 "kJ/mol" and 150 "kJ/mol" for pure and stabilized CPVC sample, respectively.

On the other side the decrease of Tg for CPVC stabilized CPVC sample can be assign to dienopholic property of PM [9,10]. This property enables PM molecules to intervene within the conjugated CPVC system. This

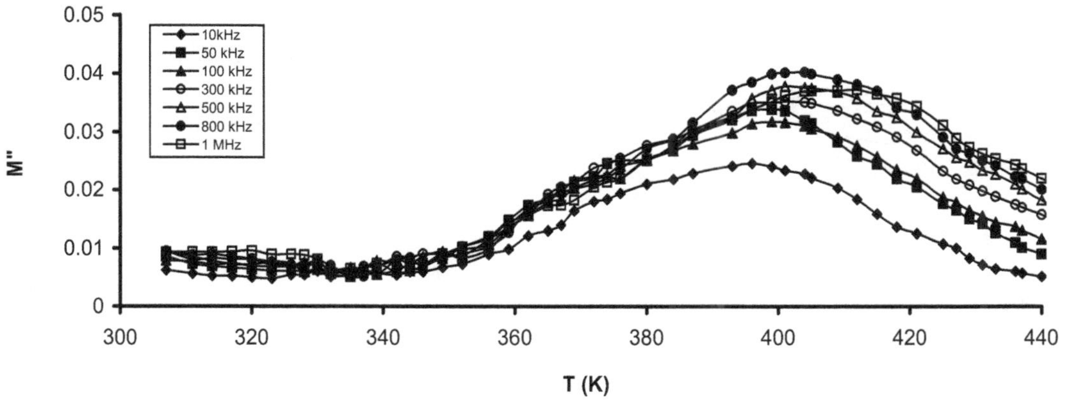

(a)

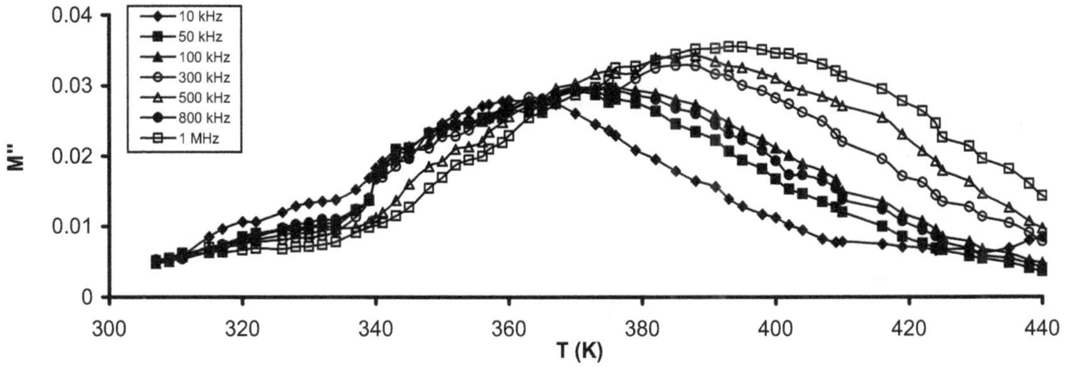

(b)

Figure 10. Temperature dependence of M'' (a) Pure CPVC and (b) CPVC stabilized with 10 wt% of PM at some constant frequencies.

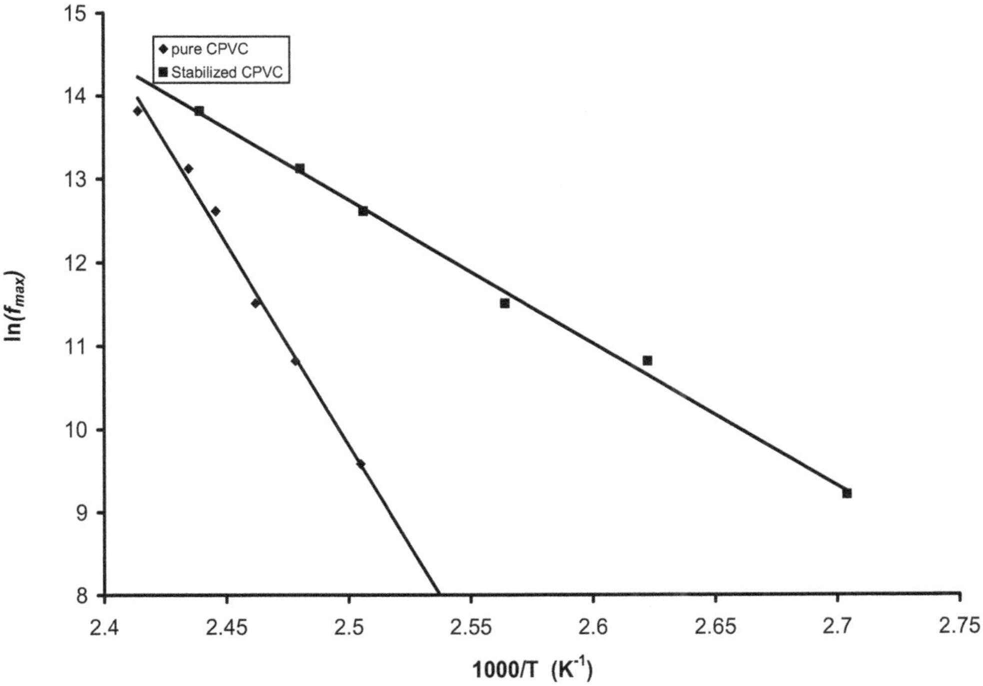

Figure 11. The relation between f_{max} and 1000/T for pure CPVC and that stabilized with 10 wt% of PM.

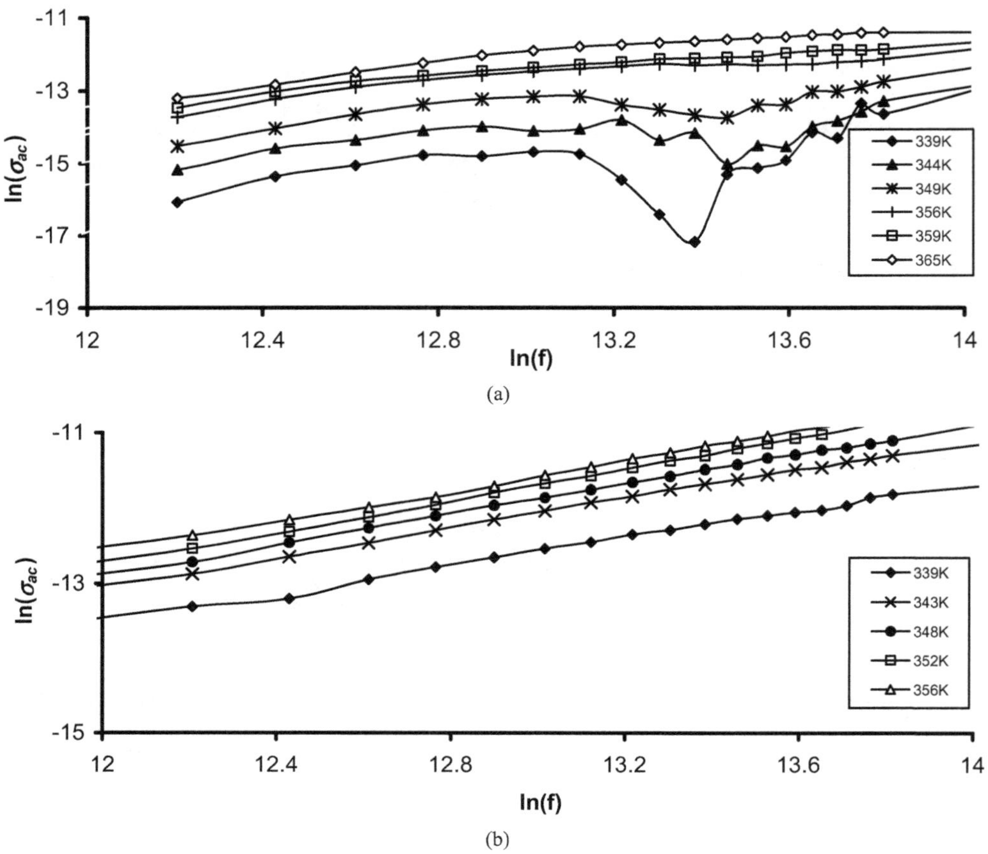

Figure 12. Temperature dependence of ln(σ_{ac}) (a) Pure CPVC and (b) CPVC stabilized with 10 wt% of PM at some constant frequencies.

could be explained by the ability of maleimide molecules to react with the polyenes of CPVC structure to give a Diels-Alder type addition reaction [9,10,32].

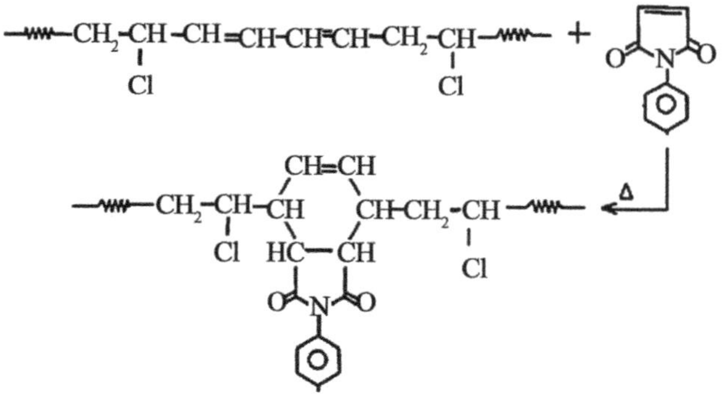

(11)

Not only Diels-Alder addition type reaction is responsible for the plasticization effect of PM molecules but also is responsible for the reduction of the number of the double bonds of CPVC structure. Consequently, PM molecules have a dual character. The first character is the stabilization effect due to the reducing of the double bonds (π-bond) and the other is the plasticization effect due to the widely spacing between the main chains of the investigated sample. So, the reduction of *Tg* of CPVC in this case can be assign to weaken of the intermolecular interaction which makes the macromolecules take mainly flexible conformation.

3.3. ac Conductivity

The frequency dependence of the conductivity at constant temperature in the amorphous materials is given by [14]:

$$\sigma(\omega) = \sigma(o) + A\omega^s$$

where $\sigma(\omega)$ is the total conductivity, $\sigma(o)$ is the frequency independent conductivity, *i.e.* σ_{dc} and *A* is constant. The exponent *s* is a constant less than one. The frequency dependent or alternating current conductivity is given by $\sigma_{ac} = A\omega^s$. Several theoretical models have been proposed to interpret the σ_{ac} for the amorphous semiconductors [20]. It is assumed that, the dielectric loss occurs because of the localized motion of charge carriers within a pair of sites [33]. Two distinct mechanisms have been proposed for relaxation phenomena quantum mechanical tunneling, QMT, of electrons or polarons through the barrier separating localized states and classical hopping over the same barrier. The frequency exponent s in the QMT model can be described by [20]:

$$s = 1 - \frac{4}{\ln(1/\omega\tau)} \quad (12)$$

This indicates that s values are temperature independent. For the correlated barrier hopping, CBH, model [33], it was assumed that single electron motion is responsible. The frequency exponent s in the CBH model can be controlled by [34]:

$$s = 1 - \frac{6kT}{W_H} \quad (13)$$

where W_H is the maximum barrier height at infinite separation, which is called the polaron binding energy, *i.e.*, the binding energy of the carrier in its localized sites. According Equation (13) s exponent decreases with increasing temperature. To specify the dominant ac conduction mechanism for pure and stabilized CPVC sample, the frequency dependence of σ_{ac} at some fixed temperatures corresponding to the main α-relaxation process has been considered. **Figures 12(a)** and **(b)** illustrates the frequency dependence of ac conductivity $\ln(\sigma_{ac})$ for pure CPVC and that stabilized with 10 wt% of PM at some fixed frequencies. It is clear that, σ_{ac} increases with increasing frequency and temperature. The calculated values of s component for all CPVC samples are listed in **Table 1**. The data of s exponent reveal that the conduction mechanism for pure CPVC sample is governed by CBH while the predominant conduction mechanism for the stabilized CPVC sample could be QMT.

4. Conclusion

Addition of phenyl maleimide at 10 wt% to the CPVC causes a reduction of its *Tg* by about 27 K, due to the plasticization effect of PM molecules. Dielectric losses and electric modulus behavior show three relaxation processes. The first process, ρ-relaxation, was assigned to MWS polarization. The second one, α'-process, was interpreted to the cooperation motion of the segmental motion of the branching of CPVC structure. The third one, α-process, is due to the micro-Brownian motion of the main polymer chain. Finally, electric modulus and ac conductivity of the investigated samples reveal that the CBH is the most probable conduction mechanism for pure CPVC. While the conduction mechanism of the stabilized CPVC sample is QMT.

Table 1. The values of the *s* exponent for pure and stabilized CPVC samples.

Temperature in K	*s* exponent	
	Pure CPVC	Stabilized CPVC
339	0.95	-
343	-	0.79
344	0.93	-
348	-	0.79
349	0.86	-
352	-	0.80
356	0.82	0.81
359	0.78	-
365	0.74	……

5. Acknowledgements

The author is very much thankful to Dr. Abir S. Abdel-Naby, Fayoum University, for preparing phenyl maleimide.

REFERENCES

[1] E. O. Elakesh, T. Hull, D. Price and P. Carty, "Effect of Stabilisers and Lubricant on the Thermal Decomposition of Chlorinated Poly(Vinyl Chloride) (CPVC)," *Polymer Degradation and Stability*, Vol. 88, No. 1, 2005, pp. 41-45. doi:10.1016/j.polymdegradstab.2004.04.027

[2] P. Carty, S. White, D. Price and L. Lu, "Smoke-Suppression in Plasticised Chlorinated Poly(Vinyl Chloride) (CPVC)," *Polymer Degradation and Stability*, Vol. 63, No. 3, 1999, pp. 465-468. doi:10.1016/S0141-3910(98)00075-5

[3] P. Carty and S. White, "The Effect of DOP Plasticizer on Smoke Formation in Poly(Vinyl Chloride)," *Polymer*, Vol. 33, No. 5, 1992, pp. 1110-1111. doi:10.1016/0032-3861(92)90033-S

[4] S. Mahrous, T. A. Hanfy and M. S. Sobhy, "Dielectric Relaxation of Chlorinated Polyvinyl Chloride (CPVC) Stabilized with Cyanoguanidine," *Current Applied Physics*, Vol. 7, No. 6, 2007, pp. 629-635. doi:10.1016/j.cap.2007.01.002

[5] S. Mahrous and T. A. Hanafy, "Dielectric Analysis of Chlorinated Polyvinyl Chloride Stabilized with Di-*n*-octyltin Maleate," *Journal of Applied Polymer Science*, Vol. 113, No. 1, 2009, pp. 316-320. doi:10.1002/app.29490

[6] N. Merah, F. Saghir, Z. Khan and A. Baz Aoune, "Effects on Fatigue Crack Growth Resistance of CPVC," *Engineering Fracture Mechanics*, Vol. 72, No. 11, 2005, pp 1691-1701. doi:10.1016/j.engfracmech.2004.12.002

[7] D. G. H. Ballard, A. N. Burgess, J. M. Dekoninck and E. A. Roberts, "The 'crystallinity' of PVC," *Polymer*, Vol. 28, 1987, pp. 3-9.

[8] R. D. Dworkin, "PVC Stabilizers of the Past, Present, and Future," *Journal of Vinyl Technology*, Vol. 11, No. 1, 1989, pp. 15-22. doi:10.1002/vnl.730110106

[9] S. B. Brown and C. M. Orlando, "Encyclopedia of Polymer Science and Engineering," Wiley, New York, 1988.

[10] M. W. Sabaa, N. A. Mohamed, E. H. Oraby and V. Yassin, "Organic Thermal Stabilizers for Rigid Poly(Vinyl Chloride) V. Benzimidazolylacetonitrile and Some of Its

derivatives," *Polymer Degradation and Stability*, Vol. 76, No. 3, 2002, pp. 367-380. doi:10.1016/S0141-3910(02)00009-5

[11] A. S. Abdel-Naby and S. A. Nouh, "Stabilization of Poly(Vinyl Chloride) against Laser Radiation with Ethyl-*N*-phenylmaleimide-4-caboxylate," *Polymer Degradation and Stability*, Vol. 76, No. 3, 2002, pp. 419-423. doi:10.1016/S0141-3910(02)00043-5

[12] M. Mujahid, D. S. Srivastava and D. K. Avasthi, "Dielectric Constant and Loss Factor Measurement of Polycarbonate, Makrofol KG Using Swift Heavy ion O^{5+}," *Radiation Physics and Chemistry*, Vol. 80, No. 4, 2011, pp 582-586. doi:10.1016/j.radphyschem.2010.12.007

[13] T. Sterzyński, J. Tomaszewska, K. Piszczek and K. Skórczewska, "The Influence of Carbon Nanotubes on the PVC Glass Transition Temperature," *Composites Science and Technology*, Vol. 70, No. 6, 2010, pp. 966-969.

[14] A. Yassin, M. Sabaa and N. Mohamed, "*N*-Substituted Maleimides as Photo-Stabilizers for Rigid Poly(Vinyl Chloride)," *Polymer Degradation and Stability*, Vol. 29, No. 3, 1990, pp. 291-303.

[15] A. Hassen, T. A. Hanafy, S. El-Sayed and A. Himanshu, "Dielectric Relaxation and Alternating Current Conductivity of Polyvinylidene Fluoride Doped with Lanthanum Chloride," *Journal of Applied Physics*, Vol. 110, No. 11, 2011, Article ID: 144119. doi:10.1063/1.3669396

[16] M. A. Ahmed, A. M. Basha, H. K. Marey and T. A. Hanafy, "Effect of Fast Neutrons and Radiation on Cobalt-Gelatin Film," *Journal of Applied Polymer Science*, Vol. 79, No. 10, 2001, pp. 1749-1755. doi:10.1002/1097-4628(20010307)79:10<1749::AID-APP20>3.0.CO;2-H

[17] A. Chelkovski, "Dielectric Physics," Elsevier Science, Amsterdam, 1980.

[18] F. Kremer and M. Arntt, "Dielectric Spectroscopy of Polymeric Materials," American chemical Society, Washington DC, 1997.

[19] M. Narisawa and K. Ono, "Polymer Interaction and Structure of PVA-Cu(II) Complex: 1. Binding of a Hydrophobic Dye toward PVA-Cu(II) Complex," *Polymer*,Vol. 30, No. 8, 1989, pp. 1540-1545.

[20] T. A. Hanafy, "Dielectric Relaxation and Alternating-Current Conductivity of Gadolinium-Doped Poly(Vinyl Alcohol)," *Journal of Applied Polymer Science*, Vol. 108, 2008, pp. 2540-2549.

[21] M. A. Ahmed and T. A. Hanafy, "Dielectric Relaxation and Poole-Frenkel Conduction in Poly(Vinyl Chloride) Blends with Bisphenol A/Egyptian Corncob Resin," *Journal of Applied Polymer Science*, Vol. 109, No. 1, 2008, pp. 182-189. doi:10.1002/app.28075

[22] S. Mahrous, "Dielectric Analysis of the α-Relaxation of PVC Stabilized with Cadmium Laurate," *Polymer International*, Vol. 40, No. 4, 1996, pp. 261-267.

[23] T. A. Hanafy, K. Elbana, S. Elsayed and A. Hassen, "Dielectric Relaxation Analysis of Biopolymer Poly(3-hydroxybutyrate)," *Journal of Applied Polymer Science*, Vol. 121, No. 6, 2011, pp. 3306-3313. doi:10.1002/app.33950

[24] R. Singh, J. Kumar, R. K. Singh, A. Kaur, R. D. P. Sinha and N. P. Gupta, "Low Frequency ac Conduction and Dielectric Relaxation Behavior of Solution Grown and Uniaxially Stretched Poly(Vinylidene Fluoride) Films," *Polymer*, Vol. 47, No. 16, 2006, pp. 5919-5928.

[25] M. Abdelaziz and M. M. Ghannam, "Influence of Titanium Chloride Addition on the Optical and Dielectric Properties of PVA Films," *Physica B*, Vol. 405, No. 3, 2010, pp. 958-964. doi:10.1016/j.physb.2009.10.030

[26] F. Bassiouni, F. Al-shamy, N. K. Madi and M. E. Kassem, "Temperature and Electric Field Effects on the Dielectric Dispersion of Modified Polyvinyl Chloride," *Materials Letters*, Vol. 57, No. 8-10, 2003, pp. 1595-1603.

[27] M. A. Ahmed, "Optical Absorption and Electrical Properties of Polyvinyl Alcohol (PVA)-Gelatin Blend," *Indian Journal of Physics*, Vol. 79, No. 10, 2005, pp. 1149-1155.

[28] G. M. Tsangaris, G. C. Psarras and N. Kouloumbi, "Evaluation of the Dielectric Behavior of Particulate Composites Consisting of a Polymeric Matrix and a Conductive Filler," *Materials Science and Technology*, Vol. 12, No. 7, 1996, pp. 533-538. doi:10.1179/026708396790166037

[29] C. N. Hampton, G. Carini, G. Dimarco and M. Lanza, "Temperature and Frequency Dependencies of the Complex Dielectric Constant of Poly(Ethylene Oxide) under Hydrostatic Pressure," *Journal of Polymer Science Part B: Polymer Physics*, Vol. 34, No. 3, 1996, pp. 425-433.

[30] G. S. Rillick and J. Runt, "A Dielectric Study of Poly(Ethylene-co-Vinylacetate)-Poly(Vinyl Chloride) Blends. I. Miscibility and Phase Behavior," *Journal of Polymer Science Part B: Polymer Physics*, Vol. 24, No. 2, 1986, pp. 279-302. doi:10.1002/polb.1986.090240206

[31] M. Ram and S. Chakrabarti, "Dielectric and Modulus Behavior of $LiFe_{1/2}Ni_{1/2}VO_4$ Ceramics," *Journal of Physics and Chemistry of Solids*, Vol. 69, No. 4, 2008, pp. 905-912. doi:10.1016/j.jpcs.2007.10.008

[32] A. A. Yassin and M. W. Sabaa, "Degradation and Stabilization of Poly(Vinyl Chloride) JMS-REV," *Macromolecular Chemistry and Physics C*, Vol. 30, No. 3-4, 1990, pp. 491-558.

[33] T. A. Hanafy, "Transient and Steady State Currents of Bisphenol A Corncobs Sample," *Advances in Materials Physics and Chemistry*, Vol. 1, 2011, pp. 99-107.

[34] F. H. Elkader, W. H. Osman, K. H. Mahmoud and M. A. F. Basha, "Dielectric Investigations and ac Conductivity of Polyvinyl Alcohol Films Doped with Europium and Terbium Chloride," *Physica B*, Vol. 403, No. 19-20, 2008, pp. 3473-3484.

2

Dielectric Permittivity of Various Cement-Based Materials during the First 24 Hours Hydration

Natt Makul
Faculty of Industrial Technology, Phranakhon Rajabhat University, Bangkok, Thailand
Email: shinomomo7@hotmail.com

ABSTRACT

The dielectric permittivity of cementitious materials during 24 hours hydration period at a frequency of 2.45 GHz using a network analyzer with open-ended probe technique was measured. Influences of water-to-cementitious ratios, cement types, pozzolans and aggregate types are taken into consideration. The results show that dielectric permittivity is strongly affected by initial water-to-cementitious ratio and the rate of hydration reaction which can be changed by fineness of cement (Types 1 and 3), pozzolan materials and aggregates (river sand with/without crushed limestone rock). Dielectric permittivity is relatively high and remains constant during the dormant period, after that it decreases rapidly when the hydration reaction resumes and continues to decrease during the acceleratory period.

Keywords: Dielectric Permittivity; Cementitious Materials; Hydration; Microwave Energy

1. Introduction

Microwave energy has been widely used as an innovative material processing for various industrial dielectric materials such as paper, wood, etc. Basically, microwave radiation interacts with the materials through dielectric permittivity resulting in rapid heating. Consequently, dipole interaction and heat generation will take place within dielectric materials which are composed of polar molecules [1].

Can microwave energy be applied to cure cementitious materials? The answer based on theoretical feasibility, is yes. Many research groups [1] have investigated, both experimentally and numerically, the accelerated curing of cements in order to gain high strength [2]. However, the actual number of successes has been very limited. A reason for this is lack of understanding of the behavior of dielectric permittivity of cement-based materials which, in fact, are greatly affected by temperature, free moisture content and hydration time. In particular, during the first 24 hours of hydration, it is critical to determine optimum conditions for high performance curing of the cementitious materials using microwave energy [3].

In this paper, the dielectric permittivity of cement-based materials during the initial period of hydration at 2.45 GHz has been investigated by using a network analyzer with an openended probe technique based on the influences of water-to-cementitious ratios, cement types, pozzolan materials and aggregates.

2. Experiments

2.1. Specimen Preparation

Three groups of 117 specimens having a cubical shape in size of 55 × 55 × 110 mm^3 were tested for dielectric permittivity, temperature rise and setting time, They were made from Types 1 and 3 Portland cements in accordance with the ASTM C150 [4] pulverized fuel ash in a class of low calcium (Type F) in accordance with the ASTM C 618 [4], and silica fume in accordance with the ASTM C1240 [4]. River sand (FM. = 2.58) and crushed limestone rock (Max. Size = 10 mm), having a grade conforming with the ASTM C33 [4], were mixed with tap water to from pastes, mortars and concretes with various proportions, as shown in **Table 1**. Superplasticizer conforming to the ASTM C494 [4] was used in a recommended dosage (500 ml per 100 kg of cementitious materials).

2.2. Test Procedures

For measuring dielectric permittivity of cementitious materials in the range of 2.45 GHz, a network analyzer with open-ended coaxial probe [5] was used. After mixing and placing into the mold, it was then wrapped by Styrofoam with 5 mm in thickness in order to protect the

Table 1. Mixing proportions of pastes, mortars and concretes.

Mix symbol	W/C	A/C	Material constitutes (air content designed = 1%)						
			Cement (kg/m³)	Pulverized fuel ash (kg/m³)	Silica fume (kg/m³)	Water (kg/m³)	Super plasticizer (L/m³)	Sand (kg/m³)	Lime stone (kg/m³)
CP1_0.38	0.38	0	1417	0	0	538	0	0	0
CP1_0.45	0.45	0	1290	0	0	579	0	0	0
CP1_0.70	0.70	0	973	0	0	679	0	0	0
CP3_0.38	0.45	0	1419	0	0	539	0	0	0
CP1_0.38SF25	0.38	0	1018	0	337	512	0	0	0
CP1_0.38PFA25	0.38	0	998	335	0	507	0	0	0
M1_0.38	0.38	2.75	561	0	0	212	0	1546	0
M3_0.38	0.38	2.75	561	0	0	212	0	1546	0
M1_0.38SUPS2.75	0.38	2.75	561	0	0	207	5.610	1544	0
M3_0.38SUPS2.75	0.38	2.75	561	0	0	207	5.610	1543	0
C1_0.38SUP	0.38	4.33	425	0	0	160	3.400	805	1030
C1_0.45SUP	0.45	4.14	425	0	0	189	3.400	771	985
C1_0.70SUP	0.70	3.48	425	0	0	296	3.400	642	830

Remarks: CP, M and C represent cement paste, mortar and concrete, respectively. 1 and 3 represent Types of Portland cement. 0.38, 0.45 and 0.70 represent water-to-cementitious materials (cement/Pulverized fuel ash/Silica fume). SF and PFA represent silica fume and pulverized fuel ash, respectively. SUP represents superplasticizer, and 2.75, 4.33, 4.14 and 3.48 represent aggregate-to-cementitious ratio.

heat loss. Both dielectric permittivity and semi-adiabatic temperature using data logger with thermo-couple (Type K) were simultaneously recorded every 180 and 15 minutes, respectively. However, in order to eliminate the effect of thermo-couple embedded in microwave radiation, we tested separately three specimens for dielectric permittivity and three for temperature rise. Furthermore, the setting time of pastes, mortars and concretes were tested by Vicat needle, modified Vicat needle and penetration resistance in accordance with the ASTM C191 [4], ASTM C807 [4], and ASTM C403 [4], respectively.

3. Results and Discussion

3.1. Effect of Water-To-Cementitious Materials

Figure 1 shows the evolution of dielectric permittivity and simultaneous temperature rise of concretes. It can be observed that the dielectric permittivity at the initial stage is relatively high in comparison with the later stage, and also increases with the increasing water content (higher w/c) in the concrete. This is due to the fact that immediately after the contact between water and cement they start to react and then dissolve Ca^{2+}, OH^- and SO_4^{2-} ions into the system. In addition, during the dormant period, the dielectric permittivity changes very little because the chemical compositions of the aqueous remains remain nearly constant [6].

3.2. Effect of Cement Types

Figure 2 shows the changes of permittivity throughout the early stage of 24 hours hydration reaction period. Permittivity remains at a high level and decreases at the end of the dormant period, approaching to a constant value when the internal structure has been stabilized. The dielectric permittivity of Type 3 pastes is higher than the Type 1 pastes because Portland cement Type 3 has finer grains of tri-calcium aluminate (C_3A) [7] than in the Type 1 causing it to dissolve with a high rate and maintaining an ion-rich system. In addition, the rate of the decrease of dielectric permittivity of Type 3 paste is higher than that of Type 1. In the acceleratory period, the Type 3 paste reacts faster than the Type I paste. This coincides with temperature rise and shorter dormant period. For setting time, the dielectric constant is maintained until the final setting time because of the high dissolution rate, however the dielectric constant drops dramatically with the high hydration rate. At the later stage after formation of the C-S-H structure, the dielectric permittivity tends to remain constant because of strong constraints imposed by its structure.

3.3. Effect of Pozzolan Materials

Effects of silica fume (SF) and pulverized fuel ash (PFA) on dielectric permittivity, temperature rises and setting time of the pastes are shown in **Figure 3**. The dielectric constant of the paste containing PFA through the first 24 hours hydration time is higher than that of the plain paste, whereas the paste containing SF is lower. Both reactions involving SF and PFA occur as secondary reactions [8]. This means, however, that the PFA can produce excessive water in the paste while increasing Si^{3+} and Ca^{2+} ions in the system. This results in an increasing dielectric

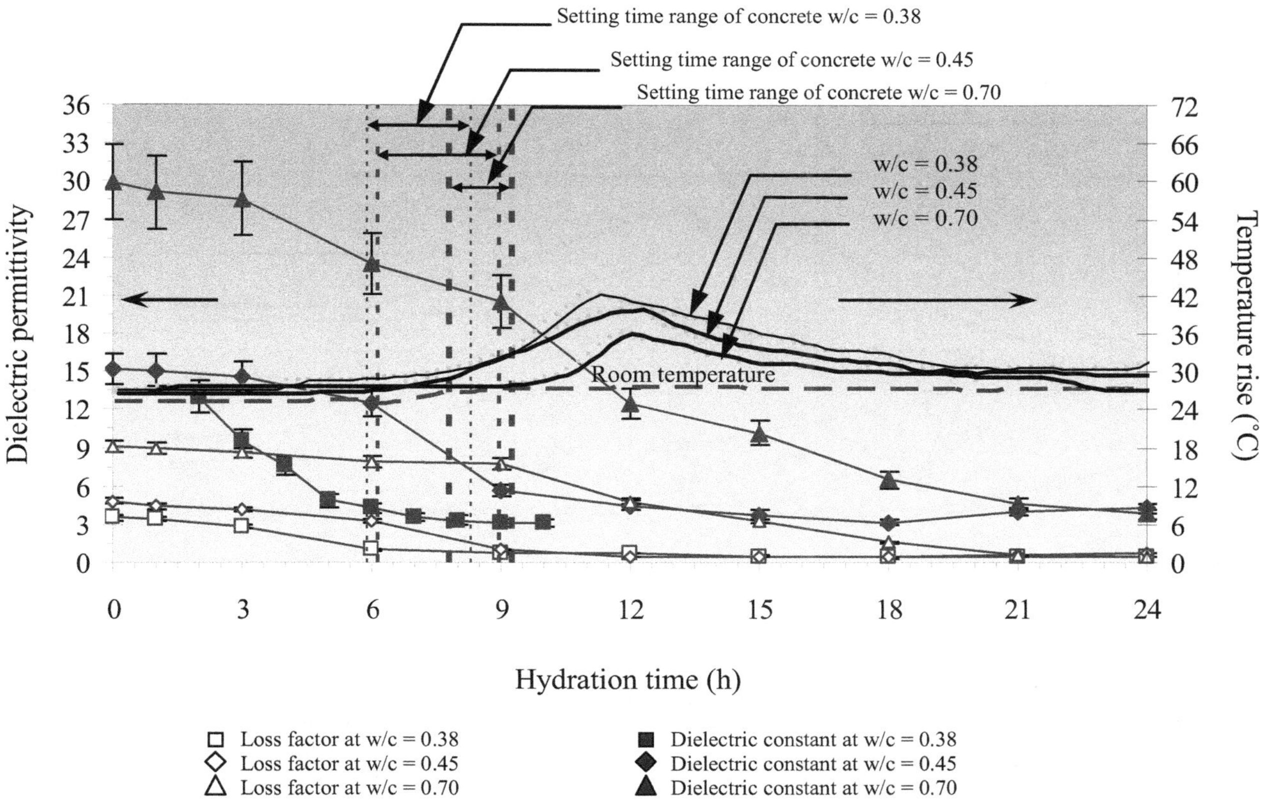

Figure 1. Dielectric permittivity of concretes with different w/c ratios.

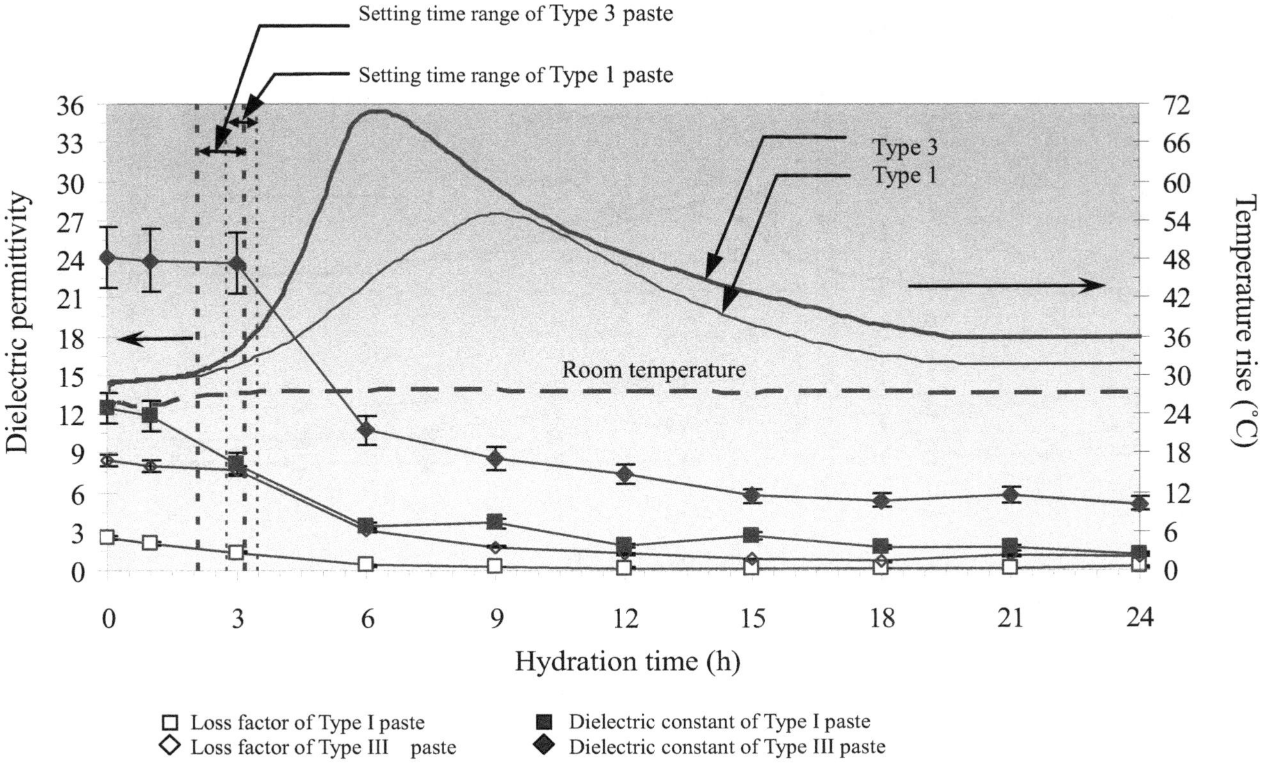

Figure 2. Dielectric permittivity of pastes with different cement types.

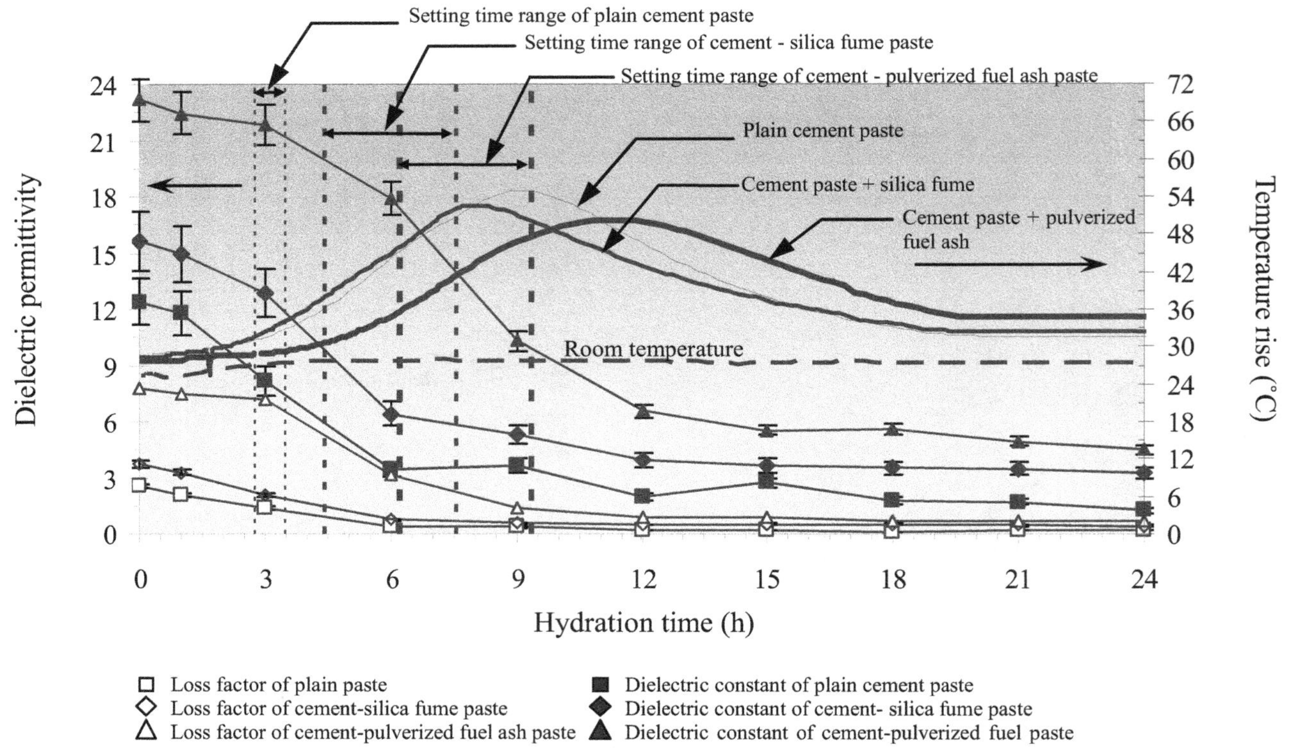

Figure 3. Dielectric permittivity of pastes with different pozzolan materials.

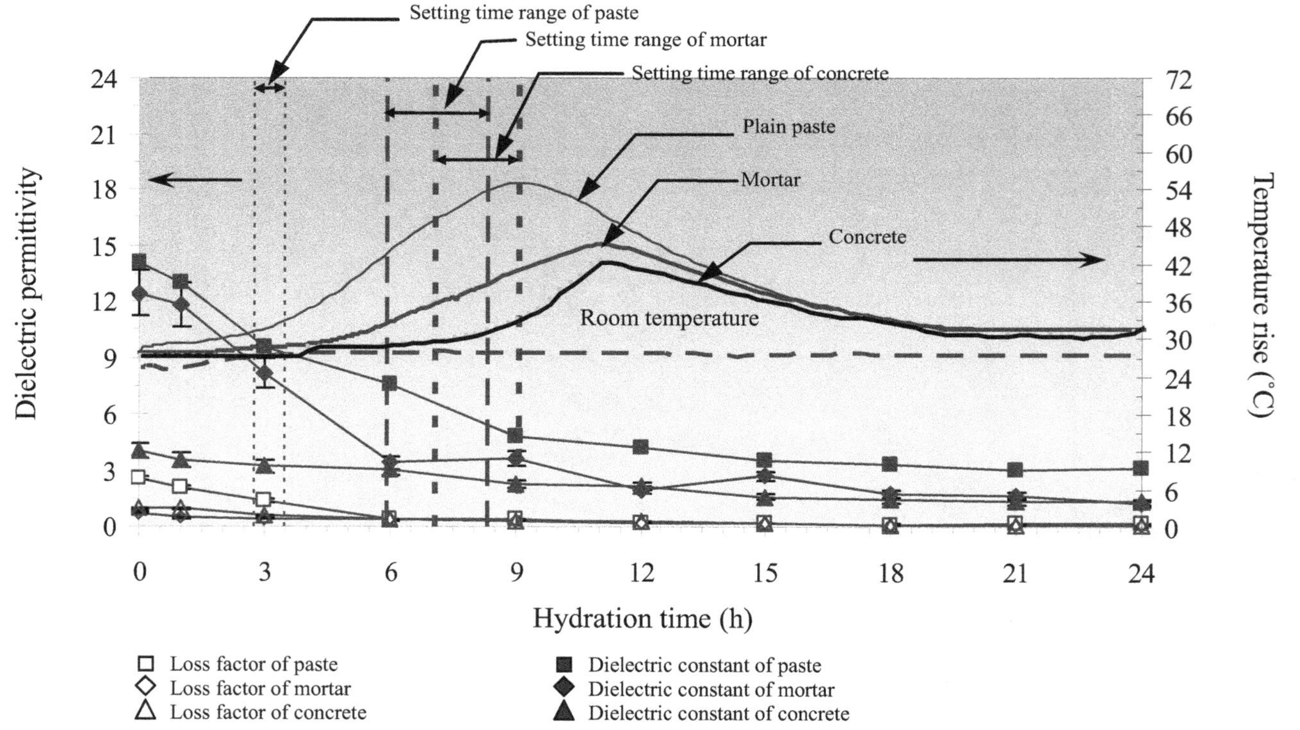

Figure 4. Dielectric permittivity of paste, mortar and concrete.

constant of the paste. Contrary to the SF paste, the com positions with PFA can dissolve it at a high rate, its fineness induces the bounding of its water molecules and ion-richness at the surface. As a result, the dielectric constant is lower than that of the plain cement paste. For loss factor evolution, the difference of loss factor is low between the conventional paste and SF-paste and large with PFA-paste; indicating that the remaining water content both during introduction and acceleratory periods in the paste has strong effect on the dielectric loss. On the other hand, the PFA particles can retain the free water in the paste system and this may raise the loss factor of the PFA-paste very high.

3.4. Effect of Aggregates

Figure 4 shows that the relative permittivity curves for the mortar and concrete are also similar to that of the conventional pastes but lower than the pastes. The decrease of cement content and the absorption of water molecules by aggregate surfaces induce a lower ions concentration in the system [9]. However, eventually, these parameters approach to constant. It can be compared in the setting time range that the slope of decrease rate of relative permittivity of the paste is somewhat higher than those of mortars and concretes. This is due to ion constraint by hydrated products which are high when compared with the aggregate-mixed ones.

4. Conclusion

Dielectric permittivity of cement-based materials is affected by initial water-to-cement ratio, cement types, pozzolan and aggregate types. However, although the volumetric fraction of water and superplasticizer in a given mixture is small, it strongly affects the dielectric permittivity of the cement due to its high dielectric permittivity. The change in the dielectric permittivity is relatively high and remains constant during the dormant period. After that it decreases rapidly when the hydration reaction resumes, and it continues to decrease during the acceleratory period.

5. Acknowledgements

The authors gratefully acknowledge the Thailand Research Fund (TRF) for supporting this research.

REFERENCES

[1] A. C. Metaxas, "Microwave Heating," *Power Engineering Journal*, Vol. 5, No. 5, 1992, pp. 237- 247. http://dx.doi.org/10.1049/pe:19910047

[2] K. Y. C. Leung and T. Pheeraphan, "Microwave Curing of Portland Cement Concrete: Experimental Results and Feasibility for Practical Applications," *Construction and Building Materials*, Vol. 9, No. 2, 1995, pp. 67-73. http://dx.doi.org/10.1016/0950-0618(94)00001-I

[3] R. G. Hutchinson, *et al.*, "Thermal Acceleration of Portland Cement Mortars with Microwave Energy," *Cement and Concrete Research*, Vol. 21, No. 5, 1991, pp. 795-799. http://dx.doi.org/10.1016/0008-8846(91)90174-G

[4] American Society for Testing and Materials, "Annual Book of ASTM Standard," Vol. 4.01, Philadelphia, 2008.

[5] Hewlett Packard Corporation, "Dielectric Probe Kit 85070 A," Research and Development Unit, Test and Measurements Laboratories, Palo Alto, 1992.

[6] P. C. Hewlett, "Lea's Chemistry of Cement and Cementitious Material," 4th Edition, John Wiley & Sons Inc., New York, 1998.

[7] F. H. Wittmann and F. Schlude, "Microwave Absorption of Hardened Cement Paste," *Cement and Concrete Research*, Vol. 5, No. 1, 1975, pp. 63-71. http://dx.doi.org/10.1016/0008-8846(75)90108-8

[8] S. Wen and D. D. L. Chung, "Effect of Admixtures on the Dielectric Constant of Cement Paste," *Cement and Concrete Research*, Vol. 31, No. 4, 2001, pp. 673-677. http://dx.doi.org/10.1016/S0008-8846(01)00475-6

[9] H. C. Rhim and O. Buyukozturk, "Electromagnetic Properties of Concrete at Microwave Frequency Range," *ACI Materials Journal*, Vol. 95, No. 3, 1998, pp. 262-271.

3

Effect of Cavity Dimensions on $TE_{01\delta}$ Mode Resonance in Split-Post Dielectric Resonator Techniques

Fang Chen[1], Sheng Mao[1], Xiaohui Wang[1], Elena Semouchkina[1,2], Michael Lanagan[2]

[1]Electrical and Computer Engineering Department, Michigan Technological University, Houghton, USA; [2]Materials Research Institute, The Pennsylvania State University, University Park, USA.
Email: esemouch@mtu.edu

ABSTRACT

The effects of cavity dimensions on the resonance frequency and resonance strength of the $TE_{01\delta}$ mode in split post-dielectric resonator (SPDR) technique are investigated by using full-wave simulations. The results of simulations provide guidance for adjusting the dimensional parameters of the set-up to ensure that a strong $TE_{01\delta}$ resonance mode is excited. The scaled designs of SPDR fixtures for operation at frequencies that are most important for applications are presented. These designs employ two sets of dielectric resonators (DRs) that can be fabricated from the standard ceramic materials. In addition, it is demonstrated that the resonance frequency of the $TE_{01\delta}$ mode in the fixture can be tuned by adjusting the gap of the split DR.

Keywords: Split Post-Dielectric Resonator; $TE_{01\delta}$ Mode; Cavity Dimensions

1. Introduction

The split-post dielectric resonator (SPDR) provides an accurate technique for the measurement of the complex permittivity of low loss dielectric materials, thin films and wideband gap semiconductors that are difficult to measure by other techniques [1-3]. SPDR typically operates at the $TE_{01\delta}$ mode that restricts the electric field component to the azimuthal direction so that the electric field remains continuous on the dielectric interfaces. The resonance mode is insensitive to the presence of air gaps perpendicular to the longitudinal axis of the fixture [4,5]. The real part of the permittivity of the sample is determined from the resonance frequency shift due to the sample insertion gap of the split post. The loss tangent can be found from the Q factors of an empty cavity and a cavity with the sample, respectively.

SPDR technique originates from the resonant post technique for measurements microwave permittivity of a cylindrical dielectric sample, which is based on excitation of a well-defined TE_{011} mode by means of metal plates on each end of the dielectric cylinder [6]. The dielectric loss is determined by subtracting the conduction Q of the metal boundaries from the measured Q for the TE_{011} resonance mode [7]. For high Q dielectrics, the metal boundary loss is a significant fraction of the measured loss so that the resonant post technique cannot be used. High Q dielectric resonators are suspended on a quartz rod in the center of a metal cavity to minimize radiation and conduction loss and the $TE_{01\delta}$ instead of the TE_{011} mode is used [8,9]. The concepts from the monolithic dielectric resonator measurements were applied to the SPDR technique [5,6].

In this paper, the effect of cavity dimensions on the $TE_{01\delta}$ resonance formation is investigated and the importance of having an appropriate cavity size, in order to accentuate the desired resonance mode in the measurements, in demonstrated.

2. Cavity Dimensions Effect on Resonance Formation

For an isolated dielectric cylinder, the resonant frequency of the $TE_{01\delta}$ mode is completely defined by the relative permittivity ε_r of the material and the dimensions of the resonator. An empirical estimation of the resonant frequency f_r for the $TE_{01\delta}$ mode of an isolated dielectric cylinder in free space is given by [9]:

$$f_r = \frac{3.4}{r\sqrt{\varepsilon_r}}\left(\frac{r}{h}+3.45\right) \quad (1)$$

where, r is the radius in mm, h is the height of the cylinder, and f_r is in the GHz units. As seen from **Figure 1**, the SPDR is a circular cylindrical dielectric resonator that is separated into two halves by an air gap, which modifies the resonant frequency given in Equation (1). A sample is placed in the gap, and coupling loop is used to excite and detect the $TE_{01\delta}$ resonance. From measurements of the resonance frequency and quality factor, the relative permittivity and loss tangent of the sample can be determined. The SPDR fixture is shielded by a metal enclosure, which contains strong evanescent electromagnetic waves not only in the air gap region between the dielectric resonators, but also in the entire cavity region for radii greater than the radius of dielectric resonators. Cavity enclosed by the metal complicates the resonance formation.

In order to investigate the effect of cavity dimensions on the resonance strength, the SPDR fixture shown in **Figure 1** has been modeled and simulated by using the CST Microwave Studio commercial software package for two sets of cavity dimensions and the same DR configuration. The two DR halves have the relative permittivity ε_d of 34.4, the diameter d_r of 25.4 mm, the height hr of 5.7 mm, and the gap distance hg is 1.3 mm. For the cavity set 1, its diameter is D = 76.3 mm, the height is L = 63.6 mm, while for the set 2 these parameters are D = 50 mm and L = 30 mm, respectively.

Cavity set 1 is an example of a design that does not provide a prominent mode peak due to large cavity size. **Figure 2(a)** presents the simulated S_{21} spectrum for the structure, which exhibits several resonance peaks. From

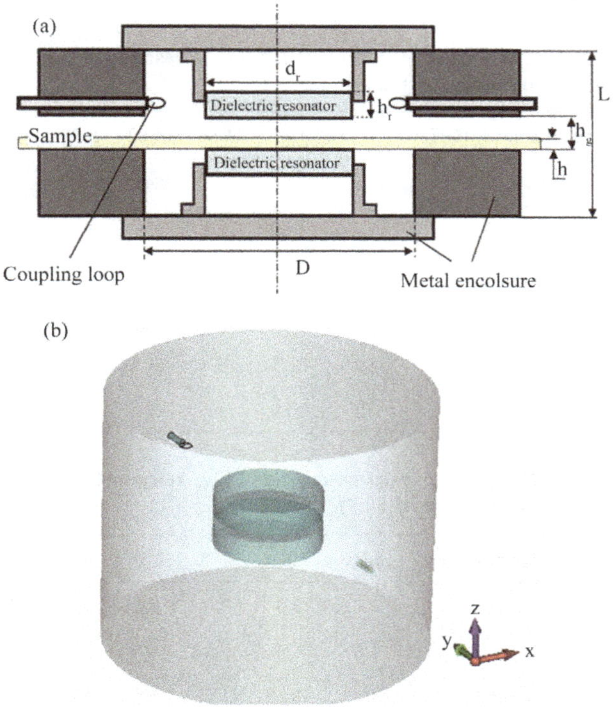

Figure 1. (a) Schematic diagram of a SPDR fixture; (b) CST simulation model of the SPDR without a sample. Parameters of the two DR halves: relative permittivity ε_d = 34.4, diameter d_r = 25.4 mm, height h_r = 5.7 mm, gap distance h_g = 1.3 mm.

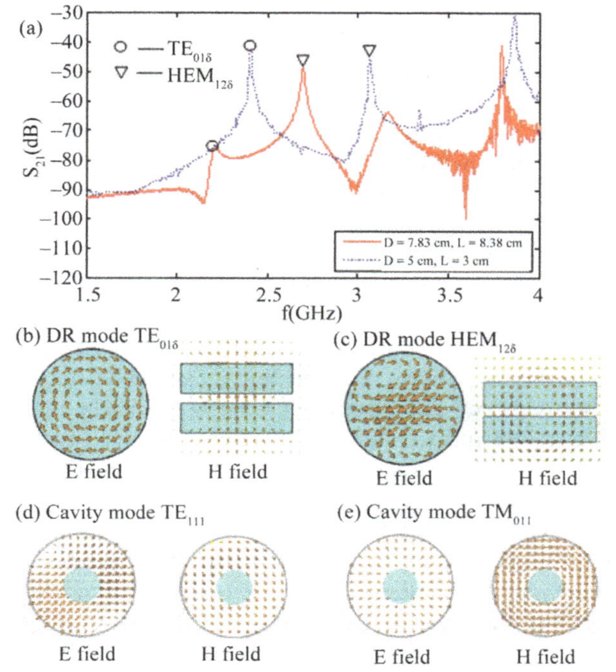

Figure 2. (a) Simulated S_{21} for a SPDR fixture for cavity set 1: D = 76.3 mm, L = 63.6 mm (in solid curve) and cavity set 2: D = 50 mm, L = 30 mm (in dotted curve); (b)-(c) Electric field distributions in the median x-y cross-section of the DR and magnetic field distributions at y-z plane of the DRs for: (b) the $TE_{01\delta}$ mode at 2.21 GHz for the cavity set 1 and at 2.40 GHz for the cavity set 2 and (c) the $HEM_{12\delta}$ mode at 2.70 GHz for the cavity set 1 and at 3.07 GHz for the cavity set 2; (d)-(e) Electric and magnetic field distributions in x-y plane in the cavity for cavity eigenmodes TE_{111} and TM_{011}, respectively.

the analysis of field distributions in the resonator at the peak frequencies, which are shown in **Figures 2(b)-(c)**, it was found that the first resonance corresponds to the $TE_{01\delta}$ mode, while the second one to the $HEM_{12\delta}$. For the third and the fourth observed resonances, respectively, it was detected that they could not be related to a resonance in the DR; instead, they were found to coincide with the cavity eigenmodes at the frequencies 3.29 GHz for TE_{111} mode and 3.82 GHz for TM_{011} mode, respectively (see **Figures 2(d)** and **(e)**). The $TE_{01\delta}$, which is employed in the SPDR technique, is observed in two DRs separated by hg at 2.21 GHz, however, the resonance is very weak, with S_{21} being at the level of –75.7 dB. The weakness of the $TE_{01\delta}$ mode could play a misleading role so that the second much stronger $HEM_{12\delta}$ resonance mode could be easily taken as being the first resonance in the experimental measurements. Cavity set 2 is an example of an appropriate design with the correctly chosen diameter and height. As seen in **Figure 2(a)**, though the same resonance modes are excited in both cavity sets, the $TE_{01\delta}$ resonance is much stronger in the cavity set 2 than in the cavity set 1.

It should be noted that the observed results are in agreement with the cavity perturbation theory [8]. According to this theory, the resonant frequency of a DR is expected to increase when the metal enclosure moves towards the DR, when the displaced magnetic energy is predominant; vice versa, it is expected to decrease when the electric energy is dominant. Therefore, in the case of a DR operating at $TE_{01\delta}$ mode, the resonant frequency should increase as metal wall moves closer to the DRs. By comparing the curves in **Figure 2(a)**, it is seen that a reduction in the cavity dimension results in an increase of the resonant frequencies. For the first $TE_{01\delta}$ mode, the resonant frequency shifts from 2.21 GHz to 2.40 GHz. In addition, the resonance strength is greatly improved for the first mode $TE_{01\delta}$ in the cavity set 2, *i.e.* S_{21} increases by about 40 dB.

The results of the performed studies assist in formulating recommendations for the designing SPDR fixtures for operation at the desired frequencies. As shown in **Figure 3(a)**, the resonant frequency of SPDR decreases by 0.11 GHz and the S_{21} parameter decreases by 35 dB, when the diameter of metal cavity is increased from 4 cm to 8 cm. **Figure 3(b)** demonstrates that when the height of metal cavity increases from 3 cm to 7 cm, the resonant frequency decreases by 0.08 GHz, while the S_{21} changes only slightly, *i.e.* by about 3 dB. Therefore, the cavity diameter produces the dominants effect on the resonance strength of the fixture, while the cavity height only causes the frequency shift. At the same time, there is a limit to which the cavity diameter could be decreased in order to strengthen the resonance, since it was found that when the probes get too close to the DRs, they distort the DR mode so that the $TE_{01\delta}$ resonance could not be excited.

3. SPDR Fixture Scaling for Operation at Different Frequencies

The SPDR fixture presented above as the cavity set 2 has been designed to operate at the frequency of 2.40 GHz, where the relative dielectric permittivity of the DR was 34.4. The design originally proposed in [5] was for operation at 5.60 GHz and used the DR permittivity of 28.9. To scale the SPDR fixture for operation at other frequencies important for practical applications based on the two original designs, the scale factors and fixture dimensions have been determined. Since the gap between the two DRs can be easily adjusted during the measurements, the effect of gap scaling has also been investigated. COMSOL Multiphysics simulation package [10] was used in the simulations of the re-scaled designs. **Table 1** and **Table 2** summarize the parameters of the SPDR fixtures that should be used for different operating frequencies with the DR having the relative permittivity of 34.4 and 28.9, respectively.

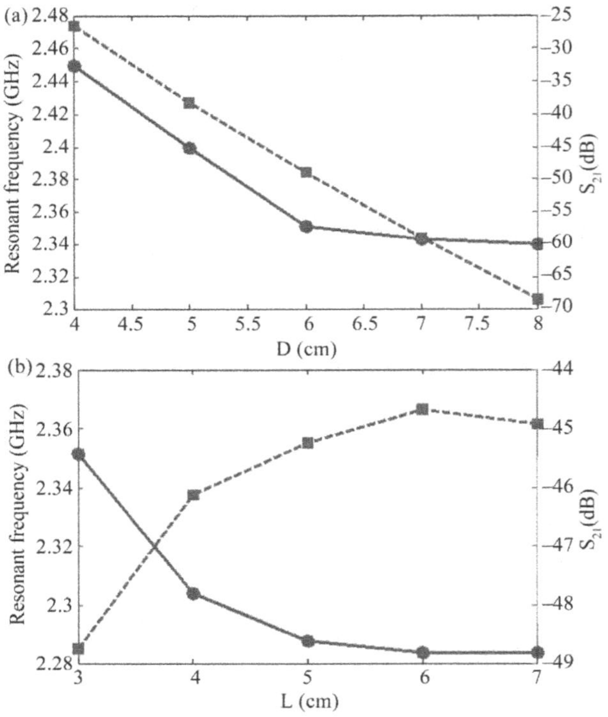

Figure 3. Dependence of the resonant frequency and S_{21} amplitude at the resonance on (a) the cavity diameter (the cavity height L is fixed at 3 cm) and (b) the cavity height (the cavity diameter is fixed at 6 cm).

Table 1. SPDR fixture parameters for $TE_{01\delta}$ resonance for different frequencies ($\varepsilon_d = 34.4$).

Resonance frequency (GHz)	2.32	1.87	1.86	5.60	5.87
Scale factor K	1.0	1.23	1.23	0.415	0.415
Air gap h_g (mm)	1.3	1.60	**1.3**	0.54	**1.3**
DR diameter d_r (mm)	25.4	31.24	31.24	10.54	10.54
DR height h_r (mm)	5.7	7.01	7.01	2.37	2.37
Cavity diameter D (mm)	50	61.5	61.5	20.75	20.75
Cavity height L (mm)	30	36.9	36.9	12.45	12.45

Table 2. SPDR fixture parameters for $TE_{01\delta}$ resonance for different frequencies ($\varepsilon_d = 28.9$).

Resonance frequency (GHz)	5.60	1.87	1.78	2.44	2.35
Scale factor K	1.0	3.0	3.0	2.3	2.3
Air gap h_g (mm)	1.6	4.8	**1.6**	3.68	**1.6**
DR diameter d_r (mm)	13.85	41.55	41.55	31.85	31.85
DR height h_r (mm)	1.98	5.94	5.94	4.55	4.55
Cavity diameter D (mm)	30.2	90.6	90.6	69.46	69.46
Cavity height L (mm)	13.0	39.0	39.0	29.9	29.9

As seen from the tables, the scale factor for the resonance frequency of the $TE_{01\delta}$ mode should be directly applied to scale the cavity size. From the comparison of two designs with the same scale factor, however, having the gap distance hg either scaled or not scaled (grey cells in the table), it could be noted that a smaller gap decreases the resonance frequency; while a bigger gap leads to a higher resonance frequency so that this frequency change could be as big as 0.26 GHz. Therefore, in practice, the frequency of the fabricated SPDR could be tuned to the desired frequency by simply adjusting the gap size.

4. Conclusion

It is demonstrated through full-wave electromagnetic simulations that the size of the cavity used in the SPDR technique to shield the dielectric resonator strongly influences the resonance frequency and the strength of the desired $TE_{01\delta}$ mode, so that the latter could be completely suppressed. Therefore, it is necessary to adjust cavity dimensions to the parameters of the dielectric resonators used in the fixture. The design parameters of SPDR fixtures providing proper operation at various frequencies important for the applications are presented. In addition, it was found that the frequency of the fixture operation can be fine tuned by adjusting the gap of the split resonator.

5. Acknowledgements

This work was supported by the National Science Foundation under Grant No. 0968850.

REFERENCES

[1] J. Baker-Jarvis, R. G. Geyer, J. H.Grosvenor, M. D. Janezic, C. A. Jones, B. Riddle, C. M. Weil and J. Krupka, "Dielectric Characterization of Low-loss Materials a Comparison of Techniques," *IEEE Transactions on Dielectrics and Electrical Insulation*, Vol. 5, No. 4, 1998, pp. 571-577. doi:10.1109/94.708274

[2] J. Sheen, "Study of Microwave Dielectric Properties Measurements by Various Resonance Techniques," *Measurement*, Vol. 37, No. 2, 2005, pp. 123-130. doi:10.1016/j.measurement.2004.11.006

[3] J. Sheen, "Comparisons of Microwave Dielectric Property Measurements by Transmission/Reflection Techniques and Resonance Techniques," *Measurement Science and Technology*, Vol. 20, No. 4, 2009, Article ID: 042001. doi:10.1088/0957-0233/20/4/042001

[4] J. Krupka, A. P. Gregory, O. C. Rochard, R. N. Clarke, B. Riddle and J. Baker-Jarvis, "Uncertainty of Complex Permittivity Measurements by Split-post Dielectric Resonator Technique," *Journal of the European Ceramic Society*, Vol. 21, No. 15, 2001, pp. 2673-2676. doi:10.1016/S0955-2219(01)00343-0

[5] J. Krupka, R. G. Geyer, J. Baker-Jarvis and J. Ceremuga, "Measurements of the Complex Permittivity of Microwave Circuit Board Substrates Using Split Dielectric Resonator and Resonant Cavity Techniques," *7th International Conference on Dielectric Materials, Measurements and Applications*, Bath, 23-26 September 1996, pp. 21-24. doi:10.1049/cp:19960982

[6] B. W. Hakki and P. D. Coleman, "A Dielectric Resonator Method of Measuring Inductive Capacities in the Millimeter Range," *IEEE Transactions on Microwave Theory Technique*, Vol. 8, No. 4, 1960, pp. 402-410. doi:10.1109/TMTT.1960.1124749

[7] Y. Kobayashi and M. Katoh, "Microwave Measurement of Dielectric Properties of Low-loss Materials by the Dielectric Rod Resonator Method," *IEEE Transactions on Microwave Theory Technique*, Vol. 33, No. 7, 1985, pp. 586-592. doi:10.1109/TMTT.1985.1133033

[8] D. C. Dube, R. Zurmuhlen, A. Bell, N. Setter and W. Wersing, "Dielectric Measurements on High Q Ceramics in the Microwave Region," *Journal of the American Ceramic Society*, Vol. 80, No. 5, 1997, pp.1095-1100. doi:10.1111/j.1151-2916.1997.tb02951.x

[9] D. Kajfez and P. Guillion, "Dielectric Resonators," Artech House, Norwood, 1986, pp. 327-376.

[10] http://www.comsol.com/products/multiphysics/

4

Dielectric and Microwave Properties of Natural Rubber Based Nanocomposites Containing Graphene

Omar A. Al-Hartomy[1,2], Ahmed Al-Ghamdi[2], Nikolay Dishovsky[3], Rossitsa Shtarkova[4], Vladimir Iliev[5], Ibrahim Mutlay[6], Farid El-Tantawy[7]

[1]Department of Physics, Faculty of Science, University of Tabuk, Tabuk, Saudi Arabia; [2]Department of Physics, Faculty of Science, King Abdulaziz University, Jeddah, Saudi Arabia; [3]Department of Polymer Engineering, University of Chemical Technology and Metallurgy, Sofia, Bulgaria; [4]Department of Chemistry, Technical University, Sofia, Bulgaria; [5]College of Telecommunications and Posts, Sofia, Bulgaria; [6]Hayzen Engineering Co., Ankara, Turkey; [7]Department of Physics, Faculty of Science, Suez Canal University, Ismailia, Egypt.
Email: dishov@uctm.edu

ABSTRACT

The development of carbon nanotubes based materials has been impeded by both their difficult dispersion in the polymer matrix and their high cost. The discovery of graphene and the subsequent development of graphene-based polymer nanocomposites is an important addition in the area of nanoscience and technology. In this study the influence of graphene nanoparticles (GNP) in concentrations from 2.0 to 10.0 phr on the dielectric (dielectric permittivity, dielectric loss angle tangent) and microwave (reflection coefficient, attenuation coefficient, shielding effectiveness) properties of nanocomposites on the basis of natural rubber has been investigated in the wide frequency range (1 - 12 GHz). The results achieved allow recommending graphene as a filler for natural rubber based composites to afford specific dielectric and microwave properties, especially when their loading with the much more expensive carbon nanotubes is not possible.

Keywords: Nanocomposites; Natural Rubber; Graphene; Dielectric and Microwave Properties

1. Introduction

Over the past five decades, industrial scale "composite materials" have been produced by adding different minerals and metals to thermosetting, thermoplastic and elastomeric polymers. As compared to the bulk polymers, these composites have shown moderate mechanical performance improvements in the mechanical properties such as Young's modulus, tensile strength, abrasion resistance and storage modulus. However, recent advances in nanoscale particle synthesis have definitely accelerated the growth of the composite industry. The capacity to synthesize and characterize atomic-level particles has produced a new generation of high-performance fillers and new class of materials—polymer nanocomposites.

Polymer nanocomposites are commonly defined as the combination of a polymer matrix and additives that have at least one dimension in the nanometer range. The additives can be one-dimensional, such as nanotubes and fibres, two-dimensional, which include layered clay minerals or graphene sheets, or three-dimensional, including spherical particles [1].

At present, nanocomposites employing carbon-based reinforcement materials are dominated by carbon nanotubes (CNTs) [2-4]. However, the development of CNT-reinforced composites has been impeded by both their difficult dispersion in the polymer matrix and their high cost.

The discovery of graphene [5] and the subsequent development of graphene-based polymer nanocomposites is an important addition in the area of nanoscience and technology. Deservedly the Nobel Prize in Physics for 2010 was awarded to Andre Geim and Konstantin Novoselov from the University of Manchester for their work on a single free-standing atomic layer of carbon (graphene). Graphene is an allotrope of carbon, whose structure is one-atom-thick planar sheets of sp^2—bonded carbon atoms packed in a honeycomb lattice. It is the basic structural element of some carbon allotropes including graphite, charcoal, carbon nanotubes and fullerenes.

Compared to carbon nanotubes, as well as its high aspect ratio and low density graphene has attracted considerable attention because of its unique and outstanding mechanical, electrical and electronic properties [6]. In addition to good thermal conductivity, remarkable mechanical stiffness and high fracture strength, graphene has been supposed to be a semiconductor with zero gap which is quite different from conventional silicon semiconductors. In graphene, electrons shoot along with minimal resistance which may allow for low-power, faster-switching transistor and become a candidate to replace silicon in the area of microchip electronics.

All these unique properties in a single nanomaterial have made physicists, chemists and material scientists exited about graphene's potential. The history, chemistry, preparation methods and possible applications of graphene are reviewed in [7-9]. Another new review focused on trends and frontiers in graphene-based polymer nanocomposites was published a year ago [10]. Recently, the trustees of Princeton University received a patent for graphene-elastomer nanocomposites where functionalized graphene sheets (FGS) had been dispersed in vulcanized natural rubber, styrene butadiene rubber, Ps-isoprene-Ps and PDMS [11]. The patented work could find a wide range of industrial applications, including food packaging, gasketing and automotive. The authors conclude that graphene-rubber nanocomposites possess qualities like those of carbon nanotube composites but are much cheaper to make.

The data about graphene influence on the microwave properties of the elastomeric composites are very scarce. Y. Chen *et al.* used functionalized graphene-epoxy composites as lightweight shielding materials for electromagnetic radiation [12]. I. M. De Rosa and coworkers have a wide expertise in the design of micro/nanocomposites based on carbon fibers and carbon nanotubes, for the realization of high performing radar absorbing screens, with tailored properties [13-15]. In a recent paper [16], the authors have accomplished a Salisbury screen, that consists of three layers. The second layer (the spacer) is a low-loss-tangent nanocomposite based on a Bisphenol-A based epoxy resin filled with GNPs at 0, 5 and 1 wt% [16]. The real and imaginary parts of the complex effective permittivity within the 8 - 18 GHz range of the nanocomposite filled with GNPs have been shown. It has been observed that the real part of the effective permittivity is nearly constant.

There are no literature data about the dielectric and microwave properties of nanocomposites comprising a higher amount of graphene nanoparticles (GNP), e.g. over 1 phr. Neither there are available reports on investigations on those properties carried out in a wider frequency range, first of all at frequencies lower than 8 GHz. Therefore the aim of this work is to study the influence that graphene nanoparticles (in amounts of 2 to 10 phr) have on the dielectric (dielectric permittivity, dielectric loss angle tangent) as well as on the microwave properties (absorption and reflection of the electromagnetic waves, the effectiveness of the electromagnetic shielding) of natural rubber based composites in a significantly greater frequency range—from 1 to 12 GHz.

2. Experimental

2.1. Characterisation of the Graphene Used

Graphene as produced by Hayzen Engineering Co., Ankara, Turkey was used in our investigation. Graphene nanoplatelets (GNP) have a "platelet" morphology, meaning they have a very thin but wide aspect. Aspect ratios for this material can range into the thousands. Each particle consists of several sheets of graphene with an overall thickness of 50 nm and average plate diameter 40 micron.

2.2. Prepapation and Vulcanization of Rubber Compounds

Table 1 summarizes the formulation characteristics of the rubber compounds (in phr) used for the investigations.

The rubber compounds were prepared on an open two-roll laboratory mill (L/D 320 × 360 and friction 1, 27) by incorporating pre-characterized graphene nanoplatelets (GNP) into a natural rubber matrix at various loadings (**Table 1**). The speed of the slow roll was 25 min^{-1}. The experiments were repeated for verifying the statistical significance. The ready compounds in the form of sheets stayed 24 hours prior to their vulcanization.

The optimal vulcanization time was determined by the vulcanization isotherms, taken on an oscillating disc vulcameter MDR 2000 (Alpha Technologies) at 150°C according to ISO 3417:2002.

These composites were evaluated for their dielectric (dielectric permittivity, dielectric loss angle tangent) and microwave (reflection coefficient, attenuation coefficient, and shielding effectiveness) properties in the 1 - 12 GHz frequency range.

Table 1. Composition of the rubber compounds studied.

	NR 1	NR 2	NR 3	NR 4
Natural rubber	100	100	100	100
Foaming agent	8	8	8	8
Stearic acid	1	1	1	1
ZnO	4	4	4	4
Processing oil	10	10	10	10
GNP	0	2	6	10
MBTS	2	2	2	2
TMTD	1	1	1	1
IPPD 4020	1	1	1	1
Sulphur	2	2	2	2

3. Measurements

3.1. Microwave Properties

3.1.1. Reflection and Attenuation

Measurements of reflection and attenuation were carried out using the measurement of output (adopted) power P_a in the output of a measuring line without losses, where samples of materials may be included. Because of the wide frequency measurement a coaxial line was used. Samples of the materials were shaped into discs with an external diameter D = 20.6 mm, equal to the outer diameter of the coaxial line and thickness of $\Delta \approx 2$ mm. The internal diameter depended on the relative dielectric permittivity of the material.

The sample reflected a part of the incident electromagnetic wave with power P_{in}. The rest of the wave with power P_p penetrated the material, so that the attenuation L depended on the coefficient of reflection $|\Gamma|$. Its module was determined by a reflect meter.

Thus the attenuation was determined by

$$L = 10\log\frac{P_a}{P_p}, dB \quad (1)$$

where

$$P_p = P_{in}\left(1-|\Gamma|^2\right) \quad (2)$$

The following scheme presents the equipment used for testing both parameters (**Figure 1**)

1) A set of generators for the whole range: HP686A and G4 - 79 to 82;

2) Coaxial section of the deck E2M Orion, with samples of material;

3) Power meter HP432A;

4) Scalar reflectance meter HP416A;

R—Reflect meter, including:

—Two directional couplers Narda 4222.16;

—Two crystal detectors Narda 4503-N.

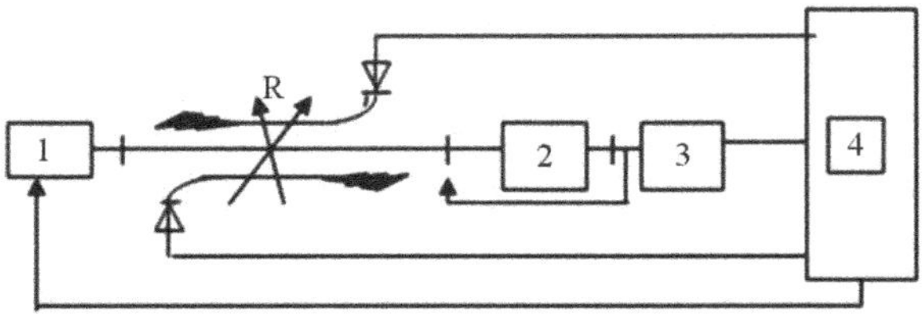

Figure 1. Scheme of the equipment for measuring the microwave properties.

3.1.2. Shielding Effectiveness (S.E.)

This parameter is defined as the sum of the reflection losses *R*, *dB* and attenuation *L*, *dB* in the material.

It can be directly measured or calculated from the measured reflectance and attenuation in the material. In the first case, as measured: incident power on the sample P_{in} and adopted after the sample P_a (**Figure 1**), S.E. is determined by

$$\text{S.E.} = 10\log\frac{P_0}{P_a}, dB \quad (3)$$

In the second, if known reflection and absorption in the material, S.E. is determined, by a definition, as

$$\text{S.E.}, dB = -20\lg|\Gamma| + L, dB \quad (4)$$

where *R*, *dB* is the attenuation due to the reflection of power at the interfaces.

In the present work the shielding effectiveness was determined by Equation (4).

3.2. Dielectric Properties

3.2.1. Complex Permittivity

The determination of complex permittivity was carried out by the resonance method, based on the cavity perturbation technique.

Having measured the resonance frequency of the empty cavity resonator f_r a measurement of the shift in resonance frequency f_ε was carried out in the presence of the sample material. Then the dielectric constant ε_r was calculated from the shift in resonance frequency, cavity and the sample cross sections S_r and S_ε, respectively

$$\varepsilon_r = 1 + \frac{S_r}{2S_\varepsilon}\cdot\frac{f_r - f_\varepsilon}{f_r}. \quad (5)$$

The sample was in the form of a disc with a diameter of 10 mm and about 2 mm thick. It was placed at the maximum electric field location of the cavity. Because the thickness of the sample was not equal to the height of the resonator, a dielectric occurred with an equivalent permittivity ε_e at the place of its inclusion. The parameter was determined by (5) and instead ε_r be saved ε_e. Then ε_r was determined by

$$\varepsilon_r \approx \varepsilon_e(k+1) - k, (\Delta \Box\ l), \quad (6)$$

where $k = l/\Delta$ and l is the distance from the disc to the top of the resonator.

3.2.2. Loss Factor tanδ

The loss factor tanδ was calculated from the quality factor of the cavity with Q_ε and without sample Q_r

$$\tan\delta = \frac{1}{4\varepsilon_r}\frac{S_r}{S_\varepsilon}\left(\frac{1}{Q_\varepsilon} - \frac{1}{Q_r}\right). \quad (7)$$

The measurement setup used several cavity resonators for the whole range, generators for the whole range, frequency meter and oscilloscope.

The following scheme presents the set used for measuring the dielectric properties (**Figure 2**):

1) Generators for the whole range: HP686A and G4 -

79 to 82;

2) Frequency meters: H 532A; FS - 54;
3) Cavity resonator;
4) Sample;
5) Oscilloscope EO 213.

4. Results and Discussion

4.1. Dielectric Properties

Figures 3 and **4** present the dependence of relative dielectric permittivity (ε_r) and dielectric loss angle tangent (tanδ) on frequency and filler concentration for the studied composites filled with graphene nanoparticles.

4.1.1. Complex Relative Dielectric Permittivity. Real Part

As seen from the plots relative dielectric permittivity increases with the increasing frequency at constant filler concentration as well as with the increasing filler concentration at constant frequency.

The real part of dielectric permittivity is the ratio of the capacity of an electric capacitor filled with the substance investigated to that of the same capacitor in vacuum, at a definite external field of frequency [17]. The dependence of the real part of the complex relative dielectric permittivity on the frequency ε_r (f) is shown in **Figure 3**. As seen in the concentration interval studied dielectric permittivity values for the non-filled and filled composites are relatively close at lower frequencies (up to 7 GHz). At frequencies higher than 7 GHz the dielectric permittivity increases with the increasing filler concentration. Moreover, there is a tendency of a more pronounced difference in the values for the non-filled and filled composites. The values we obtained for the non-filled composites are close to the ones for natural rubber at 1000 Hz reported in literature which are 2.40 - 2.70 [18]. As the figure shows our investigations also confirm the fact that the dielectric permittivity increases at higher frequencies. Of particular interest is the 9 - 12 GHz range wherein there is a relatively fast increase in the dielectric permittivity and its dependence on the GNP is the most prominent. Having in mind that the dielectric permittivity is related to the composites polarity [17], probably at frequencies higher than 9 GHz the polarization of the natural rubber matrix is hampered. Hence, the dielectric

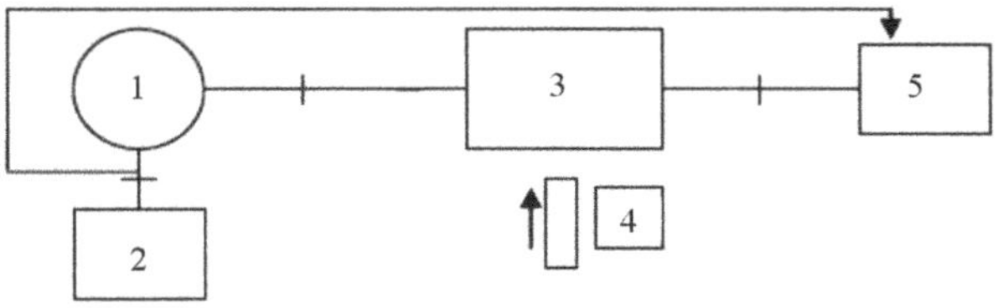

Figure 2. Scheme of the equipment for measuring the dielectric properties.

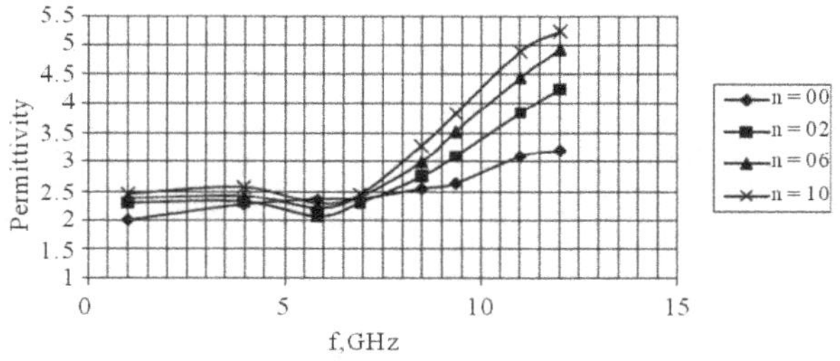

Figure 3. Frequency dependence of relative dielectric permittivity ε_r at various filler content (n-phr of graphene).

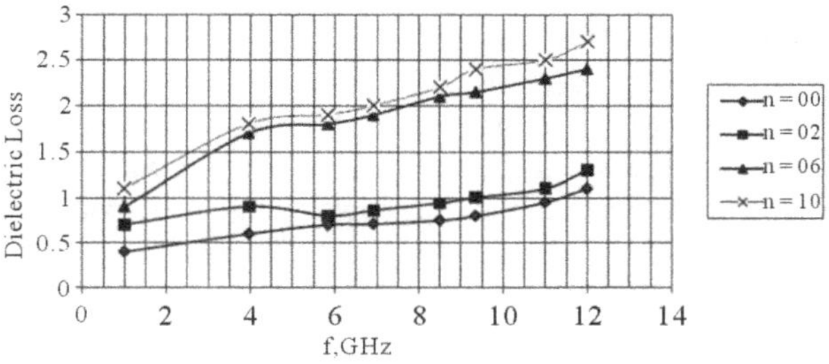

Figure 4. Frequency dependence of dielectric loss tanδ_ε at various filler content (n-phr of graphene).

permittivity increases. On the other hand, the increase in ε_r proves the lower polarization of the composite at higher frequencies.

The figure allows to calculate that the maximum difference within the range of 2 units is between the non-filled composite and the one comprising 10 phr GNP at 12 GHz. Besides, at frequencies higher than 7 GHz the frequency dependence of dielectric permittivity for the filled composites is much closer to the linear one.

4.1.2. Complex Relative Dielectric Permittivity. Imaginary Part (Dielectric Loss)

Figure 4 shows the frequency dependence of the imaginary part of the relative complex dielectric permittivity, also known as the dielectric loss angle tangent (tanδ_ε) in the 4 - 12 GHz range. As expected, the increase in frequency and filler amount leads to an increase in tanδ_ε values. The filler does not change the character of this dependence since it exists for the non-filled composite as well. Noteworthy is the fact that the addition of 2, 0 phr GNP has a relatively slight effect upon tanδ_ε values of the composite which are close to the ones of the matrix. However, tanδ_ε increases considerably at 6, 0 phr filler concentration. It is also seen that the frequency dependence of tanδ_ε both for the non-filled composite and the filled ones has an almost linear character. In the case of composites comprising a higher filler amount the frequency dependence of this parameter is more marked.

4.2. Microwave Properties

Figures 5-7 summarize the microwave properties of the

composites studied within the 1 - 12 GHz frequency range.

4.2.1. Coefficient of Reflection

The coefficient of reflection is a complex number $\dot{\Gamma}$. For the aims of the present study it is enough to estimate its modulus $|\Gamma|$ that is the reflected part of the electromagnetic wave falling upon the sample. That allows determining the absorption properties of the tested material. **Figure 5** shows the frequency dependence of $|\Gamma|$. As seen the higher the frequency and filler amount the higher $|\Gamma|$ is. The figure also reveals that $|\Gamma|$ values do not exceed 0.60 within the entire frequency range. A connection should be made between the dielectric losses mentioned above (**Figure 4**) which smoothly increase, though in a relatively small range. That to a great extent explains the small changes in the attenuation coefficient. One should keep in mind the fact that the composite comprises a foaming agent making its structure porous what lowers its attenuation coefficient values. As a whole the attenuation coefficient slightly increases with the increasing frequency and filler amount. The increase is particularly in the 6 - 12 GHz region. In the 1 - 6 GHz region the effect of frequency and filler amount on the attenuation coefficient is less pronounced. The effect is in accordance with the one upon the real part of the complex dielectric permittivity (**Figure 3**).

4.2.2. Attenuation Coefficient

The frequency dependence of the attenuation coefficient

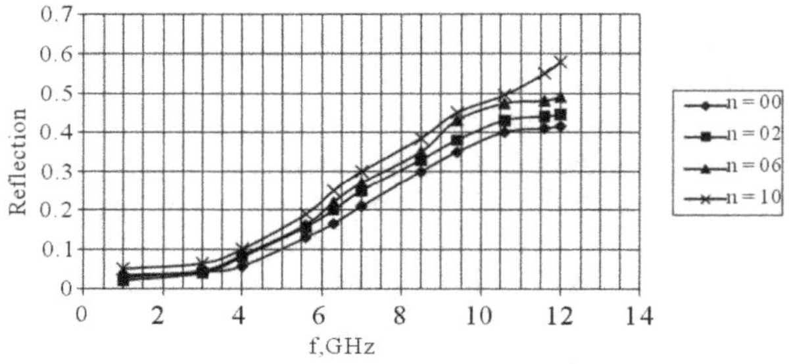

Figure 5. Frequency dependence of reflection $|\Gamma|$ at various filler concentrations (n-phr of graphene).

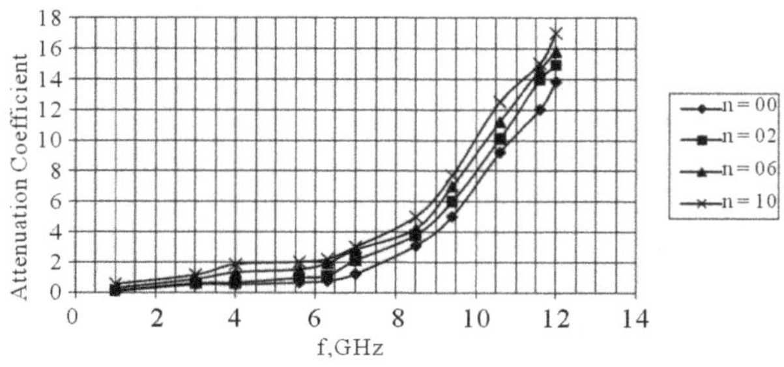

Figure 6. Frequency dependence of attenuation coefficient α of electromagnetic waves at various filler concentrations (n-phr of graphene).

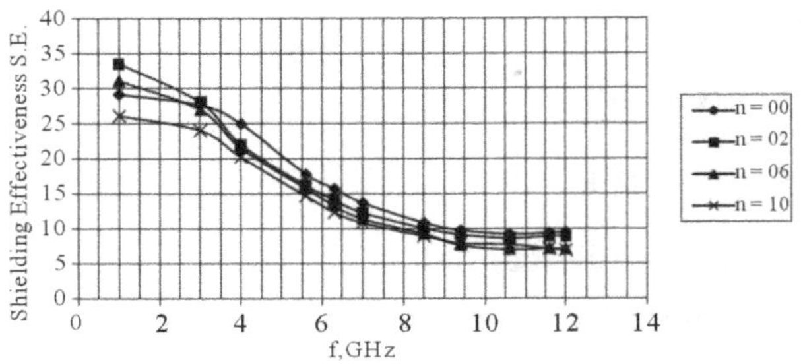

Figure 7. Frequency dependence of shielding effectiveness at various filler content (n-phr of graphene).

(α) could be studied in three frequency ranges: 1) 1 - 6 GHz; 2) 6 - 9 GHz and 3) 9 - 12 GHz (**Figure 6**). In the first range $\alpha \leq 2.2$ dB/cm; in the second $2.2 \leq \alpha \leq 6$ dB/cm and in the third $6 \leq \alpha \leq 17$. As the figures show in the first frequency range the attenuation remains almost unchanged, especially for the non-filled composites and for those with a small amount of filler.

There is a statistically significant attenuation increase with the increasing frequency and filler amount. This increase is considerable in the third frequency range especially for the filled composites. The effect is less pronounced for the non-filled ones. The comparison of the said dependence with those of the reflection coefficient and dielectric permittivity reveals a similarity that is obviously due to the same factor (the structure of grapheme and its specifics). It should be noted that the maximum attenuation of 17 dB/cm at 12 GHz is for a composite comprising 10 phr GNP.

4.2.3. Shielding Effectiveness

Figure 7 plots the frequency dependence of the electromagnetic shielding effectiveness (S.E.). At lower frequencies S.E. values are higher but decrease with the increasing frequency. In the (1 - 7) GHz the dependence has a character close to the linear one. There is also a range (7 - 12 GHz) of a relatively small effect of frequency and filler amount on S.E. values which remain in the 7 - 10 dB range. The initial values are due mainly to the return loss which are of interference nature. At the highest frequencies in the said range the attenuation in the sample is not great enough to compensate the increase in the reflection coefficient (**Figure 5**) and (**Figure 6**), hence the shielding effectiveness decreases.

5. Conclusions

1) The effect that graphene nanoparticles have upon the dielectric and microwave properties of natural rubber based composites filled at 2.0 up to 10.0 phr has been studied in the 1 - 12 GHz range.

2) It has been found that the dielectric permittivity increases slightly with the increasing frequency and filler

amount. The effect is pronounced especially in the 9 - 12 GHz range. The tendency is the same in the case of dielectric loss angle tangent, although the impact of the filler amount is less marked, *i.e.* the imaginary part of the complex relative dielectric permittivity is less sensitive to alternations of the filler amount and frequency than the real part.

3) The reflection coefficient increases with the increasing filler amount and frequency. The effects are not well pronounced. The availability of a foaming agent contributes to the low reflection coefficient.

4) The attenuation coefficient has been found to be almost frequency and filler amount independent in the 1 - 6 GHz range, then in the 6 - 9 GHz range it increases slightly, while in the 9 - 12 GHz range its values increase drastically with the increasing frequency and filler amount. The dependence is almost a linear one.

5) The frequency and filler amount have an effect upon the reflection and attenuation coefficients as well as upon the electromagnetic shielding effectiveness. S.E. values are in the 10 - 34 dB range. The attenuation is not great enough to compensate the increasing reflectance. Therefore as a whole the shielding effectiveness decreases gradually with the increasing frequency, especially at frequencies up to 7 GHz.

6) The effect of graphene nanoparticles upon the dielectric and microwave properties of the composites studied occurs mainly in the 6 - 12 GHz range at a minimal filler amount of 6 phr.

7) The results achieved allow recommending graphene as a filler for natural rubber based composites to afford specific dielectric and microwave properties, especially when their loading with the much more expensive carbon nanotubes is not possible.

6. Acknowledgements

The present research is a result of an international collaboration program between the University of Tabuk, Tabuk 71491, Kingdom of Saudi Arabia and the University of Chemical Technology and Metallurgy, Sofia, Bulgaria. The authors gratefully acknowledge the financial support from the University of Tabuk.

REFERENCES

[1] S. Thomas and R. Stephen, "Rubber Nanocomposites Preparation, Properties and Applications," Wiley, Singapore, 2010.

[2] P. M. Ajayan, "Single-Walled Carbon Nanotube-Polymer Composites: Strength and Weakness," *Advanced Materials*, Vol. 12, No.10, 2000, pp. 750-753. doi:10.1002/(SICI)1521-4095(200005)12:10<750::AID-ADMA750>3.0.CO;2-6

[3] E. T. Thostenson, "Advances in the Science and Technology of Carbon Nanotubes and Their Composites: A Review," *Composites Science and Technology*, Vol. 61, No. 13, 2001, pp. 1899-1912. doi:10.1016/S0266-3538(01)00094-X

[4] A. Krueger, "Carbon Materials and Nanotechnology," Wiley, New York, 2010.

[5] K. S. Novoselov, "Electric Field Effect in Atomically Thin Carbon Films," *Science*, Vol. 306, No. 5696, 2004, pp. 666-669. doi:10.1126/science.1102896

[6] H. J. Salavagione, G. Martínez and G. Ellis, "Graphene-Based Polymer Nanocomposites," In: S. Mikhailov, Ed., *Physics and Applications of Graphene—Experiments*, InTech, Rijeka, 2011, pp. 169-192.

[7] M. J. Allen, V. C. Tung and R. B. Kaner, "Honeycomb carbon: A Review of Graphene," *Chemical Reviews*, Vol. 110, No. 1, 2010, pp. 132-145. doi:10.1021/cr900070d

[8] Y. Zhu, S. Murali, W. Cai, X. Li, J. W. Suk, J. R. Potts and R. S. Ruoff, "Graphene and Graphene Oxide: Synthesis, Properties, and Applications," *Advanced Materials*, Vol. 22, No. 35, 2010, pp. 1-19. doi:10.1002/adma.201001068

[9] O. C. Compton and S. T. Nguyen, "Graphene Oxide, Highly Reduced Graphene Oxide, and Graphene: Versatile Building Blocks for Carbon-Based Materials," *Small*, Vol. 6, No. 6, 2010, pp. 711-723. doi:10.1002/smll.200901934

[10] P. Mukhopadhyay and R. K. Gupta, "Trend and Frontiers in Graphene-Based Polymer Nanocomposites," Plastics Engineering, 2011. www.4spe.org

[11] R. K. Prud'Homme, B. Ozbas, I. Aksay, R. Register and D. Adamson, "Functional Graphene-Rubber Nanocomposites," US Patent No. 7745528, 2010.

[12] J. Liang, Y. Huang, Y. Ma, Z. Liu, J. Cai, C. Zhang, H. Gao and Y. Chen, "Electromagnetic Interference Shielding of Graphene/Epoxy Composites," *Carbon*, Vol. 47, No. 3, 2009, pp. 922-925. doi:10.1016/j.carbon.2008.12.038

[13] I. M. De Rosa, F. Sarasini, M. S. Sarto and A. Tamburrano, "EMC Impact of Advanced Carbon Fiber/Carbon Nanotube Reinforced Composites for Next Generation Aerospace Applications," *IEEE Transactions on Electromagnetic Compatibility*, Vol. 50, No. 3, 2008, pp. 556-563. doi:10.1109/TEMC.2008.926818

[14] I. M. De Rosa, R. Mancinelli, F. Sarasini, M. S. Sarto and A. Tamburrano, "Electromagnetic Design and Realization of Innovative Fiber-Reinforced Broad-Band Absorbing Screens," *IEEE Transaction on Electromagnetic Compatibility*, Vol. 51, No. 3, 2009, pp. 700-707. doi:10.1109/TEMC.2009.2018125

[15] I. M. De Rosa, A. Dinescu, F. Sarasini, M. S. Sarto and A. Tamburrano, "Effect of Short Carbon Fibers and MWCNTs on Microwave Absorbing Properties of Polyester Composites Containing Nickel-Coated Carbon Fibers," *Composites Science and Technology*, Vol. 70, No. 1, 2010, pp. 102-109. doi:10.1016/j.compscitech.2009.09.011

[16] G. De Bellis, I. M. De Rosa, A. Dinescu, M. S. Sarto and A. Tamburrano, "Electromagnetic Absorbing Nanocomposites Including Carbon Fibers, Nanotubes and Gra-

phene Nanoplatelets," *Proceedings of the* 2010 *IEEE International Symposium on Electromagnetic Compatibility*, Fort Lauderdale, 25-30 July 2010, pp. 202-207. doi:10.1109/ISEMC.2010.5711272

[17] P. Banerjee and S. Biswas, "Dielectric Properties of EVA Rubber Composites at Microwave Frequencies," *Journal of Microwave Power and Electromagnetic Energy*, Vol. 45, No. 1, 2011, pp. 24-29.

[18] A. Kornev, A. Bukanov and O. Sheverdiaev, "Technology of Elastomeric Materials," in Russian, Istek, Moscow, 2005.

5

Nanofiller Dispersion in Polymer Dielectrics

Daniel Tan, Yang Cao, Enis Tuncer, Patricia Irwin

Dielectrics & Electrophysics Lab, GE Global Research Center, Niskayuna, Schenectady, USA.
Email: tan@ge.com, tuncer@ge.com

ABSTRACT

Nanodielectric composites have been developed in recent years attempting to improve the dielectric properties such as dielectric constant, dielectric strength and voltage endurance. Among various investigations, nanoparticle dispersion was particularly emphasized in this work. General Electric Global Research Center in Niskayuna NY USA has investigated various nanoparticles, nanocomposites and nanocomposite synthesis methods intending to understand particle dispersion and their impact on the nanocomposite dielectric properties. The breakdown strength and microstructures of the nanocomposites containing different particles were studied for projects related to capacitor and electrical insulation technologies. The nanocomposite synthesis methods either employed commerical nanoparticles or utilized nanoparticles that were self-assembled (*in-situ* precipitation) in a matrix. Our investigations have shown that nanocomposites prepared with solution chemistry were more favorable for producing uniform dispersion of nanoparticles. Structural information of nanocomposites was studied with transmission electron microscopy and the interection between particles and matrix polymers were tentatively probed using dielectric spectroscopy. In these new class of materials high energy densities on the order of 15J/cc were achievable in nanocomposites.

Keywords: Nanoparticles; Polymer; Dielectrics; Filler Dispersion; Dielectric Properties

1. Introduction

The rapid development of new advanced and renewable energy technologies as well as defence applications demand high energy density, high power density and high temperature components [1]. Among these components are advanced capacitors that greatly require the advanced dielectric materials and film processing techniques. It has been well recognized that every new generation of capacitors have been primarily credited to the innovation and engineering of new dielectric materials. The process cycle from dielectric materials (development and chemistry) all the way to capacitor fabrication (metallization, wounding and contact spray) is shown in **Figure 1**.

Conventional dielectric material development followed the paths of organic polymers, inorganic ceramics, and thin films separately, due to type of capacitors manufactured, polymeric film and multi-layer ceramic capacitors. Polymer dielectrics show very high dielectric strength (>300 kV/mm), lower dielectric losses (<0.01), and adequate mechanical flexibility in processing, which is important in thin film capacitors. However, they have low relative dielectric permittivity or constant (<4) and low operating temperatures (<200˚C) compared to ceramic materials. Ceramic dielectrics tend to have very high dielectric permittivity (>100) but relatively low dielectric strength (<50 kV/mm) and/or may be piezoelectric associated with the structure of (ferroelectric type) ceramics. The low dielectric strength in ceramics is caused by the presence of grain boundaries, intrinsic porosity, impurities, surface defects, and chemical deterioration. Thin film dielectrics are usually in nanometer to submicron in thickness with very high breakdown strength, and they require a supporting substrate. But, they are primarily useful for low voltage and microelectronics application, which are outside the scope of our current work. Increasing the dielectric breakdown strength and ensuring scalable, and reliable films are one of the challenges and goals in modern day dielectrics research in capacitor and electrical insulation applications.

Nanodielectric composites belong to a new type of materials, engineered for improved specific functionality, such as better performance in electrical insulation. Certain ceramic materials can be selected in powder form to be blended with polymers to provide synergy between the high breakdown strength polymer and high permittivity ceramic materials as shown in **Figure 2**. A number of research areas have been being actively pursued in order

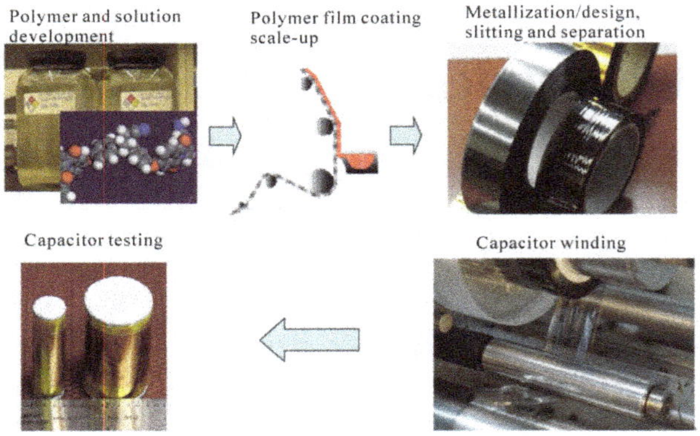

Figure 1. A sketch of capacitor manufacturing from materials to produced capacitors. Evolution of capacitor technology has a close relationship with materials development.

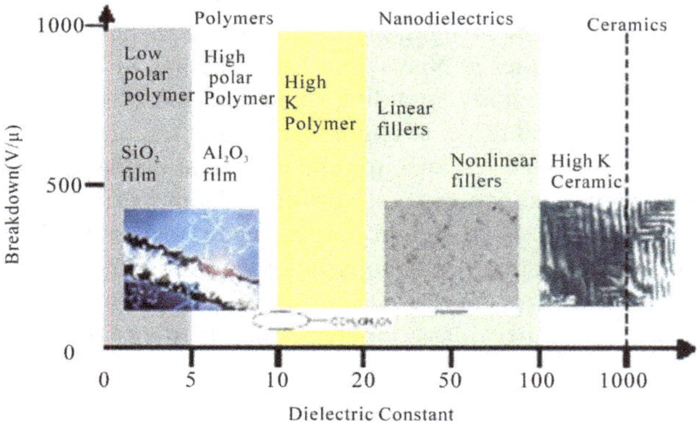

Figure 2. Relationship between breakdown strength and dielectric permittivity of various dielectric materials.

to fully explore the advantages of the functional composites for energy storage applications.

Nano-sized inorganic particles have received great attention due to their advantages in processing uniform composites and for their potential in voltage stress improvement in high voltage technology as opposed to microsized particles [2-4]. Notice that nanoparticles result in significant particle agglomeration that needs to be resolved during nanocomposite production, high interface fraction, and porosity in polymer matrices, if no special attention is given to mixing method. Porosity could result in both Maxwell-Wagner polarization at low frequencies and lower dielectric permittivities. Solving these issues will result in better nanocomposites for dielectric component and insulation applications. The current paper review the recent progresses in our labs in nanodielectric composite investigations with focus on the effect of nanofillers and its nature, size, distribution and particle-polymer interface. The particle dispersion and its importance in nanodielectric engineering is discussed. Influence of different nanocomposite synthesis methods and the resulting nanostructural features as well as the dielectric properties of selected composites are presented. The majority of the nanocomposites were prepared in wet solution method assisted with a sonication for filler dispersion. The particles were dispersed in a solvent and later introduced into a matrix polymer. For characterizations of electrical properties of nanocomposites several to tens of micrometer thick films were cast using either a solvent cast or spin coating methods. DC breakdown tests were conducted following ASTM D149 (method A) using a ball-plane electrode configuration. Structural characterization of filler shape and dispersion were studied using a transmission electron microscopy technique.

2. Need for High Energy Density Dielectric Films

Various government agencies have been actively solicitating dielectric materials for energy storage that can offer a high energy density of >20 J/cm^3 with high operating temperature capability (as high as 200˚C). Examples of applications are pulsed power and power conditioning applications for electrical grid. As a response, different methods and material approaches were studied to achieve the requirements [5,6]. For example, a homogeneous dispersion of ferroelectric $BaTiO_3$ particles were prepared in polypropylene with an in-situ polymerization process [3].

Although the permittivity was increased by almost 3*x* over polypropylene, the breakdown strength of the composite was degraded. Dielectric permittivity and energy density are related via design breakdown strength of the dielectric as shown in **Figure 3**. Note that the dielectric loss tangent in the selected material should be low enough (lower than 1%) over a broad frequency range. The relationship in **Figure 3(a)** indicates that development of better dielectric materials would be appreciated for transformational energy storage devices.

As an example, for a dielectric film to meet an energy density higher than 20 J/cm^3, the dielectric material needs to satisfy the conditions in **Figure 3(b)**, one candidate may have a dielectric permittivity of 10 - 20 and a high breakdown strength of >600 kV/mm. The lines in **Figure 3(b)** illustrate different energies densities and requirements on the dielectric properties of the candidates. If a composite material with a permittivity of higher than 60 is considered, much lower breakdown strength (<300 kV/mm) is required. Polypropylene films with relative permittivity 2.2 and design strength 800 kV/mm result in 5.6 J/cc, indicated with red symbol in **Figure 3(b)**. Other materials and composites are also labeled in **Figure 3(b)**.

One of the approaches to achieve high energy densisty has been the novel dielectric materials designed using nanoengineering and nanotechnology [4,5]. These high performance nanodielectric composites have become one of the strategic building blocks in material technology in the last decade.

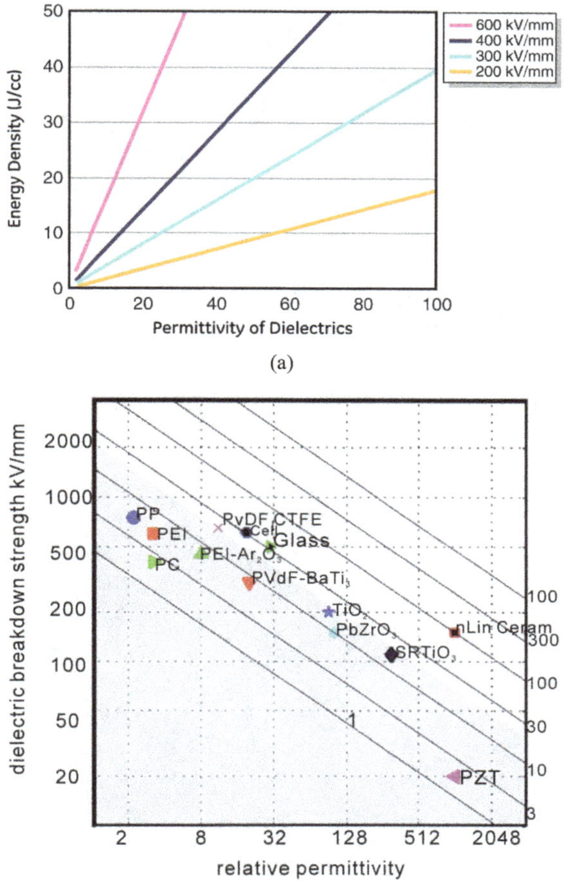

Figure 3. Energy density as a function of relative permittivity for different dielectric breakdown values. To reach the targeted energy density of >20 J/cm^3 for film material, high permittivity film or higher breakdown strength are needed. The figure on the right illustrates the equi-energy lines for 3, ..., 1000 J/cc, together with the current-state-of-the-art materials.

For this purpose, a number of research labs around the world have studied nanodielectrics to increase the dielectric permittivity of a polymers [6-9]. Even with properly selected inorganic fillers, a great success has not been achieved without retaining the breakdown strength and low dielectric loss for a broad frequency range. The large losses caused by mainly large mismatch in electric properties of constituents, relative permittivity and electrical conductivity, and interface between inorganic filler and polymeric matrix. In fact, simply filling a polymer matrix with high permittivity particles was found to yield a slight increase in composite effective permittivity. The shape of filler particles and their concentrations as well as their distribution are important factors to be considered besides the dielectric properties of the filler [10].

For example, an antiferroelectric type of ceramic material made of lead zirconate composition is able to exhibit a significantly high dielectric permittivity peak after sintering the particles at about 950°C or 1120°C, respectively (**Figure 4**). The relative permittivity of the sintered ceramic is up to several hundred at room temperature as shown in **Figure 4**. After blending 10 wt% the powder into polyetherimide (PEI) films, the composite relative permittivity was dramatically lower, about 15, compared to the relative permittivity of the powder as shown by the inset in **Figure 4**. Studies performed at General Electric Global Research Center (GE GRC) have found that high relative permittivity polymers should better be utilized in composites. However, a significant amount of effort is still required to develop materials to compete with conventional device dielectrics.

3. Nanofiller Dispersion in Polymers

One of the common challenges in blending hard (inorganic fillers) and soft (polymers) materials is the dispersion of filler particles. Usually, this has a lot to do with the filler surface energy. For example, the SiO_2 particles were often found agglomerated in polymer matrices [11]. Mixing assisted with high energy shear force generate better results via creating enough energy to the separate agglomeration of particles [10]. A high energy sonication process has been found to be an effective method to improve the particle distribution [12-14]. **Figure 5(a)** shows the TEM image of 45 nm Al_2O_3 particles of 5 vol% blended in a PEI film. Quite uniform distribution and less degree of agglomeration were obtained; observe that the particle size distribution is wide. However this method is still insufficient to disperse finer particles such as 11nm Al_2O_3 as shown in **Figure 5(b)**. The figure shows the importance of force or energy needed to separate small nanoparticles. The force is related to the particle-particle interaction and the viscosity of the medium that they were dispersed in. Utilizing surfactants or caping agents help to improve dispersion via lowering surface energy requirements.

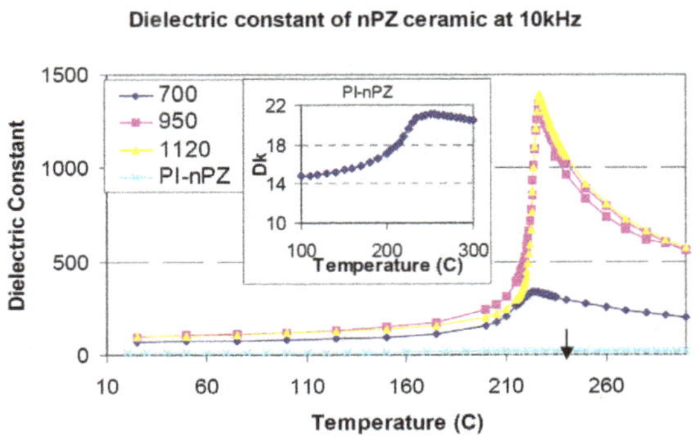

Figure 4. The temperature dependence of the relative dielectric permittivity of lead zirconate ceramics for different process conditions, and nanoparticle PZ filled polyetherimide polymer.

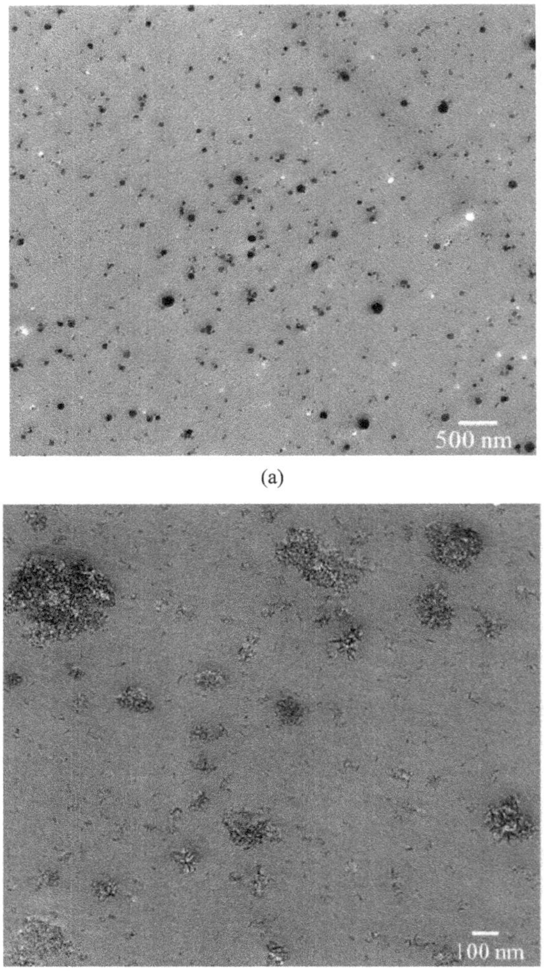

(a)

(b)

Figure 5. Transmission electron microscope images of dry nanoparticles in Ultem™. (a) 45 nm Al_2O_3 and (b) 11 nm Al_2O_3. Observe that Al_2O_3 particles are clustered and the mixing method could not break the particle agglemorations.

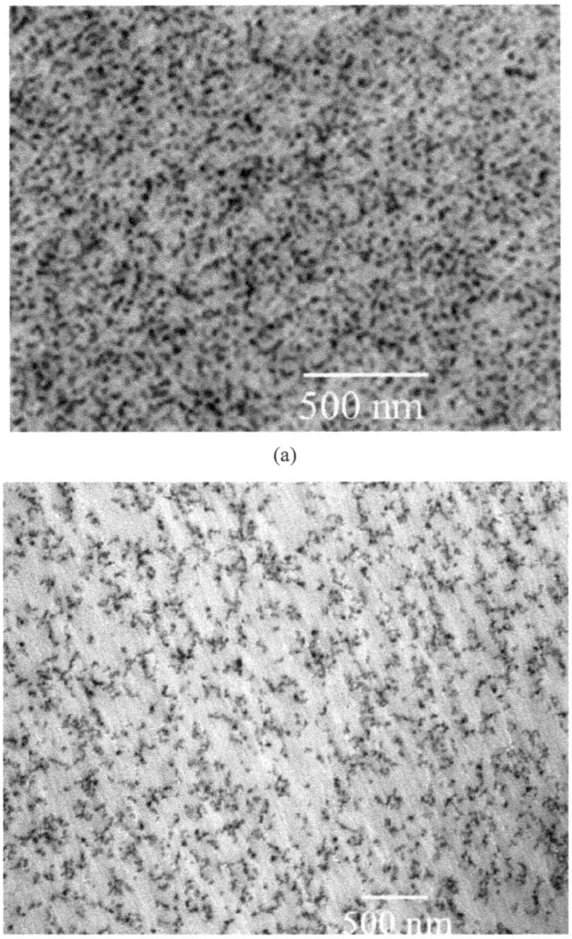

(a)

(b)

Figure 6. Transmission electron microscope images of wet Sb_2O_3 and SiO_2 nanoparticles (5 vol%) in Ultem™. The particles are better dispersed in the matrix compared to dry-powder mixed nanocomposites.

It has been shown previously that one can achieve uniform dispersion of particles when nanoparticles were synthesized using solution chemisty-particle precipitation method [12]. GE GRC found that better particle dispersions could be achievable when the nanoparticles were synthesized in-house and preserved in the solvent (*i.e.*, wet particles). The images of Sb_2O_3 and SiO_2 particles of 5 vol% that were <20 nm, and dispersed in PEI polymer were shown in **Figures 6(a)** and **(b)**. The particles are randomly distributed in the microscopic scale. Nevertheless, certain agglomeration of fine structure still remains due to lack of surfactant additives.

Much better particle dispersions were demonstrated by a small number of research labs utilizing *in-situ* polymerization process or self-assembling within a polymer matrix [15]. For *in-situ* polymerization process, the nanoparticles were first blended with the host monomer solution followed by a polymerization process. **Figure 7(a)** clearly shows the individual SiO_2 nanoparticle distributed randomly in the host polymer, where no particle agglomeration was observed [16]. The good dispersion and adhesion of modified $BaTiO_3$ in a polyamic matrix was also exhibited as shown in **Figure 7(b)** [17]. This method is effective to avoid particle agglomeration, however, the processing requires good control. This approach has not been widely adopted for nanodielectric composite processing in our labs.

Another method is *in-situ* self-assembling of nanoparticles in polymeric matrix. The *in-situ* nanoparticle precipitation method for nanocomposites has been developed for cryogenic dielectric applications by Tuncer *et al.* [18,19]. In this method, a metal salt precursor is dissolved and blended into a dissolved polymer solution. By controlling the acidity and temperature, the precursor synthesizes metal-oxide nanoparticles via chemical reac-

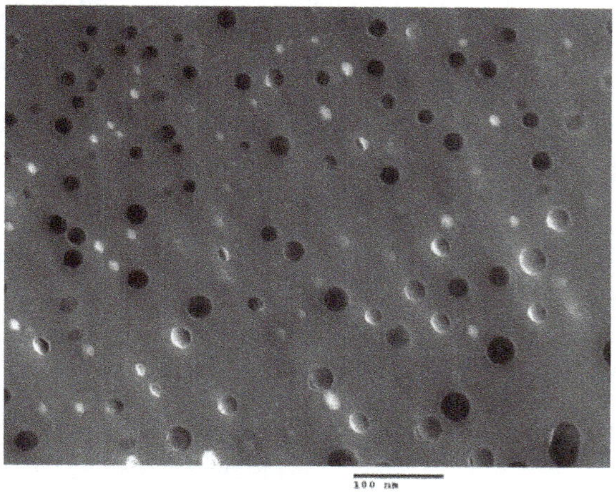

(a)

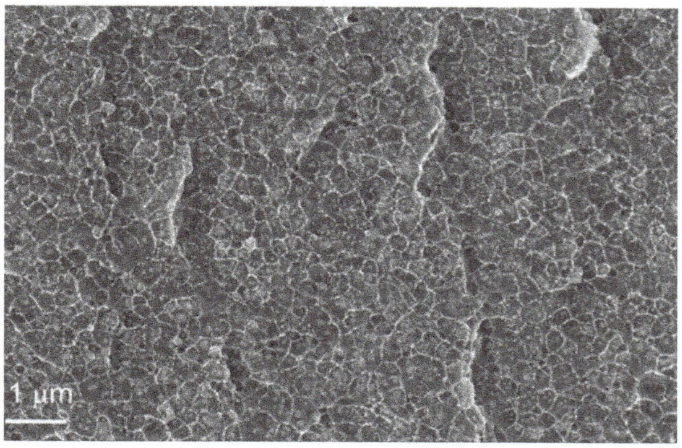

(b)

Figure 7. TEM images of nanoparticles from in-situ polermization of (a) 5 vol% SiO_2 in polyamicacid, and (b) 15 vol% 40 nm $BaTiO_3$ in a polyimid containing 2,2-Bis[4-(3,4-dicarboxyphenoxy) phenyl]propane dianhydride (BPADA) with 4,4'-oxydianiline (ODA).

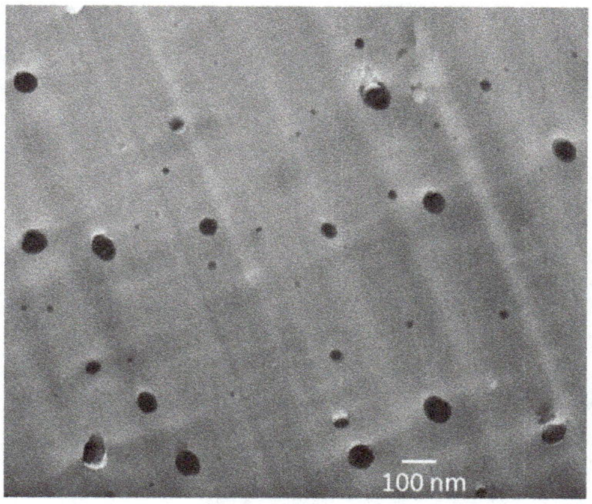

Figure 8. TEM image of 5 wt% TiO_2 nanoparticles dispersed in a PPSU film using a self-assembled *in-situ* particle precipitation process. Nanoparticles nucleated at rondom locations in the matrix that yield well-dispersion of particles without any agglemoration.

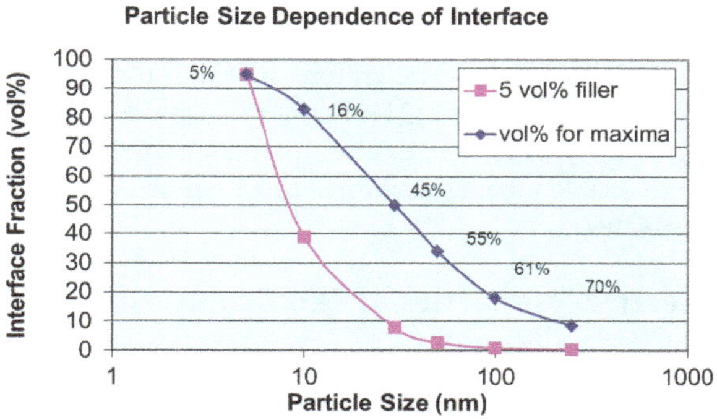

Figure 9. Particle size dependence of interface fraction of nanoparticles in a polymer matrix. The blue curve represents the volume percentage of particle required to reach the maximum interface fraction for various particle sizes.

tion. The dissolved polymer acts like surfactant for the synthesized nanoparticles which indicate good dispersion in the solution. Dissolved polymer molecules create tiny reaction chambers for nanoparticle synthesis. After solvent casting and drying films, the synthesized particles are randomly distributed, and stablized (frozen in location) in a polymer matrix. TiO_2 particle polyphenylsulphone *in-situ* nanocomposite with good nanoparticle dispersion and size distribution is shown in **Figure 8**. The cast film was 25 micron thick.

It was noted that interface fraction becomes increasingly important when the particles are much smaller. According to the computation made by Raetzke *et al.* [20], the particle interface becomes significantly high when the particle sizes are below 10 nm, see also Tuncer [18]. **Figure 9** is the re-plot of the relationship between particle interface and particle size under various filler contents. For particles of 10 nm, the composite with 5 vol% of particles contains about 40 vol% interfaces. The 5 nm particles can result in 95 vol% interface of the composites. For 10 nm particles, 15 vol% particles could result in more than 85% of interface. Tuncer *et al.* also reported very high particle density and shorter particle-particle distance at lower volume percent of particles [18]. As a result, even very small amount of particles would result in percolation where particle interconnection occurs, which would be very unlikely due to required statistical outcomes for the given size and thickness.

Experimental evidence of high interface and particle interconnection was obtained from the cyanothyl cellulose film as shown in **Figure 10**. When 20 vol% of 8 nm $BaTiO_3$ particles was added, it is clear that the particles agglomerate and form large clusters, which would be the reason for the observed high dielectric permittivity-non-spheroidal shapes in the field direction.. Particles are hard to be observed in the bulk; they introduced high interface regions. Previous investigations have shown that $BaTiO_3$ particles smaller than 15 nm can lose most

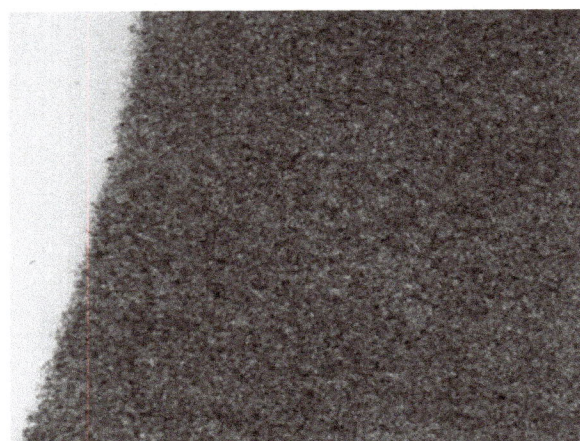

Igue08-40 vol BT in CRC_003.tit
Vol Barium Titinate in CRC
10 nm

Figure 10. TEM images of highly concentrated $BaTiO_3$ nanoparticles (20 vol%) in cyanothyl cellulose. The particles are clearly visible on the edge, otherwise it is hard to differentiate the particles in the bulk.

of their ferroelectricity, and therefore their dielectric constant is drastically lowered [21]. Further improvements in synthesis, purity and size control of nanoparticles are required to enable a better performance of the nanodielectric composites.

4. Dielectric Properties of Nanocomposites

We should concentrate how we can better utilize nanofillers and created high internal surfaces. When the particles are not interconnected, the contribution of the individual particles to the composite permittivity is limited; this is true for spherical particles. When permittivity ratio between the inclusion particles and the matrix is large, the effective permittivity of the composite would be close to the matrix permittivity below the percolation threshold [22]. At percolation threshold and above the particles would have high probability to form interconnections (clusters) union of particles act as non-spheroidal (if they were spheroidal initially) inclusions. If the union of the particles is in the direction of the applied electrical field, they would yield high polarization/depolarization values due to these effective particle shapes. In addition depending on the frequency of interest the interfacial polarization would play an additional role in the polarization, thus the effective permittivity of the composite. In 3-dimensions the percolation threshold for spherical particles is about 1/3; however, remember that the shape of inclusions will change the threshold value, which is the case when particle clusters are formed.

If we reconsider the interfacial contribution once again, embedding 5 vol% of 20 nm particles would only yield about 10% interface, if the interface has some contribution to the permittivity other than the known interfacial polarization (or Maxwell-Wagner-Sillars polarization).

GRC found that adding 5% of colloidal SiO_2 particles in a PEI film only increased the permittivity to 4 (*i.e.*, 25% higher); however due to solvent casting procedure we anticipate that there either were ionic species or residual solvent that result in a higher permittivity for the composite than the constituents-PEI and SiO_2 have relative permittivities 3.2 and 3.9, respectively. Similarly, adding 5 vol% of $BaTiO_3$ nanofillers in the cyanoethyl cellulose (CR), as described in **Figure 9**, increased the permittivity only by 25% in spite of the high permittivity of $BaTiO_3$. When the particle loading exceeds the percolation level, the permittivity was significantly increased by more than 5 times as shown in **Figure 11** [23]. Therefore, the direct contribution of the interconnected particles is most important to the increase of composite permittivity. When the nanoparticles were very small, we have found that the interconnection between particles did not cause mechanical brittleness; polymer can cover the particles depending on the size of the particles and the gyration radius of polymer chains.

Interfacial contribution, particle surface chemistry and contamination due to ionic species become important in nanocomposites that alter the electric properties of a composite, effective permittivity and conductivity, and also the dielectric breakdown strength. The interface contribution should be studied not only on its nature but also the way it affect the composite. Todd *et al.* investigated the effect of coupling agents on the interphase dielectric constant using experimental data and molecular dipole polarization calculations [24]. They have shown that the dielectric characteristics of polymer-ceramic interphase regions affect the overall dielectric response of the composite system. The interphase region depends largely on the chemical bonding at the filler interface that is capable of forming covalent or hydrogen bonds to the surface of the filler. The material in the interface has a different dielectric characteristics; it is an interphase region of a polymer-ceramic composite.

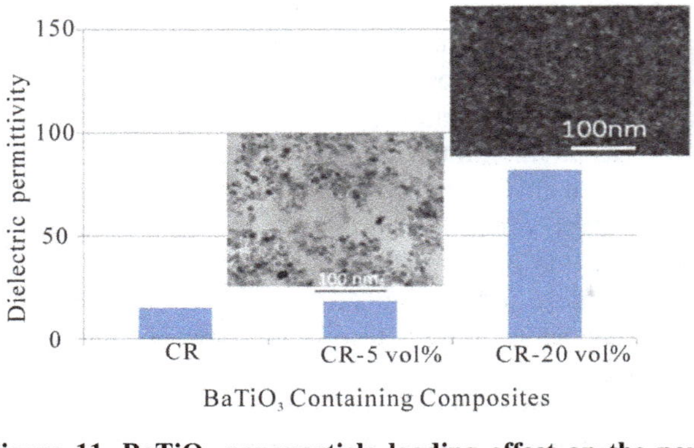

Figure 11. $BaTiO_3$ nanoparticle loading effect on the permittivity of the composites.

The dynamic response of the particle interfaces was studied by Tan *et al.* using a dielectric spectroscopy technique capable of working on a wide frequency and temperature range [25]. For the selected base polymer (PEI) containing no particles, and thus, no interfaces (interfacial polarization), the dielectric loss spectra exhibit low loss tangents of the polymer (**Figure 12**). The permittivity of the PEI is not shown since in the considered frequency and temperature ranges it has small dispersion also reflected in the loss tangent plots, which are lower than 1%. The dielectric properties of the PEI produced by extrusion and solvent (spin) casting are shown to indicate the differences in the material due to processing. However as shown in **Figure 12**, there are only slight differences at low temperatures (below 50°C) and high temperatures above 200°C. The PEI polymer is an excellent material for high temperature capacitor applications because of low loss tangent over broad temperature and frequencies.

Influence of particles and particle size on the dielectric loss tangent spectra were investigated as shown in **Figure 13**. The dielectric behavior of the polymer was altered with addition of different size of alumina particles. The particle concentration was 5 vol% in the investigations. Besides the sample with 45 nm particles, the loss tangent behavior at high temperatures and low frequencies were similar to unfilled PEI. We anticipate that the nanocomposite sample with 45 nm alumina had defects that generated high losses at high temperatures and low frequencies, which could be a results of charge conduction introduced with the defects. We discuss this later in the text. The smallest particles that were considered in the investigations were 11 nm and the composite had a loss peak at high frequencies probably due to the large surface area introduced by the alumina nanoparticles, which could be related to the interfacial polarization. The composites with large particles, 150 nm and 1 micron, illustrate that the changes in the dielectric loss tangent behaivor were higher than the unfilled PEI. The loss tangents between 150 nm and 1 um particle samples were similar with 1um particles generating slightly more loss tangents at high tempatures and low frequencies.

Size of nanoparticles has known to alter the dielectric behavior of sintered ceramics [26], such that small nanoparticles yield low dielectric permittivity sintered materials. This observation brings the question that as particle size is decreased what would be the permittivity of individual nanaparticles. Or in other words, would we observe similar dielectric behavior with similar size but

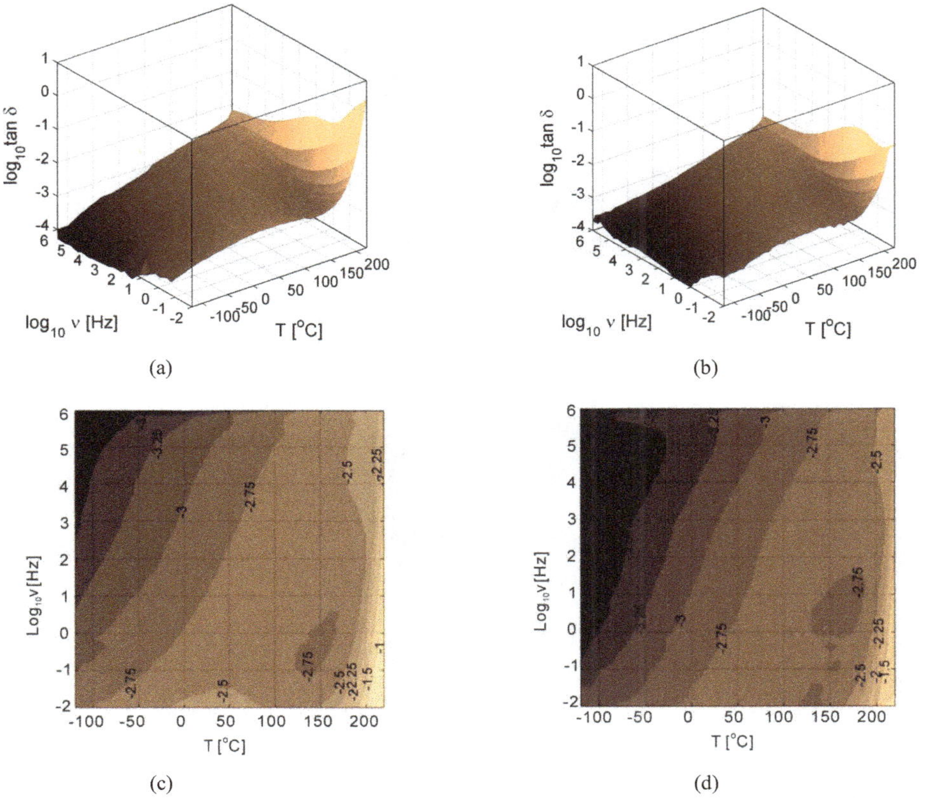

Figure 12. Dielectric loss tangent of PEI as a function of temperature and frequency. The three dimensional surface plots in (a) and (b) are shown in contour plots in (c) and (d) to help the readers to compare the loss behavior of unfilled material with those of filled material. (a) and (c) are for extruded PEI and (b) and (d) are for in-house solvent casted PEI.

different nanoparticles? Similarly one can ask what would be the electrical conductivity of the nanoparticles; size dependent conductivity. There is no data in the literature on this topic to our knowledge.

To study the type of particle and its influence on the dielectric loss tangent behavior of alumina and antimony oxide Sb_2O_3 particles were investigated for 5 vol% nanoparticle filled systems. The loss tangent is shown in **Figure 14** where the antimony oxide has higher losses than alumina. Although the relative dielectric permittivities of the two fillers are slightly similar, 10 for alumina and 12.4 for antimony trioxide [27] yielding similar permittivity ratios between the filler and the polymer-the electrical conductivities of the two fillers are different. The ratio of conductivies between PEI and alumina and PEI and antimony trioxide can be considered as 1 and 10 - 5, respectively, considering that the polymer and alumina have around 10 - 12 S/m, however antimony trioxide has 10 - 7 S/m [28]. This would generate a high loss due to the mismatch in the dielectric permittivities. While it is clear from **Figure 14** that the differences in the conductivities did not result in a significant loss tangent change in the dielectric response. Both 11 nm size fillers have a similar response considering that the behavior at high frequencies are similar. This observation indicates that perhaps the size of the particles are more important than the chemical nature of the particles.

As mentioned previously the purity of particles are important. For example, the particle surface contamination was determined using an analytical technique (XPS) [24], and concluded that the raw particles need to be cleaned up before blending into polymers. Presence of impurities on particle surfaces becomes detrimental dielectric performance under high electric fields (low dielectric breakdown strength). The data for the composite sample with 45 nm alumina shows the signs of this effect, see **Figure 13**. A cyanothyl cellulose polymer containing.

$BaTiO_3$ or $PbZrO_3$ particles was found to exhibit 50% higher breakdown strength when the particles were pretreated to remove surface contaminants [29]. Therefore, properly preparing the nanoparticles is critical for the augmentation of the dielectric properties of nanocomposites. Depending on the density of the conteminants, their influence can be detected in loss tangent via dielectric spectroscopy.

5. Discussion

To achieve improvements in the breakdown strength of nanocomposite have been mostly difficult over the years.

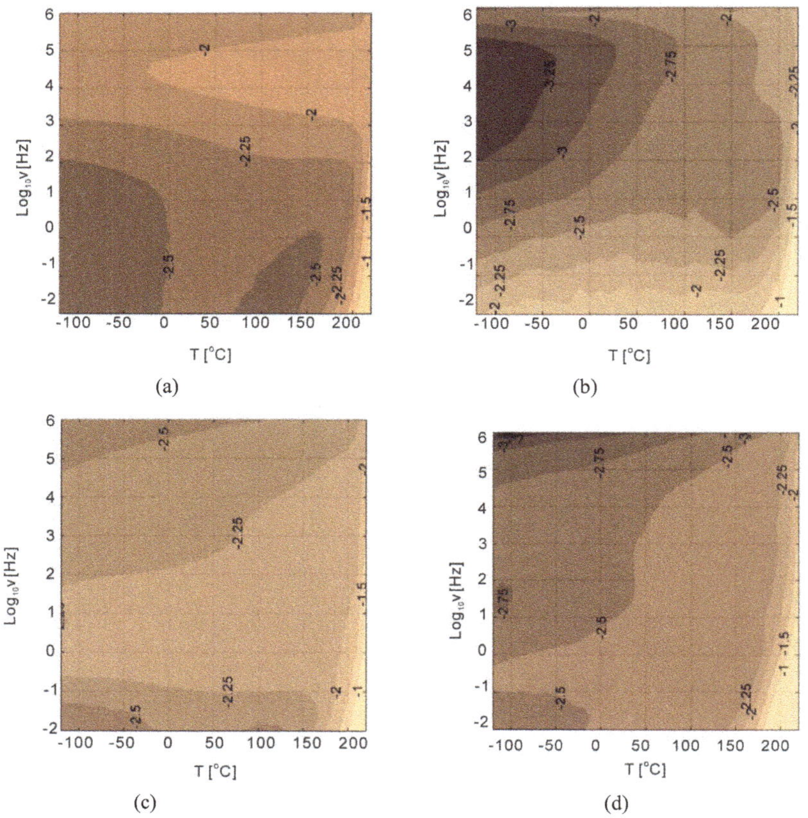

Figure 13. Dielectric loss tangent of PEI filled with alumina particles as a function of temperature and frequency. The size of particles is altered in the experiments, (a) 11 nm, (b) 45 nm (c) 150 nm and (d) 1 um. The materials are casted in-house.

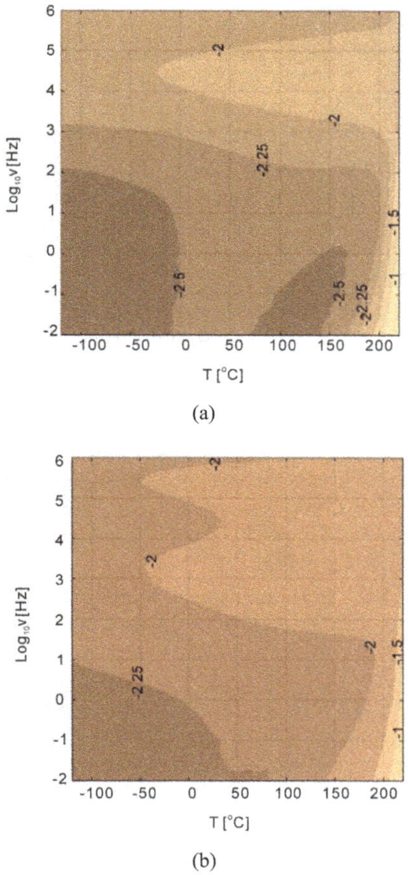

Figure 14. Dielectric loss tangent for (a) 11 nm alumina and 11 nm antimony trioxide filled PEI as functions of temperature and frequency.

In most cases, the breakdown strength equivalent to that of the base polymers were attained when the particles are well-dispersed as shown in micrographs as in **Figures 6-8** and **10** either by wet particle or in-situ particle precipitation methods. It has become a commonalty that higher loading would result in degradation in the breakdown strenght of the composites, due to increase number of particle aggregates. Therefore, leveraging low particle concentrations and high interface contribution associated with small nanoparticles would be a preferred method for equivalent or higher breakdown strength nanocomposites compared to unfilled matrix. As discussed previously, a good example of well-dispersed nanoparticles can be achieved byself-assembly of metal-oxides, see for example TiO_2 nanocompositesby Tuncer *et al.* who reported higher nanocompositebreakdown strength values than that of neat polymers [12,13]. Similar work with an epoxy nanocomposite reporteda breakdown strength increase by 20% [12,30].

Recently, breakdown strength of >700 kV/mm was achieved in thermoplastic films. These results demonstrated a promising method for reaching higher energy density materials than the conventional thermoplastics. Better engineering of nanocomposites and processing methods should be developed to achieve both higher per- mittivity and breakdown strength.

6. Conclusion

A brief review of polymeric nanodielectric composites work performed at GE GRC in Niskayuna NY was presented. The filler functionalization, distribution and dis persionhave clearly become major focus of interests in the nanodielectric research. It was found that the status of nanofiller precursors is critical to create well-defined nanostructures and desired or required dielectric and electrical properties. Successes in nanoparticle dispersion in nanocomposites require a solution chemistry and nanoparticle precursors. Examples for successful nanocomposite need wet particles, clean particle surface, suitable mixing methods such as *in-situ* and self-assembly. The on-going investigations at our Labsshow that a high dielectric permittivity and/or breakdown strength in nanodielectric composites could be achieved, if proper nanofillers and processing methods are utilized. For *in-situ* polymerization or precipitation processes, well-dispersed particles and precursors in a solution are desired; however, the discussed processes, development of their scale-up and production of nanoparticles and polymeric films still need significant efforts. For the presented self-assembled nanocomposites process, parameters and conditions are still in progress.

7. Acknowledgements

Support from DARPA DSO under a Contract FA9451-08-C-0166 is greatly acknowledged. Discussion on Barium Titanate nanoparticles with Professor Steven O'Biren and Dr. Limin Huang at Columbia University are greatly appreciated.

REFERENCES

[1] A. S. Arico, P. Bruce, B. Scrosati, J.-M. Tarascon and W. Van Schalkwijk, "Nanostructured Materials for Advanced Energy Conversion and Storage Devices," *Nature Materials*, Vol. 4, No. 5, 2005 pp. 366-377. doi:10.1038/nmat1368

[2] Y. Cao, P. Irwin and K. Younsi, "The Future of Nanodielectrics in the Electrical Power Industry," *IEEE Transactions on Dielectrics and Electrical Insulation*, Vol. 11 No. 5, 2004, pp. 797-807.

[3] N. Guo, S. A. DiBenedetto, D. K. Kwon, L. Wang, M. T. Russell, M. T. Lanagan, A. Facchetti and T. J. Marks, "Supported Metallocene Catalysis for *in Situ* Synthesis of High Energy Density Metal Oxide Nanocomposites," *Journal of the American Chemical Society*, Vol. 129, No. 4, 2007, pp. 766-767. doi:10.1021/ja0669651

[4] K. J. Nelson, "Dielectric Nanocomposite Polymers," Springer, New York, 2009.

[5] E. Tuncer, G. Polizos, I. Sauers, D. R. James, A. R. Ellis and K. L. More, "Epoxy Nanodielectrics Fabricated with *In-Situ* and *Ex-Situ* Techniques," *Journal of Experimental Nanoscience*, Vol. 7, No. 3, 2011, pp. 274-281. http://www.tandfonline.com/doi/abs/10.1080/17458080.2010.520137

[6] Y. Bai, Z.-Y. Cheng, V. Bharti, H. S. Xu and Q. M. Zhang, "High-Dielectric-Constant Ceramic-Powder Polymer Composites," *Applied Physics Letters*, Vol. 76, No. 25, 2000, p. 3804.

[7] E. A. Cherney, "Silicone Rubber Dielectrics Modified by Inorganic Fillers for Outdoor High Voltage Insulation Applications," *Annual Report Conference on Electrical Insulation and Dielectric Phenomena*, Nashville, 16-19 October 2005, pp. 1, 9.

[8] D. Q. Tan, Y. Cao and P. Irwin, "Nanostructured Dielectric Materials," *International Conference on Solid Dielectrics*, Winchester, 8-13 July 2007, pp. 411-414.

[9] P. Kim, S. C. Jones, P. J. Hotchkiss, J. N. Haddock, B. Kippelen, S. R. Marder and J. W. Perry, "Phosphonic Acid-Modified Barium Titanate Polymer Nanocomposites with High Permittivity and Dielectric Strength," *Advanced Materials*, Vol. 19, No. 7, 2007, pp. 1001-1005. doi:10.1002/adma.200602422

[10] E. Tuncer, "Structure/Property Relationship in Dielectric Mixtures: Application of the Spectral Density Theory," *Journal of Physics D: Applied Physics*, Vol. 38, No. 2, 2005, pp. 223-234. doi:10.1088/0022-3727/38/2/006

[11] S. Takahashi and D. R. Paul, "Gas Permeation in Poly (Ether Imide) Nanocomposite Membranes Based on Surface-Treated Silica. Part 2: With Chemical Coupling to Matrix," *Polymer*, Vol. 47, No. 21, 2006, pp. 7535-7547.

[12] E. Tuncer, I. Sauers, D. R. James, A. R. Ellis, M. P. Paranthaman, A. Goyal and K. L. More, "Enhancement of Dielectric Strength in Nanocomposites," *Nanotechnology*, Vol. 18, No. 32, 2007, Article ID: 325704.

[13] E. Tuncer, I. Sauers, D. R. James, A. R. Ellis, M. Pace, K. L. More, S. Sathyamurthy, J. Woodward and A. J. Rondinone, "Nanodielectrics for Cryogenic Applications," *IEEE Transactions on Applied Superconductivity*, Vol. 19, No. 3, 2009, pp. 2354-2358. doi:10.1109/TASC.2009.2018198

[14] E. Tuncer, A. J. Rondinone, J. Woodward, I. Sauers, D. R. James and A. R. Ellis, "Cobalt Iron-Oxide Nanoparticle Modified Poly(Methyl Methacrylate) Nanodielectrics: Dielectric and Electrical Insulation Properties," *Applied Physics A*, Vol. 94, No. 4, 2009, pp. 843-852. doi:10.1007/s00339-008-4881-8

[15] R. A. Vaia and E. P. Giannelis, "Polymer Nanocomposites: Status and Opportunities," *MRS Bulletin*, Vol. 26, No. 5, 2001, pp. 394-401. doi:10.1557/mrs2001.93

[16] S. H. Zhang, Y. Cao, D. Tan and P. Irwin, "Nanodielectric Ultem Films," GE Internal Report, Publishing House GE Global Research Center, Niskayuna, 2006.

[17] Y. Cao, Q. Chen, D. Q. Tan and P. C. Irwin, "Nanostructured Dielectric Materials," *International Conference on Solid Dielectrics*, Winchester, 8-13 July 2007, p. 163.

[18] E. Tuncer, A. J. Rondinone, J. Woodward, I. Sauers, D. R. James and A. R. Ellis, "Cobalt Iron-Oxide Nanoparticle Modified Poly(Methyl Methacrylate) Nanodielectrics," *Applied Physics A*, Vol. 94, No. 4, 2009, pp. 843-852. doi:10.1007/s00339-008-4881-8

[19] G. Polizos, E. Tuncer, I. Sauers and K. L. More, "Properties of a Nanodielectric Cryogenic Resin," *Applied Physics Letters*, Vol. 96, No. 15, 2010, Article ID: 152903. doi:10.1063/1.3394011

[20] S. Raetzke and J. Kindersberger, "The Effect of Interphase Structures in Nanodielectrics," *IEEJ Transactions on Fundamentals and Materials*, Vol. 126, No. 11, 2006, pp. 1044-1049. doi:10.1541/ieejfms.126.1044

[21] L. M. Huang, J. Zhang, I. Kymissis and S. O'Biren, "High K Capacitors and OFET Gate Dielectrics from Self-Assembled $BaTiO_3$ and $(Ba,Sr)TiO_3$ Nanocrystals in the Superparaelectric Limit," *Advanced Functional Materials*, Vol. 20, No. 4, 2010, pp. 554-560. doi:10.1002/adfm.200901258

[22] A. H. Sihvola, "Electromagnetic Mixing Formulas and Applications Ari Sihvola," IEE Publication Series, IEE, London, 2000.

[23] D. Q. Tan, Q. Chen, Y. Cao, P. Irwin and S. Heidger, "Polymer Based Nanodielectric Composites for Capacitors," Presentation at American Ceramic Society Electronic Materials and Applications, Orlando, Unpublished.

[24] M. G. Todd and F. G. Shi, "Characterizing the Interphase Dielectric Constant of Polymer Composite Materials: Effect of Chemical Coupling Agents," *Journal of Applied Physics*, Vol. 94, No. 7, 2003, p. 4551.

[25] D. Q. Tan, Y. Cao, P. Irwin, K. Shuman and C. McTigue, "Interfacial Study of Nanoparticle Filled Polyetherimide," Presentation at International Conference on Materials (IUMRS-ICM), July 2008.

[26] T. Tsurumi, T. Hoshina, H. Takeda, Y. Mizuno and H. Chazono, "Size Effect of Barium Titanate and Computer-Aided Design of Multilayered Ceramic Capacitors," *IEEE Transactions on Ultrasonics, Ferroelectrics, and Frequency Control*, Vol. 56, No. 8, 2009, pp. 1513-1522. doi:10.1109/TUFFC.2009.1214

[27] M. M. Abou Sekkina, "Effects of Temperature and Frequency Changes on the Dielectric Properties of Modified Chalcogenides," *Thermochimica Acta*, Vol. 120, 1987, pp. 231-239.

[28] N. Tigau, V. Ciupina, G. Prodan, G. I. Rusu, C. Gheorghies and E. Vasile, "The Influence of Heat Treatment on the Electrical Conductivity of Antimony Trioxide Thin Films," *Journal of Optoelectronics and Advanced Materials*, Vol. 5, No. 4, 2003, pp. 907-912.

[29] D. Q. Tan, "Intergrated High Energy Density Capacitors," GE Internal Report, GE Global Research Center, Niskayuna, 2010.

[30] E. Tuncer, I. Sauers, D. R. James, A. R. Ellis, M. P. Paranthaman, T. Aytug, S. Sathyamurthy, K. L. More, J. Li and A. Goyal, "Electrical Properties of Epoxy Resin Based Nanocomposites," *Nanotechnology*, Vol. 18, No. 2, 2007, Article ID: 025703. doi:10.1088/0957-4484/18/2/025703

6

Thickness Dependence of Dielectric Characteristics of $SrTiO_3$ Thin Films on $MgAl_2O_4$ Substrates

Ryuhei Kinjo, Iwao Kawayama, Hironaru Murakami, Masayoshi Tonouchi
Institute of Laser Engineering, Osaka University, Osaka, Japan
Email: tonouchi@ile.osaka-u.ac.jp

ABSTRACT

$SrTiO_3$ (STO) thin films of different thicknesses were deposited on $MgAl_2O_4$ (MAO) substrates to investigate the in-plane strain effect on the soft-mode frequency of the STO films. X-ray reciprocal space mapping (X-RSM) results indicate that there was no relaxation of the in-plane lattice strain of the STO films on MAO. Shifts in the soft-mode frequencies with a decrease in the film thickness were observed using terahertz time-domain spectroscopy (THz-TDS). However, despite the larger lattice mismatch between STO and MAO than that between STO and $DyScO_3$ (DSO), the shifts in the soft-mode frequencies of the STO films on MAO were smaller than those on DSO. The results indicate that the soft-mode frequencies of the STO films on MAO are affected by the *c*-axis (out-of-plane) lengths.

Keywords: Strontium Titanate; Ferroelectric; Soft Mode; Thin Film; Strain Effect

1. Introduction

Strontium titanate (STO) is a transition metal oxide that has a perovskite structure. Pure bulk single-crystal STO is a known quantum paraelectric that does not undergo a ferroelectric phase transition even at temperatures close to 0 K, and its permittivity is as high as 10^4 at low temperatures [1]. From theoretical predictions, lattice strain in STO changes the phase transition temperature drastically and induces ferroelectricity [2], and, in fact, Haeni *et al.* [3] have reported room-temperature ferro-electricity in a strained STO thin film deposited on a $DyScO_3$ (DSO) substrate. Since that report, various studies of strained STO thin films have been conducted, e.g., thermodynamic analysis of anisotropically strained films [4] and electric and spectrometric measurements [5-9]. However, there have only been a few reports concerning the ferroelectric and dielectric properties of STO thin films induced by large in-plane strain. In our previous study, we observed the dielectric dispersion of STO thin films on $MgAl_2O_4$ (MAO) substrates with a lattice mismatch of +3.39%, which is much larger than the lattice mismatch of +0.99% between STO and DSO. The temperature dependence of the dielectric dispersion of a 360 nm thick STO film on MAO in the terahertz region showed signs of a ferroelectric phase transition at around 170 K, much lower than that expected from the lattice mismatch [10]. We concluded, therefore, that the strain in the STO thin film on MAO used in that experiment was somewhat relaxed, and sufficient strain required for the generation of ferroelectricity at room temperature was not induced in the film. However, further experimental study is necessary to reveal the strain effects in STO thin films on MAO substrates.

In this study, we measured the thickness dependence of the dielectric dispersions of STO thin films on MAO substrates using terahertz time-domain spectroscopy (THz-TDS), and the in-plane lattice constants of the films were determined using X-ray reciprocal space maping (X-RSM). Moreover, we measured the dielectric dispersions and lattice constants of STO films on $(La_{0.3}Sr_{0.7})(AL_{0.65}Ta_{0.35})O_3$ (LSAT) and DSO substrates for comparison to elucidate the lattice strain effect on the dielectric characteristics.

2. Experimental Methods

We deposited STO films with thicknesses of 60, 120, 360, and 650 nm on MAO substrates by pulsed laser deposition. STO thin films were also deposited on LSAT and DSO substrates to compare the relationship between the in-plane strain induced by the lattice mismatch and the extension or contraction of the *c*-axis. A θ - 2θ X-ray diffraction (XRD) method was used to evaluate the crystalline quality and determine the *c*-axis lengths of the films. Only (00*l*) diffraction peaks were observed in the θ - 2θ XRD patterns, which indicated that the films were good *c*-axis orientation films. Additionally, X-RSM using a Rigaku SmartLab diffractometer was used for the 360 nm thick STO film on MAO to observe the inplane lattice constant. The dielectric dispersions of the STO films were observed using a broadband THz-TDS system with a 4-dimethylamino-*N*-methyl-4-stilbazolium tosylate (DAST) crystal as the terahertz emitter. DAST is an organic nonlinear optical crystal that can emit broad- band terahertz waves by pulsed laser irradiation except at around 1.1 THz where a large absorption peak of the DAST exists. Details of the experimental setup have been reported elsewhere [11].

3. Results and Discussion

Table 1 shows the lattice mismatch between each substrate and the STO single crystal, thicknesses, and *c*-axis lengths of the STO films on the MAO, LSAT, and DSO substrates. The *c*-axis lengths of the STO films on MAO were 0.3895, 0.3899, and 0.3900 nm for the 60 nm, 120 nm, and 360 nm thick films, respectively. The *c*-axis length of the 650 nm thick film, 0.3900 nm, is equal to that of the 360 nm thick film. The *c*-axis lengths of the STO films on the LSAT and DSO substrates were 0.3936 nm and 0.3885 nm, respectively. The *c*-axis length of the STO film on LSAT is longer than that of a bulk STO single crystal owing to in-plane compressive strain, while that on DSO was compressed owing to in-plane tensile strain. For the STO films on MAO, the *c*-axis was compressed because of in-plane tensile strain but did not change by an amount as large as that in the case of the STO films on DSO. For films on MAO thinner than 360 nm, the *c*-axis lengthens slightly with an increase in the film thickness. However, there was no extension of the *c*-axis for films on MAO thicker than 360 nm.

Figure 1 shows the results of the X-RSM measurement around the (112) reflection of the 360 nm thick STO film on MAO. The vertical axis $Q_\perp$(1/Å) corresponds to the *c*-axis direction, and the horizontal axis $Q_\parallel$(1/Å) correspond to the *a*-axis (in-plane) of the STO film. The origin in $Q_\parallel$-axis is set at the (224) reflection of MAO. The two sharp peaks in the lower part of **Figure 1** arise from the (224) reflection of MAO and are attributed to the Cu $K\alpha_1$ and $K\alpha_2$ lines of the X-ray source. The peaks are separated due to the high crystalline quality of the MAO substrate, and the tilting of the peaks indicates that the MAO substrate has some variation in its lattice constants. The broad single peak in the upper part of **Figure 1** is the (112) reflection of the STO thin film. There is no separation of the Cu $K\alpha_1$ and $K\alpha_2$ lines, and the broadness indicates the mosaicity of the film. Although the (112) reflection of the STO film was broader than that of the MAO (224), they have almost the same $Q_\parallel$ value, and no diffraction signals were seen around the calculated (112) reflection position of bulk STO (indicated by the cross mark). These results demonstrate that the in-plane lattice constant of the STO thin film on MAO corresponded to that of the MAO substrate and that the STO films were fully stained without any relaxation.

Figure 2 shows the dielectric dispersions of the four STO thin films on MAO measured by THz-TDS at room temperature. Note that the irregular dielectric dispersion observed at around 1.1 THz is not an intrinsic characteristic of the films.

We fitted the data to both the real and imaginary parts of the dielectric dispersion using a classical damping-oscillator dispersion model given by the following Equation:

$$\varepsilon(\omega) = \varepsilon_\infty + \frac{\Delta\varepsilon\omega_{TO1}^2}{\omega_{TO1}^2 - \omega^2 - i\gamma_{TO1}\omega} \quad (1)$$

where ε_∞ is the high-frequency dielectric constant, $\Delta\varepsilon$ is the dielectric strength, ω_{TO1} is the soft-mode frequency, and γ_{TO1} is the damping constant. The dashed lines in

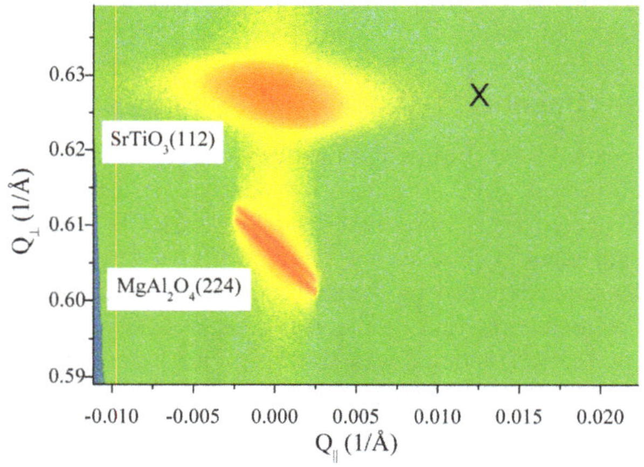

Figure 1. X-ray reciprocal space mapping around the asymmetric (112) Bragg reflection of the 360 nm thick $SrTiO_3$ thin film on a $MgAl_2O_4$ substrate.

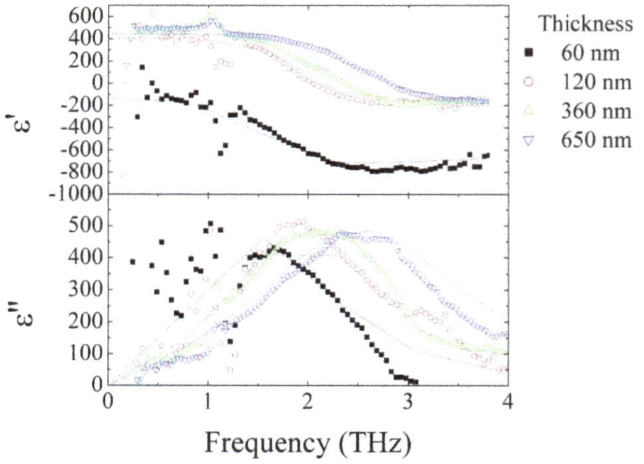

Figure 2. Thickness dependence of the dielectric dispersion of the $SrTiO_3$ thin films on $MgAl_2O_4$ substrates.

Figure 2 are the results of the fitting. The reported soft-mode frequencies of STO single crystals are around 2.7 - 3.0 THz [12-14], but the soft-mode frequency of the 650 nm thick STO thin film on MAO is 2.6 THz, which is a slightly lower frequency. The soft-mode frequency shifted to a lower frequency with a decrease in the film thickness. In the case of the 60 nm thick film, the dielectric dispersion varied widely below 1.1 THz, and the real part of the dielectric constant was noticeably smaller than the other films. This is due to the low signal-noise ratio of the 60 nm thick film, therefore the absolute value was not accurate, but it did not affect the estimation of soft-mode frequency.

Figure 3 shows the thickness dependence of the soft-mode frequency obtained from the fitting. The dashed line in the figure is the result of a linear fit to the data. Good correlation was observed between the soft-mode frequency and film thickness, and there was a tendency for the soft-mode frequency to decrease with decreasing film thickness. However, the *y*-intercept of the fit is approximately 1.6 THz, which is still much larger than the soft mode frequency of the STO film on the DSO substrate.

From the experimental results, it seems that not only the in-plane strain but also the *c*-axis length has a significant effect on the in-plane dielectric characteristics of STO films on MAO. The compressive in-plane strain in the film should induce an extension of the *c*-axis length in order to decrease the energy increased by the lattice deformation. This is implied by the lower-frequency shift of the soft-mode frequencies with a decrease in the film thickness as shown in **Table 1** and **Figure 2**. The longer extension of the *c*-axis in the STO films on MAO than that in the STO films on the other substrates might be induced by lattice defects such as oxygen and/or SrO vacancies [15], which may prevent the frequency shift of the soft mode and also an increase in the critical temperature T_c of STO thin films on MAO. Kim *et al.* [16] reported that extended *c*-axis lengths in STO films with constant in-plane lattice lengths changed the out-of-plane dielectric characteristics. Additionally, our study has implied that an extension of *c*-axis lengths also changes the in-plane dielectric characteristics of STO thin films.

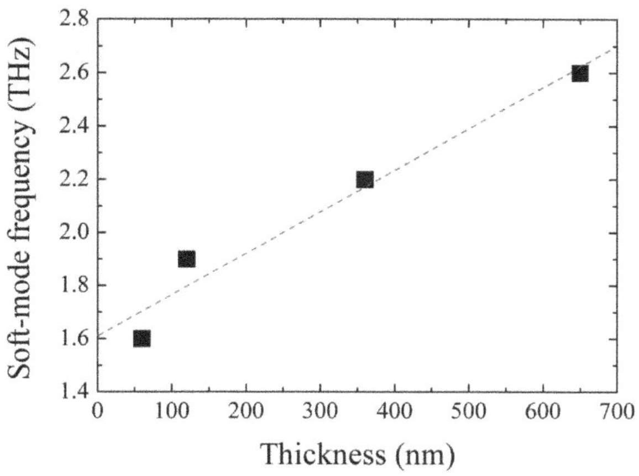

Figure 3. Thickness dependence of the soft-mode frequency of the $SrTiO_3$ thin films on $MgAl_2O_4$ substrates. The dashed line is a result of a linear fit.

Table 1. Lattice mismatch between $SrTiO_3$ and LSAT, $DyScO_3$, and $MgAl_2O_4$, thicknesses, and *c*-axis lengths of the STO films on each substrate.

Substrate	LSAT (100)	$DyScO_3$ (110)	$MgAl_2O_4$(100)		
Lattice mismatch	−0.95%	+0.99%	+3.39%		
Thickness (nm)	360	120	60	120	360, 650
c-axis length (nm)	0.3936	0.3885	0.3895	0.3899	0.3900

4. Conclusion

STO thin films of different thicknesses were deposited on MAO, LSAT, and DSO substrates. The in-plane lattice constant of the 360 nm thick STO thin film on MAO was found to correspond to that of the MAO substrate, indicating that there was no relaxation of the in-plane strain in the film. The *c*-axis lengths of the STO film on MAO were determined to be longer due to in-plane compressive strain; however, the values were smaller than that expected from the trends observed in the other substrates. The soft-mode frequencies of the films shifted to lower frequencies with a decrease in the film thickness. The minimum soft-mode frequency of a STO thin film on MAO was found to be about 1.6 THz. These results indicate that the *c*-axis length of a STO thin film affects the in-plane dielectric characteristic of the STO film and that lattice defects prevent an increase in the T_c of the films.

5. Acknowledgements

The authors received technical support from T. Tanaka and T. Takehara for the X-RSM measurement. This study was supported by a Grant-in-Aid for Scientific Research (A) (No. 22246043) and the Core-to-Core Program of the Japan Society for the Promotion of Science (JSPS).

REFERENCES

[1] H. E. Weaver, "Dielectric Properties of Single Crystals of $SrTiO_3$ at Low Temperatures," *Journal of Physics and Chemistry of Solids*, Vol. 11, No. 3-4, 1959, pp. 274, IN5-IN6, 275-277.

[2] N. A. Pertsev, A. K. Tagantsev and N. Setter, "Phase Transitions and Strain-Induced Ferroelectricity in $SrTiO_3$ Epitaxial Thin Films," *Physical Review B*, Vol. 61, No. 2,

2000, pp. R825-R829. doi:10.1103/PhysRevB.61.R825

[3] J. H. Haeni, *et al.*, "Room-Temperature Ferroelectricity in Strained $SrTiO_3$," *Nature*, Vol. 430, No. 7001, 2004, pp. 758-761. doi:10.1038/nature02773

[4] G. Sheng, *et al.*, "A Modified Landau-Devonshire Thermodynamic Potential for Strontium Titanate," *Applied Physics Letters*, Vol. 96, No. 23, 2010, Article ID: 232902. doi:10.1063/1.3442915

[5] M. D. Biegalski, *et al.*, "Relaxor Ferroelectricity in Strained Epitaxial $SrTiO_3$ Thin Films on $DyScO_3$ Substrates," *Applied Physics Letters*, Vol. 88, No. 19, 2006, Article ID: 192907. doi:10.1063/1.2198088

[6] M. D. Biegalski, *et al.*, "Influence of Anisotropic Strain on the Dielectric and Ferroelectric Properties of $SrTiO_3$ Thin Films on $DyScO_3$ Substrates," *Physical Review B*, Vol. 79, No. 22, 2009, Article ID: 224117. doi:10.1103/PhysRevB.79.224117

[7] P. Kužel, C. Kadlec, F. Kadlec, J. Schubert and G. Panaitov, "Field-Induced Soft Mode Hardening in $SrTiO_3$/$DyScO_3$ Multilayers," *Applied Physics Letters*, Vol. 93, No. 5, 2008, Article ID: 052910. doi:10.1063/1.2967336

[8] D. Nuzhnyy, *et al.*, "Soft Mode Behavior in $SrTiO_3$/$DyScO_3$ Thin Films: Evidence of Ferroelectric and Antiferrodistortive Phase Transitions," *Applied Physics Letters*, Vol. 95, No. 23, 2009, Article ID: 232902. doi:10.1063/1.3271179

[9] P. Kužel, F. Kadlec, J. Petzelt, J. Schubert and G. Panaitov, "Highly Tunable $SrTiO_3$/$DyScO_3$ Heterostructures for Applications in the Terahertz Range," *Applied Physics Letters*, Vol. 91, No. 23, 2007, Article ID: 232911. doi:10.1063/1.2822409

[10] R. Kinjo, I. Kawayama, H. Murakami and M. Tonouchi, "Strain-Induced Ferroelectricity of a $SrTiO_3$ Thin Film on a $MgAl_2O_4$ Substrate Observed by Terahertz Time-Domain Spectroscopy," *Journal of Infrared Millimeter, and Terahertz Waves*, Vol. 33, No. 1, 2012, pp. 67-73. doi:10.1007/s10762-011-9839-9

[11] R. Kinjo, *et al.*, "Observation of Strain Effects of $SrTiO_3$ Thin Films by Terahertz Time-Domain Spectroscopy with a 4-Dimethylamino-*N*-methyl-4-stilbazolium Tosylate Emitter," *Japanese Journal of Applied Physics*, Vol. 48, No. 9, 2009, Article ID: 09KA16. doi:10.1143/JJAP.48.09KA16

[12] R. A. Cowley, "Temperature Dependence of a Transverse Optic Mode in Strontium Titanate," *Physical Review Letters*, Vol. 9, No. 4, 1962, pp. 159-161. doi:10.1103/PhysRevLett.9.159

[13] A. S. Barker, Jr. and M. Tinkham, "Far-Infrared Ferroelectric Vibration Mode in $SrTiO_3$," *Physical Review*, Vol. 125, No. 5, 1962, pp. 1527-1530. doi:10.1103/PhysRev.125.1527

[14] H. Vogt, "Refined Treatment of the Model of Linearly Coupled Anharmonic Oscillators and Its Application to the Temperature Dependence of the Zone-Center Soft-Mode Frequencies of $KTaO_3$ and $SrTiO_3$," *Physical Review B*, Vol. 51, No. 13, 1995, pp. 8046-8059. doi:10.1103/PhysRevB.51.8046

[15] C. H. Park and D. J. Chadi, "Microscopic Study of Oxygen-Vacancy Defects in Ferroelectric Perovskites," *Physical Review B*, Vol. 57, No. 22, 1998, pp. R13961-R13964.

[16] Y. S. Kim, *et al.*, "Observation of Room-Temperature Ferroelectricity in Tetragonal Strontium Titanate Thin Films on $SrTiO_3$ (001) Substrates," *Applied Physics Letters*, Vol. 91, No. 4, 2007, Article ID: 042908. doi:10.1063/1.2764437

7

Crystal Growth, Optical and Dielectric Properties of L-Histidine Hydrochloride Monohydrate Nonlinear Optical Single Crystal

P. Koteeswari[1], S. Suresh[2*], P. Mani[1]
[1]Department of Physics, Hindustan Institute of Technology, Padur, India
[2]Department of Physics, Loyola College, Chennai, India
Email: *sureshsagadevan@yahoo.co.in

ABSTRACT

Optically transparent and bulk single crystal of l-histidine hydrochloride monohydrate (LHHM) was successfully grown by slow evaporation technique. The cell parameters and the crystallinity of the grown crystal were estimated by the single crystal XRD. Optical transmittance of the crystal was recorded using the UV-vis-NIR spectrophotometer. The optical band gap and optical constant of the material were determined by using transmission spectrum. The dielectric loss and dielectric constant measurements as a function of frequency and temperature were measured for the grown crystal.

Keywords: Single Crystal; Slow Evaporation Technique; XRD; UV and Dielectric Studies

1. Introduction

The organic NLO materials are attracted by many scientists due to their frequency conversion efficiency, piezo-electric, pyro-electric properties and their wide applications in the recent technologies like lasers, optical communications and data storage [1]. New materials with high optical nonlinearities are quite important due to their extensive application in harmonic generation, amplitude and phase modulation, switching and other signal processing device [2-4]. The main goal to design the molecules with the third order nonlinearities is to incorporate them into the devices used in all types of optical signal processing [5-6]. Nonlinear optical (NLO) materials have shown potential application in optical information storage, optical logic gates, laser radiation protection and phase locked laser mode. Hence the interest in searching for NLO materials has increased gradually [7]. In addition to that organic molecules also have a great attention owing to their potential application in the frontier areas such as nonlinear optics (NLO), optical switching and light emitting diodes. Thus, the potential use of organic device materials in optoelectronics has now become a serious matter [8]. The present investigation is aimed at the growth of bulk LHHM single crystal by slow evaporation technique. The grown crystal has been subjected to single X-ray diffraction analysis, UV-vis transmission spectral analysis, optical band gap measurements, and dielectric studies.

*Corresponding author.

2. Experimental Procedure

The l-histidine and hydrochloric acid were taken in equi molar ratio in double distilled water to prepare the saturated solution of l-histidine hydrochloride monohydrate (LHHM). The solution obtained is stirred well at room temperature using a temperature controlled magnetic stirrer to yield a homogenous mixture of solution. Then the solution is filtered using a Whatmann filter paper and was allowed to evaporate at room temperature. The solution is recrystallized several times in order to increase the purity of the crystal. Optically clear and good quality seed crystal is kept inside the purified saturated solution and the solution is allowed to evaporate at room temperature, which produces an improved optically high quality within a period of 30 days. The photograph of the as grown single crystal is shown in **Figure 1**.

3. Results and Discussion

3.1. Single Crystal X-Ray Diffraction

Single crystal X-ray diffraction analysis was carried out to determine the lattice parameters. The grown crystals have orthorhombic structure with $P2_12_12_1$ space group. The lattice parameter values of the grown crystals are a = 6.82 Å, b = 8.91 Å, c = 15.286 Å. The single crystal data are in good agreement with reported values [9].

3.2. Optical Transmittance Spectrum Study

The optical transmission spectrum of LHHM single crystal was recorded in the wavelength region 300 - 900 nm and is shown in **Figure 2**. For optical fabrications, the crystal should be highly transparent in the considered region of wavelength [10] and [11]. Favorable transmittance of the crystal in the entire visible region suggests its suitability for second harmonic generation [12]. The UV absorption edge for the grown crystal was observed to be around 260 nm. The dependence of optical absorption coefficient on photon energy helps to study the band structure and type of transition of electrons [13].

Optical absorption coefficient (α) was calculated from transmittance using the following relation:

$$\alpha = \frac{1}{d}\log\left(\frac{1}{T}\right) \tag{1}$$

where T is the transmittance and t the thickness of the crystal. As a direct band gap material, the crystal under study has an absorption coefficient (α) obeying the following relation for high photon energies (hv).

Figure 1. Photograph of as grown crystal.

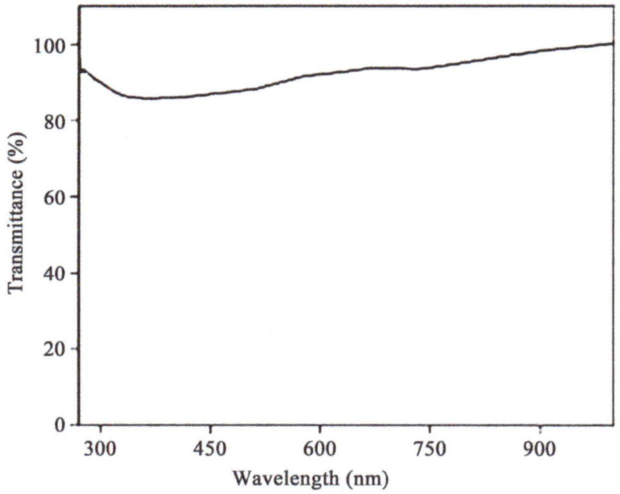

Figure 2. Transmission spectrum of the grown crystal (LHHM).

$$\alpha = \frac{A\left(hv - E_g\right)^{1/2}}{hv} \tag{2}$$

where E_g is the optical band gap of the crystal and A is a constant. A plot of variation of $(\alpha hv)^2$ versus hv is shown in **Figure 3**. E_g is evaluated using extrapolation of the linear part [14]. The energy absorption gap is of direct type and the band gap energy is found to be 3.90 eV.

3.3. Determination of Optical Constant

The dependence of optical absorption co-efficient with photon energy helps to study the band structure and the type of transition of the electron. The absorption coefficient (α) and the optical constant (n, k) are determined from the transmission (T) and reflection (R) spectrum based on the following relations, [15,16].

$$T = \frac{(1-R)^2 \exp(-\alpha t)}{1 - R^2 \exp(-2\alpha t)} \tag{3}$$

Reflectance can also be written in terms of absorption coefficient and from the *ab*

$$R = \frac{1 \pm \sqrt{1 - \exp(-\alpha t) + \exp(\alpha t)}}{1 + \exp(-\alpha t)} \tag{4}$$

and from the above equation, refractive index (n) can also be derived as

$$n = -\frac{(R+1) \pm \sqrt{3R^2 + 10R - 3}}{2(R-1)} \tag{5}$$

Figure 4 show the variation of refractive index (n) as a function of wavelength (λ), respectively. From the graphs, it is clear that refractive index (n) depend on wavelength (λ).

3.4. Dielectric Studies

The dielectric characteristics of the material are important

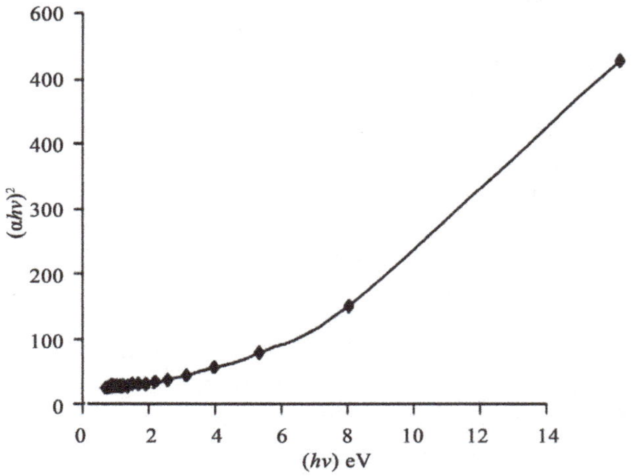

Figure 3. Plot of $(\alpha hv)^2$ vs photon energy of the title crystal.

to study the lattice dynamics in the crystal. Hence, the grown crystal was subjected to dielectric studies using a HIOKI HITESTER model 3532-50 LCR meter in the frequency range from 50 Hz to 5 MHz for different temperatures. The surface of the sample was electrode with silver paste for electrical contact. **Figure 5** shows the plot of dielectric constant (ε_r) versus log frequency. The dielectric constant has high values in the lower frequency region and then it decreases with the increase in frequency. The very high value of ε_r at low frequencies may be due to the presence of all the four polarizations, namely, space charge, orientational, electronic and ionic polarization and its low value at higher frequencies may be due to the loss of significance of these polarizations gradually. From the plot, it is also observed that dielectric constant increases with an increase in temperature, and this is attributed due to the presence of space charge polarization near the grain boundary interfaces, which depends on the purity and perfection of the sample [17].

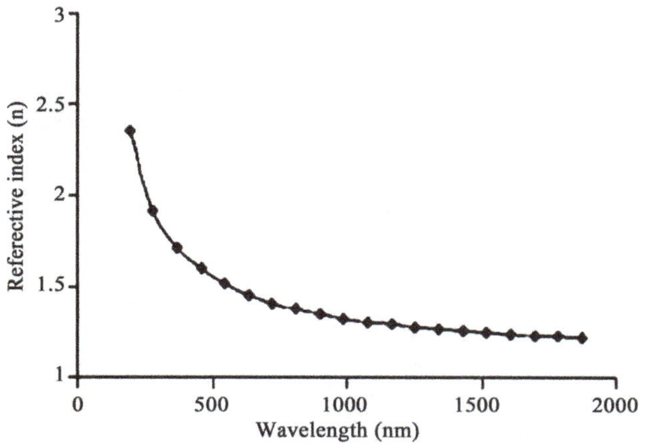

Figure 4. Variation of refractive index with wavelength.

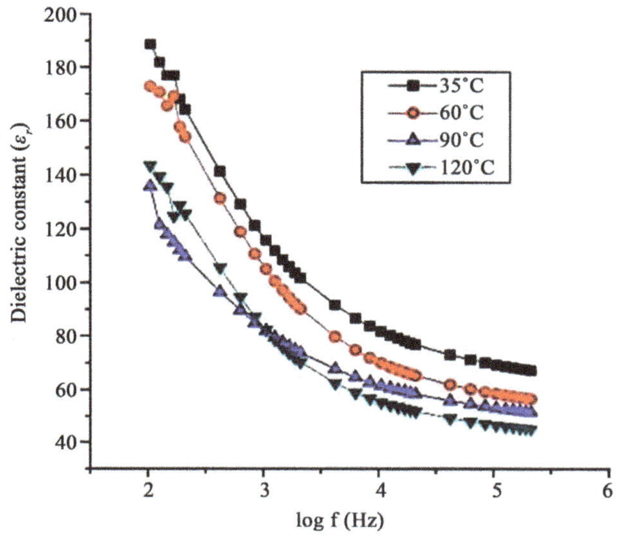

Figure 5. Variation of dielectric constant with log frequency.

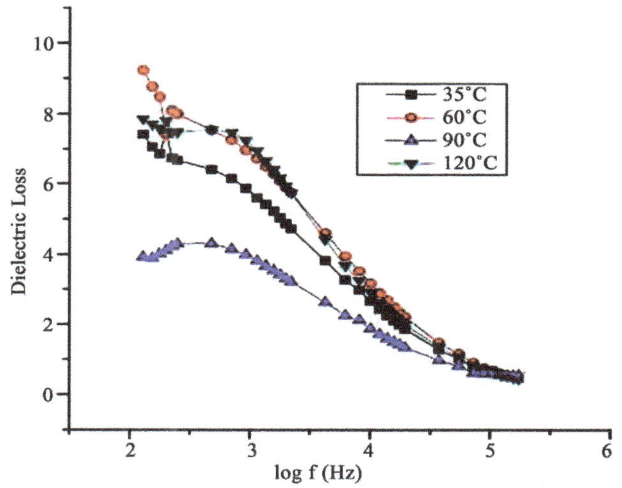

Figure 6. Variation of dielectric loss with log frequency.

The variation of dielectric loss with frequency is shown in **Figure 6**. The characteristics of low dielectric loss with high frequency for the sample suggest that it possesses enhanced optical quality with lesser defects and this parameter is of vital importance for nonlinear optical applications [18].

4. Conclusion

Bulk single crystal of semi-organic LHHM was grown from aqueous solution by a slow evaporation technique. Single crystal X-ray diffraction studies confirm that the grown crystal belongs to orthorhombic crystal system with space group $P2_12_12_1$. The optical transmission analysis indicates that LHHM has a wide transparency window in the entire visible and near IR regions with a lower cutoff wavelength at 260 nm. The band gap was estimated to be 3.90 eV. The variation of dielectric constant (ε'), dielectric loss and imaginary dielectric constant (ε'') were studied as a function of frequency at different temperatures.

REFERENCES

[1] N. B. Singh, T. Henningsen, E. P. A. Metz, R. Hamacher, E. Cumberledge, R. H. Hopkins and R. Mazelsky, "Solution Growth of Vanillin Single Crystals," *Materials Letters*, Vol. 12, No. 4, 1991, pp. 270-275. doi:10.1016/0167-577X(91)90012-U

[2] E. W. Van Stryland, H. Vanherzeele, M. A. Woodall, M. J. Soileau, A. L. Smirl, S. Guha and T. F. Bogess, "Two Photon Absorption, Nonlinear Refraction, and Optical Limiting in Semiconductors," *Optical Engineering*, Vol. 24, 1985, p. 613.

[3] T. Wei, D. J. Hagan, E. W. Van Stryland, "Sensitive Measurement of Optical Nonlinearities Using a Single Beam," *IEEE Journal of Quantum Electronics*, Vol. 26, No. 4, 1990, pp. 760-769.

[4] J. J. Rodrigues Jr., L. Misoguti, F. D. C. R. Nunes, C. R. Mendonca and S. C. Zilio, "Optical Properties of L-Threonine Crystals," *Optical Materials*, Vol. 22, No. 3, 2003, pp. 235-240. doi:10.1016/S0925-3467(02)00270-7

[5] M. Somac, A. Somac, B. L. Davies, M. G. Humphery and M. S. Wong, "Third-Order Optical Nonlinearities of Oligomers, Dendrimers and Polymers Derived from Solution Z-Scan Studies," *Optical Materials*, Vol. 21, No. 1-3, 2002, pp.485-488.

[6] L. V. Natarajan, R. L. Sutherland, V. P. Tondiaglia, T. J. Bunning, W. W. Adams and J. Nonlinear, "Electro-Optical Switching Characteristics of Volume Holograms in Polymer Dispersed Liquid Crystals," *Optical Physics and Materials*, Vol. 5, No. 1, 1996, pp. 89-98.

[7] X. Xu, W. Qiu, Q. Zhou, J. Tang, F. Yang, Z. Sun and P. Audebert, "Nonlinear Optical Absorption Properties of Two Multisubstituted p-Dimethylaminophenylethenyl Pyridiniums," *Journal of Physical Chemistry B*, Vol. 112, No. 16, 2008, pp. 4913-4917. doi:10.1021/jp7103775

[8] J. G. Breitzar, D. D. Dlott, L. K. Iwaki, S. M. Kirkpatrick and T. B. Rauchturs, "Third-Order Nonlinear Optical Properties of Sulfur-Rich Compounds," *Journal of Physical Chemistry A*, Vol. 103, No. 35, 1999, pp. 6930-6937. doi:10.1021/jp990137f

[9] J. Madhavan, S. Aruna, P. C. Thomas, M. Vimalan, S. A. Rajasekar and P. Sagayaraj, "Growth and Characterization of L-Histidine Hydrochloride Monohydrate Single Crystals," *Crystal Research and Technology*, Vol. 42, No. 1, 2007, pp. 59-64. doi:10.1002/crat.200610771

[10] V. Krishnakumar and R. Nagalakshmi, "Crystal Growth and Vibrational Spectroscopic Studies of the Semiorganic Non-Linear Optical Crystal—Bisthiourea Zinc Chloride," *Spectrochim Acta Part A: Molecular and Biomolecular Spectroscopy*, Vol. 61, No. 3, 2005, pp. 499-507. doi:10.1016/j.saa.2004.04.014

[11] V. Krishnakumar and R. J. Xavier, "FT Raman and FT-IR Spectral Studies of 3-Mercapto-1,2,4-Triazole," *Spectrochim Acta Part A: Molecular and Biomolecular Spectroscopy*, Vol. 60, No. 3, 2004, pp. 709-714. doi:10.1016/S1386-1425(03)00281-6

[12] S. A. Roshan, J. Cyriac and M. A. Ittyachen, "Growth and Characterization of a New Metal-Organic Crystal: Potassium Thiourea Bromide," *Materials Letters*, Vol. 49, No. 5, 2001, pp. 299-302. doi:10.1016/S0167-577X(00)00388-8

[13] N. Tigau, V. Ciupinaa, G. Prodana, G. I. Rusub, C. Gheorghies and E. Vasilec, "Influence of Thermal Annealing in Air on the Structural and Optical Properties of Amorphous Antimony Trisulfide Thin Films," *Journal of Optoelectronics and Advanced Materials*, Vol. 6, No. 1, 2004, pp. 211-217.

[14] A. K. Chawla, D. Kaur and R. Chandra, "Structural and Optical Characterization of ZnO Nanocrystalline Films Deposited by Sputtering," *Optical Materials*, Vol. 29, No. 8, 2007, pp. 995-998. doi:10.1016/j.optmat.2006.02.020

[15] B. L. Zhu, C. S. Xie, D. W. Zeng, W. L. Song and A. H. Wang, "Investigation of Gas Sensitivity of Sb-Doped ZnO Nanoparticles," *Materials Chemistry and Physics*, Vol. 89, No. 1, 2005, pp. 148-153. doi:10.1016/j.matchemphys.2004.08.028

[16] A. Ashour, N. El-Kadry and S. A. Mahmoud, "On the Electrical and Optical Properties of CdS Films Thermally Deposited by a Modified Source," *Thin Solid Films*, Vol. 269, No. 1-2, 1995, pp. 117-120. doi:10.1016/0040-6090(95)06868-6

[17] C. P. Smyth, "Dielectric Behaviour and Structure," McGraw-Hill, NewYork, 1965.

[18] C. Balarew and R. Duhlew, "Application of the Hard and Soft Acids and Bases Concept to Explain Ligand Coordination in Double Salt Structures," *Journal of Solid State Chemistry*, Vol. 55, No. 1, 1984, pp. 1-6. doi:10.1016/0022-4596(84)90240-8

8

Characterization and Electromagnetic Studies on NiZn and NiCuZn Ferrites Prepared by Microwave Sintering Technique

Matli Penchal Reddy[1], Il Gon Kim[1*], Dong Sun Yoo[1], Wuppati Madhuri[2], Nagireddy Ramamanohar Reddy[3], Kota Venkata Siva Kumar[2], Rajuru Ramakrishna Reddy[2]

[1]Department of Physics, Changwon National University, Changwon, South Korea; [2]Department of Physics, Sri Krishnadevaraya University, Anantapur, India; [3]Department of Materials Science and Nanotechnology, Yogi Vemana University, Kadapa, India.
Email: *igkim@changwon.ac.kr, *drlpenchal@gmail.com

ABSTRACT

The low-temperature sintered NiZn and NiCuZn ferrites with the composition of $Ni_{0.40}Zn_{0.60}Fe_2O_4$ and $Ni_{0.35}Cu_{0.05}Zn_{0.60}Fe_2O_4$ were respectively synthesized by the microwave sintering method. These powders were calcined, compacted and sintered at 950˚C for 30 min. X-ray diffraction (XRD) patterns of the samples indicate the formation of single-phase cubic spinel structure. The grain size was estimated from SEM images which increase with CuO addition. The X-ray density is higher than the bulk density in both the ferrites. The temperature variation of the initial permeability of these samples was carried out from 30˚C to 250˚C. The NiCuZn ferrite had higher initial permeability than that of the NiZn ferrite, which could be attributed to the microstructure. Saturation magnetization increases from 40 emug/g (NiZn) to 47 emug/g (NiCuZn). The dielectric constant (ε') and dielectric loss tangent $(\tan\delta)$ of NiZn and NiCuZn ferrite samples decreases with increase in frequency exhibiting normal ferrimagnetic behavior. The NiCuZn ferrite had better electromagnetic properties than the NiZn ferrite.

Keywords: Microwave Sintering; Soft Ferrites; Magnetic Properties; Dielectric Properties

1. Introduction

The use of microwave radiation in the processing of various ceramic materials such as dielectric, magnetic, superconducting, polymer and other composite materials offers several advantages over conventional processing alternatives. The unique and potential benefits that microwave energy can provide over conventional methods such as rapid, internal and selective heating have stimulated much interest in many researchers to apply this technique in ceramic processing such as firing, annealing and sintering. The direct interaction or coupling between microwaves and the material is responsible for the rapid and internal heating, which makes the energy conversion rather than energy transfer, and as a consequence volumetric heating occurs. The volumetric heating is one of the possible reasons for improvement in microstructure and hence the dielectric and magnetic properties of the materials [1,2].

The poly crystalline ferrites have very important structural, magnetic and electrical properties that are dependent on several factors such as method of preparation, substitution of cations, sintering temperature and time, chemical composition and microstructure, etc. The spinel ferrites with low loss at higher frequencies are used in pulse transformers, inductances, reflection coils, antennas and modulators, etc. The NiZn and MgZn ferrites are suitable for these applications [3]. The Ni-Zn ferrites are considered as the most versatile ferrites for their high resistivity and low eddy current losses. The high frequency applications and further miniaturization of magnetic components enable the use of NiZn or NiCuZn ferrites, because both of them have high electrical resistivity and can miniaturize magnetic components without a bobbin [4-6]. The magnetic properties of these ferrite materials are mainly determined by their chemical composition. The substitution of CuO to the NiZn ferrite could also lower its sintering temperature and modifies structural and electromagnetic properties [7,8]. However, to the best of our knowledge, the detailed comparison of low temperature sintered NiZn and NiCuZn ferrites has not yet been reported. Only limited researchers have studied the comparison of NiZn and NiCuZn ferrites with excellent magnetic and electrical properties [8-10]. Therefore, in this paper, we fixed the

*Corresponding author.

preparing process and the low-temperature sintered NiZn and NiCuZn ferrites, and compared their sintering density, microstructures, magnetic and electric properties.

2. Experimental Details

2.1. Materials

Starting materials for the synthesis of NiZn and NiCuZn ferrites were Ferric Oxide, Nickel Oxide, Zinc Oxide and Copper oxide powders. All the chemicals are purchased from Merck and used as received. All the chemicals and reagents were used without further purification.

2.2. Preparation of Soft Ferrites

In this study, a modified microwave oven (Sharp, 1.1 kW, 2.45 GHz) was used to sinter NiZn and NiCuZn ferrite samples. **Figure 1** shows the experimental settings in the microwave oven. The general configuration of the microwave furnace has been described elsewhere [11]. The NiZn and NiCuZn ferrites with the respective compositions of $Ni_{0.40}Zn_{0.60}Fe_2O_4$ and $Ni_{0.35}Cu_{0.05}Zn_{0.60}Fe_2O_4$ were prepared using the microwave sintering method. The analytical grade Fe_2O_3, NiO, ZnO, and CuO powders were weighed according to the corresponding composition, intimately mixed and the resulting powders were ball milled using a planetary ball mill (Restch PM 200, Germany) in agate bowls with agate balls in acetone medium for 20 h. The slurry was dried and loosely pressed into cakes using a hydraulic press. These cakes were pre-sintered at a temperature of 800˚C for 4 h in closed alumina crucibles. The pre-sintered cakes removed from the furnace were crushed and ball milled in an acetone medium in agate bowls with agate balls for another 30 h to obtain fine particle size. These slurries after drying were sieved to obtain a uniform particle size. The green powder thus obtained was then pressed using a suitable die in the form of toroids of dimensions 1.2 cm OD, 0.8 cm ID and 0.4 cm thickness and pellets of dimensions 1.2 cm diameter and 0.3 cm thickness with a hydraulic press at a pressure of 200 MPa using 3% PVA solution as a binder and the samples were finally sintered at 950˚C for 30 min in air at atmospheric pressure.

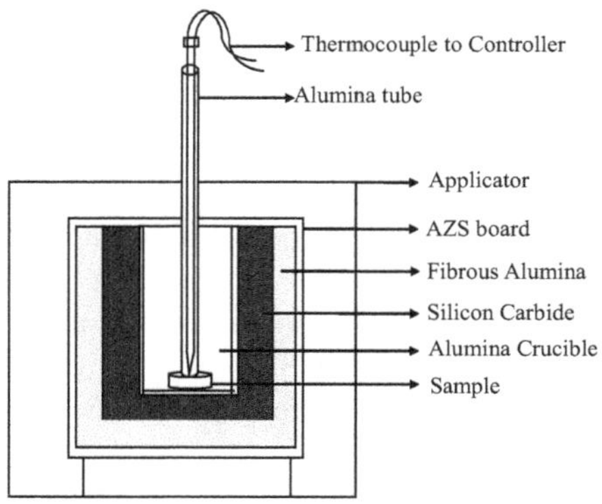

Figure 1. Schematic view of sample environment in the microwave cavity.

2.3. Characterization

The bulk density was determined using the Archimedes method using distilled water as the immersion fluid. XRD patterns of samples were obtained using an X-ray diffraction system (PM 1730, Germany) using CuK_α radiation. The specimens were polished with 0.3 μm alumina powder and thermally etched prior to examination with the help of a scanning electron microscope (SEM, CRL-ZESIS-EVO-MAI5, Japan). The DC resistivity of these ferrite pellets were evaluated using the Keithley electrometer (Model 6514, India). The initial permeability, (μ_i) of these ferrite toroids were evaluated using the standard formulae from the inductance measurements carried out at 10 kHz using a computer controlled impedance analyzer (Hioki Model 3532-50 LCR HiTester, Japan). The magnetic characteristics were measured with VSM-Lakeshore 665 Vibrating sample magnetometer (VSM). These measurements were carried out in the temperature range 30˚C - 250˚C at 10˚ temperature intervals. The dielectric constant of the samples was calculated from the capacitance (C_p) and loss factor (tanδ) values measured using impedance analyzer in the frequency range 100 Hz to 1 MHz at room temperature.

3. Results and Discussions

Figure 2 shows the XRD patterns of the NiZn and NiCuZn ferrite samples. As we can see from the figure, both the samples possesses single phase with spinel structure. **Figure 3** presents the micrographs of the NiZn and NiCuZn ferrite samples sintered at 950˚C. The average grain size is observed to be 3.8 and 4.6 μm for NiZn, NiCuZn ferrites respectively.

The Lattice constant, X-ray density, bulk density, percentage of porosity, average grain size, Curie transition

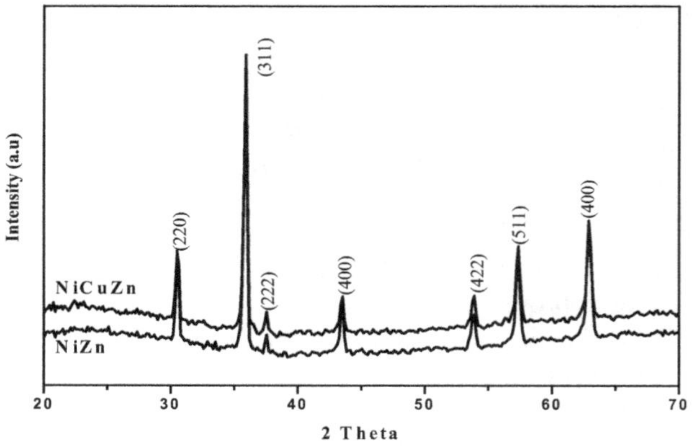

Figure 2. XRD diffraction of NiZn and NiCuZn ferrites sintered at 950˚C.

temperature and DC resistivity of sintered NiZn and NiCuZn ferrite samples are also tabulated in **Table 1**. It is seen from this table that the lattice parameter increases from 8.342 Å (NiZn) to 8.392 Å (NiCuZn) [9]. This increase is attributable to the higher ionic of Cu^{2+} (0.73 Å) compared to that of Ni^{2+} (0.69 Å). As copper content increases at the expense of Ni, the lattice seems to expand slightly to accommodate the increased number of Cu^{2+} ions of relatively larger radius resulting in increased value of lattice parameter. It can also be observed from the **Table 1**, that X-ray density of each sample (NiZn and NiCuZn ferrite) is higher than the corresponding bulk density of sintered smples. The variation of room-temperature resistivity with Cu content (**Table 1**) shows that resistivity decreases with addition of Cu content to NiZn ferrite.

The magnetic initial permeability (μ_i) as a function of temperature (at constant frequency, 10 kHz) from room temperature to Curie transition temperature was also studied. The temperature variation of magnetic initial permeability for both samples is shown in **Figure 4**. It is evident from figure that the magnetic initial permeability remains constant over a wide range for the microwave sintered sample. It can also be noticed from the figure that the microwave sintered ferrites show good thermal stability. This indicates that the shape of the permeability-temperature curves depends on the preparation conditions. It can be noted from **Figure 4** that as the temperature increases the magnetic initial permeability remains constant up to a certain temperature and increases to a peak value and then abruptly falls to a minimum value. The temperature at which this abrupt fall takes place is the magnetic Curie transition temperature (T_c). At this temperature the specimens transform from the ferrimagnetic phase to the paramagnetic phase. An increase in NiCuZn ferrite initial permeability is observed compared with NiZn-ferrite. The magnetic initial permeability for the material is expected to strongly depend on the microstructure, as the magnetic initial permeability represents the mobility of magnetic domain wall in response to the small applied field [12]. It is also seen from **Figure 2** that the NiCuZn samples result in large-grain samples leading to enhancement in permeability.

Figure 5 shows that Cu addition increases saturation magnetization, probably by increasing grain size and crystallinity at the same temperature treatment compared to NiZn-ferrite. Saturation magnetization (M_s) values are comparable to those reported by other authors [13,14]. Coercivity is related to the variation of the particle size. The substitution of nickel by copper also affects bulk density and resistivity of the ferrite (**Table 1**). The increase in density is attributed to the increase in grain size and reduction of the pores in the ferrite microstructure as seen in SEM microphotography (**Figure 3**).

The dielectric properties such as dielectric constant, dielectric loss (tanδ) are important for multilayer chip inductors used in high frequency range. The frequency dependence of dielectric constant for microwave sintered NiZn and NiCuZn ferrite samples is shown in **Figures 6(a)** and **(b)**. The plots illustrate that the dielectric constant decreases with increasing frequency reaching constant value

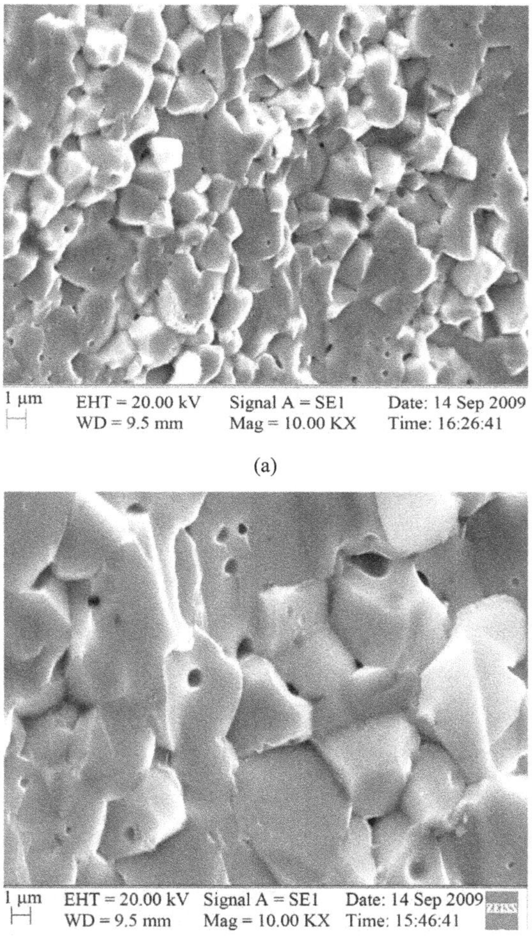

Figure 3. SEM images of the NiZn and NiCuZn ferrites sintered at 950˚C. (a) NiZn ferrite; (b) NiCuZn ferrite.

Table 1. Lattice constant, X-ray density, bulk density, porosity, average grain size, Curie transition temperature and DC resistivity for the NiZn and NiCuZn ferrites sintered at 950˚C.

Ferrites	a (Å)	d_X (gm/cm^3)	d_B (gm/cm^3)	P (%)	G_{avg} (μm)	T_c (˚C)	ρ (Ωcm)
NiZn	8.342	5.342	4.927	7.8	3.8	160	5.2×10^8
NiCuZn	8.392	5.365	5.034	6.1	4.6	150	3.7×10^8

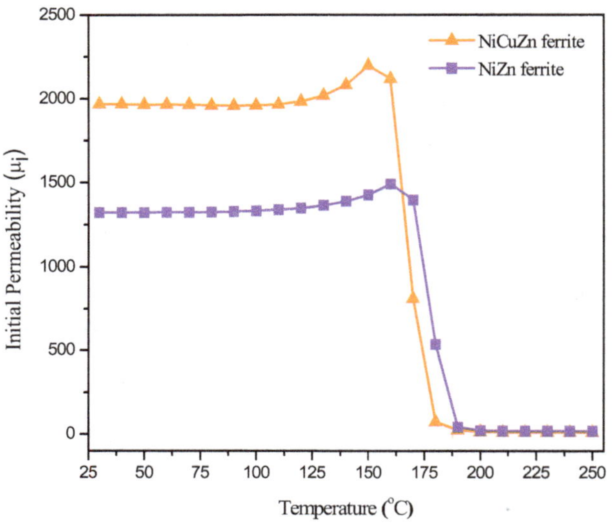

Figure 4. Temperature variation of initial permeability of NiZn and NiCuZn ferrite samples.

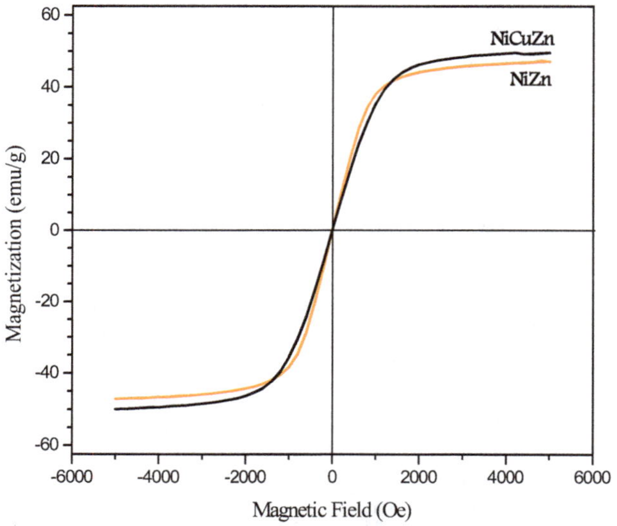

Figure 5. Room temperature M-H graphs of NiZn and NiCuZn ferrite samples.

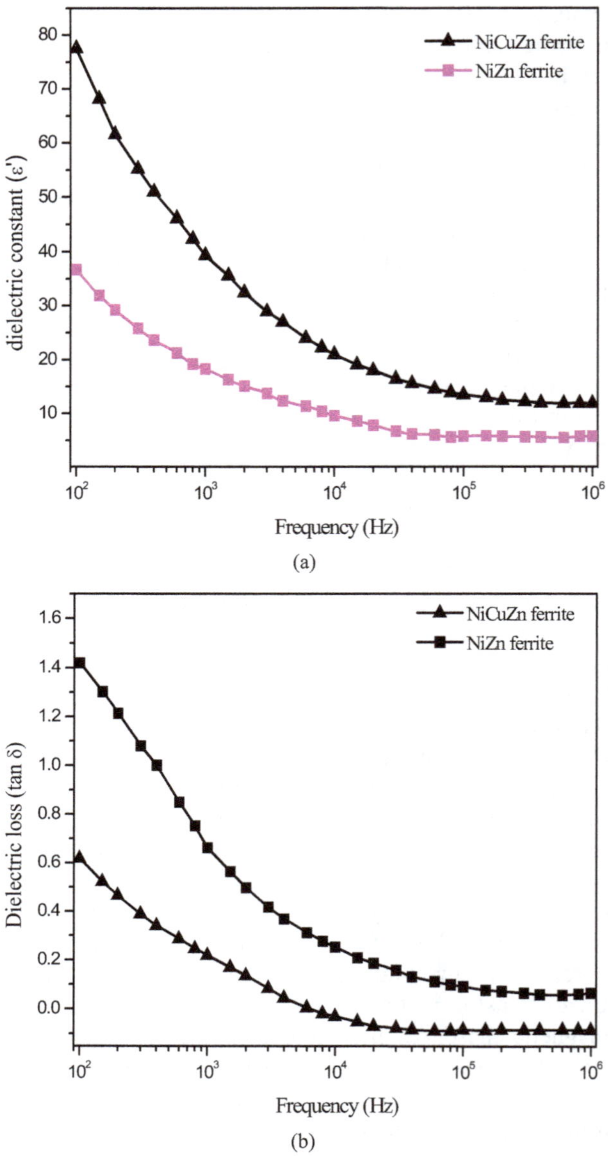

Figure 6. (a) Dielectric constant (ε') as a function of frequency of NiZn and NiCuZn ferrite samples; (b) Dielectric loss (tanδ) as a function of frequency of NiZn and NiCuZn ferrite samples.

at higher frequencies. The variation reveals the dispersion due to Maxwell-Wanger [15,16] type interfacial polarization in agreement with Koop's phenomenological theory [17]. This is the normal behavior for ferrites [18]. This is similar to the results reported earlier for NiZn [19], Ni-Cu [20] and Ni-Cu-Zn [9] ferrites. An important decrease of dielectric loss (tanδ) with frequency is observed. The low loss values (0.03) at higher frequencies show the potential of these samples for high-frequency applications. As seen in the figure, the dielectric constant for NiCuZn-ferrite is higher than for NiZn ferrite. The mechanism of polarization in polycrystalline ferrites is mainly reported to be hopping of electrons between ions of the same element but in different oxidation states [21].

4. Conclusions

In this paper, we have compared the lattice parameter, sintering density, microstructures, magnetic and dielectric properties of the low temperature sintered NiZn and NiCuZn ferrites, and the following conclusions can be drawn when the substitution of Cu in NiZn ferrite.

1) The lattice parameter increases with the copper content due to larger ionic radius of Cu^{2+} compared to Ni^{2+};

2) A remarkable increase in the value of bulk density has been found with Cu substitution for Ni in NiZn ferrite;

3) The average grain size is found to be high with the

substitution of Cu in NiZn ferrite;

4) Cu substitution in NiZn-ferrite enhances initial permeability and saturation magnetization. This is mainly attributed to the presence of Cu ions activating the sintering processes in ferrites and leading to increase in density;

5) The dielectric constant and loss tangent are found to decrease with increase in frequency. Dielectric loss ($\tan\delta$) is of the order 10^{-2}.

REFERENCES

[1] P. K. Patro, A. R. Kulkarni, S. M. Gupta and C. S. Harendranath, "Improved Microstructure, Dielectric and Ferroelectric Properties of Microwave-Sintered $Sr_{0.5}Ba_{0.5}Nb_2O_6$," *Physica B: Condensed Matter*, Vol. 400, No. 1-2, 2007, pp. 237-242. doi:10.1016/j.physb.2007.07.022

[2] Z. Xie, J. Yang, X. Huang and Y. Huang, "Microwave Processing and Properties of Ceramics with Different Loss," *Journal of European Ceramic Society*, Vol. 19, No. 3, 1999, pp. 381-387. doi:10.1016/S0955-2219(98)00203-9

[3] V. R. K. Murty and B. Vishwanathan, "Ferrites Materials: Science and Technology," Narosa Publishing House, Mumbai, 1990.

[4] Y. Matsuo, M. Inagaki, T. Tomozawa and F. Nakao, "High Performance NiZn Ferrite," *IEEE Transactions on Magnetics*, Vol. 37, No. 4, 2001, pp. 2359-2361. doi:10.1109/20.951172

[5] K. Kondo, T. Chiba, S. Yamada and E. Otsuki, "Analysis of Power Loss in Ni-Zn Ferrites," *Journal of Applied Physics*, Vol. 87, No. 9, 2000, pp. 6229-6231. doi:10.1063/1.372663

[6] A. C. F. M. Costa, E. Tortella, M. R. Morelli and R. H. G. A. Kiminami, "Synthesis, Microstructure and Magnetic Properties of Ni-Zn Ferrites," *Journal of Magnetism and Magnetic Materials*, Vol. 256, No. 1-3, 2003, pp. 174-182. doi:10.1016/S0304-8853(02)00449-3

[7] S. R. Murthy, "Low Temperature Sintering of NiCuZn Ferrite and Its Electrical, Magnetic and Elastic Properties," *Journal of Materials Science Letters*, Vol. 21, No. 8, 2002, pp. 657-660. doi:10.1023/A:1015608625798

[8] P. A. Jadhav, R. S. Devan, Y. D. Kolekar and B. K. Chougule, "Structural, Electrical and Magnetic Characterizations of Ni-Cu-Zn Ferrite Synthesized by Citrate Precursor Method," *Journal of Physics and Chemistry of Solids*, Vol. 70, No. 2, 2009, pp. 396-400. doi:10.1016/j.jpcs.2008.11.019

[9] J. C. Aphesteguy, A. Damiani, D. D. Giovanni and S. E. Jacobo, "Microwave-Absorbing Characteristics of Epoxy Resin Composites Containing Nanoparticles of NiZn- and NiCuZn-Ferrites," *Physica B: Condensed Matter*, Vol. 404, No. 18, 2009, pp. 2713-2716. doi:10.1016/j.physb.2009.06.065

[10] H. Su, H. Zhang, X. Tang, Y. Jing and Y. Liu, "Effects of Composition and Sintering Temperature on Properties of NiZn and NiCuZn Ferrites," *Journal of Magnetic Materials*, Vol. 310, No. 1, 2007, pp. 17-21. doi:10.1016/j.jmmm.2006.07.022

[11] M. Penchal Reddy, W. Madhuri, N. Ramamanohar Reddy, K. V. Siva Kumar, V. R. K. Murthy and R. Ramakrishna Reddy, "Magnetic Properties of Ni-Zn Ferrites Prepared by Microwave Sintering Method," *Journal of Electroceramics*, Vol. 28, No. 28, 2012, pp. 1-9.

[12] C. Y. Tsay, K. S. Liu and I. N. Lin, "Microwave Sintering of $(Bi_{0.75}Ca_{1.2}Y_{1.05})(V_{0.6}Fe_{4.4})O_{12}$ Microwave Magnetic Materials," *Journal of the European Ceramic Society*, Vol. 24, No. 6, 2004, pp. 1057-1061. doi:10.1016/S0955-2219(03)00401-1

[13] M. C. Dimri, A. Verma, S. Kashyap, D. C. Dube, O. P. Thakur and Ch. Prakash, "Structural, Dielectric and Magnetic Properties of NiCuZn Ferrite Grown by Citrate Precursor Method," *Materials Science and Engineering B*, Vol. 133, No. 1-3, 2006, pp. 42-48. doi:10.1016/j.mseb.2006.04.043

[14] E. Rezlescu, L. Sachelarie, P. D. Popa and N. Rezlescu, "Effect of Substitution of Divalent Ions on the Electrical and Magnetic Properties of Ni-Zn-Me Ferrites," *IEEE Transactions on Magnetics*, Vol. 36, No. 6, 2000, pp. 2841-2846. doi:10.1109/20.914348

[15] K. C. Maxwell, "Electricity and Magnetism," Oxford University Press, London, Vol. 33, 1873, p. 328.

[16] K. W. Wagner, "Zur Theorie der Unvolkommenen Dielektrika," *American Physics*, Vol. 40, 1913, p. 817.

[17] C. G. Koops, "On the Dispersion of Resistivity and Dielectric Constant of Some Semiconductors at Audio Frequencies," *Physical Review*, Vol. 83, 1953, pp. 121-125. doi:10.1103/PhysRev.83.121

[18] K. Katsmi, S. Mamoru, I. Tatrsuo and I. Katsuya, "Dielectric Behavior of Water Molecules Adsorbed on Iron(III) Oxide Hydroxides," *Bulletin of the Chemical Society of Japan*, Vol. 48, No. 6, 1975, pp. 1764-1767. doi:10.1246/bcsj.48.1764

[19] P. Yadoji, R. Peelamedu, D. Agarwal and R. Roy, "Microwave Sintering of Ni-Zn Ferrites: Comparison with Conventional Sintering," *Materials Science and Engineering B*, Vol. 98, No. 3, 2003, pp. 269-278. doi:10.1016/S0921-5107(03)00063-1

[20] D. R. Patil and B. K. Chougule, "Effect of Copper Substitution on Electrical and Magnetic Properties of $NiFe_2O_4$ ferrite," *Materials Chemistry and Physics*, Vol. 117, No. 1, 2009, pp. 35-40. doi:10.1016/j.matchemphys.2008.12.034

[21] N. Rezlescue and E. Rezlescue, "Dielectric Properties of Copper Containing Ferrites," *Physica Status Solid (a)*, Vol. 23, No. 2, 1974, pp. 575-582. doi:10.1002/pssa.2210230229

9

DC Conductivity and Dielectric Behaviour of Glassy $Se_{100-x}Zn_x$ Alloy

Mohd. Nasir, M. Zulfequar
Department of Physics, Jamia Millia Islamia, New Delhi, India
Email: nasir_sphy@yahoo.com, mzulfe@rediffmail.com

ABSTRACT

The DC conductivity and dielectric properties of glassy $Se_{100-x}Zn_x$ $2 \leq x \leq 20$ alloys have been investigated in the temperature range 303 - 487 K with frequency range 100 Hz – 1 MHz. It is observed that DC conductivity decreases and the activation energy increases with Zn content in Se-Zn system. Dielectric dispersion is observed when Zn incorporated in Se-Zn glassy system. The results are explained on the basis of DC conduction mechanism and dipolar-type dielectric dispersion.

Keywords: Glass; DC Conductivity; Activation Energy; Dielectric Constant; Dielectric Loss; DC Conduction Loss

1. Introduction

Chalcogenide Se based is very important due to its current use as photoreceptors in TV Videocon pick-up tubes [1], conventional xerographic machine and digital X-ray imaging [2,3].These types glasses are belongs to a special group of amorphous semiconductors, which include one, two, three and more chalcogenide elements S, Se, Te from the VI group of the periodic table. From the technical point of view's Se based glassy alloy is important because of their potential applications. To defeat the difficulties, confirm additives are used and mostly used of Se-Zn Se-Sb, Se-Te, Se-Ge, Se-Si and Se-ln is the great interest important properties such as greater hardness, higher sensitivity, higher conductivity and smaller aging effects as compared to pure a-Se. The chalcogenide glassy are useful semiconductors point of application in optics, electronics and optoelectronics like as holography, infrared lenses, ionic sensors, ultra fast optical sensors. It has been focused on chalcogenide glasses of Se-Zn system [4,5] as the materials have been found importance for their electrical, optical, dielectric and kinetics parameters Stable glasses which have good photosensitive properties have been produced and can be n or p type. In low field conduction, the mobility and free carrier concentration are considered to be constant with field. However, the application of high field to free carrier system may affect both the mobility and the number of charge carriers. These studies have been stimulated by the attractive possibilities of using the structural disorder in amorphous semiconductors for the development of better, cheaper and more reliable solid state devices [6,7]. Several band models have been proposed to explain the electronic structure of these materials [8,9]. In the present work, we have reported the electrical and dielectric properties of glassy $Se_{100-x}Zn_x$ alloys.

2. Experimental

Preparation of Glassy Alloys

Glassy alloys of $Se_{100-x}Zn_x$, (where $2 \leq x \leq 20$) are prepared by melt quenching method. The highly pure materials (99.999%) having the desired compositional ratio of elements (Se and Zn) are sealed in a quartz ampoules (of length 7 cm and internal diameter ~8 mm) in a vacuum of about $\sim 10^{-5}$ Torr. The sealed ampoules are kept inside a furnace where the temperature is raised to 850°C at a rate of 4°C - 5°C/min for 11 hours with frequent rocking to ensure the homogenization of the melt. After rocking the ampoules are removed from the furnace and are cooled rapidly in ice-cool water to obtain the glassy nature. This quenching is done in ice-cool water. The nature of glassy alloys is verified by X-ray diffraction. The bulk samples in the form of pellets of pellets (diameter 1.0 cm and thickness 0.2 cm) are obtained by compressing the fine powder of glassy alloys under a load of about 4.11×10^4 Pa using the hydraulic pressure. The DC conductivity and dielectric measurements are maintained under a vacuum of 10^{-3} Torr. The temperature is measured by mounting a calibrated chromel alumel thermocouple near the sample in a specially designed metallic sample holder. The current is measured with a digital picoammeter (Model DPA-111) by applied the dc voltage 1.5 Volts across the pellet sample and resulting the temperature dependent

DC conductivity is measured for the bulk samples of $Se_{100-x}Zn_x$. For the dielectric measurements the bulk samples are mounted between two steel electrodes inside a metallic sample holder. The three terminal measurements used to avoid any stray capacitance effect. All the measurements have been made on the annealed samples in vacuum of the order of 10^{-3} Torr over the entire temperature range (303 K - 487 K). A Wayne Kerr LCR meter (Model-1J4300R) is used for capacitance (C) and dissipation factor (D) measurements. The temperature dependence of dielectric parameters are studied in a heating run rate of 1 K/min. The parallel capacitance and dissipation factor are measured and then dielectric constant (ε') and dielectric loss (ε'') are calculated simultaneously. We preferred to experiment on the pellet rather than the bulk as macroscopic effects (gas bubbles, etc.) may appear in the bulk during preparation. It has been shown by Goyal *et al.* [10], both theoretically and experimentally, that bulk ingots and compressed pellets exhibit similar dielectric behaviour in chalcogenide glasses for the suspected inhomogeneities in case of compressed pellets in these materials. In this entire process the used pellets are to be an uncoated.

3. Result and Discussions

3.1. Powder X-Ray Diffraction Analysis

X-ray diffraction patterns of all samples are recorded at room temperature by using (A Panalytical (PW 3710) X-ray powder diffractometer with Cu Kα radiation (λ = 1.5405 Å). All the samples were scanned in angular range of 5° - 70° with scan speed of 0.01°/s under the similar conditions. From XRD pattern it was clear that all the four samples $Se_{98}Zn_2$, $Se_{95}Zn_5$, $Se_{90}Zn_{10}$ and $Se_{80}Zn_{20}$ are belongs to similar structure of polycrystalline in nature as shown in **Figure 1**. The crystallize size is calculated using Scherer's formula ($D = k\lambda/\beta\cos\theta_B$) of all the specimens in the Se-Zn system, where, D is thickness of crystallite, k is constant dependent on crystallite shape (0.89), λ is X-ray wavelength (1.5405Å), β is FWHM (full width at half max in radian) or integral breadth and θ_B is Bragg angle. The crystallized is found to be increases with increasing the Zn concentration in the glassy system $Se_{100-x}Zn_x$ and the values of crystallized size are given in the **Table 1**.

3.2. Scanning Electron Microscope

Figure 2 Shows the Scanning electron micrographs of all the samples and confirms the polycrystalline nature of the synthesized materials. The numbers of nanocrysts are increases with increase of Zn concentration in the complete system. It is clear from SEM micrograph at highest 20% Zn content that the nanocrysts are easy to see multi structures with the thickness in the glassy alloys $Se_{100-x}Zn_x$.

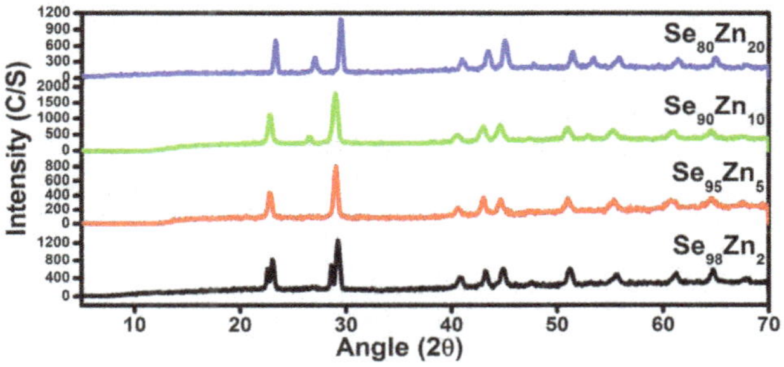

Figure 1. XRD diagrams for all the samples of $Se_{100-x}Zn_x$.

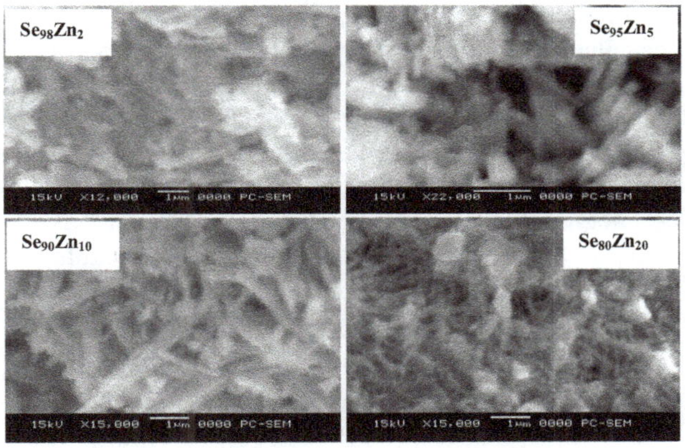

Figure 2. SEM micrograph for all the samples of $Se_{100-x}Zn_x$ glassy alloys.

Table 1. Electrical and dielectric parameters of $Se_{100-x}Zn_x$ alloys at (340 K).

Composition	σ_{DC} ($\Omega^{-1}\cdot cm^{-1}$)	ΔE (eV)	σ_0 ($\Omega^{-1}\cdot cm^{-1}$)	Crystallize Size (L) nm	T = 487 K and f = 22 KHz		ε''_{DC}
					ε'	ε''	
$Se_{98}Zn_2$	4.33×10^{-9}	0.27	3.52×10^{-5}	111.64	8.60	2.28	72.76×10^{-3}
$Se_{95}Zn_5$	2.22×10^{-9}	0.19	1.37×10^{-6}	222.36	9.21	2.61	10.94×10^{-3}
$Se_{90}Zn_{10}$	3.29×10^{-10}	0.30	6.27×10^{-6}	222.98	9.48	3.66	1.62×10^{-3}
$Se_{80}Zn_{20}$	3.69×10^{-11}	0.64	84.22×10^{-3}	223.52	9.39	2.88	3.82×10^{-3}

The thickness of synthesis nanocrysts are found in the range of (174 - 217 nm) in the system. The morphology of SEM micrograph is agreement with the powder XRD result because the numbers of clear sharp peaks are increases with increase of Zn concentration in the glassy system.

3.3. The Temperature Dependence of DC Conductivity

The temperature dependence of DC conductivity (σ_{DC}) in the bulk samples of $Se_{100-x}Zn_x$ $2 \le x \le 20$ are measured. The variation of temperature dependence DC conductivity (σ_{DC}) is in the temperature range (315K - 450K) as in the **Figure 3**. The plot of ln(σ_{DC}) vs. 10^3/T represents a straight line which shows that electrical conduction is through thermally activated process with single activation energy in these glasses [11,12]. The conductivity (σ_{DC}), activation energy (ΔE) and the pre-exponentional factor (σ_0) are represented by well known relation in case of amorphous semiconductors for all samples [8].

$$\sigma_{DC} = \sigma_0 \exp\left[-\frac{\Delta E}{kT}\right] \tag{1}$$

where (ΔE) is the activation energy for the DC conduction mechanism and "k" is the Boltzmann constant, "σ_0" is the pre-exponentional factor. The activation energy (ΔE) calculated from the slope of **Figure 3** for each sample and the values are given in the **Table 1**, which also contains the values of σ_{DC} at 340 K. DC conductivity (σ_{DC}) are decreases with increases the composition and the activation energy (ΔE) are increases with increases the composition for the glassy $Se_{100-x}Zn_x$ system as shown in the **Figure 4**. The activation energy is found to decrease slowly at 5% of zinc content and then increases for high zinc concentration in the system. The decreasing of activation energy can be attributed to shifting of Fermi level [8]. The pre-exponentional factor (σ_0) are increases nonlinearly with the composition in the glassy system.

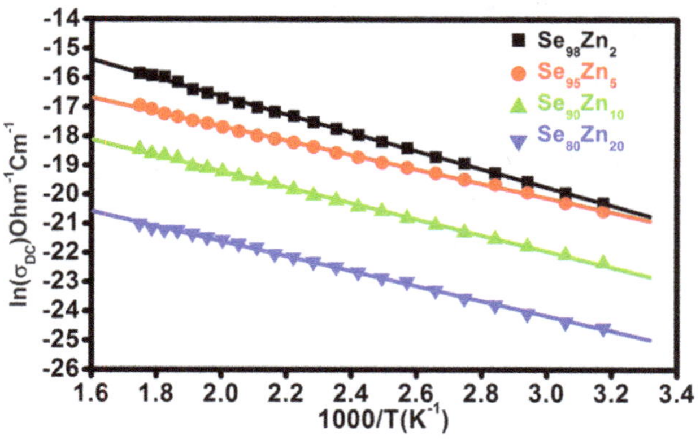

Figure 3. The plots of lnσ_{DC} vs. 10^3/T in the temperature range (315 K - 487 K) for all the samples of $Se_{100-x}Zn_x$.

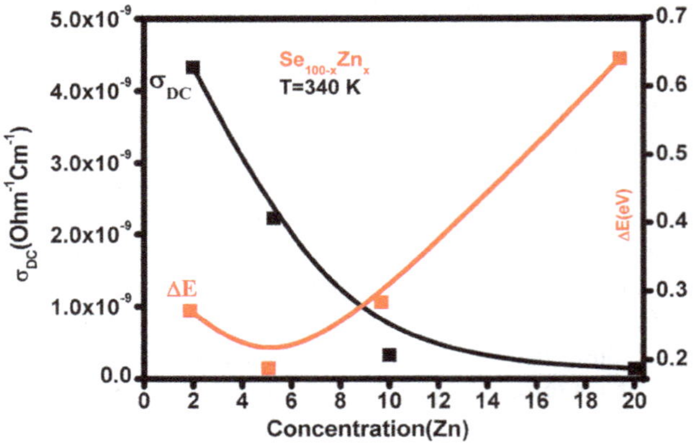

Figure 4. Conductivity (σ_{DC}) and Activation energy (ΔE) versus Zn content for all the samples of $Se_{100-x}Zn_x$.

The value of activation energy (ΔE) and (σ_0) pre-exponentional is give the information about conduction through thermally assisted tunnelling of charge careers movement in the band tails of localized states. The conduction takes place either in the extended states above the mobility gap and it can be understand on the basis of the pre-exponential factor. Mott [13] has been suggested that the pre-exponentional factor (σ_0) for conduction mechanism in the localized states should be two to three orders smaller than for conduction in the extended states, and furthermore it would still become smaller for conduction in the localized states near the Fermi level. If the conduction in extended-state, the value σ_0, reported that for Se and other Se alloys films are of the order 10^4 $\Omega^{-1}\cdot$cm^{-1} [14], However, in the present case the value of σ_0 for Se-Zn and other films are found to be less than 10^4 $\Omega^{-1}\cdot$cm^{-1}. The low value of σ_0 (10^{-6} - 10^{-3} $\Omega^{-1}\cdot$cm^{-1}) shows that a wide range of localized state in the glassy system. Therefore, the possibility of localized conduction in the band tails most likely present [15] in the system $Se_{100-x}Zn_x$. The activation energy increases with the corresponding DC conductivity decreases due to Se-Zn has blend structure both in bulk and in thin film form [15,16]. Selenium (Se) has about 40% atoms in a ring structure and 60% atoms are bonded polymeric chains with the conduction of p-type by J. Schottmiller [17]. However, the conductivity is decreases with Zn concentration that mean increase the conduction through the defect state associated with the impurity atoms [13].

3.4. Temperature and Frequency Dependent Dielectric Constant (ε') and Dielectric Loss (ε'')

The temperature dependence of the dielectric constant (ε') and the dielectric loss (ε'') is studied for various frequencies (100 Hz - 1 MHz) of the glassy $Se_{100-x}Zn_x$ ($2 \le x \le 20$) alloys. The range of temperature is (303 - 487 K) in the glassy system. The dielectric constant (ε') and the dielectric loss (ε'') for $Se_{98}Zn_2$ increases with the increasing the temperature and different for different frequencies as shown in **Figure 5**. The variation of the dielectric constant and the dielectric loss with temperature is large at lower frequencies. The temperature dependence of the dielectric constant (ε') at various frequencies for the $Se_{98}Zn_2$ system indicates that the dispersion maybe occur above 350 K. From the **Figure 6** the peaks have not been observed in the dielectric constant (ε') and dielectric loss (ε'') vs. temperature curves for different fixed frequencies. The dielectric loss increases with temperature in the glassy system. The dielectric constant (ε') for Se-Zn has a higher value as compared to a pure Se [18] and it has lower value as compared to a pure a-Se-Te, a-Se-Te-Ga [19,20] etc. These results indicate that the concentration of Zn plays an important role in the variation of dielectric parameters with temperature and frequency. This type of behavior has also been reported by various workers in chalcogenide glasses [19,21,22].The same type of graphs are observed in other samples of Se-Zn.

The **Figures 7** and **8** shows the variation of dielectric constant and dielectric loss with frequency at different fixed temperature for $Se_{98}Zn_2$. It is observed that the dielectric constant and the dielectric loss decrease with increase of frequency. The dielectric constant and the dielectric loss

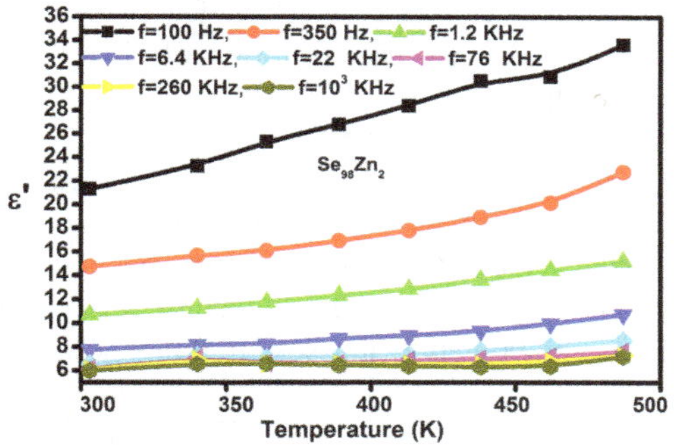

Figure 5. The plot of dielectric constant (ε') vs. Temperature at different frequency in the glassy $Se_{98}Zn_2$ alloy.

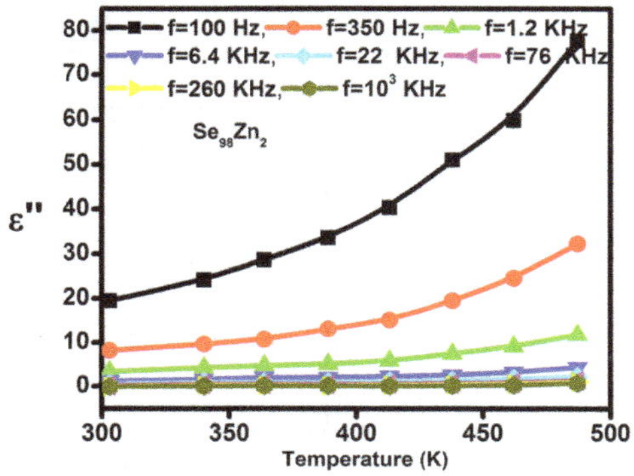

Figure 6. Dielectric loss (ε'') vs. Temperature at different frequency in the glassy $Se_{98}Zn_2$ alloy.

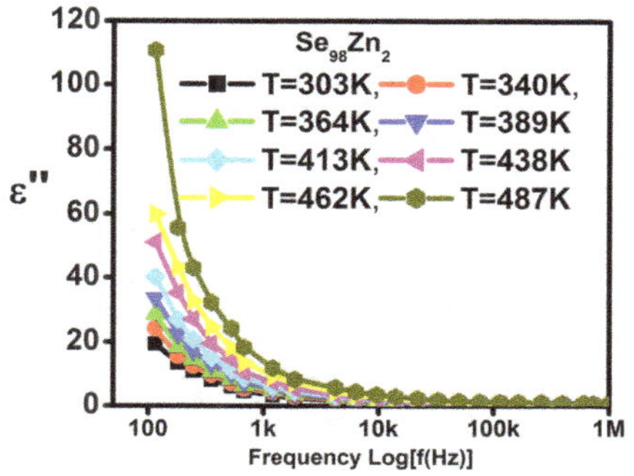

Figure 7. The plots dielectric constant (ε') and (b) dielectric loss (ε'') vs. Log of frequency at different temperatures for the glassy $Se_{98}Zn_2$ System.

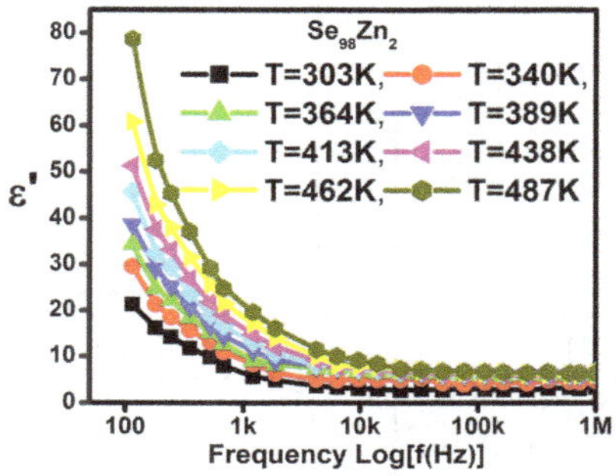

Figure 8. The plots dielectric loss (ε'') vs. log of frequency at different temperatures for the Glassy $Se_{98}Zn_2$ System.

both increase with the increase of Zn concentration in $Se_{100-x}Zn_x$ as shown in **Figure 9**. The numerical values of dielectric constant and the dielectric loss are given in **Table 1**. The higher value of the dielectric constant and the dielectric loss are observed at 10% of Zn content. The temperature dependence of the dielectric constant (ε') at various frequencies for glassy $Se_{100-x}Zn_x$ alloys indicated that the $\ln\varepsilon'$ versus $10^3/T$ plot is a straight line for all the samples as shown in **Figure 10**. This type of temperature dependence is generally observed in molecular solids where the Debye theory [23] for the viscosity dependence of relaxation time holds quite well. According to this theory, the dielectric constant (ε') should increase exponentially with temperature, and such a relationship has been observed indeed for the present samples. The above discussion indicates that dipolar type dielectric dispersion is occurring in the present samples. However, no peaks have been observed in the dielectric loss (ε'') versus ln(f) curves in the present samples as expected in the case of dipolar-type relaxation. Guintini *et al.* [24] have proposed a dipolar model for dielectric dispersion in the glassy system. This model is based on Elliott's idea [25] of hopping of charge carriers over a potential barrier between charged defect states D^+ and D^-. Each pair of sites D^+ and D^- is understood to form a dipole, which has a relaxation time depending on its activation energy [26,27],

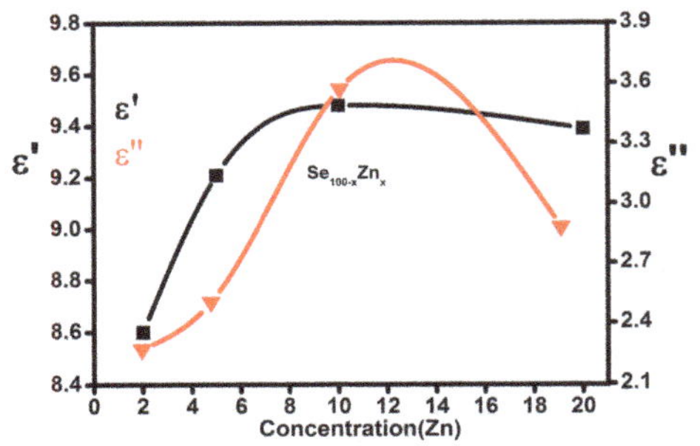

Figure 9. The variation of dielectric constant and dielectric loss with Zn concentration for $Se_{98}Zn_2$ system.

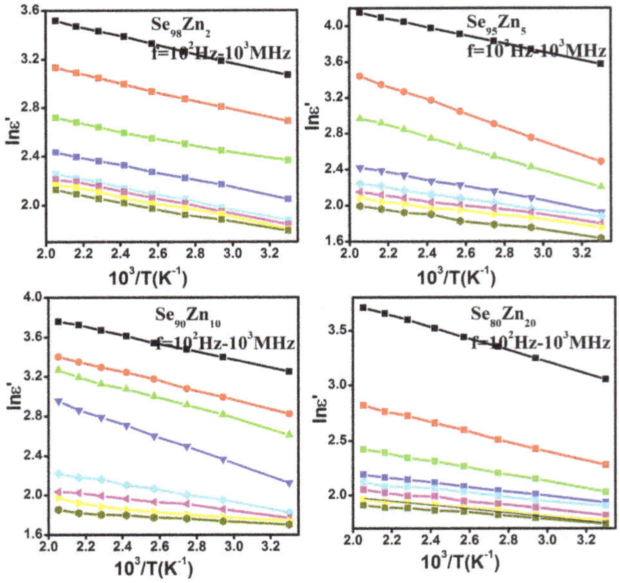

Figure 10. The plots of ln (ε″) vs. $10^3/T$ at different temperature for all the samples of $Se_{100-x}Zn_x$.

recognized to the existence of a potential barrier over which the carriers hopping occurs [28].

According to the above Guintini *et al.* [24] has also proposed a theory of dielectric loss (ε'') at a particular frequency in the temperature range of the chalcogenide glasses where the dielectric dispersion occurs, leads to the relation

$$\varepsilon''(\omega) = (\varepsilon_0 - \varepsilon_\infty) 2\pi^2 N \left[\frac{ne^2}{\varepsilon_0}\right]^3 kT \tau_0^m W_m^{-4} \omega^m \quad (2)$$

Here, m is a power of angular frequency and is negative in this case and is given by

$$m = -\left[\frac{4kT}{W_m}\right] \quad (3)$$

and where n is the number of electrons hop, N is the concentration of localized sites, ε_0 and ε_∞ are the static and optical dielectric constants, respectively, W_m is the energy required to move the electron from a site to infinity. The dielectric loss in these glasses depends upon the total number of localized states. The frequency dependent of dielectric loss ε'' is shown in **Figure 11** at different temperatures for the glassy $Se_{100-x}Zn_x$ alloy. From this figure, it is clear that ε'' is also found to decrease with increasing frequency and It is increases with increase the temperatures. The values of power (m) are calculated from the slopes of these straight lines from **Figure 11**. The values of m are negative and the magnitude of m decreases linearly with temperatures the figure is not given here. The numerical value of power (m) and energy required (W_m) are given in the **Table 2** of the glassy $Se_{100-x}Zn_x$ alloy. The values of required energy (W_m) are increases with increase of temperature in the glassy .This is consistent with theory [24] of dielectric relaxation based on the hopping of charge carriers over a potential barrier as suggested by Elliott [25] is applicable in the case of the glassy $Se_{100-x}Zn_x$ alloy. Hence, the increase of the dielectric loss with the increase of zinc concentration indicates that the density of defects states decreases on addition of zinc in the Se-Zn binary system. Schottmiller *et al.* [13] studied the effect of various elements (S, Te, Bi, As and Ge) on the structure of glassy Se by infrared and Raman Spectroscopy. They reported that in glassy Se about 40% of the atoms have a ring structure and 60% of the atoms are bonded as polymeric chains. The conducting additive in a dielectric may have a very large effect on its electrical response. The presence of additive might increase the concentration of charge carriers and there may be a shift of the Fermi level [29,30]. It is usually assumed that the addition of zinc to the Se-Zn system leads to a cross linking of the Se-Zn chains, reducing the disorder in the system, which in turn may decrease the density of defect states. It has been shown by Goyal *et al.* [10] that spurious dielectric dispersion may appear in chalcogenide glasses, if there is a poor

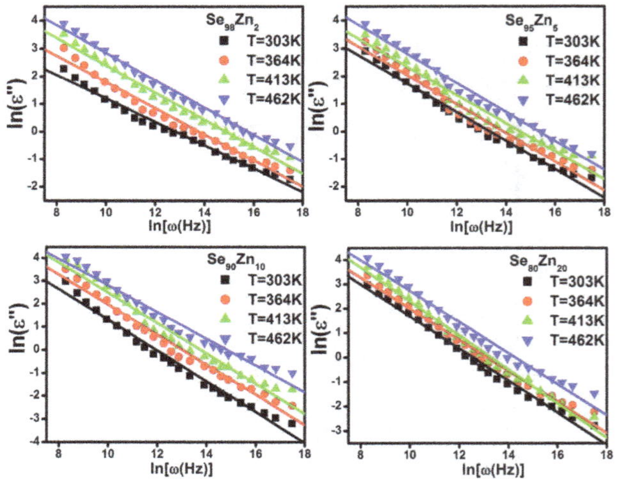

Figure 11. The plots of ln(ε″) with ln(ω) at different fixed temperature for $Se_{100-x}Zn_x$.

Table 2. Temperature dependence of slope (m) and (W_m) required energy for glassy $Se_{100-x}Zn_x$ alloy.

Temp (K)	$Se_{98}Zn_2$		$Se_{95}Zn_5$		$Se_{90}Zn_{10}$		$Se_{80}Zn_{20}$	
	(m)	W_m (eV)	(m)	W_m (eV)	(m)	W_m (eV)	(m)	W_m (eV)
303	–0.51	0.25	–0.58	0.20	–0.67	0.16	–0.70	0.04
364	–0.49	0.27	–0.52	0.24	–0.66	0.19	–0.66	0.16
413	–0.47	0.29	–0.51	0.28	–0.65	0.22	–0.64	0.21
487	–0.42	0.32	–0.50	0.30	–0.58	0.27	–0.63	0.19

electric contact between sample and the electrodes. This is likely to occur in uncoated samples. To ensure that the measured capacitance and dissipation factor represent the values of bulk samples with different thickness. Temperature and frequency dependence of capacitance and dissipation factor have been measured simultaneously. It is found that the values of capacitance and dissipation factor are thickness dependent while the corresponding dielectric constant and dielectric loss are thickness independent. These results indi- cate that the electrode polarization effect does not exist in the present case. The DC conduction loss (ε''_{DC}) is also an important parameter to understand the dielectric dispersion in chalcogenide glasses. We have also calculated the DC conduction loss (ε''_{DC}) using the following formula [20,30,31].

$$\varepsilon''_{DC} = \left[\frac{\sigma_{DC}}{\varepsilon_0(\omega)}\right] \quad (4)$$

where σ_{DC} is the DC conductivity. The calculated values of the DC conduction losses (ε''_{DC}) are shown in the **Table 1**. The **Figure 12** shows the variation of the observed dielectric loss and the calculated DC conduction loss with temperature at fixed frequency for $Se_{100-x}Zn_x$. The DC conduction losses are not dominated on observed dielectric loss because of the DC conduction loss is much smaller as compared to the observed dielectric loss. Hence, the dielectric loss cannot be attributed to the DC conduction in entire temperature range. However, the dielectric dispersion is due to the hopping of charge carriers over a potential barrier between the charged defect states D^+ and D^- for $Se_{100-x}Zn_x$ sample. Thus, these results indicate that the observed dielectric dispersion is recognized mainly to dipolar type dispersion in the present $Se_{100-x}Zn_x$ system.

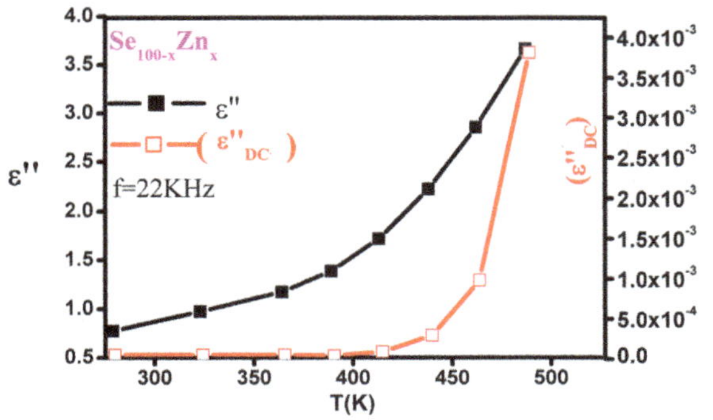

Figure 12.Variation of the observed dielectric loss (ε″), the DC conduction (ε''_{DC}) vs. temperatures at constant frequency 22 KHz for all the composition of ε″ glassy system.

4. Conclusions

On the basis of XRD and SEM micrograph picture the crystallinity is increases, crystallize size and increases the numbers of nanocryst with increases of Zn concentration in Se-Zn glassy system. The DC conductivity (σ_{DC}) is decreases and activation energy (ΔE) are increases with increases Zn composition in the system. It may be due to incorporation of Zn concentration which decreases the defect states in the mobility gap as well as increase band gap in $Se_{100-x}Zn_x$ ($2 \leq x \leq 20$) system. The conduction is most likely in the localized state.

The dielectric dispersion is occurring as dipolar-type in present glassy alloys. Dielectric constant and dielectric loss increase with temperature and as well as increases with increase of Zn content. The dielectric parameters with Zinc can be understood in terms of decrease the density of defect states on adding Zn content and increase the band gap in Se-Zn system. A detailed analysis also suggested the possibility of dipolar type relaxation in the glassy $Se_{100-x}Zn_x$ alloy, after the incorporation of Zn. A possible explanation is given in terms of hopping of charge carriers over a potential barrier between charged defect states. The potential barrier is increases with the concentration of zinc due to increases of activation energy in the system.

REFERENCES

[1] E. Maruyama, "Amorphous Built-in-Field Effect Photoreceptors," *Japanese Journal of Applied Physics*, Vol. 21, No. 2, 1982, pp. 213-223. doi:10.1143/JJAP.21.213

[2] D. C. Hunt, S. S .Kirby and J. A .Rowland, "X-Ray Imaging with Amorphous Selenium: X-Ray to Charge Conversion Gain and avalanche Multiplication Gain," *Medical Physics*, Vol. 9, No. 11, 2002, pp. 2459-2464.

[3] S. O. Kasap, "Imaging Materials," Marcel Dekker, New York, 1991, pp. 350-355.

[4] V. I. Mikla, Yu. Nagy, V. V. Mikla and A. V. Mateleshko, "The Effect of Sb Alloying on the Electro Photographic Properties of Amorphous Selenium," *Materials Science and Engineering B*, Vol. 64, No.1, 1999, pp. 1-8.

[5] N. Chaudhary and A. Kumar, "Dielectric Relaxation in Glassy $Se_{100-x}Sb_x$," *Turkish Journal of Physics*, Vol. 29, No. 3, 2005, pp. 119-125.

[6] K. W. Boer and S. R. Ovshinsky, "Electrothermal Initiation of an Electronic Switching Mechanism in Semiconducting Glasses," *Journal of Applied Physics*, Vol. 41, No. 6, 1970, pp. 2675-2681. doi:10.1063/1.1659281

[7] F. M. Collins, "Switching by Thermal Avalanche in Semiconducting Glass Film," *Journal of Non-Crystalline Solids*, Vol. 2, 1970, pp. 496-503. doi:10.1016/0022-3093(70)90163-8

[8] N. F. Mott and E. A. Davis, "Conduction in Non-Crystalline systems," *Philosophical Magazine*, Vol. 22 No. 179, 1970, pp. 903-922.

[9] M. Kastner and H. Fritzsche, "Defect Chemistry of Lone-Pair Semiconductors," *Philosophical Magazine Part B*, Vol. 37, No. 2, 1978, pp. 199-215. doi:10.1080/01418637808226653

[10] D. R. Goyal, S. Walker and K. K. Srivastava, "Errors Due to Lack of Contact in Measurements of Dielectric Relaxation Parameters for Solid Powders," *Physica Status Solidi (a)*, Vol. 64, No. 1, 1981, pp. 351-357. doi:10.1002/pssa.2210640138

[11] M. Kitao, K. Yoshii and S. Yamada, "Thermoelectric Power of Glassy $As_{40}Se_{60-x}Te_x$," *Physica Status Solidi (a)*, Vol. 91, No. 1, 1985, pp. 271-277. doi:10.1002/pssa.2210910133

[12] A. B. Gadkari and J. K. Zope, "Electrical Properties of Amorphous Semiconducting Se-Te-In System," *Journal of Non-Crystalline Solids*, Vol. 103, No. 2-3, 1988, pp. 295-299. doi:10.1016/0022-3093(88)90208-6

[13] N. F. Mott, "Conduction in Non-Crystalline Systems IV. Anderson Localization in a Disordered Lattice," *Philosophical Magazine*, Vol. 22, No. 175, 1970, pp. 7-29. doi:10.1080/14786437008228147

[14] S. Kumar, R. Arora and A. Kumar, "Space Charge Limited Conduction in a-$Ge_{22}Se_{68}M_{10}$ (M = Cd, ln, Pb, Te)," *Journal of Electronic Materials*, Vol. 22, No. 6, 1993, pp. 675-679. doi:10.1007/BF02666416

[15] R. Mach, P. Flgel and L. G. Suslina, "The Influence of Compositional Disorder on Electrical and Optical Properties of ZnS_xSe_{1-x} Single Crystals," *Physica Status Solidi (b)*, Vol. 109, No. 2, 1982, pp. 607-615. doi:10.1002/pssb.2221090219

[16] H. Mitsubas, I. Mituishi and M. Mizuta, "Coherent Growth of ZnSe on GaAs by MOCVD," *Japanese Journal of Applied Physics*, Vol. 24, 1985, pp. 578-580.

[17] J. Schottmiller, M. Tabak, G. Lucovsky and A. Ward, "The Effect of Valency on Transport Properties in Vitreous Binary Alloys of Selenium," *Journal of Non-Crystalline Solids*, Vol. 4, 1970, pp. 80-96. doi:10.1016/0022-3093(70)90024-4

[18] N. Musahwar, M. A. Majeed Khan, M. Husain and M. Zulfequar, "Dielectric and Electrical Properties of Se-S Glassy Alloys," *Physica B*, Vol. 396, No. 1-2, 2007, pp. 81-86. doi:10.1016/j.physb.2007.03.014

[19] S Kumar, M. Husain and M. Zulfequar, "Effect of Silver on Dielectric Properties of the Se-Te System," *Physica B*, Vol. 371, No. 2, 2006, pp. 193-198. doi:10.1016/j.physb.2005.09.021

[20] S. Kumar, M. Husain and M. Zulfequar, "Dielectric Relaxation in the Glassy a-Se-Te-Ga system," *Physica B*, Vol. 387, No. 1-2, 2007, pp. 400-408. doi:10.1016/j.physb.2006.04.036

[21] M. Ilyas, M. Zulfequar and M. Husain, "Anomalous Dielectric Behaviour in a-Ga_xTe_{100-x} Alloys ($0 \leq x \leq 10$)," Physica B, Vol. 271, No. 1-4, 1999, pp. 125-135. doi:10.1016/S0921-4526(99)00221-5

[22] R. S. Kundu, K. L. Bhatia, N. Kishore and P. Singh, "Effect of Addition of Zn Impurities on the Electronic Conduction in Semiconducting $Se_{80-x}Te_{20}Zn_x$ Glasses," *Philosophical Magazine Part B*, Vol. 72, No. 5, 1995, pp. 513-528.

[23] R. Arora and A. Kumar, "Dielectric Relaxation in Glassy Se and $Se_{100-x}Te_x$ Alloys," *Physica Status Solidi (a)*, Vol. 115, No. 1, 1989, pp. 307-314. doi:10.1002/pssa.2211150135

[24] J. C. Guintini, J. V. Zanchetta, D. Jullien, R. Eholie and P. Houenou, "Temperature Dependence of Dielectric Losses in Chalcogenide Glasses," *Journal of Non-Crystalline Solids*, Vol. 45, No. 1, 1981, pp. 57-62. doi:10.1016/0022-3093(81)90089-2

[25] S. R. Elliott, "A Theory of a.c. conduction in Chalcogenide Glasses," *Philosophical Magazine*, Vol. 36, No .6, 1977, pp. 1291-1304.

[26] A. E. Stern and H. Eyring, "The Deduction of Reaction Mechanisms from the Theory of Absolute Rates," *Journal of Chemical Physics*, Vol. 5, No. 2, 1937, pp. 113-113. doi:10.1063/1.1749988

[27] S. Glasstone, K. J. Laidler and H. Eyring, "Theory of Rate Processes," McGraw Hill, New York, 1941.

[28] P. Pollak and G. E. Pike, "A General Treatment of 1-Dimensional Hopping Conduction," *Physical Review Letters*, Vol. 28, No. 28, 1972, pp. 1449-14450. doi:10.1103/PhysRevLett.28.1449

[29] M. M. Malik, M. Zulfequar, A. Kumar and M. Husain, "Effect of Indium Impurities on the Electrical Properties of Amorphous $Ga_{30}Se_{70}$," *Journal of Physics: Condensed Matter*, Vol. 4, No. 43, 1992, pp. 8331-8338. doi:10.1088/0953-8984/4/43/008

[30] M. Zulfequar, A. Kumar, "Dielectric Behavior of Hot-Pressed AIN Ceramic: Effect of CaO Additive," *Journal of Electrochemical Society*, Vol. 136, No. 4, 1989, pp. 1099-1102. doi:10.1149/1.2096792

[31] M. Zulfequar and A. Kumar, "Electrical Conductivity and Dielectric Behavior of Hot-Pressed AlN," *Advanced Ceramic Matter*, Vol. 3, No. 4, 1988, pp. 332-336.

10

Effect of Copper Doping on Structural, Dielectric and DC Electrical Resistivity Properties of $BaTiO_3$

Moganti Venkata Someswara Rao[1], Kocharlakota Venkata Ramesh[2*], Majeti Naga Venkata Ramesh[2], Bonthula Srinivasa Rao[2]
[1]Department of Physics, S.R.K.R. Engineering College, Bhimavaram, India
[2]Department of Physics, GIT, GITAM University, Visakhapatnam, India
Email: *kv_ramesh5@yahoo.co.in

ABSTRACT

The modified $BaTiO_3$ ferroelectric materials are suitable for pyroelectric applications. This paper reports the structural, dielectric and electrical properties on copper influence in $BaTiO_3$ when it was substituted site "A" of perovskite structure of $BaTiO_3$. Copper has been chosen for modified $BaTiO_3$ with different concentrations with stoichiometry $Ba_{1-x}Cu_xTiO_3$, where x = 0.01%, 0.02%, 0.03% and 0.04%. The X-ray diffraction patterns of the samples doped with different composition of CuO are found to be that the positions and intensities of the diffraction peaks are similar and no secondary phases were observed. The Curie's temperature (T_c) for all CuO doped $BaTiO_3$ with were found to be in the range of 120°C to 125°C. The frequency dependence of relative permittivity (ε_r) and dielectric loss (Tanδ) of $Ba_{1-x}Cu_xTiO_3$ samples at room temperature were reported in the range 100 KHz - 1 MHz. The temperature dependence of D.C electrical resistivity studies were reported for all samples indicating that the participation of Cu^{2+}-Cu^{+} ions in the conduction process around their Curie's temperature.

Keywords: Barium Titanate; Copper Doping; Dielectric Properties; DC Electrical Resistivity; Frequency Dependence

1. Introduction

Few of crystalline materials which show electrical behavior analogous to the magnetic behavior of ferromagnets are called ferroelectrics. These materials exhibit spontaneous polarization even in the absence of external field because of their spontaneous polarization and hence hysteresis phenomenon. In ferroelectric materials this kind of behavior is observed up to a certain temperature known as Curie's temperature (T_c). This behavior is no more above this T_c. The dielectric non linearity is one of the significant characteristic of these ferroelectric materials. The structure consisting of the corner linked oxygen octahedral with a small cation filling the octahedral hole and a large cation filling the dodecahedral hole is usually regarded as a perovskite. As a result we can say that perovskite structure has a wide range of substitution of cations A and B, as well as the anions, but remember that the principles of substitution must maintain charge balance and keep sizes within the range for particular co-ordination number. Because the variation of ionic size and small displacement of atoms that lead to the distortion of the structure and the reduction of symmetry have profound effects on physical properties, perovskite structure materials play such an important role in dielectric ceramics.

Barium titanate ($BaTiO_3$) is one of the best known perovskite ferroelectric compounds ($A^{2+}B^{4+}O^{-3}$) that has been extensively studied due to the simplicity of its crystal structure, which can accommodate different types of dopents [1,2]. Because of the intrinsic capability of the perovskite structure to host ions of different size, large number different dopents can be accommodated in the lattice [3]. The dopent incorporation mechanism to $BaTiO_3$ has been extensively studied. The ionic radius is the main parameter that determines the substitution site [4]. Due to their large piezoelectric values, in addition to spontaneous polarization, reversibility of the permanent polarization by an electric field makes them more attractive for different applications. This has led to the possibility of consisting the properties [5] of doped $BaTiO_3$ for specific technological applications, such as capacitors, sensors with positive temperature coefficients of resistivity, transducers and memories [6-9].

This phenomenon is observed in polycrystalline $BaTiO_3$, with spontaneous polarization of the ferroelectric domains modifying the band-bending and hence the electronic transport across the grain boundary [10]. Doped semi conducting Barium Titanate Ceramics can be as

*Corresponding author.

positive temperature coefficient (PTC) materials. These act like a switching device [11]. These PTC materials are known to have a high temperature coefficient of resistance around and nearer to Curie temperature and having ability of self-limiting leading to useful for sensor applications.

The present study reports the physical properties with Cu doped $BaTiO_3$, with different compositions. From the known literature that the T_c of $BaTiO_3$ is about 120°C, but it can be modified to correspond with a given application by adjusting the composition and the ceramic microstructure or doping which substitutes in to either Ba or Ti sites or both. Two types of dopents, one is donor ions which have a higher valance than the ions they replace and the other is acceptor ions which have a lower valance than the ions they replace [12-14]. In the present study the divalent of copper ion is systematically replaced in the place of Barium ion of A site, which is a divalent.

The present paper reports the study of influence the CuO in the place of BaO in which no acceptor or no donor in the A site of the perovskite structure of $BaTiO_3$. Our current interest is to study the influence of copper on structural, dielectric and dc electrical resistivity properties, when it is substituted for Barium in $BaTiO_3$ ferroelectric. The modified $BaTiO_3$ ferroelectric materials are suitable for pyroelectric applications and hence copper has been chosen for modified $BaTiO_3$ with different copper concentrations. The chosen problem for is to carry out the structural, dielectric and DC electrical resistivity measurements of with $(Ba_{1-x}Cu_x)TiO_3$ with x = 0.01%, 0.02%, 0.03% and 0.04% concentrations.

2. Experimental Techniques

To characterize the prepared samples various experimental probes like XRD technique for structural properties, dielectric and dc electrical resistivity studies have been studied. Different compositions of CuO doped $BaTiO_3$ with $Ba_{1-x}Cu_xTiO_3$ (x = 0.01% to 0.04%) polycrystalline compounds were prepared by solid state reaction method. All the raw materials $BaCO_3$, TiO_2 and CuO are analytical reagent grade of Loba Chemi, India with purity 99.5%, were weighed by the stoichiometric equation $Ba_{1-x}Cu_xTiO_3$ with x = 0.01%, 0.02%, 0.03% and 0.04% respectively by using petit electronic balance MK-E series of 0.0001 gm accuracy. Mixed powders with above compositions were hand ground in Agate mortar for 10 hours thoroughly for homogeneity. The homogenous mixtures of all these compositions were then pre-heated for calcination at 900°C in air for 12 hours. After calcinations these mixtures were examined for their structural studies with X-ray diffraction (XRD) (X-ray diffractometer of make Pan analytical) Cu $K\alpha$-radiation (λ = 1.541 Å) at room temperature was used for structural studies of these samples.

The obtained mixed powders were again ground for 1 hour and granulated by adding PVA as binder, then pressed (with 250 mpa pressure) into pellets with diameter 12 mm and 2 mm thickness. Finally these pellets were sintered in air at 1100°C, for 3 hours, followed by natural cooling to the room temperature. The prepared pellet sample surfaces were polished with carborundum powder for smooth and uniform surface and then the densities for all pellet shaped samples were measured. In order to measure their dielectric and dc electrical resistivity properties, silver paste was painted on both

The polished surfaces of the samples as an electrode and fired at 500°C for 15 min. The Curie temperature (T_c) for all the prepared pellet shaped samples was measured by obtaining the corresponding capacitance with the help of sensitive LCR meter accessed with variable temperature furnace. The dielectric properties of the pellets were determined using Hewlet Packard (Model 4192A) Impedance Analyzer from 100 HZ to 13 MHZ at room temperature.

Determining the D.C. electrical resistivity of a sample is another highly informative macroscopic measurement technique. A temperature dependent study of electrical resistivity yields activation energy of the charge carriers and when coupled with suitable models, this technique can yield important information regarding the nature of the charge carriers, type of charge transport and the nature of energy states involved etc.

In the present investigations the temperature dependence of D.C. electrical resistivity studies were carried out by using a two probe technique. A muffle furnace using a super canthal wire as a heating element is used for temperature variation studies in the range 300 K - 460 K. Temperatures of the furnace as well as the sample are monitored by using a Cr-Al thermocouples. The resistance of the sample was measured using digital electrometer scientific equipment, Roorke model no DNM-121.

The temperature dependence of D.C. electrical resistivities of the samples were studied with known geometry of the samples by applying the following equation.

$$\rho = \rho_0 \exp\left(-W/K_B T\right)$$

W is the activation energy, K_B is the Boltzmann's constant and T is the absolute temperature.

3. Results & Discussion

X-ray diffraction (XRD) is a versatile and non-destructive technique that yields the detailed information about crystallographic structure of the materials.

Figures 1(a)-(d) shows the x-ray diffraction patterns of the samples doped with different composition of CuO

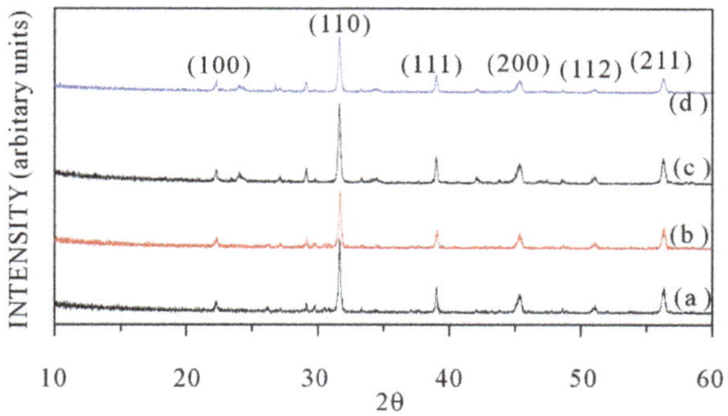

Figure 1. X-ray diffractograms of $Ba_{1-x}Cu_xTiO_3$; (a) x = 0.01; (b) x = 0.02; (c) x = 0.03; (d) x = 0.04.

content of 0.01%, 0.02%, 0.03% and 0.04% respectively. It can be found that the positions and intensities of the diffracttion peaks are similar and no secondary phases were observed. All the above patterns of XRD show the single phase tetragonal system and similar to that of standard pattern of JCPDS of pure $BaTiO_3$ with increase in CuO dopent in the place of Barium [15]. It indicates that influence of CuO did not affect the structural properties. Due to smaller concentrations of CuO doping in the place of Barium and both are of divalent ions may not affect the structural as Cu^{2+} ions replace Ba^{2+} ions. The lattice parameters of doped $BaTiO_3$ with CuO are given in **Table 1**.

The relative densities of the sintered samples are 80.35, 76.71, 75.08 and 71.26 for CuO dopent content x = 0.01%, 0.02%, 0.03% and 0.04% respectively. C. Y. Chen and W. H. Taun reported the theoretical density of pure Ba-TiO_3 is 6.02 g/cm^3 [16]. It was observed that the densities decreased with increase of CuO content even in small composition. It can be observed from the images of the scanning electron microscopy (SEM) shown in **Figure 2**. This may be due to replacement of CuO in BaO place. As the ionic radius of copper (II) is smaller than Barium (II), it leads to porosity in the samples and hence less densification. In contrast to the above, it was observed in literature that the densities of the CuO doped in $BaTiO_3$ increased when Cu (II) was doped in Ti (IV) place [17].

Figures 3(a)-(d) show the variation of capacitance of the CuO doped samples with temperature. From **Figure 3**, the Curie's temperature for each composition of CuO doped $BaTiO_3$ with x = 0.01% to 0.04% was found to be in the range of 120˚C to 125˚C. The corresponding dielectric constant values were also observed in **Figures 4 (a)-(d)** shows the variation of dielectric constant with temperature for $Ba_{1-x}Cu_xTiO_3$ system, in which the dielectric constant values are in the range of 600 to 911. The values of dielectric constant were decreased drastically when comparing with pure $BaTiO_3$, whose value of dielectric constant is around 4100 at Curie temperature

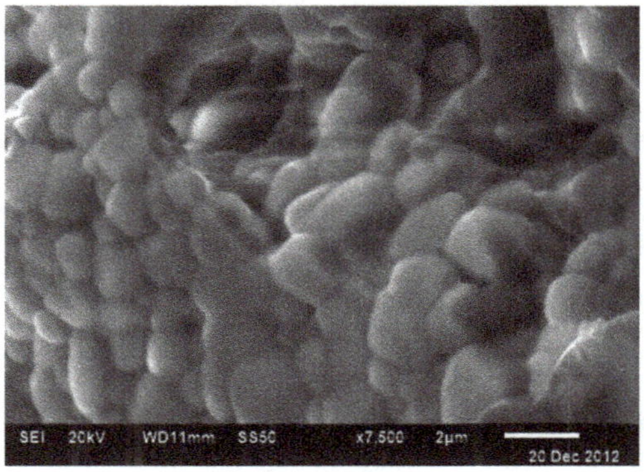

(a)

(b)

(c)

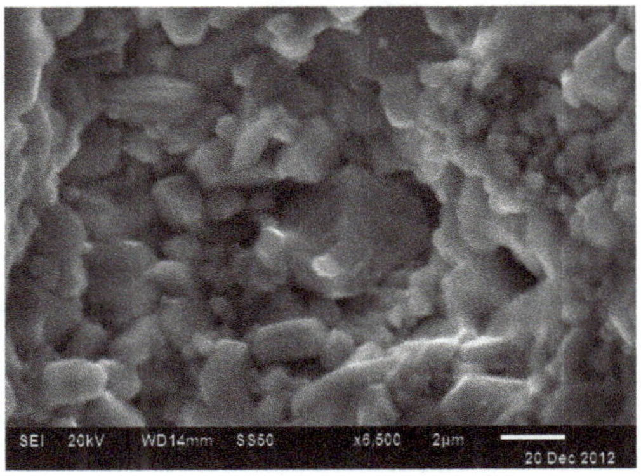

(d)

Figure 2. SEM pictures of $Ba_{1-x}Cu_xTiO_3$: (a) x = 0.01; (b) x = 0.02; (c) x = 0.03; (d) x = 0.04.

Table 1. Lattice parameters of $Ba_{1-x}Cu_xTiO_3$.

X (%)	a (Å)	c (Å)
0.01	3.9865	4.0234
0.02	3.9862	4.0237
0.03	3.9836	4.0383
0.04	3.9822	4.0412

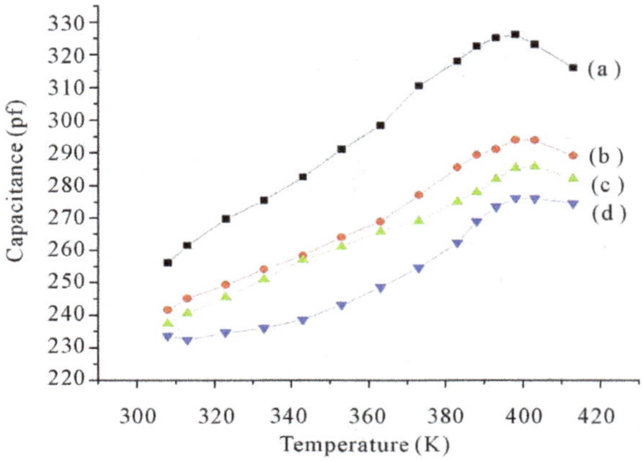

Figure 3. Variation of capacitance with temperature for $Ba_{1-x}Cu_xTiO_3$ system; (a) x = 0.01; (b) x = 0.02; (c) x = 0.03; (d) x = 0.04.

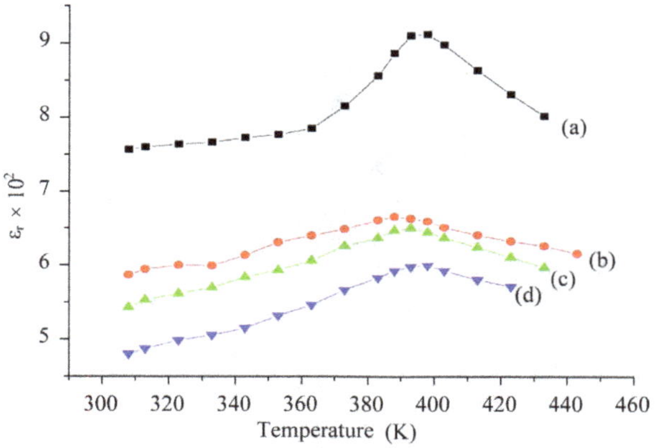

Figure 4. Variation of dielectric constant with temperature for $Ba_{1-x}Cu_xTiO_3$ system; (a) x = 0.01; (b) x = 0.02; (c) x = 0.03; (d) x = 0.04.

[17]. Even though Curie's temperatures for all the samples are maintained in the range around 120˚C, there is decrease in dielectric constant in all CuO doped samples. This might be due to the role of conducting Copper ions in $BaTiO_3$ network in the place of Barium ions. **Table 2** shows the values of dielectric constant and Curies' temperatures of CuO doped $BaTiO_3$ samples.

The frequency dependence of relative permittivity (Dielectric constant, ε_r) and dielectric loss (Tanδ) of $Ba_{1-x}Cu_xTiO_3$ samples at room temperature were given in **Figures 5** and **6** respectively. From the **Figure 5** it can

Table 2. Dielectric constant (ε_r) and Curie's temperature (T_c) of $Ba_{1-x}Cu_xTiO_3$ samples.

X (%)	ε_r	T_c (˚C)
0.01	600	125
0.02	650	120
0.03	666	115
0.04	911	125

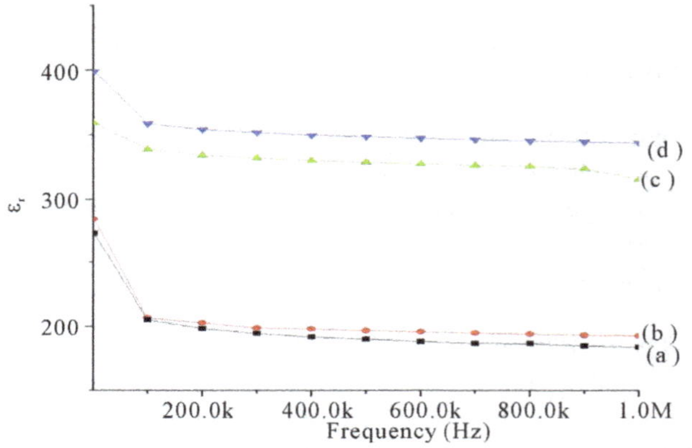

Figure 5. Frequency dependence of dielectric constant for $Ba_{1-x}Cu_xTiO_3$; (a) x = 0.01; (b) x = 0.02; (c) x = 0.03; (d) x = 0.04.

be observed that in all composition initially strong drop of dielectric constant up to 200 KHZ and then slightly reduced up to 1 MHZ. The decrease in dielectric constant is due to the dipoles which cannot follow the alternation of the applied ac electric field at higher frequencies and then total orientation polarization is less at higher frequencies.

Figure 6 shows the frequency dependence of dielectric loss (Tanδ) for all the prepared samples. It is obvious that the trend observed in dielectric loss decreases with increase of CuO content as the dielectric constant increases.

The D.C. electrical resistivities of the prepared samples were measured by using two probe techniques. The interest in D.C. electrical resistivity measurements is because of CuO doped in $BaTiO_3$, as Copper is a good conducting ion in Barium titanate which may lead to transport by way of hopping. But here Copper is substituted in the place of Barium; our interest is how far Copper is involved in the transport mechanism. **Figure 7** indicates the variation of D.C. electrical resistivity with temperature. From **Figure 7**, it can be observed that the D.C electrical resistivity decreased with increase of CuO content (from 0.01% to 0.04%) up to certain temperature and then it increased. This trend was observed to be similar in all the compositions of the samples. Initially the decrease in D.C. electrical resistivity might be due to

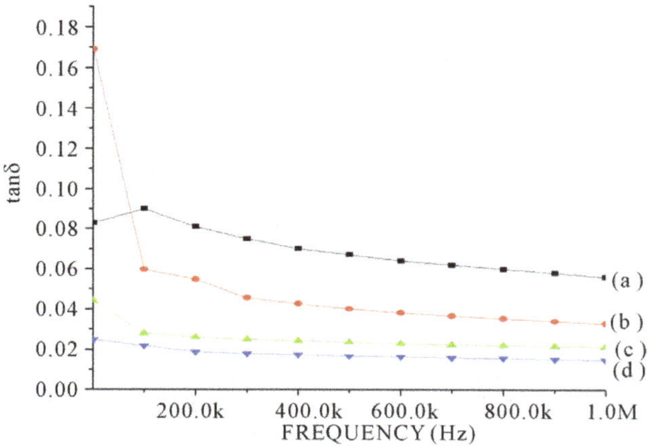

Figure 6. Frequency dependence of dielectric loss (Tanδ) for $Ba_{1-x}Cu_xTiO_3$; (a) x = 0.01; (b) x = 0.02; (c) x = 0.03; (d) x = 0.04.

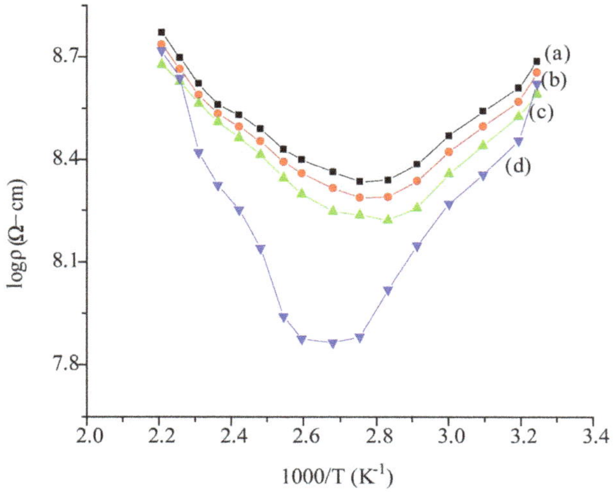

Figure 7. Temperature dependence of D.C. electrical resistivity of $Ba_{1-x}Cu_xTiO_3$; (a) x = 0.01; (b) x = 0.02; (c) x = 0.03; (d) x = 0.04.

participation of Cu^{2+}-Cu^{+} ions in the conduction process around their Curie's temperature and then increase of D.C. electrical resistivity. The increase in dc electrical resistivity at higher temperatures (*i.e.* above Curie temperature) is due to the contribution of hole conduction in all the samples which can be assumed to have been caused by the trapping of holes on Ba-O bonds. The similar trend was observed when applied to variable range hopping model mechanism also.

4. Conclusions

The present work was reports the study of preparation, experimental methodology and results in connection with structural, dielectric and DC electrical resistivity measurements of CuO modified $BaTiO_3$ ceramics. As CuO and BaO were divalent so that CuO doped in the BaO place in pure $BaTiO_3$ composition. Samples with CuO substituted $BaTiO_3$ with stiochiometry $Ba_{1-x}Cu_xTiO_3$, where x = 0.01%, 0.02%, 0.03% and 0.04% compositions were prepared via solid state route. The structural properties of the $Ba_{1-x}Cu_xTiO_3$ (x = 0.01% to 0.04%) samples were explained from the X-ray diffractograms, indicate that the diffraction peaks are similar to the pure $BaTiO_3$ and an influence of CuO in Barium titanate did not affect the structural properties.

The densities of the $Ba_{1-x}Cu_xTiO_3$ (x = 0.01% to 0.04%) samples are with less density compared to unsubstituted $BaTiO_3$. This is due to the increase in porosity of the samples with small ionic radius of Cu ion in the place of Ba ion in "A" site. Curie's temperature of the $Ba_{1-x}Cu_xTiO_3$ (x = 0.01% to 0.04%) samples were found to be in the range 120˚C to 125˚C. The corresponding dielectric constant values are in the range 600 to 910. It was observed that with increase of CuO concentration the dielectric constant also increased. The frequency dependence of dielectricconstant and dielectric loss of $Ba_{1-x}Cu_xTiO_3$ samples at room temperature were studied and they indicate that dielectric constant decreases with frequency. The dielectric loss decreases with increase of CuO content in the samples. The DC electrical resistivity of $Ba_{1-x}Cu_xTiO_3$, x = 0.01% to 0.04%) ceramics were studied in the temperature range 300 K - 460 K. These results indicate that the DC electrical resistivity decreases initially upto nearer to their T_c and then increased. It might be due to the participation of Cu^{2+}-Cu^{+} ions in the conduction process. The increase in resistivity can be attributed due to the contribution of hole conduction in all the samples, caused by the trapping of holes on Ba-O bonds.

REFERENCES

[1] D. Makovec, Z. Samadmija and M. Drofenik, "Solid Solubility of Holmium, Yttrium, and Dysprosium in $BaTiO_3$," *Journal of the American Ceramic Society*, Vol. 87, No. 7, 2004, pp. 1324-1329. doi:10.1111/j.1151-2916.2004.tb07729.x

[2] D. Maga, P. Igor and M. Sergei, "Influence of Impurities on the Properties of Rare-Earth-Doped Barium-Titanate Ceramics," *Journal of Materials Chemistry*, Vol. 10, No. 4, 2000, pp. 941-947. doi:10.1039/a909647g

[3] J. Zhi, A. Chen, Y. Zhi, P. M. Vilainho and J. L. Batptista, "Incorporation of Yttrium in Barium Titanate Ceramics," *Journal of the American Ceramic Society*, Vol. 82, No. 5, 1999, pp. 1345-1348.

[4] Y. P. Pu, W. H. Yang and S. T. Chen, "Influence of Rare Earths on Electric Properties and Microstructure of Barium Titanate Ceramics," *Journal of Rare Earths*, Vol. 25, Suppl. 1, 2005, pp. 154-157. doi:10.1016/S1002-0721(07)60546-8

[5] A. Jane, T. K. Kundu, S. K. Pradhan and D. Chakravorty, "Dielectric Behavior of Fe-Ion-Doped $BaTiO_3$ Nanoparti-

cles," *Journal of Applied Physics*, Vol. 97, No. 4, 2005, Article ID 044311. doi:10.1063/1.1846135

[6] L. E. Cross, "Ferroelectric Ceramics: Materials and Application Issues," *Ceramic Transactions*, Vol. 68, 1996, pp. 15-55.

[7] V. Dharmadhikari and W. Grannemann, "Photovoltaic Properties of Ferroelectric $BaTiO_3$ Thin Films rf Sputter Deposited on Silicon," *Journal of Applied Physics*, Vol. 53, No. 12, 1982, p. 8988. doi:10.1063/1.330456

[8] T. R. Shrout and J. P. Dougherty, "Lead Based $Pb(B_1B_2)O_3$ Relaxors vs $BaTiO_3$ Dielectrics for Multilayer Capacitors," *Ceramic Transactions*, *Ceramic Dielectrics*: *Composition*, *Processing*, *and Properties*, *American Ceramic Society*, Vol. 8, 1990, p. 3.

[9] G. H. Haertling, "Ferroelectric Ceramics: History and Technology," *Journal of the American Ceramic Society*, Vol. 82, No. 4, 1999, pp. 797-818. doi:10.1111/j.1151-2916.1999.tb01840.x

[10] W. Heywang, "Resistivity Anomaly in Doped Barium Titanate," *Journal of the American Ceramic Society*, Vol. 47, No. 10, 1964, pp. 484-490. doi:10.1111/j.1151-2916.1964.tb13795.x

[11] S. Chatterjee and H. S. Maiti, "A Novel Method of Doping PTC Thermistor Sensor Elements during Sintering through Diffusion by Vapour Phase," *Materials Chemistry and Physics*, Vol. 67, No. 1-3, 2001, pp. 294-297. doi:10.1016/S0254-0584(00)00454-5

[12] F. Batllo, E. Duverger, J. C. Jules and B. Jonnet, "Dielectric and EPR Studies of Mn-Doped Barium-Titanate," *Ferroelectrics*, Vol. 109, No. 1, 1990, pp. 113-118.

[13] V. C. S. Prasad and L. G. Kishore, "Studies on Some $BaTiO_3$ Based Commercial Electroceramics," *Ferroelectrics*, Vol. 120, No. 1, 1990, pp. 141-150. doi:10.1080/00150199008221472

[14] F. D. Morrison, D. C. Sinclair and A. R. West, "An Alternative Explanation for the Origin of the Resistivity Anomaly in La-Doped $BaTiO_3$," *Journal of the American Ceramic Society*, Vol. 84, No. 2, 2001, pp. 474-776. doi:10.1111/j.1151-2916.2001.tb00684.x

[15] W. Luan, L. Gao and J. Guo, "Size Effect on Dielectric Properties of Fine-Grained $BaTiO_3$ Ceramics," *Ceramics International*, Vol. 25, No. 8, 1999, pp. 727-729. doi:10.1016/S0272-8842(99)00009-7

[16] C. Y. Chen and W. H. Tuan, "Effect of Silver on the Sintering and Grain-Growth Behavior of Barium Titanate," *Journal of the American Ceramic Society*, Vol. 83, No. 12, 2000, pp. 2988-2992. doi:10.1111/j.1151-2916.2000.tb01671.x

[17] T. Li, K. Yang, R. Z. Xue, Y. C. Xue and Z. P. Chen, "The effect of CuO Doping on the Microstructures and Dielectric Properties of $BaTiO_3$ Ceramics," *Journal of Materials Science*: *Materials in Electronics*, Vol. 22, No. 7, 2011, pp. 838-842.

The Reaction Sequence and Dielectric Properties of $BaAl_2Ti_5O_{14}$ Ceramics

Xiaogang Yao, Wei Chen, Lan Luo
Shanghai Institute of Ceramics, Chinese Academy of Science, Shanghai, China
Email: rockyao@student.sic.ac.cn

ABSTRACT

To investigate the correct reaction sequence of BaO-Al_2O_3-5TiO_2 system, powders calcined at different temperatures are analyzed by x-ray diffraction. The results show that the source phase $BaCO_3$ decomposes below 800°C, TiO_2 and Al_2O_3 start to consume at 900 and 1100°C, respectively. $BaTi_4O_9$ phase appears at 1000°C while $BaAl_2Ti_5O_{14}$ phase starts to reveal at 1200°C. As the temperature increases, the density, dielectric constant and quality factor of the $BaAl_2Ti_5O_{14}$ ceramic increase and keep unchanged at 1350°C. The dielectric properties of $BaAl_2Ti_5O_{14}$ ceramic sintered at 1350°C for 3h are: ε_r=35.8, Q×f=5130GHz, τ_f=-6.8ppm/°C.

Keywords: Reaction Sequence; $BaAl_2Ti_5O_{14}$; Ceramics; Dielectric Properties

1. Introduction

The rapid progress in mobile communication has created a tremendous demand for the microwave dielectric materials with high dielectric constant, low dielectric loss and near-zero temperature coefficient of resonator frequency [1,2]. As a typical high permittivity system, $Ba_{6-3x}Ln_{8+2x}Ti_{18}O_{54}$ (Ln = La, Sm, Nd) has attracted plenty of attention for the high dielectric constant over 80 [3,4]. However, the shortcoming of BLT system is its relatively high dielectric loss (low quality factor) which has restricted its commercial application.

Much work was done to lower the dielectric loss of the BLT system. Zhu improved the Q × f value of $Ba_{4.2}Nd_{9.2}Ti_{18}O_{54}$ ceramic by doping with $LnAlO_3$ (Ln = La, Nd, Sm) [5]. TiO_2 was added into $Ba_{6-3x}Sm_{8+2x}Ti_{18}O_{54}$ by Ohsato and excellent dielectric properties were obtained [6]. Our previous work has revealed that the crucial point to lower the dielectric loss of BLT system is preventing the reduction of Ti^{4+} at high sintering temperatures [7]. Al_2O_3 or MgO was used as an acceptor to suppress the reduction of Ti^{4+} in $Ba_{4.2}Sm_{9.2}Ti_{18}O_{54}$ ceramic and was very effective to improve the Q×f value of the BST ceramic. In our present work, Al_2O_3 was added into Ba-Sm-Ti and Ba-Nd-Ti systems. The dielectric loss has been reduced effectively. The common results are closely related to a new phase $BaAl_2Ti_5O_{14}$ (BAT) which is observed in both systems. No relevant information about the new BAT phase was reported.

In this paper, $BaAl_2Ti_5O_{14}$ ceramic was prepared by the solid state reaction. X-ray diffraction was used to identify the crystalline phase at each calcining temperature from 800 to 1350°C. The reaction sequence of BaO-Al_2O_3-5TiO_2 system was determined and the microwave dielectric properties of $BaAl_2Ti_5O_{14}$ ceramic were measured.

2. Experiment

The BAT ceramic powders were prepared according to the desired stoichiometry of $BaAl_2Ti_5O_{14}$ by mixing the chemical grade starting materials $BaCO_3$ (99.9%), Al_2O_3 (99.9%) and TiO_2 (99.9%). After ground in deionized water with ZrO_2 balls for 24h, the mixture was dried and then calcined at different temperatures from 800 to 1350°C in air for 1h. The optimally calcined BAT powders were milled for 24h, dried at 120°C and granulated with polyvinyl alcohol (PVA). The granules were preformed and then sintered at 1275~1375°C in air for 3h with a heating rate of 5°C/min.

The bulk densities of the BAT ceramic were measured by the Archimedes method. The crystalline phases of the calcined BAT powders and sintered BAT ceramic were analyzed by a Rigaku D/max 2550V X-ray diffractometer with a conventional Cu-Kα radiation in the range of 10~70° with a step size of 0.02°. The microstructure of the BAT ceramic was examined by a Hitachi S-4800 field emission scanning electron microscope. The dielectric properties of the polished BAT samples were tested the TE_{011} mode of an Agilent E8363A PNA series network analyzer with a frequency ranges from 3 to 4GHz. τ_f was tested in the temperature ranges from 20 to 80°C and calculated by noting the change in resonant frequency as:

$$\tau_f = (f_2 - f_1)/60 f_1 \quad (1)$$

Here, f_1 and f_2 represent the resonant frequencies at 20 and 80°C, respectively.

3. Results and Discussion

3.1. Reaction Sequence of BaO-Al_2O_3-5TiO_2 ststem

Figure 1 shows the XRD patterns of the BAT powders calcined at different temperatures from 800 to 1350°C for 1h. Here, the phases identified by X-ray diffraction at each calcining temperature are listed in **Table 1**. Six phases are observed as various temperatures are used. $BaCO_3$, Al_2O_3 and TiO_2 phases are observed at 800 °C with a new phase $BaTiO_3$ accompanied. It is easy to deduce that $BaTiO_3$ is the product of the reaction of $BaCO_3$ and TiO_2. With increasing the temperature to 900°C, no

$BaCO_3$ is residual while the trace of $BaTiO_4$ is detected. Fewer $BaTiO_3$ but a predominant $BaTiO_4$ phase is found at 1000°C. Not any change is observed until the calcining temperature is increased to 1200°C. The diffraction peaks of TiO_2 and $BaTiO_4$ are getting weak while a new phase $BaAl_2Ti_5O_{14}$ appears. Only a single $BaAl_2Ti_5O_{14}$ phase exists over 1300°C.

We can easily write down the reaction sequence of the BaO-Al_2O_3-$5TiO_2$ system with the increasing of calcining temperature from 800 to 1350°C.

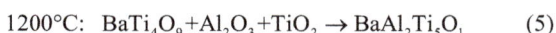

$$\text{Below 800°C:}\quad BaCO_3 \rightarrow BaO + CO_2\uparrow \tag{2}$$

$$\text{800°C:}\quad BaO + TiO_2 \rightarrow BaTiO_3 \tag{3}$$

$$\text{900~1000°C:}\quad BaTiO_3 + 3TiO_2 \rightarrow BaTi_4O_9 \tag{4}$$

$$\text{1200°C:}\quad BaTi_4O_9 + Al_2O_3 + TiO_2 \rightarrow BaAl_2Ti_5O_1 \tag{5}$$

Figure 2 shows the DSC curve of the BaO-Al_2O_3-$5TiO_2$ powder heated from room temperature to 1350°C. As shown in **Figure 2**, three exothermic peaks are observed at 825, 925 and 1300°C which correspond very well to the temperatures at which the reactions (3)-(5) happen.

3.2. Dielectric Properties of $BaAl_2Ti_5O_{14}$ Ceramic

From what has been discussed above, we can draw the conclusion that the sintering temperature of the $BaAl_2Ti_5O_{14}$ ceramic is between 1300 and 1350°C. Thus, five temperature points (1275, 1300, 1325, 1350, 1375°C) are used to study the effect of temperature on the dielectric properties.

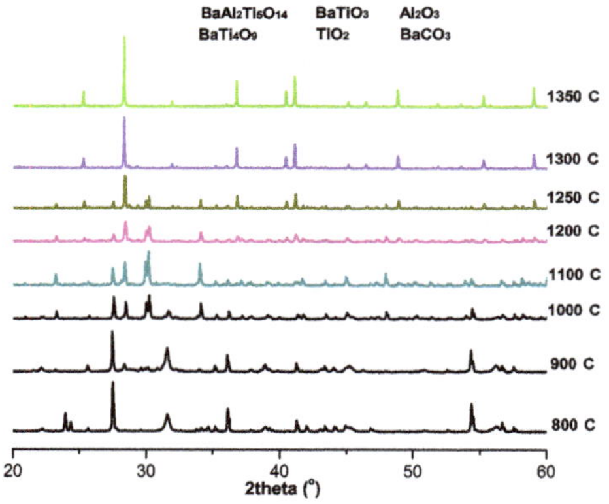

Figure 1. XRD patterns of BaO-Al_2O_3-$5TiO_2$ powders calcined at different temperatures from 800 to 1350°C for 1h.

Table 1. Crystalline phases exist of not exist at each calcining temperature.

T_c/°C	$BaCO_3$	TiO_2	Al_2O_3	$BaTiO_3$	$BaTiO_4$	$BaAl_2Ti_5O_{14}$
800	Y	Y	Y	Y	N	N
900	N	Y	Y	Y	Y	N
1000	N	Y	Y	N	Y	N
1100	N	Y	Y	N	Y	N
1200	N	Y	N	N	Y	Y
1250	N	Y	N	N	N	Y
1300	N	N	N	N	N	Y
1350	N	N	N	N	N	Y

Figure 3 shows the density of BAT ceramics at different sintering temperature. With the increasing temperature from 1275 to 1375°C, the density of BAT ceramics increases from 4.02 to the maximum value 4.17g/cm^3 at 1350°C, and then decreases slightly. We can conclude that the optimized sintering temperature of the BAT ceramics is 1350°C.

Figure 4 shows the SEM images of the BAT ceramic samples sintered at 1350°C for 3h. As we can see from **Figure 4(a)**, the BAT ceramic has a compact structure but a heterogeneous grain size. The average size is 10 μm, as shown in **Figure 4(b)**. Irregularly grown grains are seen in both images. The formation mechanism of these huge grains is still inexplicit and needs further research.

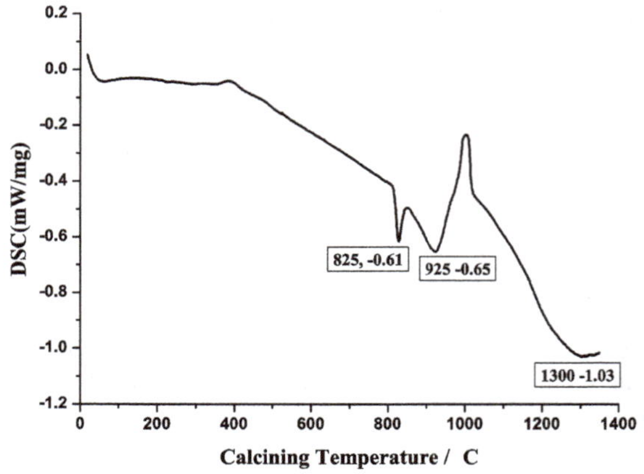

Figure 2. DSC curve of BAT powders.

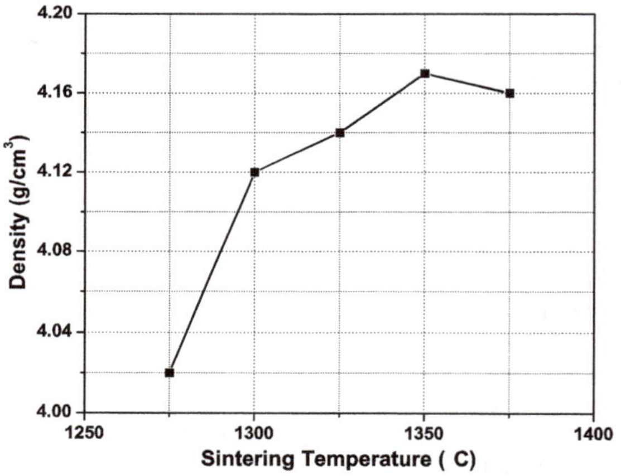

Figure 3. Density of the BAT ceramics sintered at different temperatures from 1275 to 1375°C.

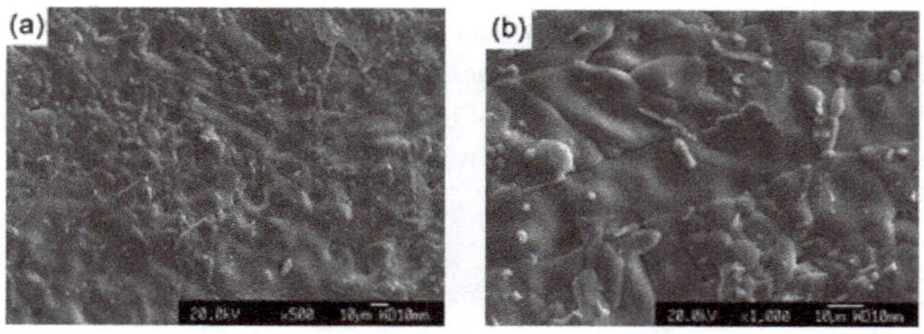

Figure 4. SEM images of the BAT ceramic sintered at 1350°C for 3h: (a) ×500; (b) ×1000.

Figure 5 shows the dielectric constant of BAT ceramics sintered at different temperatures. It is not strange that the change in dielectric constant with temperature shows the same regularity with that of density, since a more compact structure means a lower porosity. The dielectric constant of BAT ceramics reach 35.8 after sintering at 1350°C for 3h.

Figure 6 shows the Q × *f* value of the BAT ceramics sintered at different temperatures from 1275 to 1375°C for 3h. With the increasing temperature, the Q × *f* value of BAT ceramics increases from 4324GHz at 1275°C to the maximum value 5130 GHz at 1350°C. The heterogeneous grain size has very bad effect on the Q × *f* value. Huge grains can increase the dielectric loss significantly. Therefore, much work need be done to obtain a more homogenous grain distribution so as to improve the Q × *f* value of BAT ceramics.

Figure 7 shows the τ_f value of BAT ceramics sintered at different temperatures for 3h. The τ_f value of the BAT ceramics is slightly affected by the sintering temperature. BAT ceramics sintered at 1350°C for 3h has a negative τ_f value of -6.8ppm/°C.

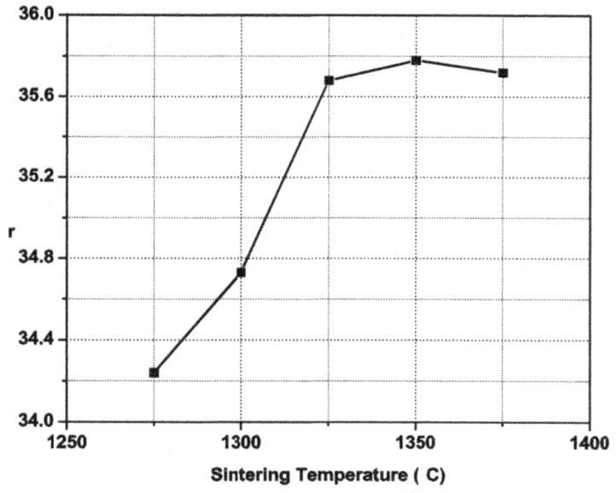

Figure 5. Dielectric constant of the BAT ceramics sintered at different temperatures from 1275 to 1375°C.

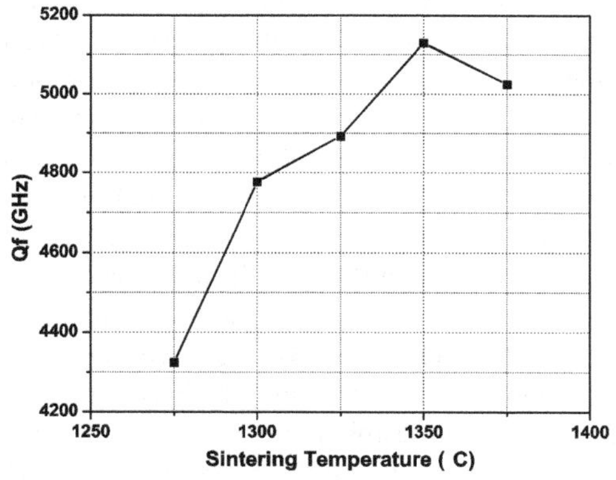

Figure 6. Q×*f* value of the BAT ceramics sintered at different temperatures from 1275 to 1375°C.

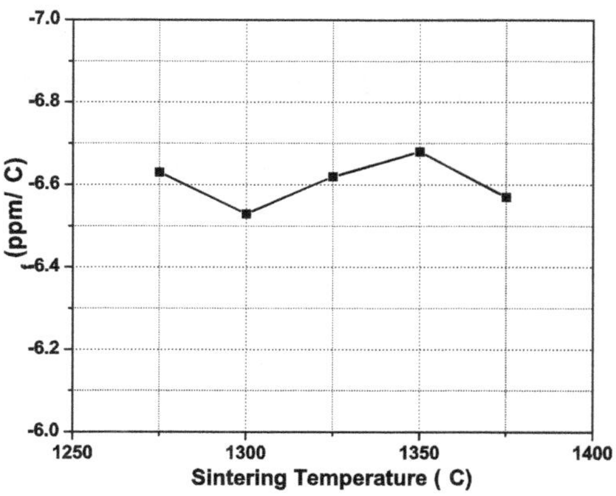

Figure 7. Resonator frequency of temperature coefficient of the BAT ceramics sintered at different temperatures from 1275 to 1375°C.

4. Conclusion

$BaAl_2Ti_5O_{14}$ ceramic is prepared by the conventional solid state reaction. The reaction sequence of BaO-Al_2O_3-$5TiO_2$ system has been established. The result shows that the ideal calcining temperature of BAT powder is 1200°C and the best sintering temperature of BAT ceramic is 1350°C. The BAT ceramic has a heterogeneous grain distribution which has very bad effect on its dielectric properties especially for the Qf value. The dielectric properties of BAT ceramic sintered at 1350°C for 3h are: ε_r=35.8, Q×*f*= 5130GHz and τ_f=-6.8ppm/°C.

REFERENCES

[1] R. Cava, "Dielectric materials for applications in microwave communications," J.Mater.Chem. pp. 54-62, 2001.

[2] I. Reaney, D. Iddles, "Microwave dielectric ceramics for resonators and filters in mobile phone networks," J.Am. Ceram. Soc.pp.2063-2072,2006.

[3] H. Ohsato, "Science of tungstenbronze-type like $Ba_{6-3x}R_{8+2x}Ti_{18}O_{54}$ (R=rare earth) microwave dielectric solid solutions," J.Euro. Ceram. Soc. pp.2703-2711,2001.

[4] H. Ohsato, M. Mizuta, "Microwave dielectric properties of tungsten bronze-type $Ba_{6-3x}R_{8+2x}Ti_{18}O_{54}$ (R = La, Pr, Nd and Sm) solid solution,"J. Ceram.Soc.Jap.pp.178-182,1998.

[5] J. Zhu, E. Kipkoech, W. Lu. "Effects of $LnAlO_3$(Ln =La, Nd, Sm) additives on the properties of $Ba_{4.2}Nd_{9.2}Ti_{18}O_{54}$ ceramics," J.Euro. Ceram.Soc.pp.2027-2030,2006.

[6] H. Ohsato, A. Komura, "Microwave dielectric properties and sintering of $Ba_{6-3x}R_{8+2x}Ti_{18}O_{54}$ (R=Sm,x=2/3) solid solution with added rutile," Jpn.J.Appl.Phys. pp.5357-5359,1998.

[7] X. Yao, H. Lin, "Antireducion of Ti^{4+} in $Ba_{4.2}Sm_{9.2}Ti_{18}O_{54}$ ceramics by doping with MgO,Al_2O_3 and MnO_2," Ceram. Int. pp.3011-3016,2012.

12

Synthesis and Electrical Properties of Polyaniline Composite with Silver Nanoparticles

Safenaz M. Reda[1*], Sheikha M. Al-Ghannam[2]
[1]Chemistry Department, Faculty of Science, Benha University, Benha, Egypt
[2]Chemistry Department, College of Girls for Science, University of Dammam, Dammam, KSA
Email: safenazr@yahoo.com

ABSTRACT

Polyaniline/silver (PANI/Ag) nanocomposite was prepared by chemical oxidative polymerization of aniline monomer in the presence of nitric acid. The formation of PANI/Ag nanocomposite was characterized by XRD, FTIR, TEM, UV-vis spectroscopy. The XRD patterns indicated that the crystalline phase of Ag is cubic with crystallite size of 93 nm. The TEM image shows that the Ag nanoparticles are well dispersed in the polyaniline matrix. Optical measurements show that the value of optical band gap of nanocomposite is lower than that of pure PANI. The DC-, AC-conductivities, dielectric permittivity (ε') and dielectric loss (ε'') of (PANI/Ag) nanocomposite and pure PANI have been measured in the temperature range from 303 to 723 K and frequency range from 10 to 10^3 kHz. The electrical conductivity of the (PANI/Ag) nanocomposite is higher than pure PANI. Temperature variation of frequency exponents in this blend suggests that AC-conductivity is attributed to correlated barrier hopping mechanism. At all frequencies, the ε' value for (PANI/Ag) nanocomposite is higher than that for pure one. The higher dielectric constant of the PANI/Ag nanocomposite indicates their better ability to store electric potential energy under the influence of alternative electric field.

Keywords: Polyaniline/Silver (PANI/Ag) Nanocomposite; DC-Electrical Conductivity; AC-Electrical Conductivity; Dielectric Permittivity; Dielectric Loss

1. Introduction

Polymers are generally insulators and to exhibit electrical conductivity they must possess, ordered conjugation with extended (pi) electrons and large carrier concentrations. Conjugated polymers are the organic compounds that have an extended (pi) orbital system and conjugated carbon system [1]. Conductive polymer with polyaromatic backbone including polypyrrole, polythiophene, polyaniline, etc. has received a great deal of attention in the last two decades [2]. The conductivity of these conjugated polymers can be controlled by the process of doping which may be carried out through a chemical route, electrochemical route or photochemical route and is characterized by charge transfer from dopant to polymer or from polymer to dopant [1]. On doping these conjugated polymers show very high conductivity similar to metals. Therefore sometimes they are also called Synthetic Metals. They combine the electrical properties of metals with the advantage of polymers such as smaller weight, greater workability, resistance to corrosion and lower cost [1]. The most exciting applications of conductive polymers are in television sets, cellular telephones, automotive dashboard displays and artificial cockpit displays, Light emitting devices (LEDs), solar cells, lightweight batteries, light emitting diodes, polymer actuators, corrosion protection agents, sensors and molecular electronic devices [1,2]. Amongst the family of conducting polymers polyaniline (PANI) is one of the most promising electrically conducting polymers due to its unique electrical, electrochemical properties, easy polymerization, high environmental stability and low cost of monomer [3]. Also, its wide applications in microelectronic devices, diodes, light weight batteries, sensors, super capacitors, microwave absorption, corrosion inhibition [3]. Moreover, PANI has attracted attention for various reasons such as high absorption coefficients in the visible part of the spectrum, high mobility of charge carriers and tremendous stability [1]. Also PANI is unique among conductive polymers in that its electrical properties could be reversibly controlled both by charge transfer doping and by protonation, which makes it a potential material for applications such as chemical and biological sensors, actuators, microelectronic devices, etc. [4]. PANI exists in a number of forms which totally differ in chemical and physical properties. The most conducting emeraldine salt has conductivity on

*Present Address: Chemistry Department, College of Girls for Science, Dammam University, Dammam, Kingdom of Saudi Arabia.

a semiconductor level of the order of 100 omh·cm^{-1}, many orders of magnitude higher than that of traditional polymers (<10^{-9} omh·cm^{-1}) but lower than that of metals (>10^4 omh·cm^{-1}) [1]. PANI possesses secondary and tertiary amines in the backbone structure that can bind metal ions; these metal ions can be released by immersing inbound PANI into a low pH solution. The ability of binding metal ions and subsequent liberation makes PANI an attractive material for environmental safety [1]. However, PANI suffers from poor processability because it is infusible and insoluble in common solvents. The properties of PANI can be tailored by changing its oxidation states, dopants or through blending it with other organic, polymeric or inorganic nanosized semiconducting particles [3]. A number of metal and metal oxide particles have been encapsulated into the conductive polymer to form nanocomposites (NCs). The NCs exhibit combination of properties like conductivity, electrochemical, catalytic and optical properties [3]. The NCs are used in applications like electrochromic devices, light-emitting diodes, electromagnetic interference shielding, secondary batteries, electrostatic discharge systems, chemical and biochemical sensors [3]. To obtain materials with syncretistic advantage between PANI and inorganic nanoparticles, various composites of PANI with inorganic naonoparticles such as Cu [5], tellurium [6], lithium [7] and TiO_2 [8] have been reported. No much work has been done on the polyaniline doped with noble metals such as silver (Ag). The incorporation of metal nanoparticles could effectively improve the electrical, optical and dielectric properties of the polyaniline composites [4]. These properties are extremely sensitive to small changes in the metal content and in the size and shape of the nanoparticles. It was reported that the nanoparticles themselves could act as conductive junctions between the PANI chains that resulted in an increase of the electrical conductivity of the composites [4]. The electrical conductivity of such composites might also depend upon the molecular structure of the conductive polymer matrix (*i.e.*, crystallinity). Since silver exhibits the highest electrical and thermal conductivities among all the metals [4], the combination of PANI with silver could yield functional materials having enhanced electrical properties.

In this study, we explore the possibility of improving the conductivity of PANI by doping it with silver atoms. PANI/Ag NCs was prepared by chemical oxidation polymerization method under UV light. The electric and dielectric properties of the PANI/Ag NCs were investigated.

2. Experimental

2.1. Synthesis

Aniline (Aldrich) was dissolved in 1 M nitric acid, as was silver nitrate (Aldrich). The solutions were mixed to start oxidation at room temperature. The concentration of aniline was 0.4 M, the silver nitrate-to-aniline molar ratio was 2:5. The reaction was slow, characterized by an induction period extending for weeks. So it was activated by exposing to UV light [9]. Green solids produced by the oxidation were collected on filter and washed several times by distilled water, and dried at room temperature.

The mechanism of synthesis of polyaniline-silver nanocomposite is given in **Scheme 1**. The amine group of aniline is transformed into protonated form by nitric acid. Nitric acid capped silver nanoparticles have negative charges on their surface. Electron negative silver nanoparticles adsorb the positively charged aniline cations through electrostatic interactions prior to polymerization and form aniline-silver complex [9].

The DC-, AC-conductivity, dielectric constant ε' and dielectric loss ε'' as a function of temperature and frequency were measured. This was carried out using programmable automatic LCR meter (PM636 Philips) at frequencies of 10 - 10^3 kHz and temperature range of 303 - 723 K.

2.2. Characterization

Structure characterization of the prepared polyaniline-silver composite was carried out by the X-ray diffraction (XRD) using Diano Corporation USA diffractometer with the Co-radiation (λ = 0.179 nm). The optical absorbance was examined by using a Perkin-Elmer Lambda 40 spectrophotometer over the spectrum range of 200 - 800 nm at room temperature. A transmission electron microscopy (TEM) examination was made to observe the surface morphology of samples using an EM10 Zeiss West Germany electron probe microanalyzer. The infrared absorption (IR) spectra, between 4000 and 400 cm^{-1}, were obtained using a Nicolet spectrometer model 670 FT-IR on KBr pellet of the samples.

3. Results and Discussion

3.1. Characterization

3.1.1. XRD Study

XRD patterns of the synthesized polyaniline-silver composite are shown in **Figure 1**. The diffraction peaks at 2θ = (38.3˚), (44.4˚), (64.6˚) and (77.5˚) to the (111), (200), (220) and (311) diffraction planes, respectively, ascribed to the cubic structure of Ag [2].

The average crystallite size of the polyaniline-silver composite was calculated using the Scherrer Equation [10]:

$$D = \frac{0.94\lambda}{\beta \cos\theta} \quad (1)$$

where D is the crystal size of Ag, λ the wavelength of

X-ray (0.179 nm), θ half diffraction angle of peak (in degrees) and β the true half peak width. The average size of the Ag determined through the (111) plane is 93 nm.

3.1.2. FT-IR Study

Figure 2 represents the FT-IR spectra of PANI/Ag. The bands at 1300, 1490, 1250, 1370, 1030 and 806 cm^{-1} correspond to polyaniline. The band corresponding to out of plane bending vibration of C-H bond of p-substituted benzene ring appears at 806 cm^{-1}. The bands corresponding to stretching vibrations of N-B-N and N=Q=N structures appear at 1490 and 1600 cm^{-1}, respectively, (where -B- and =Q= stand for benzenoid and quinoid moieties in the polyaniline backbone) [2]. The peak appeared at 1170 cm^{-1} corresponds to -N=Q-N^+-B- which is characteristic of the protonated state. The absorption band at 1250 cm^{-1} is associated with polaronic structure of PANI [2]. The bands at 1146 cm^{-1} correspond to polyaniline in the composites. The bands corresponding to vibration mode of N=Q=N ring and stretching mode of C-N bond appear at 1146 and 1300 cm^{-1}, respectively.

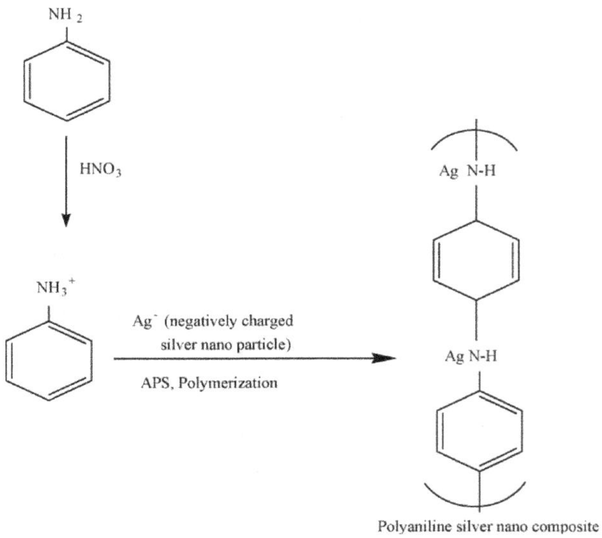

Scheme 1. Mechanism of formation of polyaniline-silver nanocomposite.

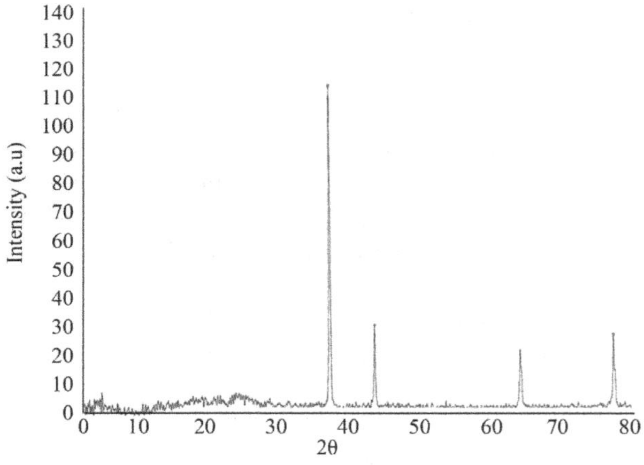

Figure 1. X-ray diffraction patterns of PANI/Ag nanocomposite.

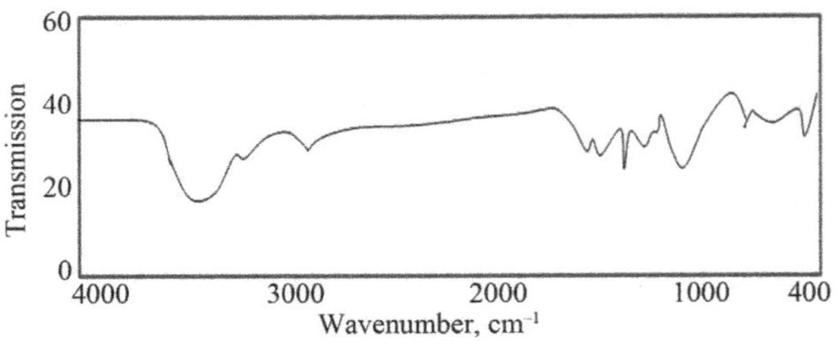

Figure 2. FTIR spectrum of PANI/Ag nanocomposite.

3.1.3. Morphological Study

Figures 3(a) and **(b)** shows TEM images for the PANI and the PANI/Ag nanocomposite, respectively. TEM shows that the samples consist of nanowires and nanorods, with width gradually decreasing from bottom to top the tip. It was found that Ag has a strong effect on the PANI's morphology; it shows a transformation in morphology of PANI particles. As shown in TEM images (**Figure 3(b)**), the silver particles are well adhered on the PANI substrate due to strong affinity of silver to nitrogen. This indicates that nano-sized inorganic particles possess a nearly spherical morphology and influence strongly the composite morphology. Since the silver nanoparticles were synthesized in the polyaniline solution, the nanoparticles got embedded into the polyaniline matrix [2]. It can also be seen that the spherically shaped Ag nanoparticles are uniformly dispersed in the PANI matrix and the particles are not clearly monodispersed. From the TEM

(a)

(b)

Figure 3. TEM of PANI (a) and PANI/Ag nanocompsite (b).

image it is observed that grains are well resolved and circular in shape and dopant particles are uniformly distributed in the polymer matrix. The formation of a relatively large cluster with Ag dispersity could be attributed to silver migration and aggregation. The migration and aggregation of silver particles might be driven by the instability of silver atoms due to their high surface free energy. Their aggregation would produce thermodynamically stable clusters [2].

3.1.4. UV-Visible Absorption Study

The photon absorption in many amorphous materials is found to obey the Tauc's relation [2]:

$$(\alpha h\upsilon) = \left(h\upsilon - E_g\right)^n \quad (2)$$

where (E_g) the optical band gap, (α) the absorption coefficient, and ($h\upsilon$) the energy of the incident photon. The index n has discrete values such as 1/2, 3/2, 2 or higher depending on whether the transition is direct or indirect and allowed or forbidden. In the direct and allowed cases, the index n is 1/2, whereas for the direct but forbidden cases it is 3/2. But for the indirect and allowed cases n = 2 and for the forbidden cases it is 3 or higher. In the present case the photon energy ($h\upsilon$) is plotted against $(\alpha h\upsilon)^2$ for n = 0.5, **Figure 4**. It gives a straight line fit, which implies that the samples undergo direct transition. Then the band gap has been extracted by extrapolating the straight portion of the graph on $h\upsilon$ axis at $\alpha = 0$. The calculated values of the optical band gap for PANI and PANI/Ag are 2.7 and 2.25 eV, respectively. The value of optical band gap is found to decrease after doping. Since the optical absorption depends on the short range order in the amorphous state and defect state associated with it, the decrease in the optical band gap in the present system may be due to reduction in the disorder of the system and increase in the density of defect states. From this it is clear that the silver doping has a potential effect on the optical properties of PANI. Optical conductivity of polyaniline increases in presence of silver nanoparticles due to the decrease in optical band gap and this type of behavior is also observed in the electrical conductivity measurements.

3.2. DC-Electrical Conductivity

To illustrate the effect of Ag nanoparticles on the DC-conductivity of polyaniline material, a comparison of pure PANI and PANI/Ag nanocomposite was made. **Figure 5** shows the temperature dependence of DC-conductivity in the temperature range 303 - 723 K for undoped PANI and PANI/Ag. From the figure, it is evident that the DC-electrical conductivity of the composite is higher than that of polyaniline. It is clear from the figure that in all the samples, the plots of ln σ_{DC} vs 1000/T are nearly straight lines, indicating that the conduction in these samples through an activated process having single activation energy in the temperature range 303 - 723 K. The activation behavior of the samples are studied by using Arrhenius Equation [1],

$$\sigma_{DC} = \sigma_O e^{\frac{-E_a}{2kT}} \quad (3)$$

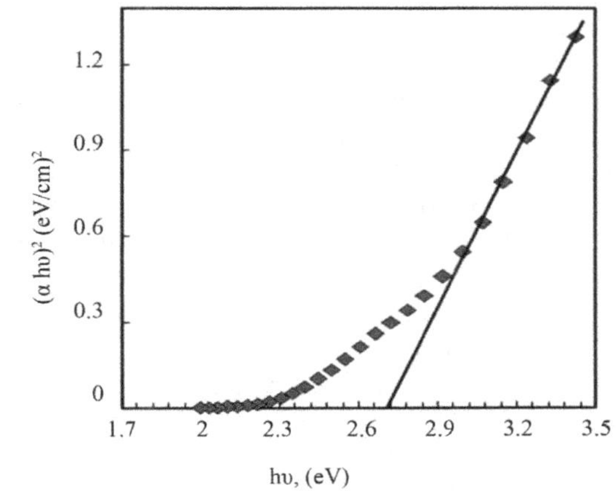

(a)

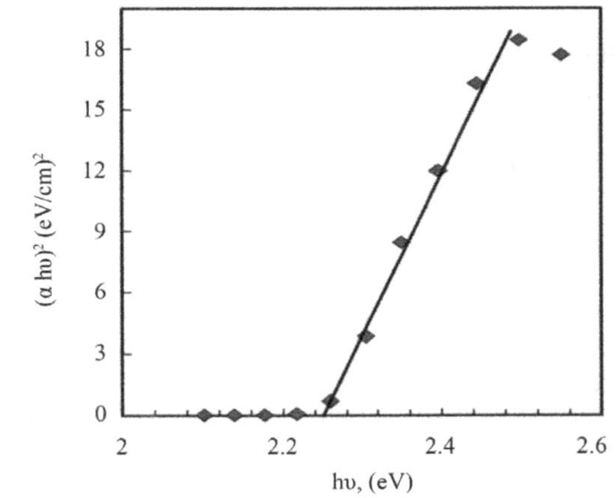

(b)

Figure 4. Determination of band gap energy for PANI (a) and PANI/Ag nanocomposite (b).

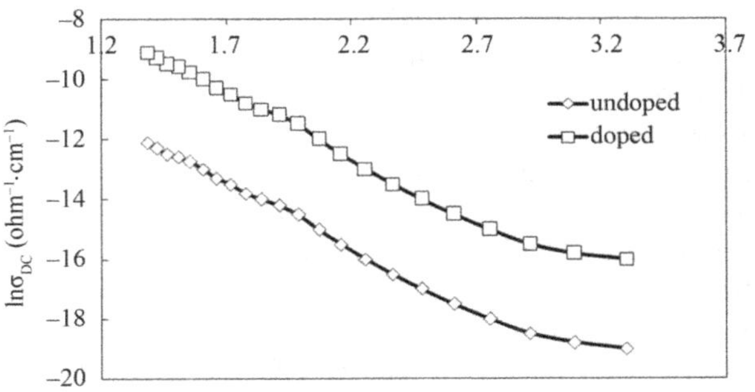

Figure 5. Effect of temperature on DC-electrical conductivity for all investigated samples.

where k is the Boltzman's constant and σ_O is the conductivity at infinity temperature.

The values of activation energy calculated from **Figure 5** and are given in **Table 1**. It was found that the activation energy for pure PANI is higher than PANI/Ag composite. An increase in DC-conductivity with corresponding decrease in activation energy is found to be associated with a shift of Fermi level in doped samples [1]. From a single value of activation energy it is clear that the conduction is through the carrier concentration at the Fermi level. But the activation energy (E_a) alone does not provide any information whether the conduction takes place in extended states or by hopping in localized states. This can be explained on the basis of the values of pre-exponential factor (σ_o). According to Mott and Davis [11] a value of σ_o in the range 10^3 - 10^4 ohm·cm^{-1} indicates that the conduction takes place mostly in extended states. A smaller value of σ_o indicates a wide range of localized states and the conduction is taking place by the hopping process. In our case, the values of σ_o are found to be of the order of 10^{-3} for PANI and 10^{-2} for PANI/Ag, therefore the conduction takes place by the hopping process due to the wide range of localized states present in the sample. From the above results we can conclude that the hopping mechanism is responsible for an increase in the conductivity of the samples. A smaller value of σ_o indicates that the density of defect states increases in the sample and further supports our argument that the conduction mostly takes place by the hopping process in the Ag doped polyaniline. The formation of polarons and bipolarons can also be used to explain the conduction mechanism [1]. Polarons and bipolarons play a leading role in determining the charge injection, optical and transport properties of conductive polymers. These are self-localized particles like defects associated with characteristic distortions of the polymer backbone and with quantum states deep in the energy gap due to strong lattice coupling.

3.3. AC-Electrical Conductivity

The frequency-temperature dependences of AC-electrical conductivity of PANI and PANI/Ag nanocomposite are shown in **Figure 6**. As shown in this figure, the AC-electrical conductivity of the PANI/Ag nanocomposite is higher than that of the pure PANI. The lower conductivity of the pure PANI could be ascribed to low level of protonation of the PANI chains. The improvement of AC-conductivity for PANI/Ag nanocomposite comes from the effective dispersion of Ag nanoparticles in the PANI matrix (shown in TEM images), which favors better electronic transport. It is evident that the AC-conductivity is both frequency and temperature dependent and enhanced with an increase of both the frequency and the temperature. This indicates that there may be charge carriers, which can be transported by hopping through the defect sites along the polymer chain [4].

It can be seen that the conductivity of PANI/Ag nanocomposite increases as the frequency increases. The frequency dependence of the AC-conductivity is considered to be a result of interface charge polarization (or Maxwell-Wagner-Sillars effect) and intrinsic electric dipole polarization [4]. This phenomenon appears in heterogeneous systems like metal-polymer composites due to the accumulation of mobile charges at the interfaces and the formation of large dipoles on metal particles or clusters. Several theoretical models have been proposed for the AC-conduction in amorphous semiconductors such as classical hopping and quantum mechanical tunneling of

Table 1. DC- and AC-conductivity data for all investigated samples.

Samples	E_a (eV)	#Temp. Range (K)	$^{\$}\sigma_{DC}$ ohm^{-1}·cm^{-1}	$W_M \times 10^3$ (eV)	$^{*}\sigma_{AC}$ ohm^{-1}·cm^{-1}
PANI	0.32	370 - 450	3.1×10^{-6}	0.25	3.3×10^{-3}
PANI/Ag	0.10	370 - 450	6.1×10^{-5}	0.01	5.2×10^{-2}

$^{\$}\sigma_{DC}$ at 563 K; $^{*}\sigma_{AC}$ at 563 K and frequency of 100 kHz; #Temperature range at which the activation energy; E_a, was calculated.

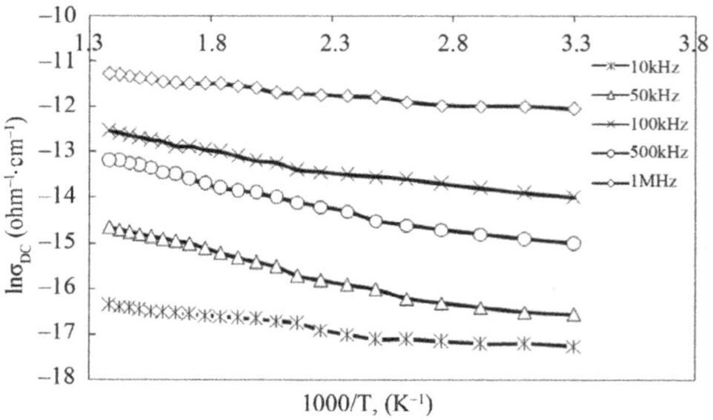

(a)

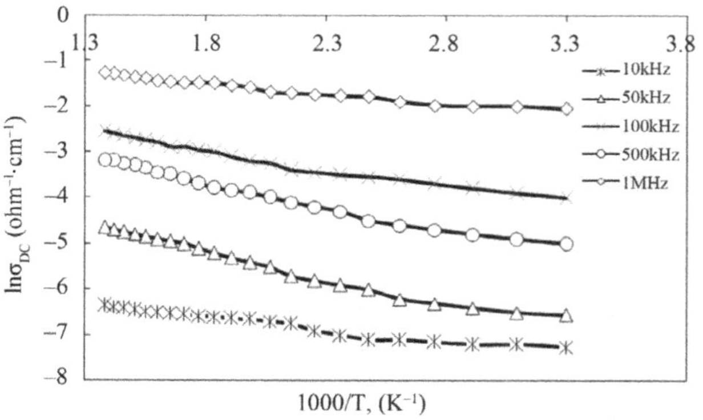

(b)

Figure 6. Effect of temperature on AC-electrical conductivity for PANI (a) and PANI/Ag nanocomposite(b).

charge carriers over the potential barrier separating two energetically favorable centers in a random distribution [12]. In electron tunneling model, s is independent of T. In case of small polaron tunneling, s decreases as T increases. The variation of s with T for the PANI/Ag is shown in **Figure 7**, where a decrease in the values of s becomes higher as the temperature is increased. In the correlated barrier hopping (CBH) model, the charge carrier hops between the sites over the potential barrier separating them. The existence of Coulomb interaction between individual barriers causes a barrier reduction. The frequency exponent s for such model is given by the following Equation [12]:

$$S = 1 - \left\{ \frac{6kT}{W_M} \right\} \tag{4}$$

where W_M is the effective barrier height at infinite interstice separation R. From **Table 1**, it can be seen that the highest values of W_M are assigned to PANI compared with the PANI/Ag. The high potential barrier exhibited in this sample acts to hinder charge carrier transport and this is responsible for its lower conductivity.

3.4. Temperature Dependence of Dielectric Permittivity and Dielectric Loss

Figures 8 and **9** show the temperature dependence of the dielectric constants and dielectric losses of PNAI and PNAI/Ag in the frequency range of 10 - 10^3 kHz and temperature range of 303 - 723 K. In these figures, higher dielectric constants and losses of PANI/Ag nanocomposite than those of pure PANI are observed. It can be seen that the incorporation of Ag enhanced the dielectric constant of the PANI/Ag nanocomposite about 3 times of that recorded for the pure PANI under the same conditions, **Figure 8**. At the same time, there is about a four orders of magnitude increase in dielectric loss upon incorporation of Ag in the polyaniline, **Figure 9**. From the results described above, it has been conclude that the enhanced dielectric constants and losses might originate from the increased conductivity. The better dielectric constant of the PANI/Ag nanocomposite indicates their better ability to store electric potential energy under the influence of alternative electric field. The frequency dependency of dielectric constant is found to be increased in presence of Ag in the composite (**Figure 8**). This is thought to be a result of the slow dielectric relaxation of the PANI matrix and the interface of the composite. The dielectric constant and loss are very high due to the AC-conduction loss that dominates over other types of losses present in the polymer e.g. dipole segmental losses and orientation of the polar group [4].

4. Conclusion

We have synthesized polyaniline and polyaniline-silver nanocomposite by chemical oxidative method. The formation of Ag nanoparticles and their presence in the prepared nanocomposite were confirmed by XRD, FTIR, TEM, UV-Vis spectroscopy. The XRD patterns indicated that the crystalline phase of Ag is cubic with crystallite size of 93 nm. TEM analysis showed uniform dispersion of the spherically shaped Ag nanoparticles in the PANI matrix. The optical studies indicated that the absorption mechanism is due to direct allowed transition and the optical band gap of polyaniline is higher than that of

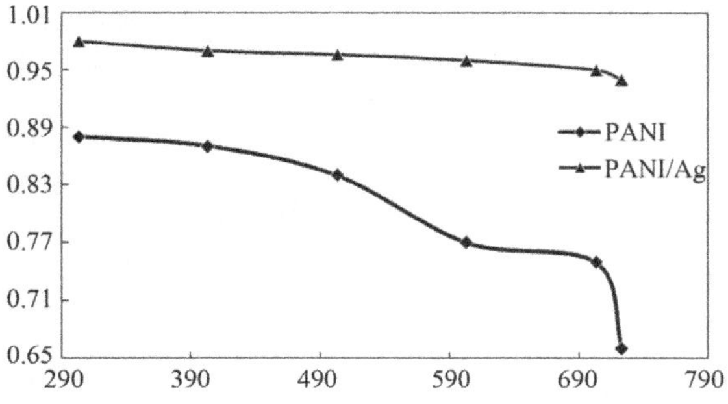

Figure 7. Effect of temperature on s value for all investigated samples.

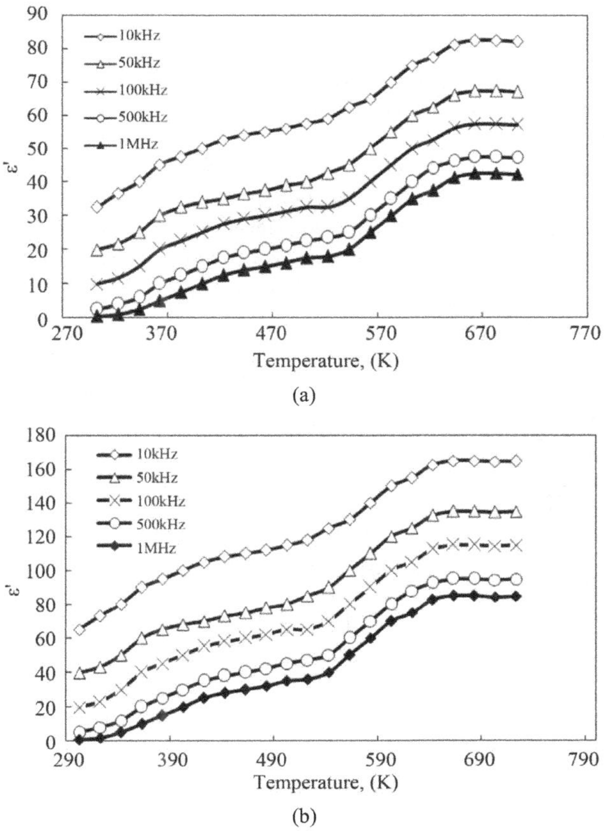

Figure 8. Effect of temperature on dielectric constant for PANI (a) and PANI/Ag nanocomposite (b).

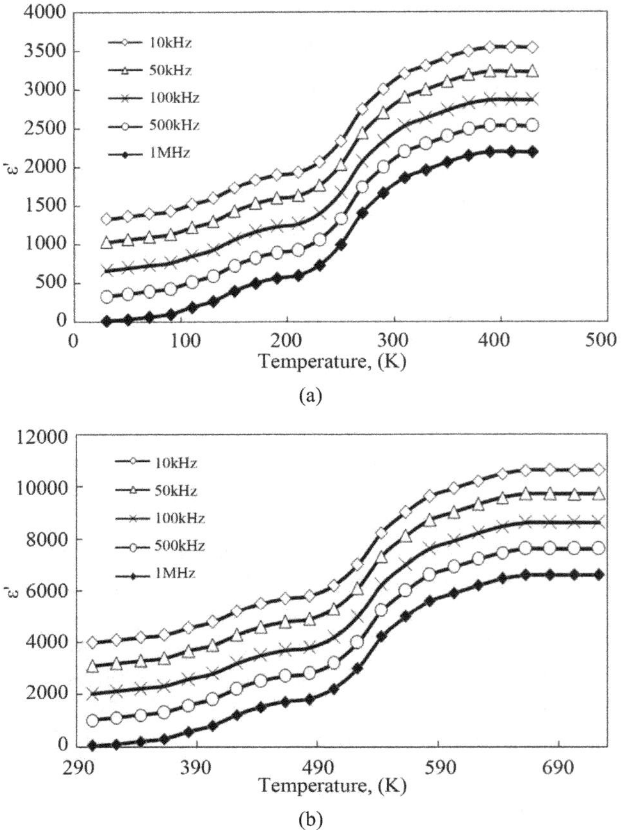

Figure 9. Effect of temperature on dielectric loss for PANI (a) and PANI/Ag nanocomposite (b).

polyaniline/silver (PANI/Ag) nanocomposite. Conductivity and dielectric properties such as dielectric constant and dielectric loss have been measured and the conduction mechanism has also been investigated in terms of Mott and Davis. It is suggested that conduction is taking place by the hopping process due to a wide range of localized states present in the sample. The PANI/Ag nanocomposite exhibit remarkable improvement of electrical conductivity and dielectric properties when compared with pure PANI. So, this is a simple way by which optical and electrical properties of other conductive polymers may be enhanced by using different nanoparticles.

5. Acknowledgements

The authors wish to gratefully acknowledge the support from Chemistry Department, College of Science for Girls, Dammam University, Dammam, Kingdom of Saudi Arabia.

REFERENCES

[1] G. B. Shumaila, V. S. Lakshmi, M. Alam, A. M. Siddiqui, M. Zulfequar and M. Husain, "Synthesis and Characterization of Se Doped Polyaniline," *Current Applied Physics*, Vol. 11, No. 2, 2010, pp. 217-222. doi:10.1016/j.cap.2010.07.010

[2] K. Gupta, P. C. Jana and A. K. Meikap, "Optical and Electrical Transport Properties of Polyaniline-Silver Nanocomposite," *Synthetic Metals*, Vol. 160, No. 13-14, 2010, pp. 1566-1573.

[3] S. S. Umare, B. H. Shambharkar and R. S. Ningthoujam, "Synthesis and Characterization of Polyaniline-Fe_3O_4 Nanocomposite: Electrical Conductivity, Magnetic, Electrochemical Studies," *Synthetic Metals*, Vol. 160, No. 17-18, 2010, pp. 1815-1821. doi:10.1016/j.synthmet.2010.06.015

[4] A. Choudhury, "Polyaniline/Silver Nanocomposites: Dielectric Properties and Ethanol Vapour Sensitivity," *Sensors and Actuators B: Chemical*, Vol. 138, No. 1, 2009, pp. 318-325. doi:10.1016/j.snb.2009.01.019

[5] V. Ali, R. Kaur, N. Kamal, S. Singh, S. C. Jain, H. P. S. Kang, M. Zulfequar and M. Husain, "Use of Cu^{+1} Dopant and It's Doping Effects on Polyaniline Conducting System in Water and Tetrahydrofuran," *Journal of Physics and Chemistry of Solids*, Vol. 67, No. 4, 2006, pp. 659-664. doi:10.1016/j.jpcs.2005.10.172

[6] S. Kazim, V. Ali, M. Zulfequar, M. M. Haq and M. Husain, "Electrical, Thermal and Spectroscopic Studies of Te Doped Polyaniline," *Current Applied Physics*, Vol. 7, No. 1, 2007, pp. 68-75. doi:10.1016/j.cap.2005.11.072

[7] K. S. Ryu, B. W. Moon, J. Joo and S. H. Chang, "Characterization of Highly Conducting Lithium Salt Doped Polyaniline Films Prepared from Polymer Solution," *Polymer*, Vol. 42, No. 23, 2001, pp. 9355-9360. doi:10.1016/S0032-3861(01)00522-5

[8] J. C. Xu, W. M. Liu and H. L. Li, "Titanium Dioxide Doped Polyaniline," *Material Science Engineering: C*, Vol. 25, No. 4, 2005, pp. 444-447.

[9] N. V. Blinova, J. Stejskal, M. Trchova, I. Sapurina and G. C. Marjanovic, "The Oxidation of Aniline with Silver Nitrate to Polyaniline-Silver Composites," *Polymer*, Vol. 50, No. 1, 2009, pp. 50-56. doi:10.1016/j.polymer.2008.10.040

[10] M. Ristić, M. Ivanda, S. Popvić and S. Musić, "Dependence of Nanocrystalline SnO_2 Particle Size on Synthesis Route," *Journal of Non-Crystalline Solids*, Vol. 303, No. 2, 2002, pp. 270-280. doi:10.1016/S0022-3093(02)00944-4

[11] N. F. Mott and E. A. Davis, "Electronic Process in Non-Crystalline Materials," Clarendon Press, Oxford, 1979.

[12] S. Ebrahim, A. H. Kashyout and M. Soliman, "Ac and Dc Conductivities of Polyaniline/PolyVinyl Formal Blend Films," *Current Applied Physics*, Vol. 9, No. 2, 2009, pp. 448-454. doi:10.1016/j.cap.2008.04.007

13

Magnetooptical Properties of Layer-by-Layer Deposited Ferromagnet—Dielectric Nanocomposites*

Viktoria Evgenèvna Buravtsova[1#], Elena Alexandrovna Ganshina[1], Sergey Alexandrovich Kirov[1], Yuriy Egorovich Kalinin[2], Alexandr Viktorovich Sitnikov[2]

[1]Physics Faculty, Lomonosov Moscow State University, Moscow, Russia; [2]Voronezh State Technical University, Voronezh, Russia.
Email: [#]v.e.buravtsova@gmail.com

ABSTRACT

The present work was initiated to investigate how technology of preparation of nanocomposites $(Co_{45}Fe_{45}Zr_{10})$ $Z(Al_2O_3)_{1-Z}$ affects their magneto-optical (MO) properties. The spectral, magnetic field and concentration dependences of the transversal Kerr effect (TKE) have been studied either for bulk or layer-by-layer deposited nanocomposites within a wide range of the ferromagnetic (FM) phase concentrations and for various thicknesses of layers. It was found that the MO response of the layer-by-layer deposited nanocomposites with compositions inside the percolation interval differs essentially from the one of the bulk composites and depends on the layer thicknesses. With decreasing thicknesses of layers the percolation threshold has been shifted towards the lower contents of the FM phase. In addition, it has been established that the size and shape of the granules inside the nanocomposite layer also depends on the layer thickness as well as the microstructure of the layer-by-layer sputtered composites considerably differs from the microstructure of the bulk nanocomposite.

Keywords: Magnetooptics; Nanostructure; Composites

1. Introduction

The steady current interest to nanostructures is caused by possibility to modify and fundamentally change properties of the known materials when they turn into a nanocrystalline state. New nanodimensional magnetic materials exhibit a number of extraordinary properties: a giant magnetoresistance (GMR), giant magnetic impedance (GMI) [1], anomalous Hall effect (AHE) [2], strong magnetooptical (MO) response [3] and anomalous optical effects [4]. All these phenomena open vast prospects both for fundamental investigations and most promising possibilities for their applications. Wide use of nanocomposites as magnetic layers in multilayer structures ferromagnet/semiconductor [5,6] for reduction of diffusion on the metal-semiconductor interfaces requires both optimization of their compositions and technologies of manufacturing film nanocomposites and investigation how their microstructure affects their electrical, magnetic and optical properties.

The properties of nanocomposites depend critically on their composition and microstructure, particularly on the size of the granules, their distribution by volume of the sample, on the magnetic phase concentration and interface properties. That is why optical and MO investigation techniques are of much interest, since they possess a number of advantages, the main of which is that these methods are sensitive to presence of magnetic heterogeneities, to changes of particle shape and size, their three-dimensional distribution and appearance of new magnetic phases. This is confirmed by recent investigations of optical and MO spectra of granular systems [4,7-13], that revealed many particularities of linear and non-linear optical and MO Kerr effect and many other properties.

In the present publication we compare MO properties of the thin-film nanocomposites, prepared by evaporation on a fixed substrate (bulk composites) and the nanocomposites, prepared by layer-by-layer deposition of thin layers.

First a brief description of MO properties of bulk nanocomposites. Studies of MO properties of bulk nanocomposites with various chemical compositions, both ferromagnetic granules ($Co_{84}Nb_{14}Ta_2$, $Co_{40}Fe_{40}B_{20}$, $Co_{45}Fe_{45}Zr_{10}$, Co, FePt), and a dielectric matrix (SiO_2, Al_2O_3 LiNbO) [14-17] have shown that MO spectra substan-

*This work was partly supported by the Russian Foundation for Basic Research and Russian Ministry of Education and Science (grant No. 16.516.12.6019).
#Corresponding author.

tially differ from witnesses' spectra, FM alloys and Co. The TKE spectra demonstrate a strong nonmonotonous dependence on the metallic volume fraction with a cross-over at the percolation threshold. In the near IR range we observed appreciable enhancement of the TKE in the granulated system which occurs near to the percolation threshold.

Studies of MO properties of amorphous metal-dielectric bulk nanocomposites have shown that changes of MO properties in all our systems with increasing concentration proceed virtually identically [14,15]. The TKE spectra of nanocomposites of all systems demonstrate similar frequency dependencies, varying only in details, namely by magnitude of the effect, as well as position of maximums and zero points of the effect **Figure 1(A)** example of spectra of nanocomposite $(Co_{86}Nb_{12}Ta_2)_Z(SiO_2)_{100-Z}$ in the wide concentration range, B—spectra of nanocomposites with different compositions, the FM phase concentration being near the percolation threshold). Particularly large variations of $\delta(h\nu)$ are found in energy range $h\nu < 1.5$ eV. While in pure alloys ($Co_{84}Nb_{14}Ta_2$, $Co_{40}Fe_{40}B_{20}$, $Co_{45}Fe_{45}Zr_{10}$) and in Co with decrease of light energy the TKE magnitude decreases to zero ($\delta \approx 0$ at $h\nu = 0.7$ eV), in nanocomposites the effect changes sign and reaches extreme negative values in the energy range 0.7 - 1.2 eV. By their absolute value the TKE of nanocomposites in this spectral interval is several times larger, than in the witness, *i.e.* pure ferromagnetic alloy. At the same time it should be remembered that in amorphous nanocomposites, demonstrating the largest TKE magnitudes, the magnetic phase content is virtually two times less as compared with the ferromagnetic alloy.

The TKE dependence on content of the FM phase is nonmonotonic (**Figure 1(C)**) shows the TKE dependence on concentration for the nanocomposite $(Co_{86}Nb_{12}Ta_2)_Z(SiO_2)_{100-Z}$). The largest variations of TKE demonstrate the samples with magnetic phase concentration corresponding to the relevant percolation threshold. The effect magnitude essentially depends on chemical composition of the granules. Maximum magnitudes of MO response are achieved in the systems with Zr as amorphizator, at Z = 47 at%, in the system with Boron content at Z = 58 at% and in the system with niobium and tantalum at Z = 46 at% [15]. Concentration position of the maximum is determined by geometrical particularities of the composites, which are near percolation threshold—minimal thickness of the dielectric barrier, through which the tunneling of polarized electrons is going between the FM granules.

Evolution of the TKE magnetic field dependence with increase of the FM phase content is also similar in all the systems studied [14,15]. As example in **Figure 2** are

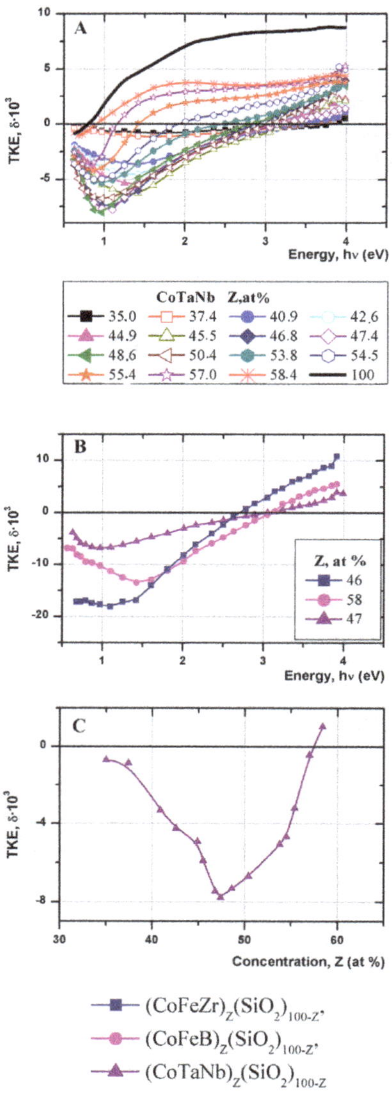

Figure 1. (A), (B)—spectral, (C)—concentration dependences of TKE for nanocomposites: $(Co_{45}Fe_{45}Zr_{10})_Z(SiO_2)_{100-Z}$, $(Co_{40}Fe_{40}B_{20})_Z(SiO_2)_{100-Z}$, $(Co_{84}Nb_{14}Ta_2)_Z(SiO_2)_{100-Z}$.

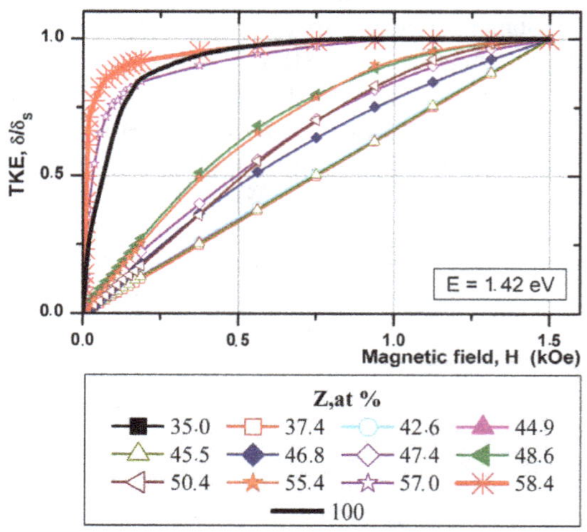

Figure 2. Field dependences of TKE for $(Co_{86}Nb_{12}Ta_2)_Z(SiO_2)_{100-Z}$.

shown the TKE magnetic field dependence of a series of nanocomposites $(Co_{86}Nb_{12}Ta_2)_Z(SiO_2)_{100-Z}$, measured at $h\nu = 1.42$ eV (normalized). Three concentration regions can be seen which correspond to different magnetization processes. The first group of samples with low content of metallic phase Z < 46 at% demonstrates linear growth of TKE with increase of magnetic field up to $H = 1.5$ kOe. Such behavior of the $\delta(H)$ curves testifies to superparamagnetic type of magnetization, what is typical for samples up to percolation threshold. For these samples the size of equilibrium granules is 2 - 4 nm. For samples of the second group with Z within 46 - 55 at%, the $\delta(H)$ dependence changes, acquiring features typical for bulk ferromagnets (witnesses). At that the size of the granules in this structure increases up to 5 - 7 nm. Nanocomposites of the third group, that are beyond percolation threshold (Z > 55 at%), demonstrate magnetization of a ferromagnetic type. For them magnetization process ends mainly in the field up to 1 kOe. Such behavior of the $\delta(H)$ dependences may be explained by microstructure change dynamics of our nanocomposites: in structures with low ferromagnet content (Z ≤ 46 at%) the metallic part of the nanocomposite represents independent granules, non-interacting with each other, their magnetic behavior close to the superparamagnetic type; with increase of the ferromagnetic phase content the dielectrical interlayers, which divide these regions, become thinner, thus interaction between granules becomes possible and the ferromagnetic contribution into MO response of the structure increases. The increase of the ferromagnetic contribution may be also caused by growing size of ferromagnetic granules. For ferromagnet content around Z ~ 46 at%, *i.e.* in the percolation threshold, the ferromagnetic component of the sample is magnetized as a large single cluster, but apparently, there are also small separate magnetic clusters, which do not interact with each other, and add a superparamagnetic contribution into magnetooptical response of the nanostructure.

Investigation of magnetoresistance and magnetostriction [15] of the same samples have shown a correlation between maximum magnitudes of the giant magnetoresistance (GMR), transversal Kerr effect (TKE) and magnetostriction of constituting FM granules (**Figure 3**). With linear increase of saturation magnitudes of magnetostriction of ferromagnetic inclusions, when passing from CoNbTa to CoFeB and further to CoFeZr both GMR and TKE grow linearly.

The observed correlations between saturation magnetostriction of the ferromagnetic phase, maximum magnitudes of the giant magnetoresistance and the TKE are caused by one and the same mechanism and may be related to increasing contribution of *d*-electrons and magnitude of the spin-orbital interaction in the series of granular nanocomposites CoNbTa →CoFeB→ CoFeZr. It

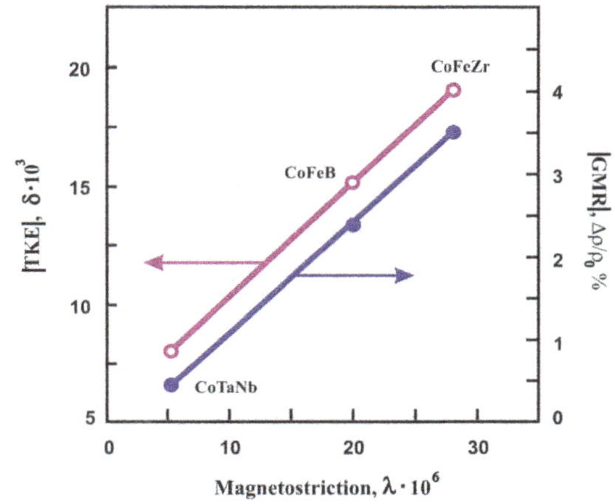

Figure 3. The correlation between the GMR, TKE and magnetostriction of FM granules for nanocomposites $(Co_{45}Fe_{45}Zr_{10})_{57}(SiO_2)_{43}$, $(Co_{40}Fe_{40}B_{20})_{60}(SiO_2)_{40}$ and $(Co_{84}Nb_{14}Ta_2)_{61}(SiO_2)_{39}$.

was also established that in nanocomposites of this type the dominant charge transfer process from granula to granula is by mechanism of hopping conduction of electrons across the dielectrical barrier [14-17]. Density of states of the polarizerd *d*-electrons near Fermi level depends on the granule material and it grows in the series of granular nanocomposites CoNbTa→CoFeB→CoFeZr, what as a consequence, leads to increase of magnetoresistance, magnetostriction and MO effects. Thus, MO methods allow to follow the changes of a nanocomposite structure and determine percolation threshold in nanocomposites by maximum magnitudes of the effect in IR region, and by deviation from linearity of the $\delta(H)$ dependence.

Calculations of the MO spectra using different effective medium approximations show that the observed changes in the MO spectra can be described using the optical and MO data typical for the bulk materials, both of grains and the matrix, but taking into account the size and shape of the particles. Thus, MO response of a composite medium depends on the form and size of granules, and its amplification near the percolation threshold is caused by changes of optical and MO parameters under influence of topology and modifications of nanocomposite microstructure [14].

Since the composite layers were supposed to be used as magnetic layers with thickness of the order of several nanometers in multilayer films of nanocomposite/semiconductor type, so there emerged a problem to follow the changes of MO properties of the nanocomposite itself with variation of its thickness. For this purpose we have studied MO properties of bulk nanocomposites $(Co_{45}Fe_{45}Zr_{10})_Z(Al_2O_3)_{1-z}$ and layer-by-layer deposited nanocomposites, prepared by the same technology as the multilayer films $[(Co_{45}Fe_{45}Zr_{10})_X/(a\text{-}Si)_Y]$ in [5].

2. Samples. Preparation and Structure

Bulk granular amorphous films containing nano-dimensional clusters of the $Co_{45}Fe_{45}Zr_{10}$ alloy randomly distributed in the insulating amorphous matrix Al_2O_3 were prepared in argon atmosphere by ion-beam sputtering of compound targets, containing both the ferromagnetic and the dielectric components, on pyroceram substrates. The choice of a rather complicated composition of the granules $Co_{45}Fe_{45}Zr_{10}$ was determined by the requirement to stabilize the amorphous structure of the ferromagnet at room temperatures.

During simultaneous sputtering of both the metallic alloy and the dielectric from the compound target a fragmented structure forms composed of metallic amorphous granules imbedded inside the dielectric matrix, a wide and continuous range of the ferromagnetic phase concentrations x from 30 to 60 at% being obtained. The manufactured film samples were about 4 μm thick. When the concentration of dielectric was maximum the average size of ferromagnetic granules sputtered on a static substrate was 2 - 4 nm. With decrease of the dielectric fraction content the size of ferromagnetic granules increases, and in nanocomposites with large concentrations of the metallic phase (x = 50 - 60 at%) it becomes 5 - 7 nm (**Figure 4**). Deposition of solid nanocomposites was performed in vacuum 1×10^{-5} Torr onto non-cooled fixed pyroceram substrates.

The layered granular amorphous films were deposited layer by layer onto rotating substrate by the same technology as the multilayer films [5]. Varying the rotation speed of a special device ("carousel"), on which the substrate is fixed, it is possible to obtain various thicknesses of nanocomposite layers. Number of "carousel" revolutions corresponds to the number of deposited layers, what allows controlling the total film thickness. The thicknesses of the layers in the layered granular films were 1 - 2.7 nm for the 1st series and 2.7 - 5.8 nm for the 2d one. Number of layers in the 1st series structures was 101, and in the 2d series –40.

The magnetooptical properties have been studied by the transversal Kerr effect (TKE). It consists in relative change of the intensity of the linearly polarized light (p-wave) reflected from the sample in the case when magnetization vector is parallel to the surface and at the same time lies perpendicular to the incident light plane (the transversal geometry):

$$\delta = \frac{I_H - I_{H=0}}{I_{H-0}}.$$

Here I_H and $I_{H=0}$ are intensities of the light reflected by the magnetized and non-magnetized sample correspondingly. The measurements of TKE were performed using a dynamic method developed for measuring MO-effects [18] in which the remagnetization of the sample by the AC magnetic field leads to a modulation of the reflected light intensity, the depth of which determines the magnitude δ of the MO effect. The TKE spectra $\delta(h\nu)$ and the magnetic field dependence curves $\delta(H)$ were measured in the 0.5 - 4.0 eV photon energy range for the incidence angle φ = 70° in the AC magnetic field of 78 Hz frequency and amplitude up to 2.5 kOe.

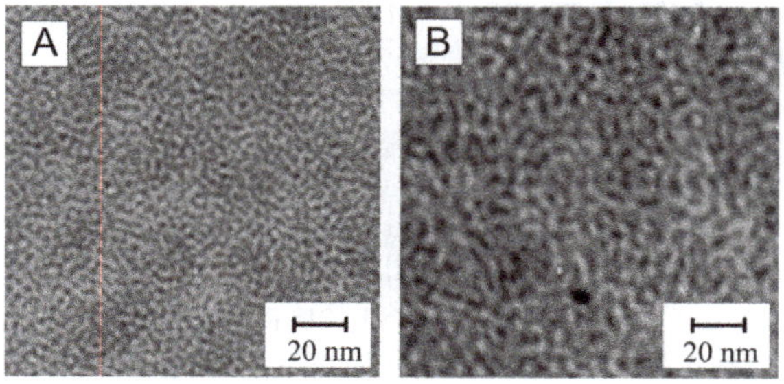

Figure 4. Photomicrographs of bulk nanocomposites $(Co_{45}Fe_{45}Zr_{10})_Z(SiO_2)_{100-Z}$: (A) for Z = 35 at%; (B) Z = 45 at%.

3. Experiment and Discussion

The TKE spectral and magnetic field dependence in layer-by-layer deposited structures of 1 and 2 series and the bulk composite are shown in **Figures 5** and **6**. Similar

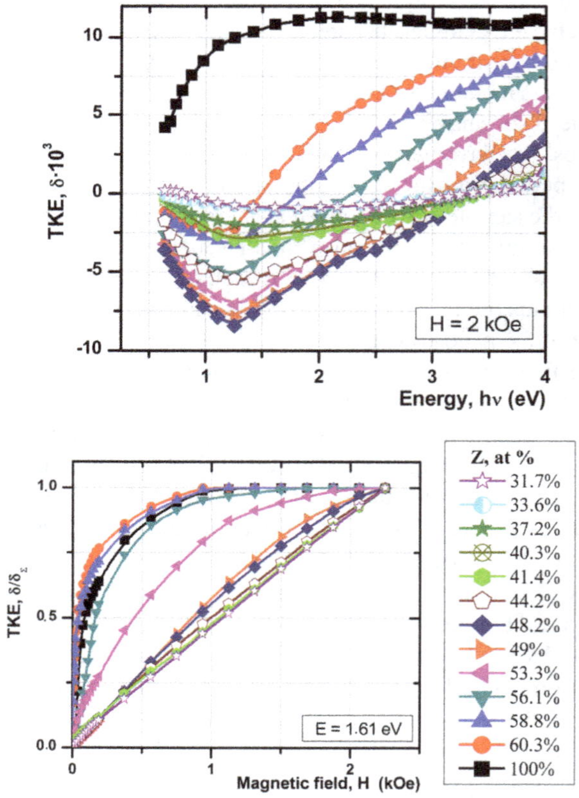

Figure 5. Spectral and field dependences of TKE for bulk nanocomposites $(Co_{45}Fe_{45}Zr_{10})_Z(Al_2O_3)_{100-Z}$.

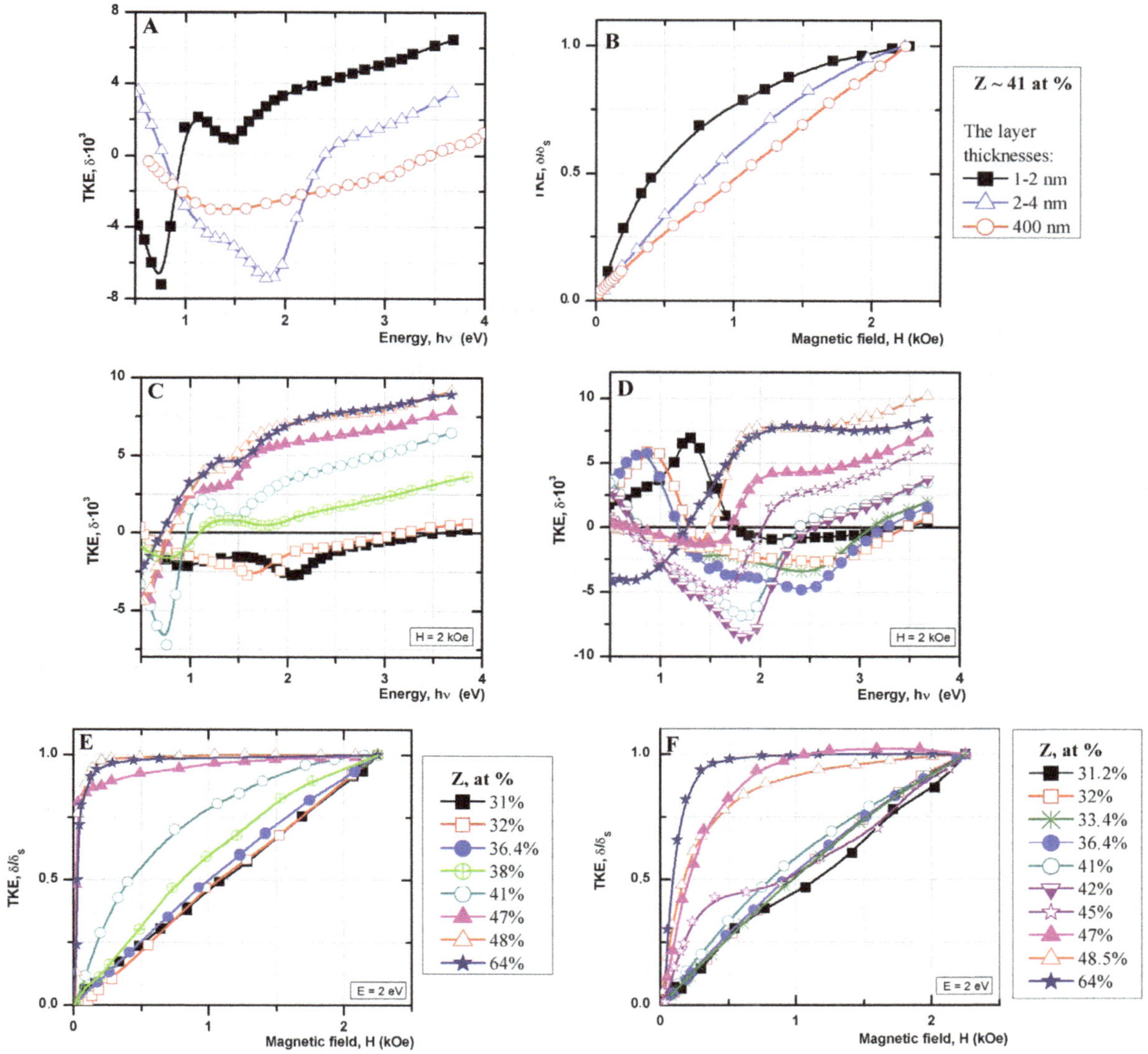

Figure 6. The comparison of spectral (A) and field (B) dependences for bulk and "layer by layer" nanocomposite $(Co_{45}Fe_{45}Zr_{10})_Z(Al_2O_3)_{100-Z}$ with different thickness of forming layers; (C)-(F) spectral and field dependences for the 1st and 2d series "layer by layer" nanocomposite $(Co_{45}Fe_{45}Zr_{10})_Z(Al_2O_3)_{100-Z}$.

to the previously investigated nanocomposites the bulk composites demonstrate substantial modification of the TKE spectra shape and magnitude according to the FM phase concentration and appearance of a big negative effect in the energy range below 2 eV, with concentrations near percolation threshold and smaller. Yet the form of spectra curves substantially changes when we pass to the layer-by-layer deposited nanocomposites.

Thus in the series 1 samples the TKE is negative in the near IR and visible range (0.5 - 3.25 eV) and only on condition that $Z \leq 33.1\%$, what is similar to the TKE spectra of the bulk nanocomposite with the same concentration Z. With increasing FM phase content ($Z \geq 36.4$ at%) the shape of the spectra appreciably changes and becomes similar to the TKE spectrum of the bulk $(Co_{45}Fe_{45}Zr_{10})_Z(Al_2O_3)_{100-Z}$ with concentration $Z \sim 64$ at%: the negative sign of the effect was observed in the energy range up to 1 eV, and in the rest part of the spectrum (1 - 3.5 eV) the TKE is positive.

In the TKE spectra of the 2 series samples up to concentration $Z = 36.4$ at% in the visible light range a negative effect was observed, what corresponds to the spectra of bulk composite films, but in the IR spectral range there is seen a positive peak of the effect uncharacteristic for bulk nanocomposites. At the same time the maximum magnitude of the TKE in the IR range is several times larger than TKE amplitude in a similar content FM phase samples of bulk nanocomposites with increasing the FM phase concentration the positive effect in the IR range sharply falls, and in the rest of the spectrum the effect magnitude grows. At Z = 41 at% the 2 series samples demonstrate the maximum magnitude of the effect, twice as large as in the bulk sample with the same concentration. With increasing concentration Z up to 64 at% the TKE spectra of both series, deposited by layer-by-layer procedure, become analogous to each other and similar to $\delta(h\nu)$ in the bulk nanocomposite with Z ~ 60 at%.

Changes of the TKE magnetic field dependences of the layer-by-layer deposited samples show the same tendency as the bulk nanocomposite—with increase of Z content the magnetization curves evolve from a superparamagnetic type to the ferromagnetic one (**Figure 6**). Yet if deviation from the linear (superparamagnetic) type is observed in the bulk nanocomposite with Z ≤ 48 at%, then in the layer-by-layer deposited composites it takes place at Z ≤ 36.4 at% for the 1st series and 41 at% for the 2d one. Such behavior of the magnetic field TKE dependences, together with the observed TKE growth in the near IR region and modifications of the TKE curves shape—all this points to the fact that percolation threshold in the layer-by-layer deposited composites moves into the region of lesser FM phase concentrations. So, taking into consideration all the above resemblances and differences between the layer-by-layer deposited samples and the bulk nanocomposite, one can say that not only the metallic phase concentration affects the MO response of the systems, but as well the thickness of the composite layers essentially affects the magnetooptical properties of nanostructures.

The changes in the preparation technology of the structures lead to changes in microstructure (topology). This results in another grain sizes, different number of contacts between grains, and, therefore, the MO response also changes (**Figure 7**).

The film thickness is predetermined by the substrate rotation velocity around sputtering targets [5]. Between depositions of each layer some time elapses, during which the thinner film will be cooled faster than the thicker one. The bulk composite was deposited onto a fixed substrate, so the structure formation temperatureremained the same. Supposing that during cooling down of thin layers formation of the granules proceeds differently than in the bulk composite, it will be logical to as sume that the size of the granules in the layer-by-layer deposited samples is smaller than their characteristic size in the bulk composite. Thus, the probability of the granules to contact each other becomes greater when sizes of the granules diminish. That is the system with most thin layers should have the percolation threshold at smallest Z values among all the studied systems, which is what we observe in experiment.

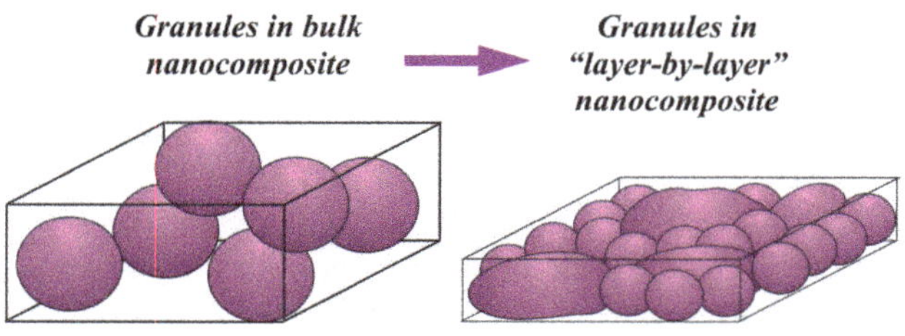

Figure 7. Model of the microstructure of bulk and "layer-by-layer" nanocomposites.

Figure 8 shows the TKE-concentration dependence for all our systems. It is seen that in the near IR region (E ≤ 2.55 эB) all the systems demonstrate a non-monotonic TKEdependence on FM phase concentration, the greatest changes taking place in the percolation region. Yet if for the bulk nanocomposite the shape of the curves does not depend on the light wavelength and all the curves show peaks at $Z = Z_{per}$ (where Z_{per} was obtained from magnetic field dependence), then for the layer-by-layer nanocomposite the shape of the curves $\delta(Z)$ depends strongly on the light wavelength. At the same time for energies $E \geq$ 1.22 eV the series 1 near Z_{per} exhibits only a local minimum on the curve $\delta(Z)$ that grows monotonically with increase of concentration Z, while for series 2 the local extremum at $Z = Z_{per}$ is observed in the energy range 1.22 eV ≤ E ≤ 1.73 eV, at E = 0.76 eV and 2.55 eV the local extremum shifts into the lesser concentrations region, and at $Z = Z_{per}$ a zero crossover is seen.

More complicated nature of the concentration curves $\delta(Z)$ for the layer-by-layer nanocomposites is an evidence of their greater heterogeneity. This is also witnessed by dependence of the $\delta(H)$ curve shape on the light wavelength and appearance of anomalous TKE magnetic field dependences measured in the near IR range. **Figure 9** demonstrates anomalous field depend

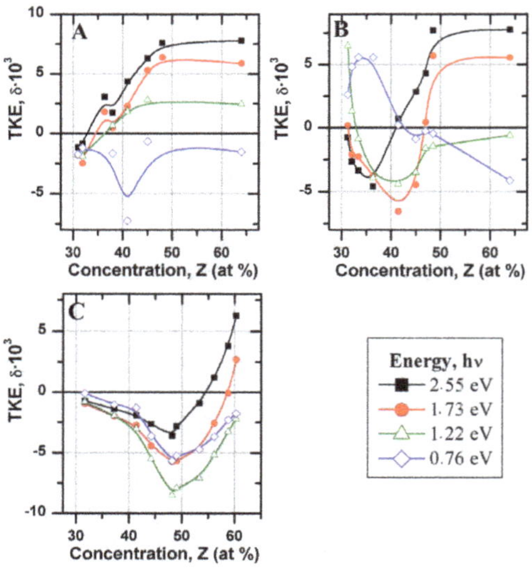

Figure 8. Concentration dependences of TKE for nanocomposite $(Co_{45}Fe_{45}Zr_{10})_Z(Al_2O_3)_{100-Z}$: (A) 1st series (the thickness of forming layers ~1 – 2.3 nm); (B) 2d series (the thickness of forming layers 2.7 - 6 nm); (C) bulk nanocomposite.

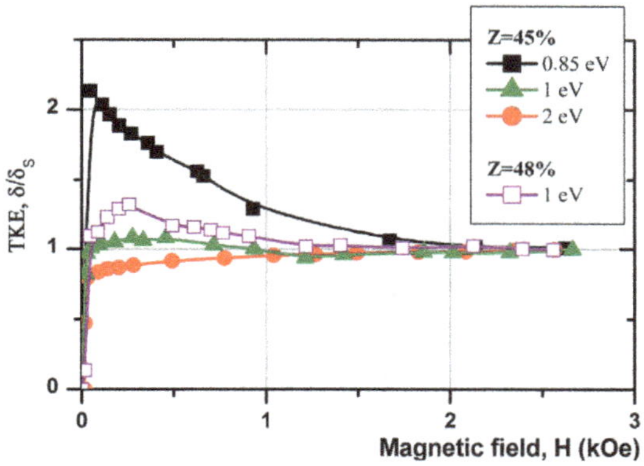

Figure 9. Anomalous field dependencies of TKE for samples 1st series (the values are normalized to the value of TKE at maximum field).

ences for several samples of the series 1. The appearance of these anomalies can be explained if we assume that during fabrication process there are formed not only magnetic granules with the size smaller, than in the bulk composite, but also rather big FM clusters, which add their contribution to the MO response of the whole system. In other words we can say that anomalous curve $\delta(H)$ is a sum of contributions of two magnetically different phases: a ferromagnetic and superparamagnetic one, at that the superparamagnetic granules produce TKE of one sign, and the ferromagnetic ones of the opposite. Thus in small field there appears a sharp saturation for the ferromagnetic phase, and with increasing magnetic field a superparamagnetic phase begins to saturate, for which the TKE have opposite sign, what results in reduction of the cumulative magneto-optical response. Taking into consideration that in the whole investigated energy range the TKE in CoFeZr is positive, and in the nanocomposite $(Co_{45}Fe_{45}Zr_{10})_Z(Al_2O_3)_{100-Z}$ in the region before percolation—negative, we can suppose that FM contribution in weak fields is associated with remagnetization of big CoFeZr clusters, and superparamagnetic contribution (in stronger fields)—with magnetization of the whole nanocomposite. By competition of these two contributions we can apparently explain a shift of the TKE spectral dependences for the 1 system into the positive region at Z > 36 at%, relatively the bulk composite (**Figure 6**). For the 2 system with increase of the deposited layers thickness the contribution of big CoFeZr clusters becomes smaller, and a more complicated dependence of concentration vs. wavelength is observed.

To sum up, the experimental data obtained for the systems of nanocomposites deposited by thin layers, proves that shape and dimensions of the granules in nanocomposite layer depend on the layer thickness; the microstructure of layers is essentially different from the microstructure of the bulk nanocomposite. This should be taken into account in analysis of experimental data for multilayer nanocomposite/semiconductor systems.

4. Conclusion

We have carried out studies of magnetic and magnetooptical properties of $(Co_{45}Fe_{45}Zr_{10})_Z(Al_2O_3)_{100-Z}$ nanocomposites, prepared by layer-by-layer sputtering of thin layers, over a wide range of layers thickness and concentrations of the ferromagnetic phase. There has been found a strong influence of preparation technology and technological parameters (the resulting layer thickness, ferromagnetic phase concentration) on magnetic and magnetooptical properties. Evolution of the spectra and found anomalies of the TKE magnetic field dependences for a number of samples strongly suggest that the films are magnetically heterogenous, and along with small isolated $Co_{45}Fe_{45}Zr_{10}$ granules in the layer-by-layer deposited nanocomposite there are rather big FM clusters. In the layer-by-layer deposited composites, there was found a shift of the percolation threshold in the direction of lesser FM phase concentrations, the greater shift found in films with lesser layers thickness. It was proved that both size and shape of the granules in the nanocomposite layer depend on layer thickness, and microstructure of the layer-by-layer deposited composites considerably differs from the microstructure of the bulk nanocomposites.

REFERENCES

[1] S. Mitani, H. Fujimori, K. Takanashi, K. Yakusiji, J. G. Ha, S. Takanashi, S. Maekawa, S. Ohnuma, N. Kobayashi, T. Masumoto, M. Ohnuma and K. Hono, "Tunnel-MR and Spin Electronics in Metal-Nonmetal Granular Systems," *Journal of Magnetism and Magnetic Materials*, Vol. 198-199, 1999. pp. 179-182. doi:10.1016/S0304-8853(98)01041-5

[2] J. C. Slonczewski, "Conductance and Exchange Coupling of Two Ferromagnets Separated by Tunneling Barrier," *Physical Review B*, Vol. 39, No. 10, 1989, pp. 6995-7002. doi:10.1103/PhysRevB.39.6995

[3] E. Ganshina, A. Granovsky, B. Dieny, R. Kumaritova and A. Yurasov, "Magneto-Optical Spectra of Discontinuous Multilayers Co/SiO_2 with Tunnel Magnetoresistance," *Physica B*, Vol. 229, No. 3-4, 2001, pp. 260-264. doi:10.1016/S0921-4526(01)00476-8

[4] A. B. Granovsky, I. V. Bykov, E. A. Gan'shina, V. S. Gushchin, M. Inoue, Y. E. Kalinin, A. A. Kozlov and A. N. Yurasov, "Magnetorefractive Effect in Magnetic Nanocomposites," *Journal of Experimental and Theoretical Physics*, Vol. 96, No. 6, 2003, pp. 1104-1112. doi:10.1134/1.1591221

[5] A. V. Ivanov, Yu. E. Kalinin, V. N. Nechaev, A. V. Sitnikov, "Electrical and Magnetic Properties of $[(CoFeZr)_x$-$(Al_2O_3)_{1-x}/(\alpha\text{-Si:H})]_n$ Multilayer Structures," *Physics of the Solid State*, Vol. 51, No. 12, 2009, pp. 2474-2479. doi:10.1134/S1063783409120087

[6] L. Förster, M. Karski, J. M. Choi, A. Steffen, W. Alt, D.

Meschede, A. Widera, E. Montano, J. H. Lee, W. Rakreungdet and P. S. Jessen, "Microwave Control of Atomic Motion in Optical Lattices," *Physical Review Letter*, Vol. 103, No. 23, 2009, pp. 233001-233005. doi:10.1103/PhysRevLett.103.233001

[7] V. G. Kravets, A. K. Petford-Long and A. F. Kravets, "Optical and Magneto-Optical Properties of $(CoFe)_x$-$(HfO_21)_{1-x}$ Magnetic Granular Films," *Journal of Applied Physics*, Vol. 87, No. 4, 2000, pp. 1762-1768. doi:10.1063/1.372089

[8] E. Gan'shina, A. Granovsky, V. Gushin, M. Kuzmichev, P. Podrugin, A. Kravetz and E. Shipil, "Optical and Magneto-Optical Spectra of Magnetic Granular Alloys," *Physika A*, Vol. 241, No. 1-2, 1997, pp. 45-51. doi:10.1016/S0378-4371(97)00057-5

[9] E. Gan'shina, M. Kochneva, M. Vashuk, A. Vinogradov, A. Granovsky, V. Guschin, P. Scherbak, Ch.-O. Kim and Ch. G. Kim, "Magneto-optical properties of Magnetic Nanocomposites," *The Physics of Metals and Metallography*, Vol. 102, Suppl. 1, 2006, pp. S32-S35. doi:10.1134/S0031918X06140079

[10] H. Akinaga, M. Mizuguchi, T. Manado, E. Gan'shina, A. Granovsky, I. Rodin, A. Vinogradov and A. Yurasov, "Enhanced Magnetooptical Response of Magnetic Nanoclusters Embedded in Semiconductor," *Journal of Magnetism and Magnetic Materials*, Vol. 242-245, 2002, pp. 470-472. doi:10.1016/S0304-8853(01)01067-8

[11] T. Fukumura, Y. Yamada, K. Tamura, K. Nakajima, T. Aoyama, A. Tsukazaki, M. Sumiya, S. Fuke, Y. Segawa, T. Chikyow, T. Hasegawa, H. Koinuma and M. Kawasaki, "Magneto-Optical Spectroscopy of Anatase TiO_2 Doped with Co," *Japanese Journal of Applied Physics*, Vol. 42, 2003, pp. 105-107. doi:10.1143/JJAP.42.L105

[12] E. Gan'shina, R. Kumaritova, A. Bogorodisky, M. Kuzmichov and S. Ohnuma, "Magneto-Optical Spectra of Insulating Granular System Co-Al-O," *Journal of Magnetism and Magnetic Materials*, Vol. 203, No. 1, 1999, pp. 241-243. doi:10.1016/S0304-8853(99)00275-9

[13] T. V. Murzina, E. A. Gan'shina, V. S. Guschin, T. V. Misuryaev and O. A. Aktsipetrov, "Non-Linear Magnetooptical Kerr Effect and Second Harmonic Generation Interferometry in Co-Cu Granular Films," *Applied Physics Letters*, Vol. 73, No. 25, 1998, pp. 3769-3771. doi:10.1063/1.122889

[14] E. A. Gan'shina, M. V. Vashuk, A. N. Vinogradoy, A. B. Granovsky, V. S. Gushchin, P. N. Shcherbak, Yu. E. Kalinin, A. V. Sitnikov, C.-O. Kim and C. G. Kim, "Evolution of the Optical and Magnetooptical Properties of Amorphous Metal-Insulator Nanocomposites," *Journal of Experimental and Theoretical Physics*, Vol. 98, No. 5, 2004, pp. 1027-1036. doi:10.1134/1.1767571

[15] V. E. Buravtsova, V. S. Guschin, Yu. E. Kalinin, S. A. Kirov, E. V. Lebedeva, S. Phonghirun, A. V. Sitnikov, N. E. Syr'ev and I. T. Trofimenko, "Magnetooptical Properties and FMR in Granular Nanocomposites $(Co_{84}Nb_{14}Ta_2)_x$-$(SiO_2)_{100-x}$," *Central European Science Journals*, Vol. 2, No. 4, 2004, pp. 566-578. doi:10.2478/BF02475564

[16] E. Gan'shina, K. Aimuta, A. Granovsky, M. Kochneva, P. Sherbak, M. Vashuk, K. Nishimura and M. Inoue, "Optical and Magneto-Optical Properties of Magnetic Nanocomposites FePt-SiO_2," *Journal of Applied Physics*, Vol. 95, No. 11, 2004, pp. 6882-6884. doi:10.1063/1.1687537

[17] V. A. Vyzulin, V. E. Buravtsova, V. S. Gushchin, E. A. Ganshina, E. V. Lebedeva, N. E. Syriev, S. Phonghirun, Yu. E. Kalinin and A. V. Sitnikov, "Magnetic and Magnetooptical Properties of Ferromagnetic-Ferroelectric Nanocomposites," *Bulletin of the Russian Academy of Sciences: Physics*, Vol. 70, No. 7, 2006, pp. 1075-1078.

[18] G. S. Krinchik and V. S. Guschin, "Investigation of Inter-Band Transitions in Ferromagnetic Metals and Alloys by Means of Magnetooptical Method," *Journal of Experimental and Theoretical Physics*, Vol. 56, 1969, pp. 1833-1842.

14

Consolidation of Ancient Raw Materials Using a Reversible, Elastic, Soft Polymer

Juan Manuel Navarrete[1], Gustavo Leonardo Martínez[2†]

[1]Inorganic and Nuclear Chemistry Department, Faculty of Chemistry, National University of Mexico, Mexico City, Mexico
[2]National Coordination to Restore the Cultural Inheritance, National Institute of Anthropology and History, Mexico City, Mexico
Email: jmnat33@unam.mx

ABSTRACT

Aging of ancient raw materials usually finish with disintegration, which starts on surface of walls to progress toward the inside mass of a huge variety of mineral compounds. This is particularly harmful when antique buildings keep mural paintings, which suffers destruction before the wall itself. Same case appears on sculptures and monuments, whose surfaces are often attacked by living organisms which start a deterioration process previous to complete disintegration. The main factor to produce these unwanted effects is humidity, either rain for materials exposed to open air, or underground humidity going up by capillarity of minerals, in this case represented by porosity of associated salts forming the material. This paper describes a method to measure easily the relative porosity of diverse raw materials at laboratory level, by using a radioactive labeled solution, and also a procedure to reduce their porosity of those minerals. The efficiency of this procedure is measured in the same way, and so the results obtained at laboratory level have encouraged its use at real scale, where it has been quite successful for a number of materials in a limited span of five years.

Keywords: Consolidation; Preservation; Humidity; Ancient; Materials

1. Introduction

Ancient raw materials are usually attacked from two fronts: one is from outside, mainly by rain, snow, dust, wind, light and severe temperature changes, other is from inside, mainly by underground humidity which goes up through wall porosity, as a solution carrying on a great variety of salts. When both fronts meet, destructive conditions for raw materials become most powerful, but just one front is enough to deteriorate and finally destroy bricks, volcanic stones, rocks, feldspars, clays, adobes and so on. Even at controlled room conditions of temperature and humidity, such as those in museums, just at the time goes by materials surface start a more or less visible process of aging. However, as humidity and unstable temperature are the most common causes of destruction, and underground humidity is very difficult to control or shield efficiently in ancient buildings, it is proposed here to use at laboratory level a very easy and simple method to measure first the absorption capacity or porosity in samples of Mexican ancient raw materials, in order to determine a decreasing factor when the samples have been permeated with an organic gelatin solution, added with two food preservatives (sodium benzoate and potassium sorbide) plus a very strong organic preservative (37% formaldehyde solution). These additives guarantee that gelatin will not become a fungus or bacteria culture as time goes by, and by other part form a tougher polymer, much more insoluble in water than plain gelatin, but still soluble in hot water (60˚C) [1]. Applied at real scale, this simple technique has showed till now a surprising efficiency in every material tested during a limited span of five years.

2. Materials and Methods

As sodium sulfate is a very common salt found in solid deposits on ancient walls [2], this salt was chosen to be labeled with radioactive ^{22}Na (half life 2.6 years, annihilation peak of 0.511 MeV of energy). A radioactive solution was made, 0.1% Na_2SO_4 concentration and ^{22}Na radioactivity about 30 Bq/ml. Samples were collected from debris found in archeological sites of interest such as: Teotihuacan, Tula, Palenque, Veracruz and Bonampak. They were cut in smaller samples prisms of about 3 cm height and 1 cm wide, which were cleaned and weighed.

Then, prisms were placed on a disposable tray divided by squares, where 2 ml per square of the labeled solution had been previously deposited, in order to get approximately 1 - 2 cm over the prism base (**Figure 1**). In half an hour samples were totally wet by the labeled solution. They were isolated from the radioactive solution and dried at room temperature during 24 hours. Sodium sulfate salt was deposited on the surfaces. They were carefully cleaned wearing gloves and manipulating a spatula, before being conditioned to be detected either in test tubes or plastic bags (**Figures 2** and **3**). Samples were detected and counts accumulated during 10 to 30 minutes. Detector used was a well type 3 × 3" solid scintillator NaI (Tl), Bicron trade mark, coupled to a PC charged with the Maestro program for radioactive detection. Counts accumulated per time and weight unit represent in relative terms their different porosity from one to another material, and in consequence its ability to absorb humidity, either from soil, or in open air conditions from the environment, both equally destructive. These differences were as much as a factor of ten for different materials (porous volcanic stones and feldspar), and till 25% for different samples of same material, because materials are far from being quite homogeneous. After the relative measurement of porosity in different raw materials was performed, same type of prism samples were permeated with a warm solution (5%) (60°C) of organic powdered gelatin, so called French gelatin, plus two food preservatives A.R.: sodium benzoate and potassium sorbate (2.5% each) (**Figure 4**).

After 3 hours, when the gelatin is not solid yet, and the material looks humid and brilliant, by using one hand sprinkler, gloves and a mask protector, was added 37% formaldehyde aqueous solution, one tenth to one fifth of the gelatin volume solution previously made. Tougher gelatin is formed overnight. When next day the porosity of same materials was tested by means of the radioactive solution, in exactly the same way that samples with no gelatin treatment, after carefully cleaning from external gelatin residues, the difference was dramatic. From 70% to 0% of counts accumulated, that is to say, from 30% to 100% of porosity reduction, according the nature of the material tested. An inverse relation was found between the porosity of samples and efficiency with gelatin treatment. So, plain dry mud, adobe, calcareous and porous volcanic stones, showed a total blockade to labeled solution at laboratory conditions, while less porous material such as dolomite and feldspar showed an absorption rate about 30% lower than before treatment. Some other raw

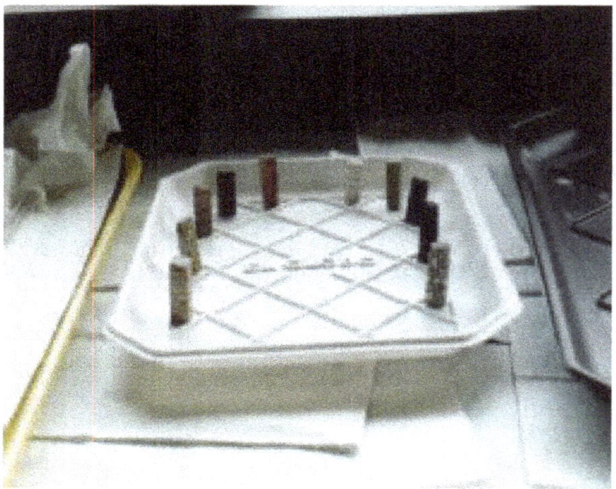

Figure 1. Samples of different materials conditioned to absorb the labeled solution.

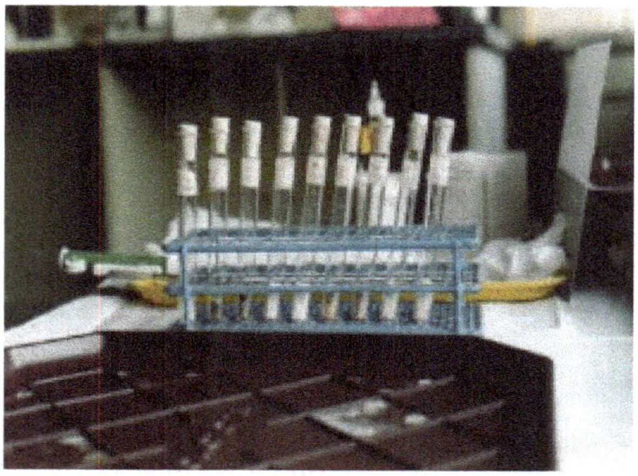

Figure 2. Samples conditioned in test tubes to be detected.

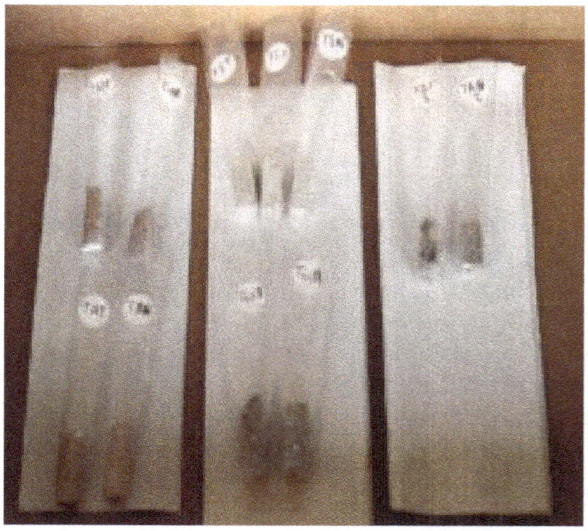

Figure 3. Samples conditioned in polyethylene bags to be detected.

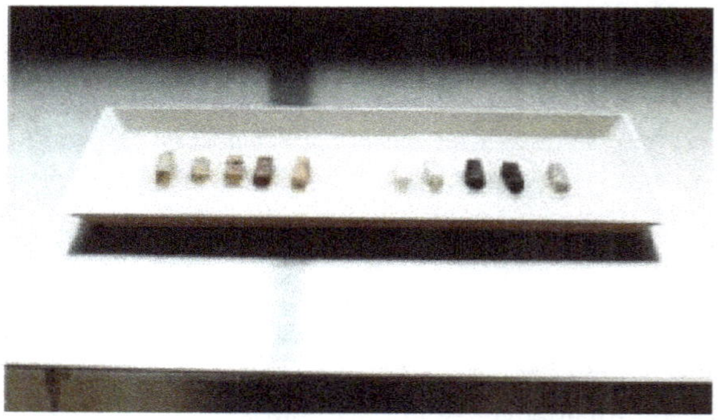

Figure 4. Samples absorbing gelatin.

materials such as brick, clay, limestone and assorted volcanic stones showed intermediate results. A piece of gelatin obtained in the described way, was immersed in plain water at room temperature. After one month begun to be slightly soluble, but after two years was not completely dissolved. On the contrary, by using water at 60˚C - 70˚C it was soluble in one hour. Also, another test was made at laboratory level, using a colored methyl red aqueous solution. Two adobe bricks were permeated with the colored solution in identical conditions: one treated with the gelatin on just one surface, that opposite to liquid permeation, while the other brick was untreated.

The non treated brick got wet till the top in few minutes, becoming dark red with the solution, sticking easily on it a piece of tissue paper, while the other one remained one month with periodical replacement of solution lost by evaporation, with no color on the top surface and of course non sticking any piece of tissue paper (**Figure 5**). Nevertheless, the application of the technique at real scale is quite another matter, since conditions of volume, time and humidity persistence, either external or internal by underground humidity, are enlarged beyond any proportion compared with those of laboratory level. In this way, permeation of gelatin has been used in some Mexican monuments and whole objects always following the next recipe:

1) First, to clean up the surface in the usual way with water, brush and soft detergent. Once dry, sprinkle the gelatin solution (5%) plus food preservatives, sodium benzoate and potassium sorbide (2.5% each) at a temperature about 60˚C - 70˚C. Application must be as homogeneous as possible on the surface, avoiding slipping down of the solution, which means to wait some seconds to allow the warm liquid permeates the porosity of material. Sprinkler managed by hand was preferred to those running by a compressor, because it is easier to avoid the slipping down of the solution, and warm temperature of it is much better kept, which means larger absorption in the material. Yielding of solution is a function of material porosity, but a suitable average seems to be 2 $L \cdot m^{-2}$.

2) After about three hours, gelatin begins to solidify and material looks humid and brilliant. Then it must be sprinkled with 37% formaldehyde solution, in a proportion 10% to 20% in volume applied of gelatin solution. The hand sprinkler was used again, wearing a protective mask, to avoid inhalation of formaldehyde vapors. Both solutions can not be mixed at the start of process, because then gelatin does not become solid at any time. Instead, a tougher polymer, more insoluble in cold water is formed, just adding the formaldehyde when gelatin begins to condensate. Treatment is finished by cooling overnight. Material surface is protected by an interior layer of tough gelatin, against humidity from rain at open air conditions, as well as underground humidity and its dissolved salts, being quite inert to any change in a reasonable time span.

3. Results

First, the method has been applied on a plastered wall located in Teotihuacan archeological zone [3].

It is roofed, but extremes and front are open. Conditions of maintenance were rather precarious (**Figure 6**). After longer than five years, salts deposition has been completely interrupted, and not fungus mushrooms, or some other culture of organic material has appeared on it. A small rabbit made out from volcanic stone during the Aztec empire period, at open air conditions for indefinite time, was treated and placed as before [4]. Conservation has lasted longer than five years with no problem (**Figures 7** and **8**). Five calcareous stone Maya pieces sent to be exposed in European countries, showed an evident loss of powder material when they were manipulated and conditioned for the trip, even when they have always been at museum conditions. After treatment, manipulation was made with no material loss and treatment was quite unnoticeable (**Figures 9** and **10**).

Figure 5. Bricks with and without treatment, absorbing colored solution.

Figure 6. Wall in Teotihuacan region under treatment.

Figure 7. Aztec little rabbit after years at open air conditions.

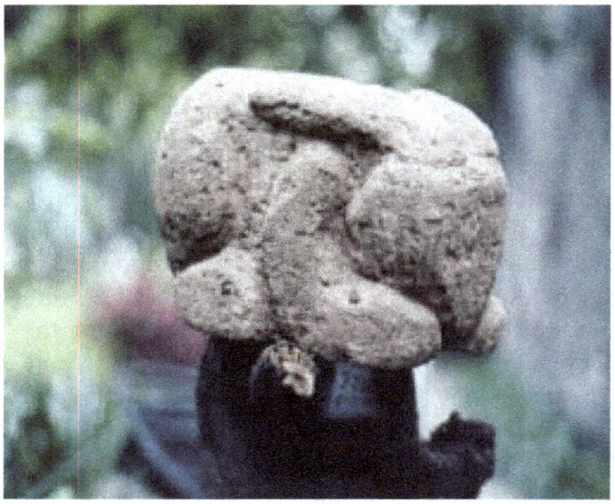

Figure 8. Aztec little rabbit after longer than five years treatment.

Figure 9. Mayan representation after treatment for travelling to Europe.

Figure 10. Another Mayan piece ready to be transported after treatment.

A jaguar sculpted on one piece of very porous volcanic stone, in the Tula region, lost its right paw in unknown circumstances, but its left paw became thinner and very fragile, even when it has been at museum conditions for very long time. Disintegration of material with just hand touch was evident. After treatment, it can be manipulated just with some care, left paw remains thin but much firmer than before, and it seems that at museum conditions it is going to be well preserved for many years (**Figure 11**). In the same archeological region and museum, an oven or furnace, probably used at underground conditions, was untouchable because the plain dried mud of its material was then quite loosed, being impossible to transport it to some other location for the great risk of total disintegration. The method was applied with success, and now it is just another museum piece perfectly transportable (**Figures 12** and **13**). Finally, an image of our Lady of Guadalupe, painted in 1885 on a plastered wall, under roof but in an open corridor at both extremes, was almost completely destroyed by internal humidity, both underground and broken pipes. Pigments lost cohesion with the plaster, and plaster with the wall. So, gelatin was used in every step of restoration. It was possible to cut out the whole picture from the wall, in order to be transported to restoration workshop, plaster was replaced and painted on it, finishing was also made with gelatin solution, and now it is in the same place, but in a niche behind a glass (**Figures 14** and **15**).

4. Discussion

Experimental results obtained at laboratory level for relative humidity absorption by materials and the way to avoid it, have been very useful when the proposed me-

Figure 11. Tula region's jaguar under treatment.

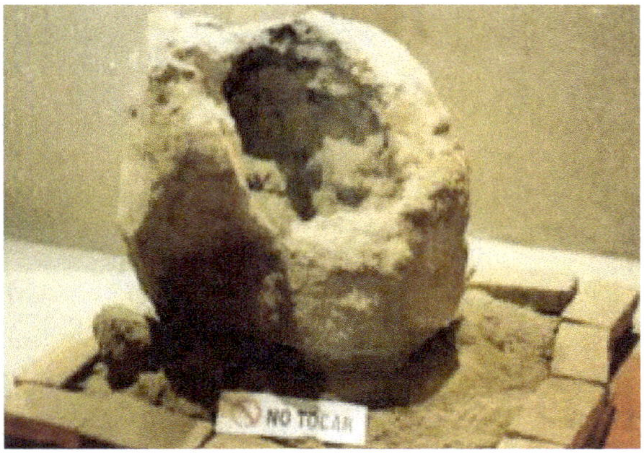

Figure 12. Tula region's oven before treatment.

Figure 13. Tula region's oven under treatment.

thod has been applied at real scale, even when five years is a very short span to test it. However, the fact that it can be applied periodically, and also that gelatin can be removed easily by hot water (60°C - 70°C), if its evolution were undesirable, which has not been the case till now, seem to be so attractive characteristics as the easiness,

Figure 14. Virgin of Guadalupe's image before treatment.

Figure 15. Virgin of Guadalupe's image after treatment.

almost non visible and time saving to perform it. As a matter of fact, it seems that clogging porosity with a soft, elastic and compressible material such as gelatin, charged with two food preservatives and one very strong preservative of organic material, such as formaldehyde, differs greatly from clogging porosity with a tough, insoluble and in some way as biodegradable as the material under conservation.

5. Conclusion

The proposed method has worked all right, either to avoid underground and environmental humidity, as to consolidate loose materials. Promising results as they are, suggest trying on for size the method in truly big restoration projects. Just to mention few examples: recently discovered painted walls of Cacaxtla in Mexico, as well as Hagia Sophia in Turkey, for underground humidity, and Xian's warriors in China, to consolidate material.

REFERENCES

[1] Merck and Co. Inc., "The Merck Index," 9th Edition, Whitehouse Station, Hunterdon County, 1976, p. 4096.

[2] H. Rosch and J. Schwarz, "Damaged to Frescoes Caused by Sulphate-Bearing Salts: Where Does the Sulphur Come from?" *Studies in Conservation*, Vol. 38, No. 4, 1993, pp. 224-230. doi:10.2307/1506367

[3] G. L. Martinez and J. M. Navarrete, "Use of a Radiotracer to Test and Reduce the Porosity and Humidity Absorption from the Soil in Pre-Hispanic Raw Materials," *Journal of Radioanalytical and Nuclear Chemistry*, Vol. 263, No. 1, 2005, pp. 35-38. doi:10.1007/s10967-005-0008-2

[4] G. L. Martinez and J. M. Navarrete, "A New Technique to Preserve Raw Materials of Ancient Monuments against the Humidity and Its Test Using ^{22}Na Labelled Solutions," *Journal of Radioanalytical and Nuclear Chemistry*, Vol. 274, No. 3, 2007, pp. 651-655. doi:10.1007/s10967-007-7097-z

15

Dielectric Behavior of Some Vinyl Polymers/Montmorillonite Nanocomposites on the Way to Apply Them as Semiconducting Materials

Amal Amin[1], Eman H. Ahmed[1], Magdy W. Sabaa[1], Magdy M. H. Ayoub[1], Inas K. Battisha[2]
[1]Polymers and Pigments Department, National Research Center (NRC), Giza, Egypt
[2]Solid State Physics Department, National Research Center (NRC), Giza, Egypt
Email: aamin_07@yahoo.com, aamin_2011@yahoo.com

ABSTRACT

Some vinyl polymers/montmorillonite nanocomposites were prepared via *in-situ*-atom transfer radical polymerization (ATRP) in presence of clay. Methyl methacrylate, styrene and n-butyl methacrylate were involved in the formation of such polymeric nanocomposites. Their dielectric properties were extensively studied to invest them in the a.c. power applications. Several dielectric parameters such as dielectric constant loss (ε'') and a.c. conductivity (σ) were measured at both different frequencies (0.1 Hz to 100 KHz) and temperature ranged from (20°C to 90°C). From the dielectric results, it was realized that the dielectric a.c. conductivity was enhanced by increasing the temperature for the four prepared polymer nanocomposites.

Keywords: Nanocomposites; Atom Transfer Radical Polymerization; Insulation; Dielectric Measurements; Intercalation; Polarization

1. Introduction

At the last decades, polymers/clay nanocomposites (PCNs) became an interesting area of research where PCNs have evoked an intense industrial and academic research, due to their outstanding mechanical, thermal, chemical and electrical properties over the pure polymers [1]. Several kinds of nanofillers were used such as layered silicates which include montmorillonite as the widely used nanoclay [2,3]. Organic polymers, loaded with small amounts of montmorillonite (MMT) clay produce polymer-clay nanocomposites (PCNs) having a good resemblance in dielectric properties with respect to their structural properties [4]. Generally, the research work on novel polymeric materials in the field of power engineering and high voltage technology is of great significance both nationally and internationally due to the increasing demands of more cost effective and environmentally better materials for high voltage equipment [5]. Traditionally, additive agents and fillers are often used for improving dielectric and mechanical properties of materials [6]. Nowadays, polymeric nanocomposite materials are attracting attention of many researchers in the field of dielectric [7] where polymeric nanocomposites are thought to be the state of the art of dielectric materials with improved dielectric performance [8]. Polymeric semiconducting materials are widely applied to power apparatuses and cables [9]. Therefore, recent preliminary work has been already done to investigate the dielectric performance of polymeric nanocomposite materials regarding the reduction of space charge, variation of conductivity and increase of electric strength where these studies were carried out in case of various polymeric materials, doped with nanofillers [10-12]. Generally, three methods are commonly used for the synthesis of PCNs including *in-situ* polymerization [13], solvent intercalation [14] and polymer melt intercalation [15]. For *in-situ* technique, free radical polymerization is likely the most important process [16]. However, conventional free radical polymerization methods lack the control due to chain transfer and termination reactions. Therefore, living controlled polymerization techniques (LCR) appeared as the vital alternative including nitroxide mediated polymerization (NMP) [17], reversible addition fragmentation process (RAFT) [18] and metal mediated atom transfer radical polymerization (ATRP) [19]. ATRP appeared as the most attractive living controlled polymerization techniques tolerant to various monomers, solvents, temperatures and catalytic systems. The aim of the current work was to prepare some vinyl polymeric/montmorillonite clay nanocomposites via *in-situ*-atom transfer radical polymerization where their dielectric behavior was successively studied by measuring different related parameters such as dielectric constant (ε'), dielectric constant loss (ε'') and a.c. conductivity (σ) [20-22].

2. Experimental Section

2.1. Materials

Montmorillonite (CEC = 88 meq/100 g) was received from Sigma-Aldrich chemical company (USA). Other chemicals-otherwise mentioned were used as received from Sigma-Aldrich without further purification. The monomers such as styrene (St), methyl methacrylate (MMA), and n-butyl methacrylate (n-BuMA) were purified by filtration through an activated basic alumina column and then stored under argon in the fridge.

2.2. Characterization

Gel permeation Chromatography (GPC) was used to determine number-average molecular weight ($\bar{M}_{nGPC}$) and polydispersity ($\bar{M}_w/\bar{M}_n$) of the polymers by using Agilent-1100 GPC-technologies-Germany. The refractive index detector was G-1362 A with 100-10^4-10^5 A°, using polystyrene (PS) as a standard and tetrahydrofuran (THF) as the eluent. The structures of the formed polymers were determined via proton nuclear magnetic resonance spectrometry (^{1}H-NMR) which was carried out with Jeol-ECA 500 MHz, using tetramethylsilane (TMS) as internal standard and $CDCl_3$ as the main solvent. X-ray Diffraction (XRD) measurements were carried out by using Phillips X Pert X-ray generator with Cu Kα radiation at 40 kV. Dynamic mechanical thermal analyses (DMTA) were carried out with a diffractometer (type PW 1390) with employing Ni-filtered CuK. A typical sample weight was about 8 - 10 mg and the analyses were performed at a heating rate of 10°C/min from 50 to 600°C under Helium atmosphere. Transmission electron microscopy (TEM) was used to determine the morphology of the formed polymers/clay nanocomposites by using transmission electron microscopy (TEM-JEOL JX 1230) with micro-analyzer electron probe and magnification up to 600 kx, giving a resolution down to 0.2 nm. The measurements were performed at an accelerating voltage of 100 kV. The dielectric measurements were performed using Computerized LRC-bridge (Hioki model 3531 zHi Tester). The dielectric constant ε' for the investigated samples was studied at both different temperatures and frequencies ranging from (20°C to 90°C) and (0.1 Hz up to 100 KHz). The samples used in the dielectric measurements were in disc form, having 10 mm in diameter and 3 mm in thickness, pressed using a pressure of 10 ton at room temperature. Then, silver paste was coated to form electrodes on both sides of the sintered ceramic specimens in order to ensure good contacting. The electric measurements were carried out by inserting the sample between two parallel plate conductors forming cell capacitor. Then, the whole arrangement was placed in non-inductive furnace for heating the samples with constant rate. The relative dielectric permittivity was calculated using the relations:

$$\varepsilon' = \mathrm{Cd}/\varepsilon_0 A \quad (1)$$

$$\varepsilon'' = \varepsilon' \tan\delta \quad (2)$$

where, ε' is the permittivity, ε'' is the dielectric loss and tan δ is the loss tangent and A is the area of the electrode. The a.c. resistivity of the prepared samples was estimated from the dielectric parameters. As long as the pure charge transport mechanism is the major contributor to the loss mechanism, the resistivity (ρ) can be calculated using the following relation:

$$(\rho) = 1/(\mathrm{w}\varepsilon_0\varepsilon' \tan\,\delta)\,\Omega Cm \quad (3)$$

where $\omega = 2\pi f$, ω is the angular frequency and f is the frequency of the applied electric field in Hertz.

$$\sigma = 2\pi f d C \tan\,\delta/\mathrm{A} \quad (4)$$

where σ is the a.c. conductivity, f is the operating frequency, d is the thickness of the dielectrics, tanδ is the dielectric loss, C is the capacitance and A is the area of the electrode.

2.3. General Polymerization Procedure

2.3.1. Organic Modification of Clay with Cationic Surfactant (CTAB) [23]

The montmorillonite clay (MMT, 20 gm) was dispersed in 500 ml distilled water containing certain amount of cationic surfactant, as cetyltrimethyl ammonium bromide (CTAB) (6 gm, 0.0165 mol) at room temperature. The temperature was increased to 80°C with vigorous stirring for 6 - 8 h. The clay was separated by filtration and washed several times with distilled water where the filtrate was tested with $AgNO_3$ solution (1 gm $AgNO_3$ in 100 ml distilled H_2O) to obtain the modified clay without CTAB residuals. The resulting modified clay with CTAB (MMT-CTAB) was dried under vacuum at 60°C for 24 h. MMT-CTAB was grinded and characterized via XRD and TEM.

2.3.2. Preparation of the MMA or St Homopolymers/ Montmorillonite Nanocomposites [24]

MMT-CTAB (0.15 g, 4%filler) was placed in a test tube, then bipyridine (bpy) (0.15 g, 4.9×10^{-4} mol) and CuBr (I)

(0.07 g, 4.9×10^{-4} mol) were added. The tube was closed and purged with argon for 15 minutes. 5.2 ml of MMA or 5.6 ml of St (4.9×10^{-2} mol) were successively injected into the reaction mixture under argon atmosphere. The tube was immersed in an oil bath with continuous stirring at 90°C. The reaction mixture was left for 10 minutes, then, 1 ml (4.9×10^{-4} mol) of ethyl-2-bromoisobutyrate was added to the reaction mixture as a sacrificial initiator. After 6 h, the vial was opened and the polymerization was terminated by adding THF to the reaction mixture. The polymers/clay nanocomposites were washed with aqueous solution of disodium salt ethylene-diamine-tetra-acetic acid (EDTA-disodium salt) to remove the catalysts and dried in vacuum at 40°C. For GPC measurements, the free polymer was separated by precipitation of THF solution in n-hexane, then filtration, re-dissolving it in THF and passing through alumina column to get the free polymer.

2.3.3. Preparation of Co- and Ter-Polymers/Montmorillonite Nanocomposites

The co- and ter-polymerization experiments were carried out as previously described in case of homo-polymers but in different amounts as:

2.3.4. In Case of Copolymers

0.3 g of MMT-CTAB, 0.306 g (9.8×10^{-4} mol) of bpy and 0.14 g (4.9×10^{-4} mol)of CuBr (I) were added. 5.6 ml (4.9×10^{-2} mol) of St were added under argon atmosphere with continuous stirring. After 4 h, the vacuum was applied to the reaction mixture to remove the remaining of the first monomer, then, 5.2 ml (4.9×10^{-2} mol) of MMA were added and the reaction was left for additional 20 h.

2.3.5. In Case of Ter Polymerization

0.45 g of MMT-CTAB, 0.45 g (9.8×10^{-4} mol) of bpy and 0.21 g (4.9×10^{-4} mol) of CuBr (I) were added. Then, the vessel was closed for 30 min. Then, the first and second monomers were sequentially added as previously described. St and MMA were added as in case of co-polymerization. 7.7 ml of n-BuMA (4.9×10^{-2} mol) were added after 24 h. The reaction was left for additional 24 h and terminated by adding THF as previously described.

3. Results and Discussion

Several polymers/MMT nanocomposites were prepared via *in-situ*-atom transfer radical polymerization (ATRP) of some vinyl monomers. On that way, MMT was modified with CTAB cationic surfactant where the basal spacing (d) increased from 1.210 nm at 2 theta of 7.5 for MMT to 1.837 nm at 2 theta value of 5 for MMT-CTAB. That extension in the interlayer distance between the clay platelets was referred to the intercalation of the CTAB molecules between the clay layers which caused widening of the interlayer distance. MMA and St/MMT nanocomposites were prepared via in-situ ATRP in presence of MMT-CTAB. Block copolymer/MMT nanocomposite was obtained by in-situ ATRP of St and applying vacuum to get rid of the monomer residuals, then adding MMA as the second monomer. In case of block terpolymer, n-BuMA was added as the third monomer after successive additions of St and MMA, respectively and by removing the excess of each monomer by vacuum before proceeding in further new polymerization step. The resulting polymers/MMT nanocomposites such as PMMA, PSt, PSt-b-MMA and PSt-b-MMA-b-n-BuMA/MMT were characterized via XRD, TGA, DSC and TEM. As shown in **Figure 1**, XRD indicated d values of 1.174 nm and 1.019 nm for PMMA/MMT and PSt-b-PMMA/MMT at 2 theta values of 7.7 and 8.7, respectively, with lower intensity than recorded for MMT-CTAB. The values of basal spacing d for both PSt/MMT and PSt-b-PMMA-b-P-n-BuMA/MMT were 8.251 and 1.894 nm at 2 theta values of 10.7 and 5.7, respectively. The formed polymers/MMT nanocomposites were characterized via TGA where as shown in **Figure 2**, PMMA lost about 88% from its initial mass where the degradation started at 170°C with slight weight loss (2.0%) till 310°C. Then, it lost 85.55% of its weight from 310 up to 455°C where the sample completely decomposed at 600°C. Pst sample lost about 42.2% from its initial mass where it started to decompose at 120°C and it lost 2.11% of its weight till 150°C then it lost 12.78% up to 260°C followed by weight loss of 17.97% up to 410°C where the complete decomposition occurred at 600°C. On the other hand, in case of PSt-b-PMMA, the sample stayed unaffected till 200°C where it began to lose about 8 % of its weight up to 340°C, and then gradual decomposition happened where it lost 45.11% up to 440°C and finally complete degradation was recorded at 600°C. Generally, it seemed that PSt-b-PMMA lost about 54% from its initial mass. In case of PSt-b-PMMA-b-P-n-BuMA, it lost about 73.94% from its initial mass where it started to decompose at 30°C up to 130°C with a weight loss 14.6%, and then the sample gradually degraded up to 415°C with weight loss of 47.94% and 11.4% at two stages but the sample completely degraded at 600°C.

DSC measurements indicated T_g values of 90°C, 117°C, 156°C and 180°C for PMMA, PSt, PSt-b-PMMA and PSt-b-PMMA-b-P-n-BuMA polymers inside their nanocomposites with MMT. Some of the prepared nanocomposites were characterized by TEM as shown in **Figure 3**. TEM showed that the stacks of the intercalated clay platelets are embedded in the polymer matrices showing good distribution of the MMT particles within the polymer matrices. The structure of the resulting polymers was confirmed by ^{1}H-NMR [25]. ^{1}H-NMR of

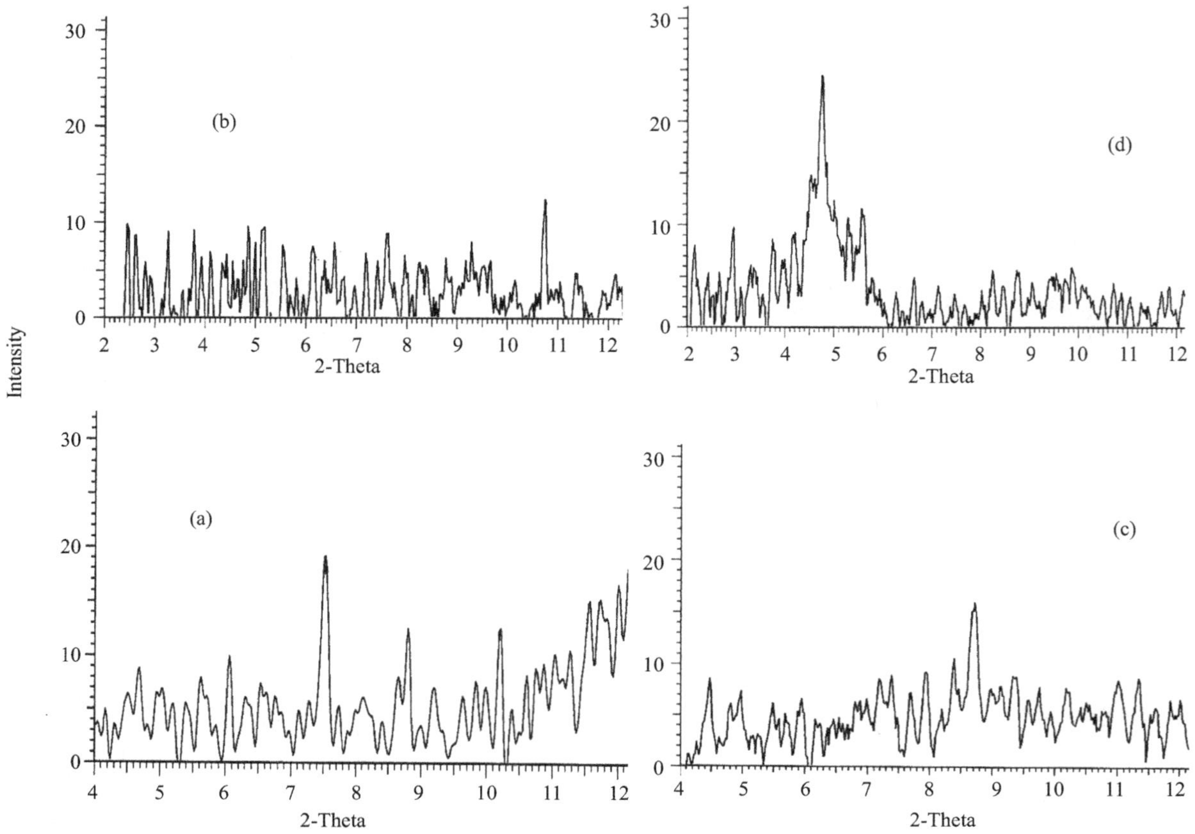

Figure 1. XRD of (a) PMMA/MMT, (b) PSt/MMT, (c) Pst-b-PMMA/MMT, (d) PSt-b-PMMA-b-P-n-BuMA/MMT nanocomposites.

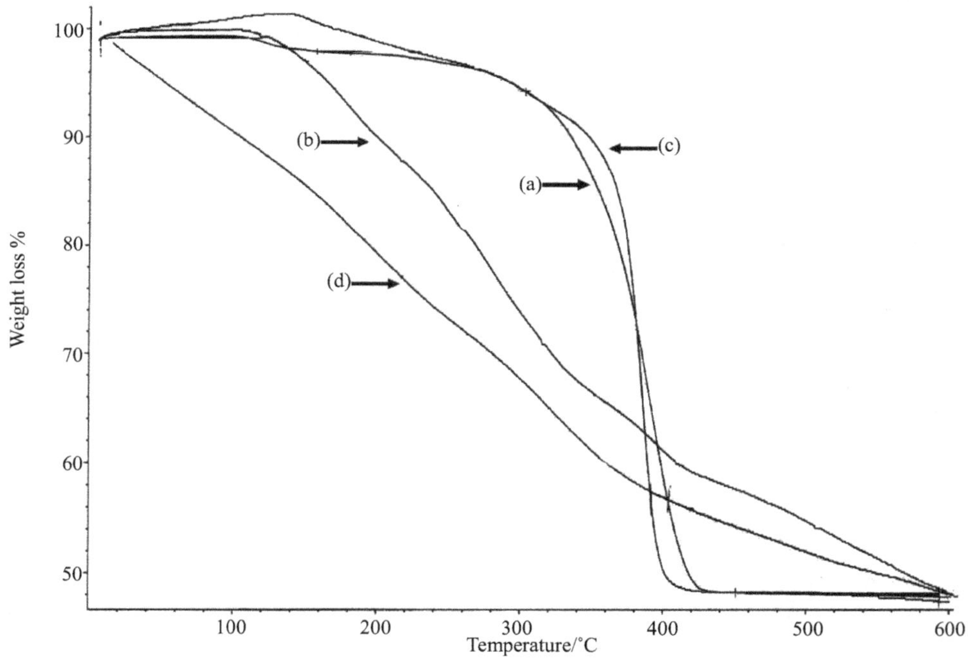

Figure 2. TGA of (a) PMMA/MMT; (b) PSt/MMT; (c) Pst-b-PMMA/MMT; (d) PSt-b-PMMA-b-P-nBuMA/MMT nanocomposites.

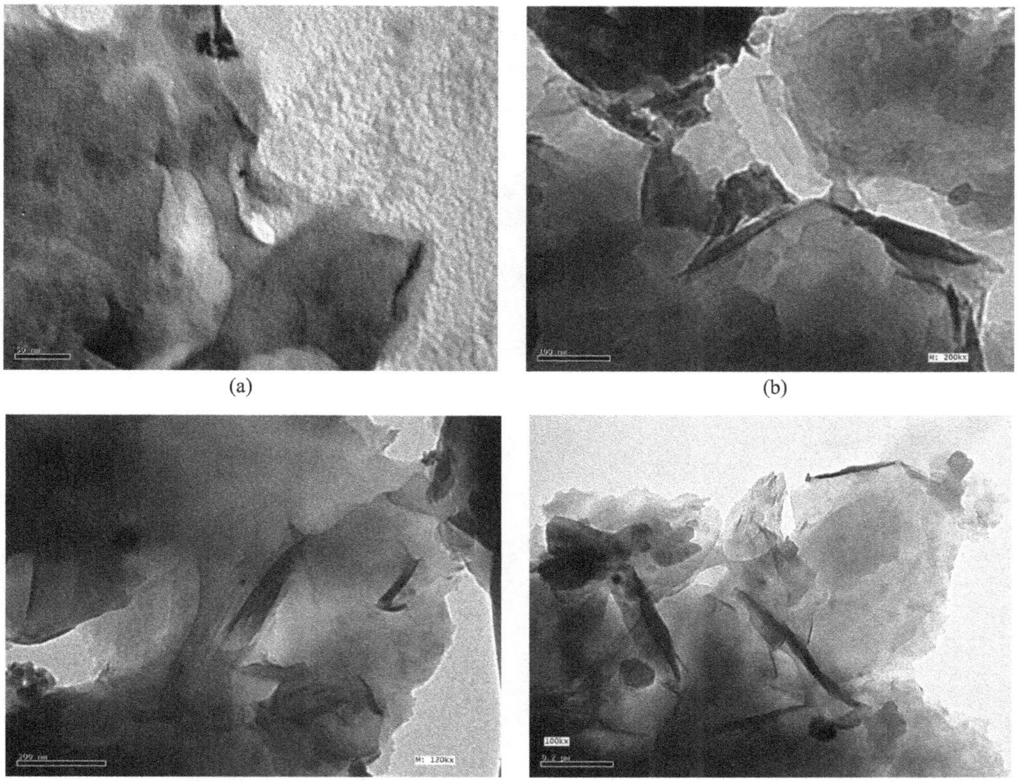

(a) (b) (c) (d)

Figure 3. TEM of (a) PMMA/MMT; (b) PSt/MMT; (c) Pst-b-PMMA/MMT; (d) PSt-b-PMMA-b-P-nBuMA/MMT nanocomposites.

MMA polymer, indicated the appearance of characteristic bands at chemical shifts (δ, ppm) at (0.8 - 1.3) for (CH_3CH_2, C-CH_3), 1.5 - 1.9 for (CH_3CH_2) and at 3.5 for (OCH_3). However, for St polymer, dense bands appeared in high intensity at $\delta = 6.5 - 7.5$ which were referred to the phenyl groups inside the skeleton of the polymer. On the other hand, (Br-CH-ph) appeared at $\delta = 4$ ppm. In case of PSt-b-PMMA copolymer, typical bands for both MMA and St units were observed. Similarly, in case of PSt-b-PMMA-b-P-n-BuMA, characteristic bands for St. and MMA were noticed in addition to several bands at 2 - 2.9 which were attributed to several (CH_2) groups inside the terpolymer skeleton in occasion of the presence of n-BuMA moiety.

Dielectric Measurements of Polymers/Clay Nanocomposites [26,27]

The dielectric constant loss ε" and a.c. conductivity (σ) were studied in the temperature range (20°C - 90°C) and frequency range (0.1 Hz up to 100 KHz) for all polymers/clay nanocomposites samples. **Figures 4(a)-(d)** shows the variation of the dielectric loss ε" of poly (a, PMMA/ MMT), (b, PSt/MMT), (c, PSt-b-PMMA/MMT) and (d, PSt-b-PMMA-b-P-n-BuMA/MMT) nanocomposites sam- ples as a function of frequency at various temperatures. From these graphs, it was noticed that the dielectric loss decreased by increasing the frequency for all investigated samples. This behavior for all polymers/clay nanocomposites prepared samples can be described by the Debye dispersion relation [28],

$$\varepsilon'' = \omega\tau\left[\left(\grave{e}_s - \grave{e}_\infty / 1 + \omega^2\tau^2\right)\right],$$

where the interfacial polarization dominates [29] at relatively low frequency range which competes the mentioned normal behavior (the rise of ε" with increasing frequency). This leads to diminish the low frequency behavior. Tareev *et al.*, [30] reported that, as the frequency increases, many types of losses get to be reduced because the field frequency begins to exceed their characteristic natural frequency. This is ascribed to the initial high value of dielectric parameters for polar materials, but as the frequency of the field is raised, the value begins to drop which may be due to the fact that the dipoles not being able to follow the field variation at higher frequencies and also due to the polarization effects.

Figures 5(a)-(d) show the dependence of a.c. conductivity on the frequency at different temperatures for (a, PMMA/MMT), (b, PSt/MMT), (c, PSt-b-PMMA/MMT) and (d, PSt-b-PMMA-b-P-n-BuMA/MMT) nanocomposites. The a.c. conductivity behavior of all prepared sam-

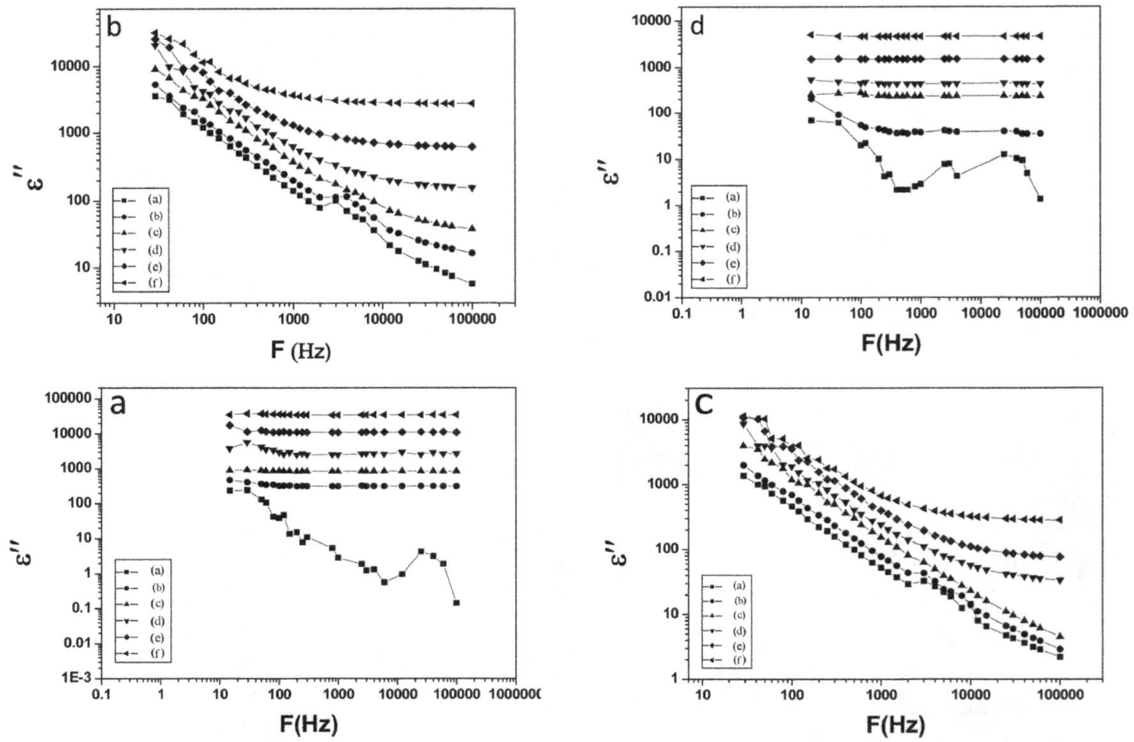

Figure 4. The dielectric constant loss ε'' as a function of frequency, (a) PMMA/MMT, (b) PSt/MMT, (c) Pst-b-PMMA/MMT, (d) Pst-b-PMMA –b-P-n-BuMA/MMT, where a = 20°C, b = 40°C, c = 60°C, d = 70°C, e = 80°C, f = 90°C.

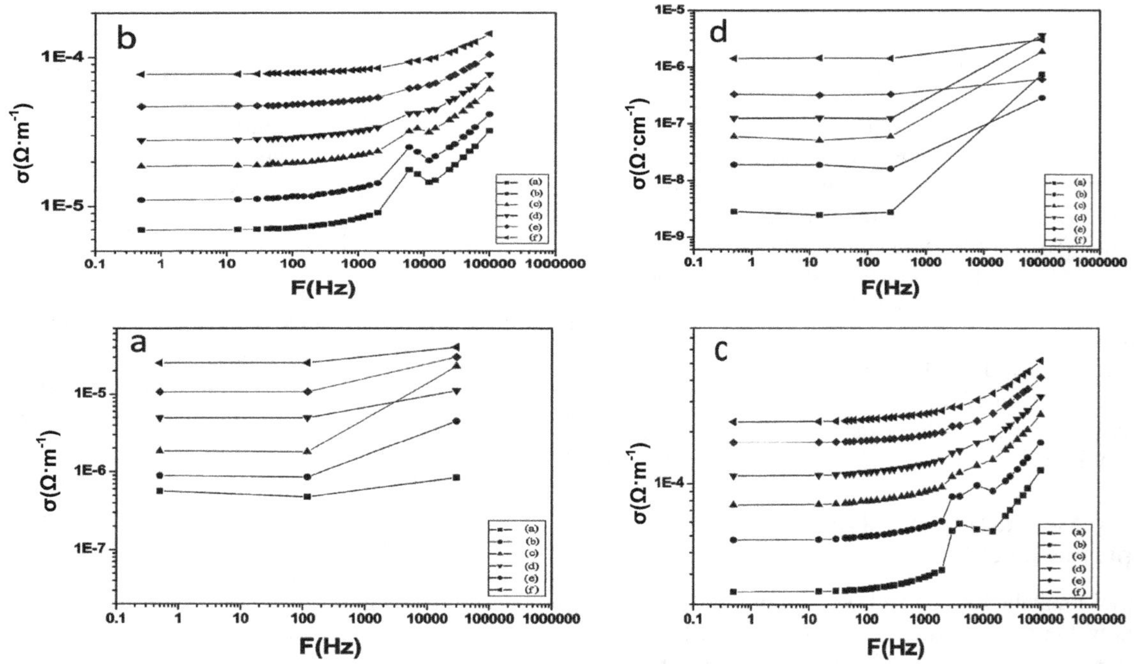

Figure 5. F as a function of a.c. conductivity $\sigma_{a.c.}$ of (a) PMMA/MMT, (b) Pst/MMT ,(c) Pst-b-PMMA/MMT, (d) Pst-b-PMMA-b-P-nBuMA/MMT, where a = 20°C, b = 40°C, c = 60°C, d = 70°C, e = 80°C, f = 90°C.

ples was investigated at both frequency and temperature ranges from (0.1 Hz up to 100KHz) and (20 up to 90°C), respectively. It was found that the a.c. conductivity for all the prepared samples remarkably increased by increasing the frequency, exhibiting nearly similar behaveior. The maximum values of the a.c. conductivity of (b, PSt/MMT) and (c, PSt-b-PMMA/MMT) exhibited higher conductivities when compared with (a, PMMA/MMT) and (d, PSt-b-PMMA-b-P-n-BuMA/MMT) nanocomposites, as shown in **Figures 5(b)** and **(c)** as a result of good intercalation of PSt and PSt-b-PMMA in montmorillonite clay and tremendous increase of the mobility of charge carriers. For the four systems, the a.c. conductivities were enhanced by increasing the temperature from 20 up to 90˚C. The samples exhibited better a.c. conduction at high temperature where the intercalation of polymers within clay interlayers was advantageous to the movement of Li^+ ions inside these interlayers which, resulted in the increase of a.c. conductivity. The lower values of a.c. conductivities at lower temperatures were attributed to two reasons [31]; 1) a few of un-intercalated polymers remained in the grain boundary of montmorillonite clay in addition to the excessive polymer in the exterior of montmorillonite which caused decrease in the a.c. conductivity at low temperature and 2) the number of charge carriers, had high relaxation time due to higher energy barrier at lower temperature and might be less in number with low barrier at higher temperature causing an increase in the a.c. conductivity [32].

4. Conclusion

Vinyl polymers/montmorillonite clay nanocomposites were prepared via in-situ-atom transfer radical polymerization (ATRP). Their dielectric properties were extensively studied to invest them in the a.c. power applications. Intercalated vinyl polymers/MMT nanocomposites gave enhanced values of the dielectric parameters such as a.c. conductivity (σ), and dielectric constant loss (ε''). For the four systems, the a.c. conductivities were enhanced by increasing the temperature from 20˚C up to 90˚C. The samples exhibited better a.c. conduction (σ) at high temperature where increasing temperature facilitated the intercalation of polymers inside clay layers which was advantageous to Li^+ ions movement inside these layers which results in the increase of a.c. conductivity.

REFERENCES

[1] M. Biswas and S. S. Ray, "Recent Progress in Synthesis and Evaluation of Polymer-Montmorillonite nanocomposites," *Advances in Polymer Science*, Vol. 155, 2001, 167-221. doi:10.1007/3-540-44473-4_3

[2] S. Pande, H. Swaruparani, M. D. Bedre, R. Bhat, R. Deshpande and A. Venkataraman, "Synthesis, Characterization and Studies of PANI-MMT Nanocomposites," *Nanoscience and Nanotechnology*, Vol. 2, No. 4, 2012, pp. 90-98. doi:10.5923/j.nn.20120204.01

[3] W. Zhai, C. B. Park and M. Kontopoulou, "Nanosilica Addition Dramatically Improves the Cell Morphology and Expansion Ratio of Polypropylene Heterophasiccopolymer Foams Blown in Continuous Extrusion," *Industrial & Engineering Chemistry Research*, Vol. 50, No. 12, 2011, pp. 7282-7289. doi:10.1021/ie102438p

[4] I. L. Hosier, A. S. Vaughan and S. G. Swingler, "On the Effects of Morphology and Molecular Composition on the Electrical Strength of Polyethylene Blends," *Journal of Polymer Science Part B: Polymer Physics*, Vol. 38, No. 17, 2000, pp. 2309-2322.

[5] F. Frutos, M. Acedo, M. Mudarra, J. Belana, J. Òrrit, J. A. Diego, J. C. Cañadas and J. Sellarès, "Effect of Annealing on Conductivity in XLPE Mid-Voltage Cable Insulation," *Journal of Electrostatics*, Vol. 65, No. 2, 2007, pp. 122-131.

[6] P. Dubois and M. Alexandre, "Performant Clay/Carbon Nanotube Polymer Nanocomposites," *Journal of Advanced Engineering Materials*, Vol. 8, No. 3, 2006, pp. 135-221.

[7] T. Tanaka, M. Kazako and N. Fuse, "Dielectric Nanocomposites with Insulating Properties," *IEEE Transactions on Dielectrics and Electrical Insulation*, Vol. 12, 2005, pp. 669-681.

[8] C. C. Reddy and T. S. Ramu, "Investigation of Space Charge Distribution and Volume Resistivity of XLPE/MgO Nanocomposite Material under DC Voltage Application," *Transactions on Dielectrics and Electrical Insulation*, Vol. 15, No. 1, 2008, pp. 310- 322.

[9] M. C. Lança, M. Fu, E. Neagu, L. A. Dissado, J. M. Mendes, A. Tzimas and S. Zadeh, "Space Charge Analysis of Electrothermally Aged XLPE Cable Insulation," *Journal of Non-Crystalline Solids*, Vol. 353, No. 47-51, 2007, pp. 4462-4466.

[10] M. Mukherjee, B. Mukherjee, Y. Choi, K. Sim, J. Do and S. Pyo, "Investigation of Organic n-Type Field Effect Transistor Performance on the Polymeric Gate Dielectrics," *Synthetic Metals*, Vol. 160, No. 5-6, 2010, pp. 504-509.

[11] S. H. Sonawane, "Removal of Brilliant Green from Wastewater Using Conventional and Ultrasonically Prepared Poly(acrylic acid) Hydrogel Loaded with Kaolin Clay: A Comparative Study," *Ultrasonics Sonochemistry*, Vol. 20, No. 3, 2013, pp. 914-923.

[12] Y. Maekawa, T. Yamanaka, T. Kimura, Y. Murat, S. Katakai, "The Hitachi Densen," 2002.

[13] M. F. J. Solomon, Y. Bhole and A. G. Livingston, "High Flux Hydrophobic Membranes for Organic Solvent Nano-Filtration (OSN)—Interfacial Polymerization, Surface Modification and Solvent Activation," *Journal of Membrane Science*, Vol. 434, 2013, pp. 193-203.

[14] P. S. Suchithra, L. Vazhayal, A. P. Mohamed and S. Ananthakumar, "Mesoporous Organic-Inorganic Hybrid Aerogels through Ultrasonic Assisted Sol-Gel Intercalation of Silica-PEG in Bentonite for Effective Removal of Dyes, Volatile Organic Pollutants and Petroleum Products from Aqueous Solution," *Chemical Engineering Journal*, Vol. 200-202, 2012, pp. 589-600. doi:10.1016/j.cej.2012.06.083

[15] Z. Chen, X. Chen, Z. Shao, Z. Yao and L. T. Biegler, "Parallel Calculation Methods for Molecular Weight Distribution of Batch Free Radical Polymerization," *Compu-*

ters and Chemical Engineering, Vol. 48, 2013, pp. 175-186. doi:10.1016/j.compchemeng.2012.09.002

[16] S. Sadhu and A. K. Bhowmick, "Preparation and Characterization of Styrene Butadiene Rubber Based Nanocomposites and Study of their Mechanical Properties," *Advanced Engineering Materials*, Vol. 6, No. 9, 2004, pp. 738-742.

[17] B. Charleux, "Review Nitroxide-Mediated Polymerization," *Progress in Polymer Science*, Vol. 38, No. 1, 2013, pp. 63-235. doi:10.1016/j.progpolymsci.2012.06.002

[18] A. B. Gomide, C. H. Thome, G. A. dos Santos and G. A. Ferreira, "Disrupting Membrane Raft Domains by Alkyl-Phospholipids," *Biochimica et Biophysica Acta*, Vol. 1828, No. 5, 2013, pp. 1384-1389.

[19] G. D. Fu, Z. Shang, L. Hong, E. T. Kang and K. G. Neoh, "Preparation of Cross-Linked Polystyrene Hollow Nanospheres via Surface-Initiated Atom Transfer Radical Polymerizations," *Macromolecules*, Vol. 38, No. 18, 2005, pp. 7867-7871. doi:10.1021/ma0509098

[20] K. Terashima, "Insulation of Cables Using Pulsed Electroacoustic Method," *Transactions on Power Delivery*, Vol. 13, 1998, pp. 7-16. doi:10.1109/61.660837

[21] Y. Murata, Y. Sekiguchi, Y. Inoue and M. kanaoka, *Electrical Insulating Materials*, Vol. 2, 2005, pp. 650-653.

[22] Y. Cao *et al.*, "The Future of Nanodielectrics in the Electrical Power Industry," *IEEE Transactions on Dielectrics and Electrical Insulation*, Vol. 11, No. 5, 2004, pp. 797-807.

[23] M. Xu, Y. S. Choi, Y. K. Kim, K. H. Wang and I. J. Chung, "Synthesis and Characterization of Exfoliated Poly(styrene-co-methyl methacrylate)/Clay Nanocomposites via Emulsion Polymerization with AMPS," *Polymer*, Vol. 44, No. 12, 2003, pp. 6387-6395.

[24] H. Datta, N. K. Singha and A. K. Bhowmick, "Beneficial Effect of Nanoclay in Atom Transfer Radical Polymerization of Ethylacrylate: A One Pot Preparation of Tailor Made Polymer Nanocomposites," *Macromolecules*, Vol. 41, No. 1, 2010, pp. 50-57. doi:10.1021/ma071528s

[25] A. Amin, R. Sarkar, C. N. Moorefield and G. R. Newkome, "Synthesis of Polymer-Clay Nanocomposites of Some Vinyl Monomers by Surface Initiated Atom Transfer Radical Polymerization," *Designed Monomers and Polymers*. doi:10.1080/15685551.2013.771304

[26] S. Singha and M. J. Thomas, "Permittivity and Tan Delta Characteristics of Epoxy Nano Composites," *IEEE Transactions on Electrical Insulation*, Vol. 15, No. 1, 2008, pp. 2-11. doi:10.1109/T-DEI.2008.4446731

[27] N. M. Renukappa and G. Swaminathan, "Experimental Investigation of the Influence of Clay on Dielectric Properties of Epoxy Nanocomposites," *Proceedings of the 9th International Conference on Properties and Applications of Dielectric Materials*, 19-23 July 2009, Harbin, pp. 833-836.

[28] P. Debye, "Polar Molecules," Dover, New York, 1945.

[29] A. K. Jonscher, "The 'Universal' Dielectric Response," *Nature*, Vol. 267, No. 5613, 1977, pp. 673-679. doi:10.1038/267673a0

[30] B. Tareev, "Physics of Dielectric Materials," Mir Publications, Moscow, 1979.

[31] W. Chen, Q. Xu and R. Z. Yuan, "Effect on Ionic Conductivity with Modification of Polymethylmethacrylate in Poly(ethylene oxide)-Layered Silicate Nanocomposites (PLSN)," *Materials Science and Engineering*, Vol. 77, No. 1, 2000, pp. 15-18. doi:10.1016/S0921-5107(00)00448-7

[32] I. I. Prepechko, "An Introduction to Polymer Physics," Mir Publications, Moscow, 1981.

Approximate Electromagnetic Cloaking of a Dielectric Sphere Using Homogeneous Isotropic Multi-Layered Materials

Hany M. Zamel[1], Essam El Diwany[1], Hadia El Hennawy[2]

[1]Electronics Research Institute (ERI), Microwave Engineering Department, Cairo, Egypt; [2]Faculty of Engineering, Ain Shams University, Cairo, Egypt.
Email: res_ass@yahoo.com

ABSTRACT

In cloaking, a body is hidden from detection by surrounding it by a coating consisting of an unusual anisotropic nonhomogeneous material. The permittivity and permeability of such a cloak are determined by the coordinate transformation of compressing a hidden body into a point or a line. The radially-dependent spherical cloaking shell can be approximately discretized into many homogeneous anisotropic layers; each anisotropic layer can be replaced by a pair of equivalent isotropic sub-layers, where the effective medium approximation is used to find the parameters of these two equivalent sub-layers. In this work, the scattering properties of cloaked dielectric sphere is investigated using a combination of approximate cloaking, where the dielectric sphere is transformed into a small sphere rather than to a point, together with discretizing the cloaking material using pairs of homogeneous isotropic sub-layers. The back-scattering normalized radar cross section, the scattering patterns are studied and the total scattering cross section against the frequency for different number of layers and transformed radius.

Keywords: Approximate Cloaking; Dielectric Sphere; Cloaking by Layered Isotropic Materials

1. Introduction

Recently, the concept of electromagnetic cloaking has drawn considerable attention concerning theoretical, numerical and experimental aspects [1-8]. One approach to achieve electromagnetic cloaking is to deflect the rays that would have struck the object, guide them around the object, and return them to their original trajectory, thus no waves are scattered from the body [1]. In the coordinate transformation method for cloaking, the body to be hidden is transformed virtually into a point (3D or spherical configuration) or a line (2D or cylindrical configuration), and this transformation leads to the profile of ε, μ in the cloaking coating. Some components of the electrical parameters of the cloaking material (ε, μ) are required to have infinite or zero value at the boundary of the hidden object. This requires the use of metamaterials which can produce such values, however, they are narrow band since they rely on using array of resonant elements (as split ring resonators) [9-12].

Approximate cloaking can be achieved by transforming the hidden body virtually into a small object rather than a point or a line as shown in **Figure 1**, which eliminates the zero (point transformed) or infinite (line transformed) values of the electrical parameters [13,14]. This, however, leads to some scattering, since the hidden body is virtually transformed into a small object rather than a point or a line, and the scattering decreases as the transformed sphere radius is smaller.

The radially-dependent spherical cloaking shell can be approximately discretized into many homogeneous anisotropic layers, provided that the thickness of each layer is much less than the wavelength, and this discretization decreases the level of scattering as the number of layers increases. Each anisotropic layer can be replaced by a pair of equivalent isotropic sub-layers, where the effective medium approximation is used to find the parameters of these two equivalent sub-layers [15,16]. Near the boundary of the hidden sphere, the values of the radial components of ε, μ of the cloaking material are nearly zero. This makes the values of ε, μ in one of the pair of isotropic layers to be very small, and can be implemented using metamaterials [11,17,18]. Two approaches are possible in choosing the material properties in the pair of sub-layers. The first is to take the smaller values of ε, μ to be in one layer and the larger values to be in the second layers [19], whereas the other possibility is to take the smaller value of ε together with the larger value of μ in one sub-layer, and the opposite combination in the other sub-layer [20]. The scattering properties of cloaked conducting sphere [21] and conducting cylinder [22] were investigated using a combination of approximate cloaking, where the conducting sphere is transformed into a small sphere rather than a point and the conducting cylinder is transformed into small cylinder rather than a line, together with discretizing the cloaking material using pairs of homogeneous isotropic sub-layers.

In this work, the scattering properties of cloaked dielectric sphere is investigated using a combination of approximate cloaking, where the dielectric sphere is transformed into a small sphere rather than a point, together with discretizing the cloaking material using pairs of homogeneous isotropic sub-layers. The solution is obtained by rigorously solving Maxwell equations using Mie series expansion. The back-scattering normalized radar cross section, the scattering cross section patterns are studied and the total scattering cross section against the frequency for different numbers of layers and the transformed radius.

2. Design Parameters of the Approximate Spherical Cloak

Perfect spherical cloak can be constructed by compressing the electromagnetic fields in a spherical region $r' \le R_2$ into a spherical shell $R_1 \le r \le R_2$ as shown in **Figure 1**. The coordinate transformation is $r' = f(r)$, with $f(R_1) = 0$ for perfect cloaking or $f(R_1) = c$ for approximate cloaking and $f(R_2) = R_2$ [14]. The radial and transverse permittivity and permeability of the spherical cloak, depending on r, are given as [23]:

$$\frac{\varepsilon_r}{\varepsilon_0} = \frac{\mu_r}{\mu_0} = \frac{f(r)^2}{r^2 f'(r)}, \frac{\varepsilon_\theta}{\varepsilon_0} = \frac{\mu_\theta}{\mu_0} = \frac{\varepsilon_\varnothing}{\varepsilon_0} = \frac{\mu_\varnothing}{\mu_0} = f'(r) \quad (1)$$

A linear transformation is usually used, given for approximate cloaking by (for ideal cloaking c =0) [14], [24]:

$$f(r) = r' = \frac{1}{(R_2 - R_1)}\left[r(R_2 - c) + R_2(c - R_1)\right] \quad (2)$$

Thus, the permittivity and permeability of the approximate spherical cloak are given from the above equations by:

$$\frac{\varepsilon_r}{\varepsilon_0} = \frac{\mu_r}{\mu_0} = \frac{\left[r(R_2 - c) + R_2(c - R_1)\right]^2}{r^2 (R_2 - R_1)(R_2 - c)} \quad (3)$$

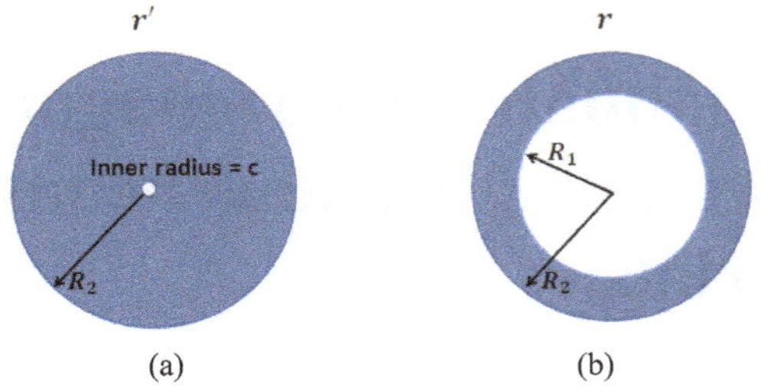

Figure 1. (a) Virtual domain, (b) Actual domain.

$$\frac{\varepsilon_t}{\varepsilon_0} = \frac{\mu_t}{\mu_0} = \frac{(R_2 - c)}{(R_2 - R_1)} \quad (4)$$

at $r = R_1$,

$$\frac{\varepsilon_r}{\varepsilon_0} = \frac{c^2 (R_2 - R_1)}{R_1^2 (R_2 - c)} \quad (5)$$

Thus, the permittivity at $r = R_1$ decreases as the cloaked radius c and the shell thickness decrease with values less than unity. At $r = R_2$,

$$\frac{\varepsilon_r}{\varepsilon_0} = \frac{(R_2 - R_1)}{(R_2 - c)} \quad (6)$$

3. The Discretization of an Anisotropic Nonhomogeneous Spherical Cloak Using Isotropic Layered Materials

We first discretize the cloaking anisotropic and nonhomogeneous shell into $2M$ homogenous layers with identical thickness ($d_B = d_A = d$) as shown in **Figure 2**, and then the outer radius of each sub-layer is:

$$r_{i+1} = R_1 + i\frac{R_2 - R_1}{2M}, \cdots i = 1, 2, \cdots, 2M \quad (7)$$

An anisotropic layer can be replaced by a pair of equivalent isotropic sub-layers, where the effective medium approximation is used to find the parameters of these two equivalent sub-layers, **Figure 2** [19,20]. The radial parameter σ_r (relative ε or μ) in the anisotropic layer can be considered as a series combination of the parameters of the equivalent sublayers σ_1, σ_2 (considered as capacitors) related by:

$$\frac{1}{\sigma_r} = \frac{1}{2\sigma_1} + \frac{1}{2\sigma_2} \quad (8)$$

The tangential parameters σ_t in the anisotropic layer can be considered as a parallel combination of the parameters of the equivalent sub-layers σ_1, σ_2 related by:

$$\sigma_t = (\sigma_1 + \sigma_2)/2 \quad (9)$$

One can obtain the equivalent medium parameters for the isotropic sub-layers as:

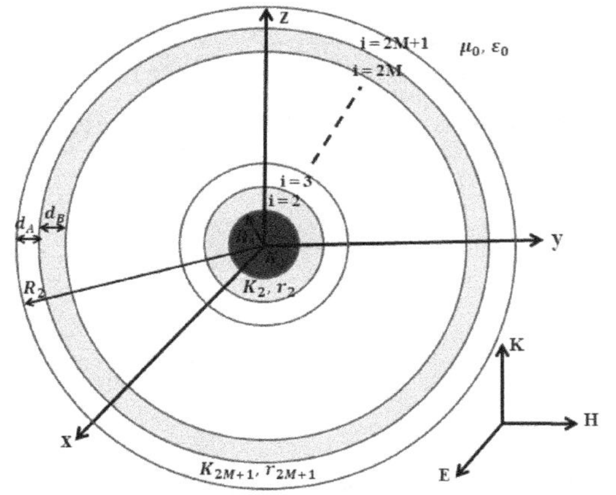

Figure 2. Plane wave scattering by a multi-layer dielectric sphere.

$$\sigma_1 = \sigma_t + \sqrt{\sigma_t^2 - \sigma_t\sigma_r} \tag{10}$$

$$\sigma_2 = \sigma_t - \sqrt{\sigma_t^2 - \sigma_t\sigma_r} \tag{11}$$

where σ_r, σ_t are given by Equations (3) and (4). σ_2 is small, particularly near the conducting sphere where σ_r is nearly zero. The value of r for a discretized layer is taken at the average radius of this layer *i.e.* at the interface of layers A and B.

Two approaches are possible in choosing the material properties in the pair of sub-layers. The first is to take the larger values of ε, μ to be in one sub-layer, *i.e.* given by Equation (11), and the smaller values in the other sub-layer from Equation (10) (case I). For this case, the impedances in the layers are the same, but the refractive index (the square root of the product of the relative ε, μ), suffers from strong jumps at the successive layers. The other possibility is to take the smaller value of μ, Equation (11), together with the larger value of ε, Equation (10), in one sub-layer, and the opposite combination in the other sub-layer (case II) [20]. In this case, the impedances of the successive layers suffer from jumps, but the refractive index profile changes continuously.

4. Scattering from a Dielectric Sphere with Multi-Layered Coating

The configuration for electromagnetic scattering by a dielectric sphere coated by $2M$ layers is shown in **Figure 2**. The external radius, permittivity, and permeability of the core and the layers are denoted by a_i, ε_i and μ_i (i = 1, 2,$\cdots$, $2M+1$), respectively. **Figure 2** shows an E_x polarized plane wave with amplitude E_0, $E^i = E_o e^{-jK_0 z}\bar{x}$, incident upon the coated sphere along the $\hat{z}$ direction. $k_0 = \omega\sqrt{\mu_0\varepsilon_0}$ is the wave number in free space. The time dependence $e^{j\omega t}$ is suppressed.

The fields in the different regions are expanded in terms of spherical harmonics of TE_r and TM_r modes w.r.t. the radial directions. The field (E or H) with only transverse components (θ, $\varnothing$) is expressed by the harmonics m, whereas the other field having the three components is expressed by the harmonics n [25,26].

$$m_e^o = \pm\frac{1}{\sin\theta}J_n(kr)P_n^1(\cos\theta)\begin{matrix}\cos\\ \sin\end{matrix}\varnothing\hat{\theta} - J_n(kr)\frac{\partial P_n^1}{\partial\theta}\begin{matrix}\sin\\ \cos\end{matrix}\varnothing\hat{\varnothing} \tag{12}$$

$$n_e^o = \frac{n(n+1)}{kr}J_n(kr)P_n^1(\cos\theta)\begin{matrix}\sin\\ \cos\end{matrix}\varnothing\hat{r} + \frac{1}{kr}\left[krJ_n(kr)\right]'\frac{\partial P_n^1}{\partial\theta}\begin{matrix}\sin\\ \cos\end{matrix}\varnothing\hat{\theta} \pm \frac{1}{kr\sin\theta}\left[krJ_n(kr)\right]'P_n^1(\cos\theta)\begin{matrix}\cos\\ \sin\end{matrix}\varnothing\hat{\varnothing} \tag{13}$$

where $J_n(kr)$ is a spherical Bessel function of order n of any kind and $P_n^1(\cos\theta)$ is a Legendre polynomial of the first kind, first order and n[th] degree. The $\varnothing$-dependnce is taken as sin$\varnothing$ or cos$\varnothing$ (odd, *o*, and even, *e*) to conform with the incident plane wave.

The incident plane wave can be expressed in terms of spherical harmonics with Bessel functions of the first kind as:

$$E^i = E_o e^{-jK_0 z}\bar{x} = \sum\nolimits_{n=1} G_n\left(m_o^i + jn_e^i\right) \tag{14}$$

$$H^i = -\sum\nolimits_{n=1}\frac{G_n}{Z_0}\left(m_e^i - jn_o^i\right) \tag{15}$$

where

$$G_n = (-j)^n\frac{2n+1}{n(n+1)}E_o \tag{16}$$

and $Z_0 = \sqrt{\mu_0/\varepsilon_0}, j = \sqrt{-1}$.

The scattered fields can be expanded in terms of spherical harmonics with spherical Hankel functions $h_n^2(k_0 r)$ representing scattered outgoing waves with unknown coefficients e^s, d^s of the TE, TM parts, respectively.

$$E^s = \sum\nolimits_{n=1} G_n\left(e^s m_o^{sH} + jd^s n_e^{sH}\right) \tag{17}$$

$$H^s = -\sum\nolimits_{n=1}\frac{G_n}{Z_0}\left(d^s m_e^{sH} - je^s n_o^{sH}\right) \tag{18}$$

We can write the fields in the ith Layer as:

$$E^{ci} = \sum\nolimits_{n=1} G_n\left(d^{cH^1 i}m_o^{cH^1} + je^{cH^1 i}n_e^{cH^1} + d^{cH^2 i}m_o^{cH^2} + j\,e^{cH^2 i}n_e^{cH^2}\right) \tag{19}$$

$$H^{ci} = -\sum_{n=1} \frac{G_n}{Z_0} \begin{pmatrix} e^{cH^1 i} m_e^{cH^1} - j d^{cH^1 i} n_0^{cH^1} \\ + e^{cH^2 i} m_e^{cH^2} - j d^{cH^2 i} n_o^{cH^2} \end{pmatrix} \quad (20)$$

where the modes in the cloak region are represented by Hankel functions of the first and second kinds with arguments ($k_i r_i$), where $k_i = \omega\sqrt{\mu_i \in_i}$, and unknown coefficients d, e. The boundary conditions at the interface between layers i, $i + 1$ leads to two equations relating the TM coefficients d^{cH^1}, d^{cH^2} of the two layers, and two equations relating the TE coefficients e^{cH^1}, e^{cH^2} of the two layers. The finiteness of the field in the dielectric core leads to the following ratios in the dielectric core [27]:

$$\frac{d^{cH^2(1)}}{d^{cH^1(1)}} = 1, \frac{e^{cH^2(1)}}{e^{cH^1(1)}} = 1 \quad (21)$$

The ratios $d^{cH^2(i)}/d^{cH^1(i)}$ and $e^{cH^2(i)}/e^{cH^1(i)}$ in the successive larger layers can be obtained iteratively from the following equations [27]:

$$\frac{d^{cH^2(i+1)}}{d^{cH^1(i+1)}} = -\frac{H_n^1(K_{i+1}r_i) - R_H^i H_n^{1'}(K_{i+1}r_i)}{H_n^2(K_{i+1}r_i) - R_H^i H_n^{2'}(K_{i+1}r_i)}, \quad (22)$$
$$i = 1, 2, \cdots 2M$$

$$\frac{e^{cH^2(i+1)}}{e^{cH^1(i+1)}} = -\frac{H_n^1(K_{i+1}r_i) - R_E^i H_n^{1'}(K_{i+1}r_i)}{H_n^2(K_{i+1}r_i) - R_E^i H_n^{2'}(K_{i+1}r_i)}, \quad (23)$$
$$i = 1, 2, \cdots 2M$$

where $H_n^1(KR) = KRh_n^1(KR)$ and $H_n^2(KR) = KRh_n^2(KR)$ are the Riccati-Hankel functions.

$$R_H^i = \sqrt{\frac{\mu_{i+1}\epsilon_i}{\epsilon_{i+1}\mu_i}} \frac{H_n^1(K_i r_i) + \frac{d^{cH^2(i)}}{d^{cH^1(i)}} H_n^2(K_i r_i)}{H_n^{1'}(K_i r_i) + \frac{d^{cH^2(i)}}{d^{cH^1(i)}} H_n^{2'}(K_i r_i)}, \quad (24)$$
$$i = 1, 2, \cdots 2M$$

$$R_E^i = \sqrt{\frac{\mu_i\epsilon_{i+1}}{\epsilon_i\mu_{i+1}}} \frac{H_n^1(K_i r_i) + \frac{e^{cH^2(i)}}{e^{cH^1(i)}} H_n^2(K_i r_i)}{H_n^{1'}(K_i r_i) + \frac{e^{cH^2(i)}}{e^{cH^1(i)}} H_n^{2'}(K_i r_i)}, \quad (25)$$
$$i = 1, 2, \cdots 2M$$

Finally, the boundary conditions between the outer layer and air lead to the scattering coefficients $b_n = d^s$(TM part) and $a_n = e^s$(TE part):

$$b_n = -\frac{j_n(K_0R_2) - R_H^{2M+1} j_n'(K_0R_2)}{H_n^2(K_0R_2) - R_H^{2M+1} H_n^{2'}(K_0R_2)} \quad (26)$$

$$a_n = -\frac{j_n(K_0R_2) - R_E^{2M+1} j_n'(K_0R_2)}{H_n^2(K_0R_2) - R_E^{2M+1} H_n^{2'}(K_0R_2)} \quad (27)$$

The scattering cross section σ_s and the normalized radar cross sections $Q_{(\theta,\varphi)}$ are given by [28,29]:

$$\sigma_s = 4\pi r^2 \frac{|E^s|^2}{|E^i|^2} \quad (28)$$

$$Q_{(\theta,\varphi)} = \frac{\sigma_s}{\pi R_1^2} = \frac{4}{(K_0R_1)^2}\left(|S_1(\theta)|^2 \sin^2\varphi + |S_2(\theta)|^2 \cos^2\varphi\right) \quad (29)$$

where $S_1(\theta)$ and $S_2(\theta)$ are defined by :

$$S_1(\theta) = \sum_n \frac{(2n+1)}{n(n+1)}\left[a_n\pi_n(\theta) + b_n\tau_n(\theta)\right] \quad (30)$$

$$S_2(\theta) = \sum_n \frac{(2n+1)}{n(n+1)}\left[b_n\pi_n(\theta) + a_n\tau_n(\theta)\right] \quad (31)$$

In the above two equations $\pi_n(\theta)$ and $\tau_n(\theta)$ describe the angular scattering patterns of the spherical harmonics used to describe S_1 and S_2 and follow from the recurrence relations [29,30]:

$$\pi_n = \frac{P_n^1(\cos\theta)}{\sin\theta} = \frac{2n-1}{n-1}\cos\theta \cdot \pi_{n-1} - \frac{n}{n-1}\pi_{n-2} \quad (32)$$

$$\tau_n = \frac{dP_n^1(\cos\theta)}{d\theta} = n \cdot \cos\theta \cdot \pi_n - (n+1)\pi_{n-1} \quad (33)$$

starting with the initial values: $\pi_0 = 0$, $\pi_1 = 1$.

The total scattering normalized cross section Q_{sca} follows from the integration of the scattered power over all directions θ, φ, given by [31]:

$$Q_{sca} = \frac{2}{(K_0R_1)^2}\sum_{n=1}^{\infty}(2n+1)\left(|a_n|^2 + |b_n|^2\right) \quad (34)$$

The backscattering normalized radar cross section Q_b, applicable to monostatic radar, is given by [31]:

$$Q_b = \frac{1}{(K_0R_1)^2}\left|\sum_{n=1}^{\infty}(-1)^n(2n+1)(a_n - b_n)\right|^2 \quad (35)$$

The mode series is truncated at the mode number $n_m = K_0R_2 + 4 \times (K_0R_2)^{1/3} + 2$ [29,30].

5. Results

To check the above analysis, the scattering pattern and the total scattering cross section are calculated for scattering by a dielectric sphere and compared with References [27,32], and for a lossy dielectric sphere coated by a lossy dielectric layer and compared with Reference [28], leading to identical results.

The scattering properties of cloaked dielectric spheres (ε_d = 2, 5) coated with isotropic homogenous layers are investigated concerning the normalized total scattering cross section versus the normalized frequency k_0 and for the bistatic radar cross section (RCS) with varying the reduced radius c and the number of isotropic cloaking layers with $R_2 = 2R_1$.

5.1. Normalized Total Scattering Cross Section

For the ideal cloaking case (c = 0) the results of the total scattering for dielectric spheres (for both case I and case II) with relative permitivities ε_d = 2, ε_d = 5, are found to be greatly identical with the results for a cloaked conducting sphere, shown in **Figure 3** (φ = 0 [20,21]). This can be attributed to the fact that the cloaking properties for ideal cloaking (c = 0) is independent of the properties of the cloaked object (conductor or dielectric), since the incident field cannot penetrate to the cloaked object [23].

On the other hand, for approximate cloaking $(c \neq 0)$ the cloaked body is not transformed to a point (as in ideal cloaking) but to a finite sphere. For a conducting sphere with $c = R_1/2$ the variation of the total scattering with frequency takes nearly the shape of the scattering by a conducting sphere but with reduced level as shown in **Figure 3** [21]. As c decreases ($c = R_1/5$, $R_1/10$, $R_1/40$, the scattering approaches that of the ideal case (c = 0)).

For the ideal profile (c = 0), $\varepsilon_r/\varepsilon_0 = 0$ at $r = R_1$, Equation (5). As c increases, $\varepsilon_r/\varepsilon_0$ at $r = R_1$ increases. For $c = R_1/40$, $\varepsilon_r/\varepsilon_0 \cong 1/3200$. For $c = R_1/10$, $\varepsilon_r/\varepsilon_0 \cong 1/200$.

Figures 4 and **5** show the effect of the reduced radius c on the normalized total scattering cross section versus frequency for cloaked dielectric spheres with relative permitivities ε_d = 2 and ε_d = 5, respectively, for case II with $2M$ = 40. Compared with the cloaking for a conducting sphere, **Figure 3**, the scattering from a dielectric sphere shows the presence of some peaks, which decrease as c decreases. Such peaks may be due to the presence of multiple reflections effects in a dielectric sphere, compared with a conducting sphere [28]. The behavior for ε_d = 2 and ε_d = 5 for $c = R_1/40$ is nearly identical with the conducting sphere, since it approaches the ideal profile, and has variations for $c = R_1/10$, **Figures 3-5**. As the frequency increases the total scattering increases because the ratio of the layer thickness with respect the wavelength increases. At the lower frequencies, the scattering decreases as c decreases, which is the most useful region with low scattering. As the frequency increases the cases with larger c show less scattering, but this region is not useful because the level of scattering is high. This behavior with the variation of c is due the multiple reflections and the interaction effects [20].

Figures 6 and **7** show the effect of the number of layers

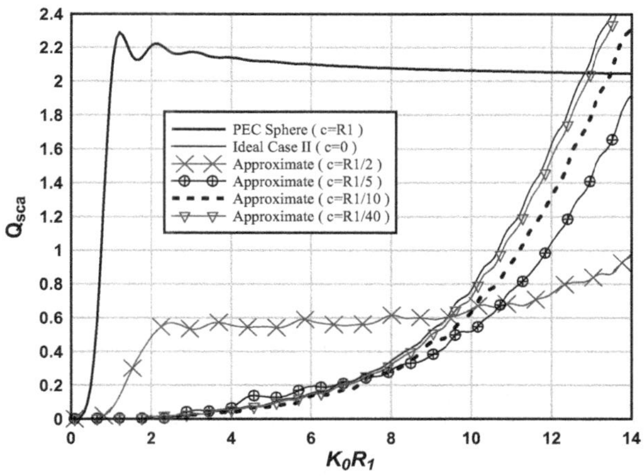

Figure 3. The normalized total scattering cross section of cloaked conducting sphere for different values of c (40-Layers) [21].

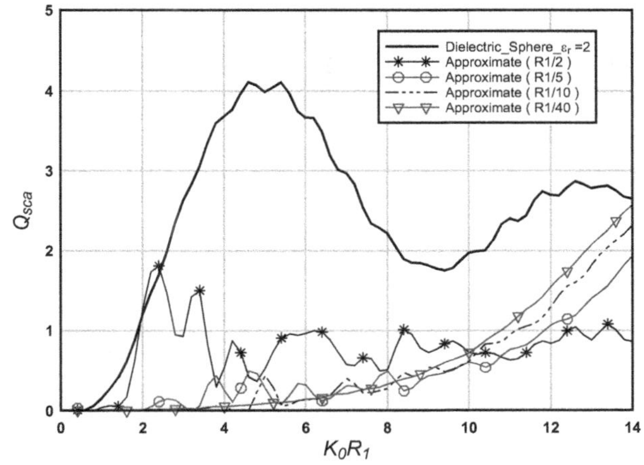

Figure 4. The normalized total scattering cross section of cloaked dielectric sphere ε_d = 2, with multi-layered isotropic structure for different values of c, case II, $2M$ = 40.

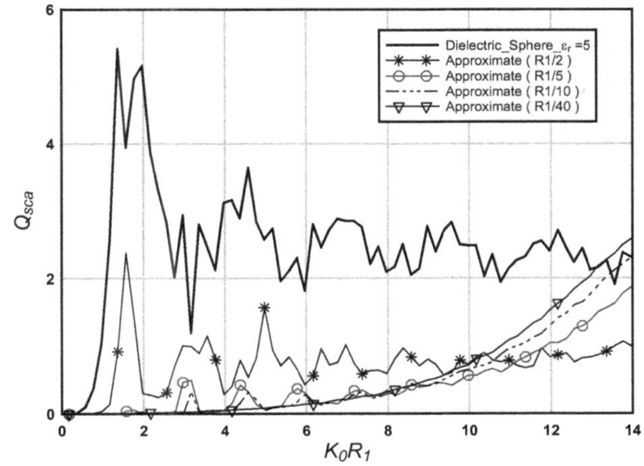

Figure 5. The normalized total scattering cross section of cloaked dielectric sphere ε_d = 5, for different values of c, case II, $2M$ = 40.

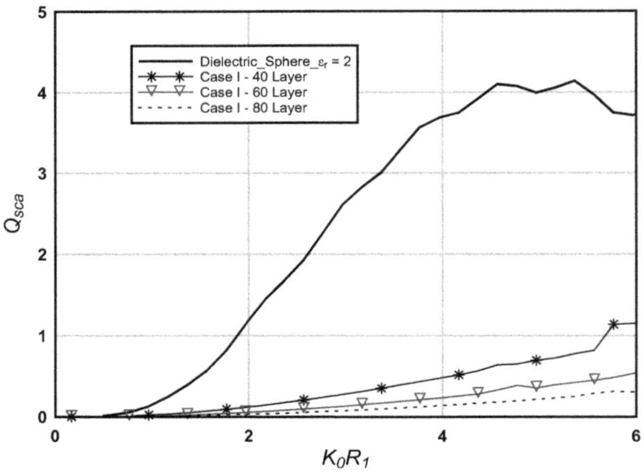

Figure 6. The normalized total scattering cross section for cloaked dielectric sphere with ε_d = 2 for different numbers of layers, case I , $c = R_1/40$.

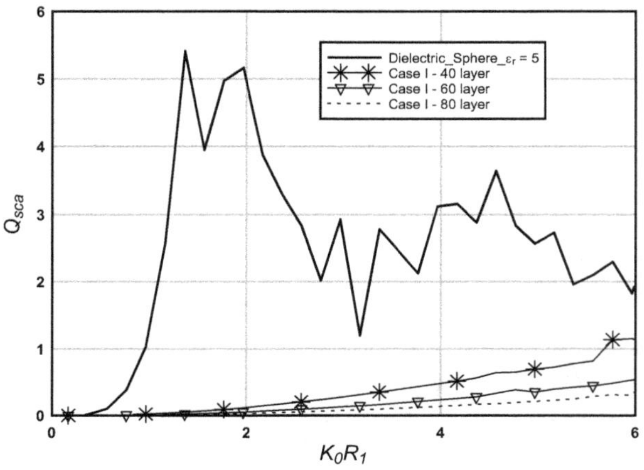

Figure 7. The normalized total scattering cross section for cloaked dielectric sphere with ε_d = 5 for different numbers of layers, case I , $c = R_1/40$.

on the normalized total scattering cross section versus frequency for case I, for different values of the dielectric constant $\varepsilon_d = 2$ and $\varepsilon_d = 5$, respectively, and $c = R_1/40$. The total scattering level is high compared with case II, **Figures 4** and **5**, since the total scattering results mainly from the forward scattering, which is high for case I, **Figures 8** and **9** [20]. The reduction of scattering does not extend beyond $k_0R_1 = 6$. As the number of layers increases the scattering decreases. The behavior for $\varepsilon_d = 2$, $\varepsilon_d = 5$ is nearly the same.

5.2. Bistatic RCS

Figures 8 and **9** show the effect of the reduced radius c on the bistatic RCS for $\varepsilon_d = 2,5$, respectively, for cases I, II, $R_1 = \lambda$ and $f = 2$ GHz. The scattering pattern for case II for the ideal profile $c = 0$ is identical with that for the conducting sphere, **Figure 10**, and deviates slightly for

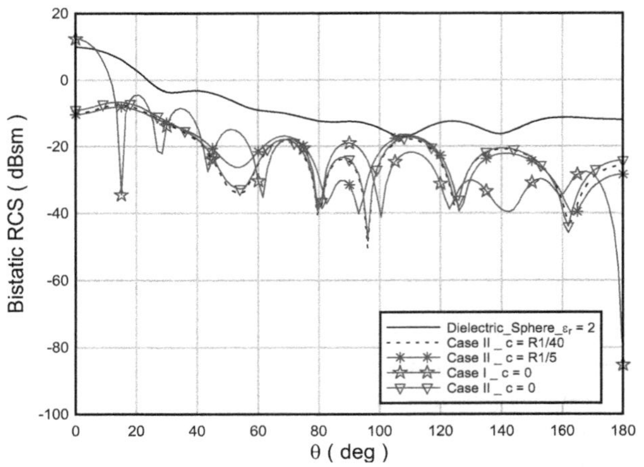

Figure 8. Bistatic RCS of cloaked dielectric sphere with ε_d = 2 for different values of c, $2M = 40$.

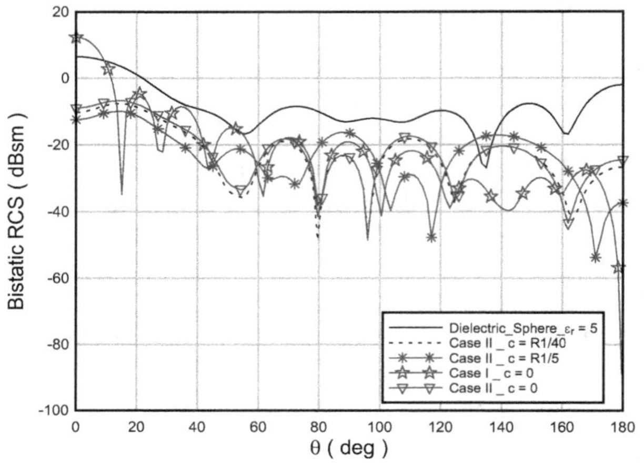

Figure 9. Bistatic RCS of cloaked dielectric sphere with ε_d = 5 for different values of c, $2M = 40$.

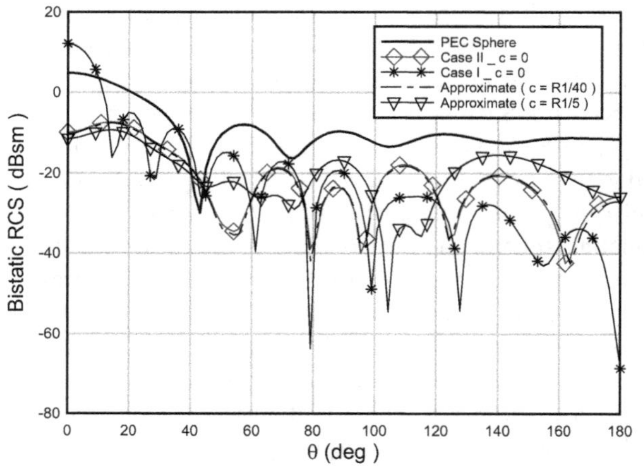

Figure 10. Bistatic RCS of cloaked conducting sphere for different values of c, $2M = 40$ [21].

$c = R_1/40$ and deviates more for $c = R_1/5$, particularly at the large angles (at and near back scattering, $\theta = 180°$).

Case I with constant impedance in the layers, shows clear end-fire scattering behavior (forward, $\theta = 0°$) corresponding to phases of radiating elements as the phase of the incident wave [33]. For case I the scattering pattern for ideal cloaking profile deviates from that of the conducting sphere for $\theta > 110°$. For case II the forward scattering ($\theta = 0°$) is lower than case I [20]. On the other hand the back scattering is lower for case I than case II. As shown in **Figures 11** and **12** for case I, the pattern for $c = R_1/40$ deviates from the ideal profile also for $\theta > 110°$.

Figures 13 and **14** show the effect of the number of layers on the scattering patters for case II, for $\varepsilon_d = 2, 5$, respectively, $c = R_1/40$. The results are nearly identical with the conducting sphere [21]. It can be seen that, when the number of layers increases (each isotropic layer is thinner), the scattered field decreases.

6. Conclusion

The scattering properties of a cloaked dielectric sphere

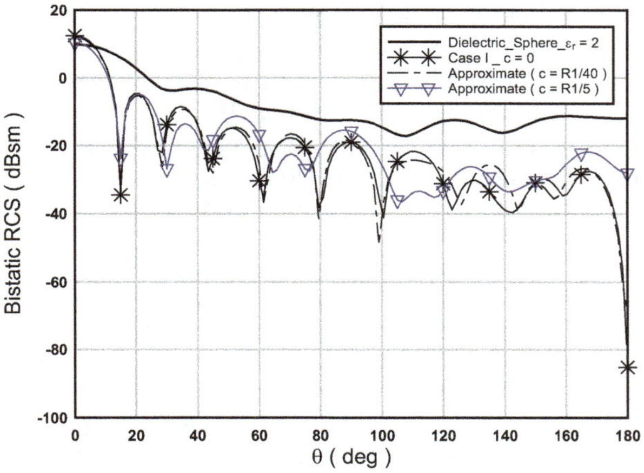

Figure 11. Bistatic RCS of cloaked dielectric sphere for different values of c for case I at $\varepsilon_d = 2$, $2M = 40$.

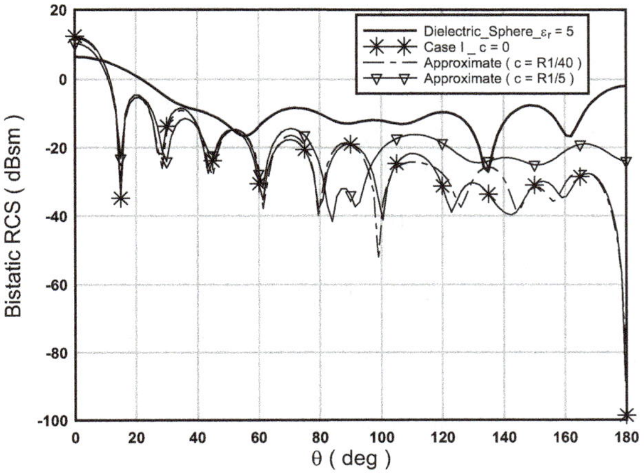

Figure 12. Bistatic RCS of cloaked dielectric sphere for different values of c for case I at $\varepsilon_d = 5$.

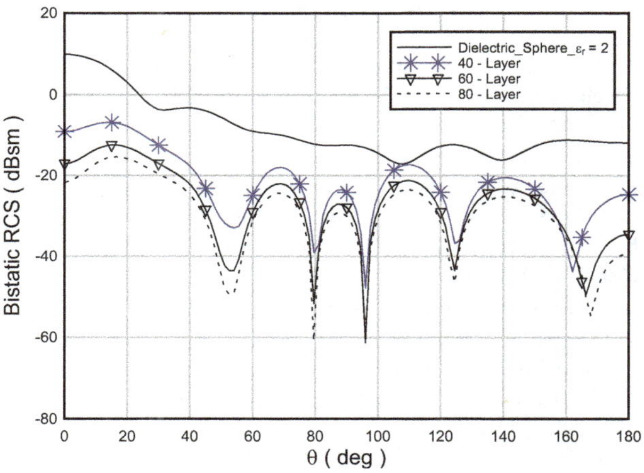

Figure 13. Bistatic RCS for different numbers of layers for case II ($\varepsilon_d = 2$) at $c = R_1/40$.

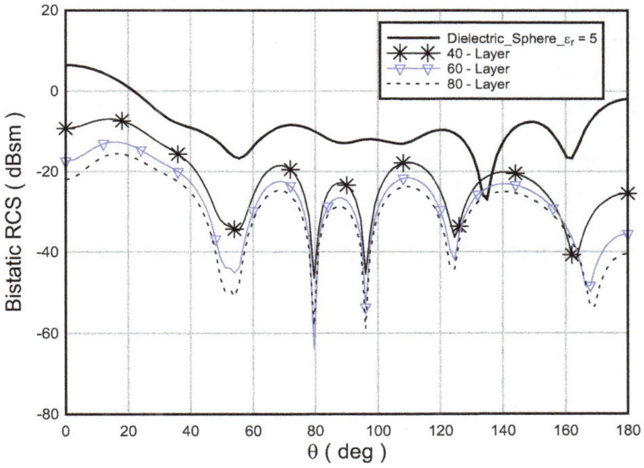

Figure 14. Bistatic RCS for different numbers of layers for case II ($\varepsilon_d = 5$) at $c = R_1/40$.

with isotropic homogenous cloaking layers are investigated concerning the normalized total scattering cross section versus the normalized frequency k_0R_1 and for the bistatic RCS for both case I and case II and with varying the transformed radius, where the dielectric sphere is transformed into a small sphere rather than to a point, together with discretizing the cloaking material using pairs of homogeneous isotropic sub-layers with different numbers. The solution is obtained by rigorously solving Maxwell equations using Mie series expansion. When the number of layers increases (each isotropic layer is thinner), the scattered field decreases. As the transformed radius decreases the total scattering decreases, and approaches the ideal layered cloaking as the transformed radius tends to zero. The discretization has an effect on the scattering at the higher frequencies. The results are compared with the case of cloaked conducting sphere, showing a number of similarities, particularly for small transformed radius.

REFERENCES

[1] J. B. Pendry, D. Schurig and D. R. Smith, "Controlling Electromagnetic Fields," *Science*, Vol. 312, No. 5781, 2006, pp. 1780-1782. http://dx.doi.org/10.1126/science.1125907

[2] Q. Cheng, W. X. Jiang and T. J. Cui, "Investigations of the Electromagnetic Properties of Three-Dimensional Arbitrarily-Shaped Cloaks," *Progress in Electromagnetics Research*, Vol. 94, 2009, pp. 105-117. http://dx.doi.org/10.2528/PIER09060705

[3] J. J. Yang, M. Huang, Y. L. Li, T. H. Li and J. Sun, "Reciprocal Invisible Cloak with Homogeneous Metamaterials," *Progress in Electromagnetics Research M*, Vol. 21, 2011, pp. 105-115. http://dx.doi.org/10.2528/PIERM11090904

[4] A. Shahzad, F. Qasim, S. Ahmed and Q. A. Naqvi, "Cylindrical Invisibility Cloak Incorporating PEMC at Perturbed Void Region," *Progress in Electromagnetics Research M*, Vol. 21, 2011, pp. 61-76. http://dx.doi.org/10.2528/PIERM11061302

[5] X. X. Cheng, H. S. Chen and X. M. Zhang, "Cloaking a Perfectly Conducting Sphere with Rotationally Uniaxial Nihility Media in Monostatic Radar System," *Progress in Electromagnetics Research*, Vol. 100, 2010, pp. 285-298. http://dx.doi.org/10.2528/PIER09112002

[6] J. Zhang and N. A. Mortensen, "Ultrathin Cylindrical Cloak," *Progress in Electromagnetics Research*, Vol. 121, 2011, pp. 381-389. http://dx.doi.org/10.2528/PIER11091205

[7] Y. B. Zhai and T. J. Cui, "Three-Dimensional Axisymmetric Invisibility Cloaks with Arbitrary Shapes in Layered-Medium Background," *Progress in Electromagnetics Research B*, Vol. 27, 2011, pp. 151-163.

[8] D. Schurig, J. J. Mock, B. J. Justice, S. A. Cummer, J. B. Pendry, A. F. Starr and D. R. Smith, "Metamaterial Electromagnetic Cloak at Microwave Frequencies," *Science*, Vol. 314, No. 5801, 2006, pp. 977-980. http://dx.doi.org/10.1126/science.1133628

[9] J. B. Pendry, A. J. Holden, D. J. Robbins and W. J. Stewart, "Magnetism from Conductors and Enhanced Nonlinear Phenomena," *IEEE Transactions on Microwave Theory and Techniques*, Vol. 47, No. 11, 1999, pp. 2075-2084. http://dx.doi.org/10.1109/22.798002

[10] G. V. Eleftheriades and K. G. Balmain, "Negative Refraction Metamaterials—Fundamental Principles and Applications," John Wiley, Hoboken, 2005.

[11] N. Engheta and R. W. Ziolkowski, "Metamaterials: Physics and Engineering Explorations," Wiley-IEEE Press, Hoboken, 2006. http://dx.doi.org/10.1002/0471784192

[12] J. Wang, S. Qu, J. Zhang, H. Ma, Y. Yang, C. Gu, X. Wu and Z. Xu, "A Tunable Left-Handed Metamaterial Based on Modified Broadside-Coupled Split-Ring Resonators," *Progress in Electromagnetics Research Letters*, Vol. 6, 2009, pp. 35-45. http://dx.doi.org/10.2528/PIERL08120708

[13] H. Liu, "Virtual Reshaping and Invisibility in Obstacle Scattering," *Inverse Problems*, Vol. 25, No. 4, 2009, pp. 1-10. http://dx.doi.org/10.1088/0266-5611/25/4/045006

[14] T. Zhou, "Electromagnetic Inverse Problems and Cloaking," Ph. D. Thesis, Washington University, St Louis, 2010.

[15] Y. Huang, Y. Feng and T. Jiang, "Electromagnetic Cloaking by Layered Structure of Homogenous Isotropic Materials," *Optics Express*, Vol. 15, No. 18, 2007, pp. 1-4.

[16] C. Qiu, L. Hu and S. Zouhdi, "Isotropic Non-Ideal Cloaks Providing Improved Invisibility by Adaptive Segmentation and Optimal Refractive Index Profile from Ordering Isotropic Materials," *Optics Express*, Vol. 18, No. 14, 2010, pp. 14950-14959. http://dx.doi.org/10.1364/OE.18.014950

[17] C. Simovski and S. He, "Frequency Range and Explicit Expressions for Negative Permittivity and Permeability for an Isotropic Medium Formed by a Lattice of Perfectly Conducting Ω Particles," *Physics Letters A*, Vol. 311, No. 2-3, 2003, pp. 254-263. http://dx.doi.org/10.1016/S0375-9601(03)00494-8

[18] C. Simovski and B. Sauviac, "Toward Creating Isotropic Microwave Composites with Negative Refraction," *Radio Science*, Vol. 39, No. 2, 2004, pp. 1-18.

[19] C. W. Qiu, L. Hu, X. Xu and Y. Feng, "Spherical Cloaking with Homogenous Isotropic Multilayered Structures," *Physical Review E*, Vol. 79, 2009, pp. 1-4.

[20] C. M. Ji, P. Y. Mao and F. D. Ning, "An Improved Method of Designing Multilayered Spherical Cloak for Electromagnetic Invisibility," *Chinese Physics Letters*, Vol. 27, No. 3, 2010, pp. 1-4.

[21] H. Zamel, E. El-Diwany and H. El-Hennawy, "Approximate Electromagnetic Cloaking of a Conducting Sphere using Homogeneous Isotropic Multi-Layered Materials," *2nd Middle East Conference on Antennas and Propagation*, 29-31 December 2012, Cairo.

[22] W. Song, X. Yang and X. Sheng, "Scattering Characteristic of 2-d Imperfect Cloaks with Layered Isotropic Materials," *IEEE Antennas and Wireless Propagation Letters*, Vol. 11, 2012, pp. 53-56. http://dx.doi.org/10.1109/LAWP.2011.2182590

[23] M. Yan, W. Yan and M. Qiu, "Invisibility Cloaking by Coordinate Transformation," *Progress in Optics*, Vol. 52, 2009, pp. 261-304. http://dx.doi.org/10.1016/S0079-6638(08)00006-1

[24] H. Zamel, E. El-Diwany and H. El-Hennawy, "Approximate Electromagnetic Cloaking of Spherical Bodies," *29th National Radio Science Conference (NRSC)*, Cairo, 10-12 April 2012, pp. 19-28. http://dx.doi.org/10.1109/NRSC.2012.6208502

[25] J. A. Stratton, "Electromagnetic Theory," McGraw-Hill, Boston, 1941.

[26] R. F. Harrington, "Time Harmonic Electromagnetic Fields," McGraw-Hill, Boston, 1961.

[27] J. Jin, "Theory and Computation of Electromagnetic Fields," John Wiley, Hoboken, 2010. http://dx.doi.org/10.1002/9780470874257

[28] G. T. Ruck, D. E. Barrick, W. D. Stuart and C. K. Krich-

baum, "Radar Cross Section Handbook," Kluwer Academic, Boston, 1970.

[29] O. Pena and U. Pal, "Scattering of Electromagnetic Radiation by a Multilayered Sphere," *Computer Physics Communications*, Vol. 180, No. 11, 2009, pp. 2348-2354. http://dx.doi.org/10.1016/j.cpc.2009.07.010

[30] L. Kai and P. Massoli, "Scattering of Electromagnetic Plane Waves by Radially Inhomogeneous Spheres: A Finely Stratified Sphere Model," *Applied Optics*, Vol. 33, No. 3, 1994, pp. 501-511. http://dx.doi.org/10.1364/AO.33.000501

[31] A. Aden and M. Kerker, "Scattering of Electromagnetic Waves from Two Concentric Spheres," *Journal of Applied Physics*, Vol. 22, No. 10, 1951, pp. 1242-1246. http://dx.doi.org/10.1063/1.1699834

[32] N. Tsitsas and C. Athanasiadis, "On the Scattering of Spherical Electromagnetic Waves by a Layered Sphere," *Journal of Applied Mathematics and Mechanics*, Vol. 59, No. 1, 2005, pp. 55-74. http://dx.doi.org/10.1093/qjmam/hbi031

[33] E. Jordan and K. Balmain, "Electromagnetic Waves and Radiating Systems," Prentice-Hall, Upper Saddle River, 1968.

Microstructure and Dielectric Properties of Bi Substituted PLZMST Ceramics

Hayet Menasra, Zelikha Necira, Karima Bouneb, Abdelhak Maklid, Ahmed Boutarfaia*

Applied Chemistry Laboratory, Exact and Natural and Life Sciences Faculty, Materials Science Department, Mohamed Kheider University of Biskra, Biskra, Algeria.
Email: *hayetmenasra@yahoo.com

ABSTRACT

Bismuth (Bi) and lanthanum (La) doped lead manganese antimoine zirconate titanate (PZMST) ceramic powders have been synthesized by high temperature solid-state reaction method. Preliminary X-ray structural analysis of the compounds shows the formation of tetragonal structure. Scanning electron micrographs (SEM) shows a uniform grain distribution and grain size of the order of ~2.28 μm. Detailed dielectric studies of the $Pb_{0.95}(La_{1-z}Bi_z)_{0.05}[(Zr_{0.6}Ti_{0.4})_{0.95}(Mn_{1/3}Sb_{2/3})_{0.05}]O_3$ samples as a function of the temperature (from 25˚C to 450˚C) at frequency 1 kHz suggest that the compounds undergo a diffuse phase transition. The transition temperature shifts increase with increasing the Bi ratio. Diffusivity (γ) study of phase transition of these compounds provided its value from 1.59 to 1.78 indicating the degree of the disordering in the system.

Keywords: PLZT Ceramics; Grain Size; Dielectric Properties; Diffusivity (γ)

1. Introduction

Since its discovery, the lead zirconate titanate (PZT) ceramic system has been widely used in transducer design due to its notable electromechanical features [1-3]. One of the main characteristics of the phase diagram of this perovskite compound is the existence of a Morphotropic Phase Boundary (MPB), approximately at Zr/Ti ~ 53/47 wish divides the rombohedral Zr-rich from tetragonal Ti-rich one. Samples with Zr/Ti ratio near the MPB show the highest dielectric and piezoelectric responses [4].

On the other hand, PZT ceramics have been extensively modified (doped) with small amount of different additives that make them more attractive for any specific application. Such kinds of modification are classified as "soft" or "hard" by differentiating cases where the dopantion has, respectively, higher or lower valence than the targetion on the ABO_3 perovskite cell [5]. Addition of donor dopants like La enhanced the electro optical and mechanical proprieties of ceramics [6,7]. In this formula La^{+3} ions goes to the A-site and vacancies are created on the B-site to maintain change balance. The influence of soft doping (trivalent Bi [8-13]) and hard doping (acceptors Mn and Sb [14-16]) in PLZT have been reported to have high electromechanical properties. However, the relationship between morphological and electrical properties with reference to these combinatorial ceramic compositions with hard (acceptor Mn and Sb B-site) and soft (trivalent Bi A-site) doping in PLZT has not been addressed.

The aim of the present work is to study the morphological and dielectric properties of the $Pb_{0.95}(La_{1-z}Bi_z)_{0.05}[(Zr_{0.6}Ti_{0.4})_{0.95}(Mn_{1/3}Sb_{2/3})_{0.05}]O_3$ ceramics. By using X-ray, SEM and the observed dielectric behavior and the temperature value will be interpreted in terms of a Ferroelectric-Paraelectric (FE-PE) phase transition.

2. Experimental Procedure

The ceramic samples were prepared by solid state reaction. High purity raw materials (Pb_3O_4, ZrO_2, TiO_2, La_2O_3, Bi_2O_3, MnO_2 and Sb_2O_3) were stoichiometrically weighed according to the composition $Pb_{0.95}(La_{1-z}Bi_z)_{0.05}[(Zr_{0.6}Ti_{0.4})_{0.95}(Mn_{1/3}Sb_{2/3})_{0.05}]O_3$ via the chemical route for z = 0.0, 0.2, 0.4 and 0.6. From now on, we shall refer to this compound as PLBZMST or 100/0, 80/20, 60/40 and 40/60 according to the La and Bi ratio. The batch powders were dispersed in acetone and mixed by a magnetic stirrer during two hours. The obtained paste is being dried at 80˚C, and then crushed in a mortar

*Corresponding author.

out of a glass during 4 hours. Powders were calcined at 900°C for 120 minutes with a heating rate of 2°C/min. calcined powders were crushed in a similar manner to the first crushing but with 6 hours, for better agglomerate size reduction. A 5% polyvinyl alcohol (PVA) water solution was used as binder to increase the plasticity of the powders. The weight ratio between the PVA solution and the powders was 1:20. Powder and PVA solutions were mixed in a mortar and then uniaxially pressed into pellets with a pressure of 2000 kg/cm^3 in a cylindrical stainless steel dies using a hydraulic press. The size of those pellets was 13 mm in diameter; while the thickness is 1 mm. Pellets were packed into covered alumina crucibles. The inner space of the crucibles was filled up with the powders of $PbZrO_3$, in order to prevent intensive evaporation of the lead during the sintering. A typical sintering schedule consisted of heating rate of 2°C/min to 1100°C, 1150°C, 1180°C and 1200°C, for 120 minutes and natural cooling in the furnace.

Powder X-ray diffraction was recorded by X-ray powder (Philips) diffractometer using CuKα radiation (λ = 1.5406 Å) in a wide range of Bragg angles (20° ≤ 2θ ≤ 60° at a scanning rate of 2°/min. Densities of sintered pieces were calculated from the sample dimensions and weights. Microstructural features such as a grain size and pores were characterized by means of scanning electron microscopy (SEM). Sintered pellets were electrified by silver paste, and fired at 750°C for forty five minutes, before using for any electrical measurements. The dielectric permittivity and loss tangent of the samples were measured as a function of frequency at different temperature (room temperature to 500°C) using LCR meter (Good Will Instrument Co., LTD).

3. Results and Discussion

X-ray diffraction patterns of PLBZMST (100/0, 80/20, 60/40; and 40/60) ceramics sintered at 1180°C are shown in "**Figure 1**". The PZT phase could be identified from X-ray peaks in a range of $2\theta \sim 42° - 47°$ [17], wish cor responded to $(200)_T$ - $(002)_T$ peaks of tetragonal (T) structure. The associated changes in lattice parameter and degree of tetragonality (c/a) were calculated and listed in **Table 1**.

The calculated tetragonality of all the samples indicates that the addition of Bi to PLZMST does not cause mush variation in tetragonality, except that a pyrochlore peak is observed near the perovskite (110) for all the com-

Table 1. Lattice and dielectric parameters of PLBZMST samples.

Lattice parameters in (Å)	Composition Bi%			
	0	2	4	6
a = b	3.9615	3.9936	3.8928	3.9666
C	4.0725	4.1937	4.068	4.0619
c/a	0.9727	0.9522	0.9659	0.9765
V(Å^3)	63.91	66.88	61.64	63.91
Physical parameters				
ε_m	13059.73	13275.19	21728.37	28272.79
T_C (°C)	350	360	360	410
ΔT_m (°C)	28	26	19	42
γ	1.59	1.63	1.78	1.73

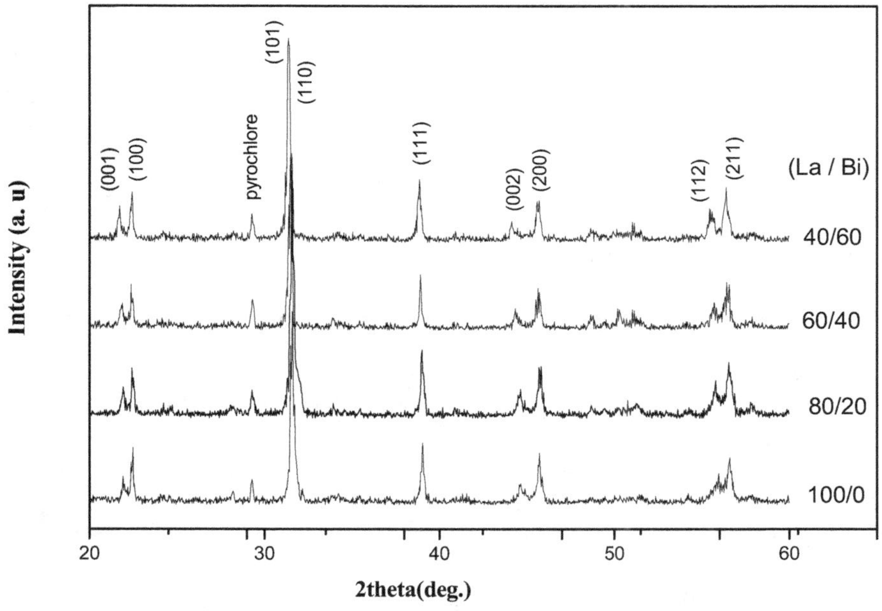

Figure 1. XRD patterns of PLBZMST sintered at 1180°C.

positions. The relative amount of the pyrochlore phase to the perovskite phase was esteemed using the following peak area ration equation [18].

$$\text{pyrochlore\%} = \frac{A_{\text{pyro.}}}{A_{\text{pyro.}} + A(100)} \times 100 \qquad (1)$$

where $A_{\text{pyro.}}$ and A (110) are the area under the pyrochlore peak and the (110) perovskite phase. The pyrochlore value is increased from 3% to 6% with increasing Bi doping, which is small and agreement with the value reported in doped PZT [19]. Hence, this composition can also be considered as a single phase wish confirms the homogeneous diffusivity of Bi^{3+} in PLZMST perovskite. The donors ions Bi^{3+} partially substitute Pb^{2+} at A-site and acceptors Mn^{4+} and Sb^{3+} partially substitutes Zr^{4+}/ Ti^{4+} at B-site due to the approximately close and similar ionic radii of respective cations at A-site and B-site in PLZT lattice.

Figure 2(a) shows the apparent density of undoped and Bi modified PLZMST ceramics at different sintering-temperatures. This curve shows the similar variation trend with increasing sintering temperature. The density of the compositions sintered at 1180˚C showed the maximum value of 7.56 mg/cm^3 at ratio La/ Bi (60/40). The apparent densities (as a percentage of theoretical density) of the sintered pellets are shown in **Figure 2(b)**. This figure shows that the composition doped with 4% of Bi has highest density values at each temperature sintering.

SEM micrographs of the pellet prepared from PLBZMST powder and sintered at 1180˚C are shown in **Figure 3**. The sintered pellets have been found to have a grain size of the order of ~2.28 μm and uniform grain distribution, which is in accordance with the high density value as can be seen from **Figures 2(a)** and **(b)**. The average grain size was determined directly from the SEM micrographs by using the classical linear interception method.

Permittivity and AC conductivity were analysed in the ferro and paraelectric phase. **Figure 4(a)** shows the variation of the dielectric constant (ε_r) as a function of temperature at frequency of 1 kHz. It is observed that ε_r increases with the increase of temperature up to the onset of phase transition. A significant improvement in values of ε_r has been observed with the increase of sintering temperatures. The value of dielectric constant (ε_{max} = 21728.37) of the sample doped with Bi (60/40) at a transition temperature (T_c = 360˚C) is a high as 2 times that of the other samples. On the other hand, while at room temperature; permittivity increased with dopants concentration mainly because of the lower values of porosity. This fact can be explained by using the Bruggeman model for inhomogeneous media [20].

Figure 4(b) shows the variety of dielectric loss (tan*δ*) as a function of temperature at 1 kHz; for all the com-

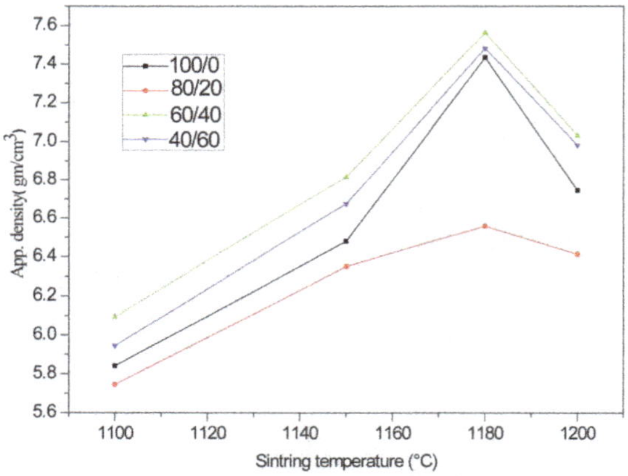

(a)

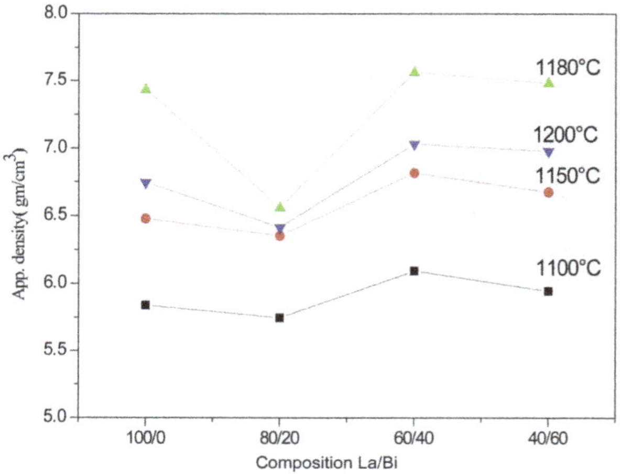

(b)

Figure 2. (a) Variation of apparent density of PLBZMST with sintered temperature; (b) Variation of apparent density of sintered PLBZMST samples with Bi addition.

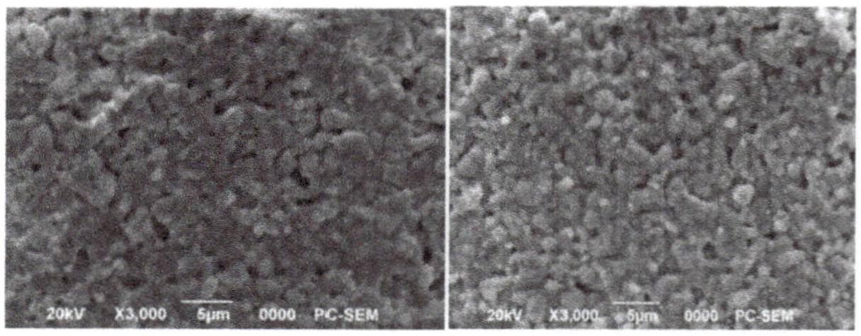

(a) (b)

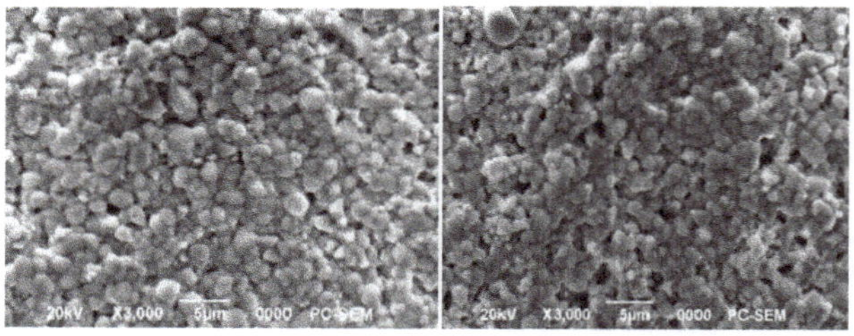

(c) (d)

Figure 3. SEM micrographs of PLBZMST pellets (a) 100/0; (b) 80/20; (c) 60/40; (d) 40/60 sintered at 1180˚C.

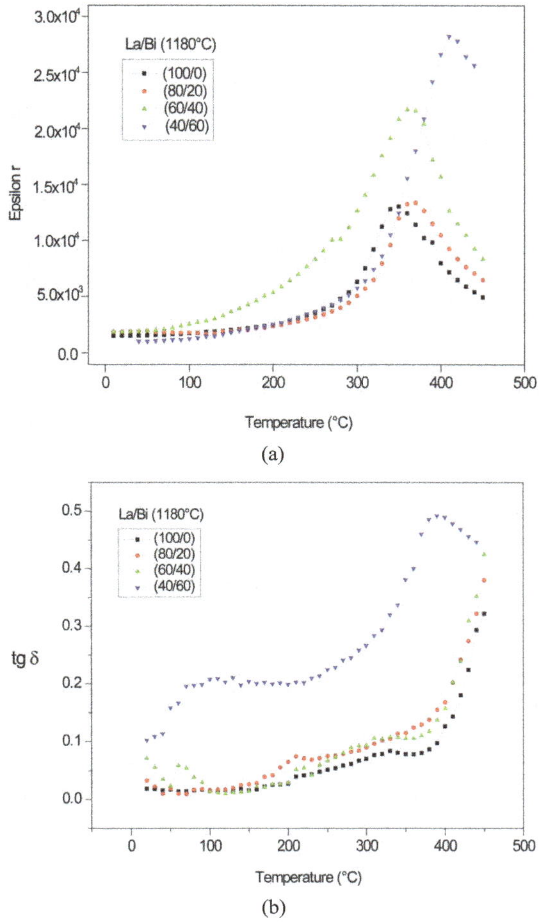

Figure 4. Variation of (a) Dielectric constant and (b) dielectric loss respectively with temperature at 1 kHz for all samples sintered at 1180˚C.

positions, as temperature increase, loss tangent is almost constant up to 300˚C, but then it starts increasing with temperature. This increase in (tanδ) may be due to an increase in the electrical conduction of the residual current and absorption current [21].

It is known that the dielectric constant ε of a normal ferroelectric, above the curie temperature follows the Curie-Weiss Law [22] described by:

$$\frac{1}{\varepsilon}=\frac{T-T_0}{C} \quad (T>T_c) \tag{2}$$

where T_0 is the Curie-Weiss temperature and C is the Curie-Weiss constant. **Figure 5(a)** shows the plot of temperature versus inverse dielectric constant (at 1 kHz) fitted to the Curie-Weiss law for PLZMST ceramics.

$$\Delta T_m = T_{cw} - T_c \tag{3}$$

The parameter ΔT_m, to illustrate the degree of deviation from the Curie-Weiss law, was defined as where Tcw denotes the temperature from which the permittivity starts to deviate from the Curie-Weiss law, and Tm represents the temperature of the dielectric constant maximum. At 1 kHz, it can be seen from **Table 1** that the ΔT_m decreases slightly with increased Bi addition.

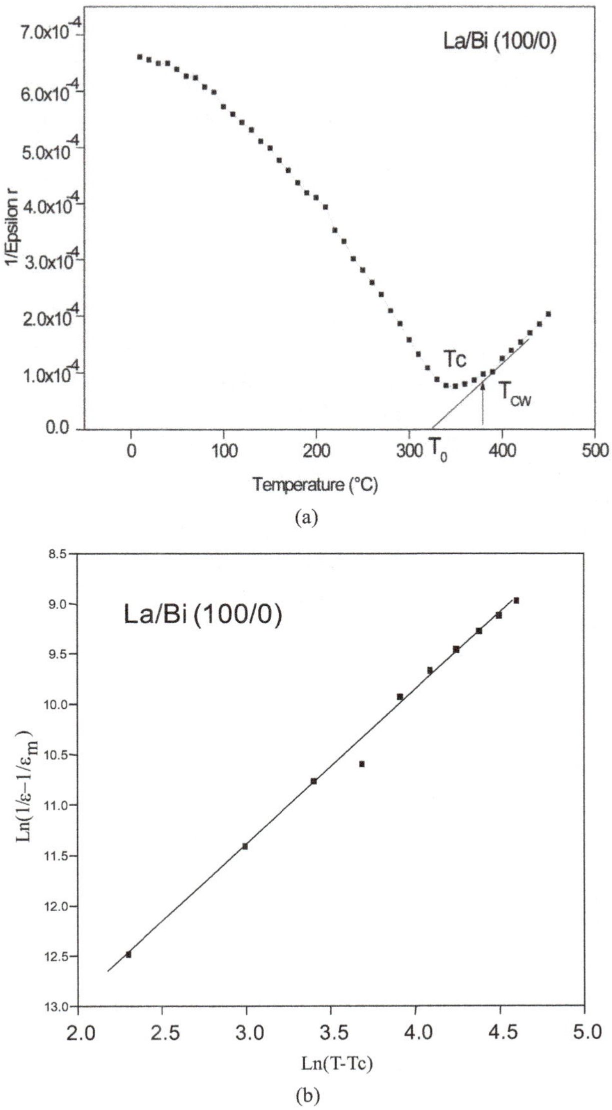

Figure 5. (a) The inverse ε as a function of temperature at 1 kHz for PLBZMST; (b) ln(1/ε − 1/ε_m) vs. ln(T − T_c) of PLZMST at 1 kHz.

The dielectrics characteristics of relaxor ferroelectrics are known to deviate from the typical Curie-Weiss behavior and can be described by a modified Curie-Weiss relation-ship [23].

$$\frac{1}{\varepsilon}-\frac{1}{\varepsilon_m}=\frac{(T-T_m)^\gamma}{C} \quad 1\le\gamma\le 2 \tag{4}$$

where γ and C are assumed to be constant. The parameter γ give information on the character of the phase transition: For $\gamma = 1$, a normal Curie-Weiss law is obtained, $\gamma = 2$ describe a complete diffuse phase transition [24]. **Fig-**

ure 5(b) shows the plot of $\ln((1/\varepsilon) - (1/\varepsilon_m))$ versus $\ln(T - T_m)$ at 1 kHz of PLZMST sample. Linear relationships were observed. The slopes of the fitting curve were used to determine the parameter γ. The values of γ are listed in **Table 1**. It can be seen the increase in value of γ with Bi_2O_3 content (**Table 1**) indicates an increase in diffusivity.

4. Conclusion

The novel PLBZMST ceramics have been prepared by conventional solid-state reaction route. All the compounds are crystallized in pure perovskite phase with the predominant tetragonal phase. Grain size was obtained in the order of ~2.28 μm and uniform grain distribution wish is in accordance with the density. The increase in core temperature and dielectric constant can be attributed to the emerging lead vacancies creation by donor cations (Bi), and acceptor to reduce the oxygen vacancies mobility to balance charge in the modified PLZMST. The diffusivity (γ) study of phase transition of these compounds gives the values from 1.59 to 1.78 indicating the degree of the disordering in the system.

REFERENCES

[1] B. Jaffe, W. R. Cook Jr. and H. Jaffe, "Piezoelectric Ceramics," Academic Press, New York, 1971.

[2] G. H. Haertling, "Ferroelectric Ceramics: History and Technology," *Journal of the American Ceramic Society*, Vol. 82, No. 4, 1999, pp. 797-818. doi:10.1111/j.1151-2916.1999.tb01840.x

[3] B. Jaffe, R. S. Roth and S. Marzullo, "Piezoelectric PropErties of Lead Zirconate-Lead Titanate Solid-Solution Ceramics," *Journal of Applied Physics*, Vol. 25, No. 6, 1954, pp. 809-810. doi:10.1063/1.1721741

[4] F. Agullo-Lopez, J. M. Carbrera and F. Agullo-Rueda, "Electrooptics Phenomena, Materials and Applications," Academic Press INC, San Diego, 1994, pp. 146-149.

[5] R. Rai, S. Sharma and R. N. P. Choudhary, "Effect of Al Doping on Structural and Dielectric Properties of PLZT Cramics," *Journal of Materials Science*, Vol. 41, No. 13, 2006, pp. 4259-4265. doi:10.1007/s10853-005-5455-1

[6] H. Tamura, T. Knolle, Y. Sakable and K. Wakino, "Improved High-Q Dielectric Resonator with Complex Perovskite Structure," *Journal of the American Ceramic Society*, Vol. 67, No. 4, 1984, pp. C59-C61.

[7] K. Wakino, K. Minal and H. Tamura, "Microwave Characteristics of $(Zr, Sn)TiO_4$ and $BaO-PbO-Nd_2O_3-TiO_2$ Dielectric Resonators," *Journal of the American Ceramic Society*, Vol. 67, No. 4, 1984, pp. 278-281. doi:10.1111/j.1151-2916.1984.tb18847.x

[8] S. Dutta, R. N. P. Choudhary and P. K. Sinha, "Ferroelectric Phase Transition in Bi-Doped PLZT Ceramics," *Materials Science and Engineering: B*, Vol. 98, No. 1, 2003, pp. 74-80. doi:10.1016/S0921-5107(02)00612-8

[9] R. N. P. Choudhary, "Phase Transition in Bimodified PLZT Ferroelectrics," *Materials Letters*, Vol. 54, No. 2, 2002, pp. 175-180. doi:10.1016/S0167-577X(01)00559-6

[10] P. Goel, K. L. Yadav and A. R. James, "Double Doping Effect on the Structural and Dielectric Properties of PZT Ceramics," *Journal of Physics: Applied Physics*, Vol. 37, No. 22, 2004, pp. 3174-3179. doi:10.1088/0022-3727/37/22/019

[11] S. Dutta, R. N. P. Choudhary and P. K. Sinha, "Ferroelectric Phase Transition in Sol-Gel Derived Bi-Doped PLZT Ceramics," *Journal of Materials Science*, Vol. 39, No. 9, 2004, pp. 3129-3135. doi:10.1023/B:JMSC.0000025842.46451.64

[12] P. Goel and K. L. Yadav, "Substitution Site Effect on Structural and Dielectric Properties of La-Bi Modified PZT," *Journal of Materials Science*, Vol. 42, No. 11, 2007, pp. 3928-3935. doi:10.1007/s10853-006-0416-x

[13] R. Rai, S. Sharma and R. N. P. Choudhary, "Structural and Dielectric Properties of Bi Modified PLZT Ceramics," *Solid State Communications*, Vol. 133, No. 10, 2005, pp. 635-639. doi:10.1016/j.ssc.2005.01.005

[14] Y. K. Gao, K. Uchino and D. Viehland, "Rare Earth Metal Doping Effects on the Piezoelectric and Polarization Properties of $Pb(Zr,Ti)O_3$-$Pb(Sb,Mn)O_3$ Ceramics," *Journal of Applied Physics*, Vol. 92, No. 4, 2002, pp. 2094-2099. doi:10.1063/1.1490617

[15] Z. G. Zhu, Z. J. Xu, W. Z. Zhang and Q. R. Yin, "Effect of PMS Modification on Dielectric and Piezoelectric Properties in xPMS-(1 − x) PZT," *Journal of Physics: Applied Physics*, Vol. 38, No. 9, 2005, pp. 1464-1469. doi:10.1088/0022-3727/38/9/021

[16] R. Rai, S. Mishra and N. K. Singh, "Effect of Fe and Mn Doping at B-Site of PLZT Ceramics on Dielectric Properties," *Journal of Alloys Compounds*, Vol. 487, No. 40180, 2009, pp. 494-498. doi:10.1016/j.jallcom.2009.07.161

[17] K. Kakegawa, J. Mohri, T. Takahashi, H. Yammamura and K. Shirasaki (Solid State Communication), "A Compositional Fluctuation and Properties of $Pb(Zr,Ti)O_3$," *Solid State Communications*, Vol. 24, No. 11, 1977, pp. 769-772. doi:10.1016/0038-1098(77)91186-3

[18] A. Garg and D. C. Agarwal, "Effect of Rare Earth (Er, Gd, Eu, Nd and La) and Bismuth Additives on the Mechanical and Piezoelectric Properties of Lead Zirconate Titanate Ceramics," *Materials Science and Engineering: B*, Vol. 86, No. 2, 2001, pp. 134-143. doi:10.1016/S0921-5107(01)00655-9

[19] S. R. Shanningrahi, F. E. H. Tay, K. Yao and R. N. P. Choudhary, Effect of Rare Earth (La, Nd, Sm, Eu, Gd, Dy, Er and Yb) Ion Substitutions on the Microstructural and Electrical Properties of Sol-Gel Grown PZT Ceramics," *Journal of the European Ceramic Society*, Vol. 24, No. 1, 2004, pp. 163-170. doi:10.1016/S0955-2219(03)00316-9

[20] A. Pelaiz-Barranco, "Ferroelectric Properties and Conduction Mechanisms in the Modified PZT Ceramic System," Ph.D. Thesis, University of Havana, Havana, 2001.

[21] B. Tareev, "Physics of Dielectric Materials," Mir Publisher, Moscow, 1979, p. 157.

[22] M. E. Lines and A. M. Glass, "Principles and Application

of Ferroelectric and Related Materials," Clarendon Press, Oxford, 1977.

[23] V. Koval, C. Alemany, J. Briančin and H. Bruncková, "Dielectric Properties and Phase Transition Behavior of xPMN-(1 − x)PZT Ceramic Systems," *Journal of Electroceramics*, Vol. 10, No. 1, 2003, pp. 19-29.

[24] G. A. Smolenskii, "X-Ray Scattering and the Phase Transition of $KMnF_3$ at 184 K," *Journal of the Physical Society of Japan*, Vol. 28, No. 2, 1970, pp. 26-37. doi:10.1023/A:1024023823871

18

Influence of Cr^{3+} Ion on the Dielectric Properties of Nano Crystalline Mg-Ferrites Synthesized by Citrate-Gel Method

M. Raghasudha[1], D. Ravinder[2*], P. Veerasomaiah[3]

[1]Department of Chemistry, Jayaprakash Narayan College of Engineering, Mahabubnagar, India; [2]Department of Physics, Nizam College, Basheerbagh Osmania University, Hyderabad, India; [3]Department of Chemistry, Osmania University, Hyderabad, India.
Email: *ravindergupta28@rediffmail.com

ABSTRACT

Mixed Mg-Cr Nano ferrites having the compositional formula $MgCr_xFe_{2-x}O_4$ (where x = 0.0, 0.1, 0.3, 0.5, 0.7, 0.9 and 1.0) were synthesized using Citrate-Gel auto combustion method. Structural characterization was carried out by XRD Analysis which confirmed the formation of single phase cubic spinel structure without any impurity peak. The dielectric properties such as Dielectric constant (ε'), Dielectric Loss tangent (tan δ) and AC conductivity (σ_{AC}) of Mg-Cr nano ferrites were studied at room temperature in the frequency range of 2 Hz - 2 MHz using Agilent E4980A Precision LCR meter. The dielectric constant, loss tangent and AC conductivity shows a normal behavior with frequency. A qulitative explanation is given for composition and frequency dependance of the dielectric constant, dielectric loss tangent and AC conductivity of the nano ferrite. The loss tangent for the synthesized samples was found to be decreased from 0.09 to 0.054 in higher frequency region showing the potential applications of these materials in high frequency micro wave devices. On the basis of these results the explanation of dielectric mechanism in Mg-Cr ferrites is suggested.

Keywords: Dielectric Constant; Dielectric Loss Tangent; Mg-Cr Nano Ferrites; LCR Meter

1. Introduction

Nano ferrites have many applications in high frequency devices and they play a useful role in technological applications. One important characteristic of ferrites is their high values of resistivity, low magnetic and dielectric losses [1] which make them ideal for high frequency applications. Owing to the dielectric behavior, they are sometimes called multiferroics. They are commercially important because they can be used in, many devices such as Phase Shifter, high frequency transformer cores, switches, resonators, computers, TVs and mobile phones [2,3]. Dielectric properties of ferrites are dependent upon several factors, including method of preparation, chemical composition and grain size [4,5]. Since $MgFe_2O_4$ and related ferrites are widely used components in micro wave family due to their high electrical resistivity, low magnetic and dielectric losses [6,7] we investigated the dielectric behavior of these ferrites over wide range of frequencies at room temperature. In this study, we prepared nano sized $MgCr_xFe_{2-x}O_4$ compounds containing different levels of Cr with the assumption that dielectric properties would be improved by substitution of Fe^{3+} ions with Cr^{3+} ions. Substitution of Magnesium Ferrites with Cr^{3+} ions at B site should be effective in enhancing the electrical resistivity. In the present work the aim of Cr^{3+} ion substitution for Fe^{3+} ions is to reduce dielectric loss. In this article we report the influence of Cr substitution on structural and dielectric properties of $MgCr_xFe_{2-x}O_4$ ferrites synthesized by Citrate-gel auto combustion method as a function of frequency and composition at room temperature.

2. Experimental

2.1. Synthesis of Mg-Cr Ferrites

Ferrites with chemical formula $MgCr_xFe_{2-x}O_4$ (x = 0.0, 0.1, 0.3, 0.5, 0.7, 0.9 and 1.0) have been prepared by the Citrate-gel auto combustion method using high purity AR grade of Magnesium Nitrate—($Mg(NO_3)_2 6H_2O$), Ferric Nitrate—($Fe(NO_3)_2 9H_2O$), Chromium Nitrate—($Cr(NO_3)_2 9H_2O$), Citric acid—($C_6H_8O_7 \cdot H_2O$), Ammonia

*Corresponding author.

—(NH_3) as starting materials for the synthesis. Required quantities of metal nitrates were dissolved in a minimum quantity of distilled water and mixed together. Aqueous solution of Citric acid was then added to the mixed metal nitrate solution. Ammonia solution was then added with constant stirring to maintain P^H of the solution at 7. The resulting solution was continuously heated on the hot plate at 100˚C up to dryness with continuous stirring. A viscous gel has resulted. Increasing the temperature up to 200˚C lead the ignition of gel. The dried gel burnt completely in a self propagating combustion manner to form a loose powder. The burnt powder was ground in Agate Mortor and Pistle to get a fine Ferrite powder. Finally the burnt powder was calcined in air at 500˚C temperature for four hours and cooled to room temperature.

2.2. Characterization

The structural characterization of powders was carried out by X-Ray Diffractometer to analyze the phase and crystallite size. For dielectric measurements the powders were added with a small amount 2% PVA as a binder to press the powders into circular pellets of diameter 13mm and thickness 1 mm applying a pressure of 5 tons. The prepared pellets were sintered at 500˚C for four hours in air in muffle furnace for the densification of the sample. For dielectric measurements silver paint was applied on both sides of the pellets and air dried to have good ohmic contact. The dielectric measurements were made using Agilent E4980A Precision LCR meter at room temperature in the frequency range 20 Hz to 2 MHz.

Using LCR meter the dielectric parameters such as Capacitance of the pellet, tan δ (loss tangent) and Capacitance of air with the same thickness as the pellet were measured.

The real part of the dielectric constant (ε') was determined from the following formula [8]

$$\varepsilon' = \frac{C_p}{C_{Air}}$$

where ε' = Real part of dielectric constant; C_p = Capacitance of the Pellet in Faraday; C_{Air} = Capacitance of Air in Faraday.

The imaginary part of the dielectric constant (ε'') or dielectric loss was measured by using the following relation [9]

$$\varepsilon'' = \varepsilon' \cdot \tan\delta$$

The Ac conductivity was calculated using the values of frequency (f) and loss tangent factor as [9]

$$\sigma_{AC} = 2\pi f \varepsilon_0 \varepsilon' \tan\delta$$

where ε_0 = Constant permittivity of free space = 8.854 × 10^{-12} F/m; ε' = Real part of dielectric constant; tan δ = loss tangent.

3. Results and Discussion

3.1. Structural Characterization

The structural characterization of all the nano-ferrites was carried out by X-ray Diffraction. From the analysis the 2 theta and intensity data were used and graphs were plotted as shown in **Figure 1**. From the X-ray diffraction pattern, crystalline phases were identified by comparison with reference data from the ICSD card No. 71-1232 for Magnesium ferrites. All Bragg reflections have been indexed which confirms the formation of a well defined single phase cubic spinel structure without any impurity peak. The strongest reflection comes from (311) plane that indicates spinel phase. The XRD patterns of all the Chromium substituted Magnesium ferrites showed the homogeneous single phased cubic spinel belonging to the space group Fd3m (confirmed by ICSD Ref 71-1232).

The crystallite size (D) was calculated for all the compositions using high intensity peak (311) from Scherrer's formula [10] and was in the range of 7 to 23 nm.

Lattice parameter (a) of the individual composition was calculated by using the following formula

$$a = d\sqrt{h^2 + k^2 + l^2}$$

where a = lattice parameter; d = interplanar spacing; hkl = the miller indices.

The lattice parameter was found to decrease linearly with increase of Cr^{3+} ions in the Mg-Cr nano-ferrites system indicating that the system obeys Vegard's law [11].

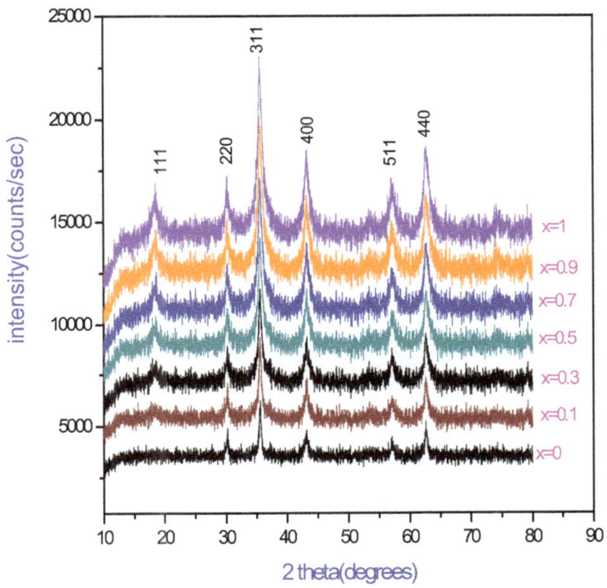

Figure 1. XRD patterns of $MgCr_xFe_{2-x}O_4$ (x = 0.0, 0.1, 0.3, 0.5, 0.7, 0.9 and 1.0).

3.2. Dielectric Properties

3.2.1. Dielectric Constant (ε' and ε'')

The frequency dependence of the real and imaginary part of dielectric constant (ε' and ε'') for all the samples was studied at room temperature in the range of 20 Hz to 2 MHz. **Figures 2** and **3** depict the variation of the real and imaginary part of the dielectric constant (ε' and ε'') as a function of frequency for mixed ferrites $MgCr_xFe_{2-x}O_4$ with different compositions (x = 0.0, 0.1, 0.3, 0.5, 0.7, 0.9 and 1.0). It is observed that all the samples have higher dielectric constant at lower frequency 20 Hz and there is a decreasing trend in value with increasing frequency from 1000 Hz to 2 MHz which is a normal behavior of ferromagnetic materials. The decrease in ε' is sharp initially from 20 Hz to 1000 Hz (lower frequency) and then ε' value decreases slowly with the increase in frequency and showed almost frequency independent behavior at high frequency regions [12]. Similar behavior was observed in our publications on Mg-Zn Ferrites (Ravinder and Latha, 1999), Li-Cd ferrites (Radha and Ravinder, 1995). This normal dielectric behavior of spinel ferrites was attributed to the lagging of existing charge carriers (hopping electrons) between Fe^{2+} and Fe^{3+} ions at localized sites, where the Fe^{2+} ions were formed in the samples during the sintering process at high temperature, responsible for polarization behind the applied field as its frequency increases [13,14]. The data revealed that none of the samples exhibit any anomalous behavior of peaking. The variation of dielectric constant with frequency may be explained on the basis of space-charge polarization phenomenon [15]. According to this, dielectric material has well conducting grains separated by highly resistive grain boundaries. On the application of electric field, space charge accumulates at the grain boundaries and voltage drops mainly at grain boundaries [16]. Koops proposed that grain boundary effect is more at low frequencies [16]. As the frequency increased beyond a certain limit the electron exchange between Fe^{2+} and Fe^{3+} ions does not follow the variations in applied field, so the value of dielectric constant becomes constant. According to Maxwell and Wagner [17,18] two layer model, space-charge polarization is because of the inhomogeneous dielectric structure of the material. It is formed by large well conducting grains separated by thin poorly conducting intermediate grain boundaries. Rabinkin and Novikova [19] pointed out that polarization in ferrites is similar to that of conduction. The electron exchange between Fe^{2+} and Fe^{3+} ions results in local displacement of electrons in the direction of applied field that determines polarization. Polarization decreases with increasing value of frequency, and then reaches a constant value. It is due to the fact that beyond a certain frequency of external field, the electron exchange $Fe^{2+} \leftrightarrow Fe^{3+}$ cannot follow the alternating field. The high value of dielectric constant at lower frequency is due to the predominance of the species like Fe^{2+} ions, oxygen vacancies, grain boundary defects, etc. [18] while the decrease in dielectric constant with frequency is natural that is any species contributing to the polarizability is found to show the applied field lagging behind at higher frequencies [20].

3.2.2. Loss Tangent (tan δ)

The value of tan δ measures the loss of electrical energy from the applied electric field into the samples at different frequencies. It is observed that the tan δ shows a decreasing trend with increasing in frequency. The loss tangent (tan δ) is defined as the ratio of the loss or resistive current to the charging current in sample. Also it is known that there is strong correlation, between the conduction

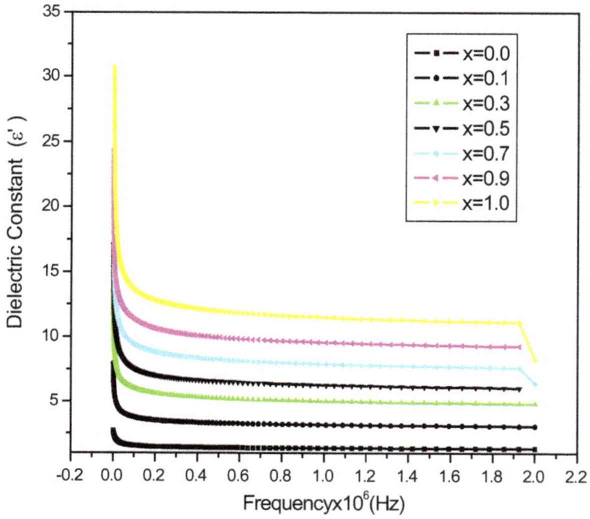

Figure 2. Variation of real part of dielectric constant ε' with frequency.

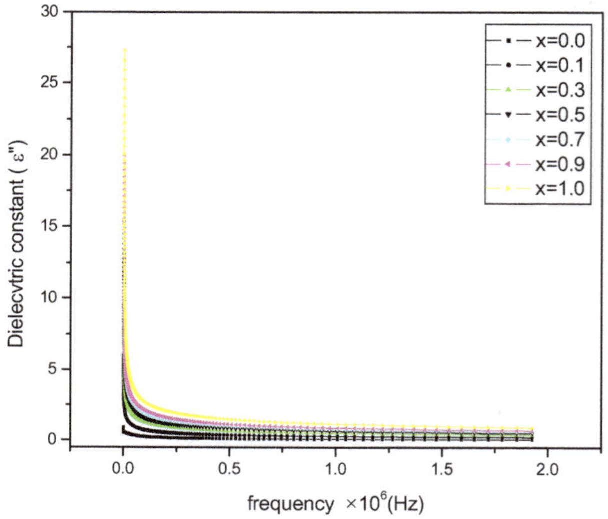

Figure 3. Variation of imaginary part of dielectric constant ε' with frequency.

mechanism and the dielectric constant behavior (Polarization mechanism) in ferrites. Variation of tan δ (the loss tangent) as a function of frequency (20 Hz to 2 MHz) at room temperature for all compositions is shown in **Figure 4**. The tan δ decreases with the increasing frequency which is normal behavior of any ferrite materials. The values of dielectric loss tangent decrease from 0.09 (x = 0.1) to 0.054 (x = 0.9) at 2 MHz. This shows that with increase in Cr concentration the energy losses decrease at high frequencies. The low loss values at higher frequencies show the potential applications of these materials in high frequency micro wave devices. From the figure it is clear that the loss decreases rapidly in the low frequency region while the rate of decrease is slow in high-frequency region and it shows an almost frequency independent behavior in high frequency region. The behavior can be explained on the basis that in the low frequency region, which corresponds to a high resistivity (due to the grain boundary), more energy is required for electron exchange between Fe^{2+} and Fe^{3+} ions, as a result the loss is high. In the high frequency region, which corresponds to a low resistivity (due to the grains), small energy is required for electron transfer between the two Fe ions at the octahedral site. Moreover, the dielectric loss factor also depends on a number of factors such as stoichiometry, Fe^{2+} content, and structural homogeneity which in turn depend upon the composition and sintering temperature of the samples [21].

3.2.3. AC Conductivity (σAC)

Conductivity is the physical property of a material which characterizes the conducting power inside the material. The electrical conductivity in ferrites is mainly due to the hopping of electrons between the ions of the same element presented in more than one valence state. **Figure 5** shows the variation of the AC conductivity (σ_{AC}) of mixed Mg-Cr ferrites of all compositions as a function of frequency in the range of 20 Hz to 2 MHz at room temperature. At low frequency range the AC conductivity was nearly independent of the frequency and showed an increasing trend with increase in frequency for all the samples. This behavior is akin to Maxwell-Wagner type. The dielectric structure of ferrites is given by Koops phenominologics theory and Maxwell-Wagner theory [18,22]. At lower frequencies the conductivity was found low due to the grain boundaries that are more active which acts as hindrance for mobility of charge carriers and hence the hopping of Fe^{2+} and Fe^{3+} ions is less at lower frequencies. As the frequency of applied field is increased, the conductive grains become more active thereby promoting the hopping between Fe^{2+} and Fe^{3+} ions and also responsible for creating charge carriers from different centers. These charge carriers take part in the conduction phenomenon thereby increasing the AC conductivity. The linear increase in conductivity was observed with frequency that confirms the polaron type of conduction. The frequency dependent conduction is attributed to small polarons [23]. At higher frequency where conductivity increases greatly with frequency, the transport is dominated by contributions from hopping infinite clusters.

3.2.4. Compositional Dependence of Dielectric Parameters (ε', ε'', tan δ and σ_{AC})

Figures 6-9 represent the variation of dielectric parameters (ε', ε'', tan δ and σ_{AC}) as a function of Cr composition at selected frequencies (20 Hz, 1000 Hz, 2000 Hz, 3000 Hz, 1 MHz and 2 MHz) respectively. It can be seen that all the dielectric parameters ε', ε'', tan δ and σ_{AC} increase upto 10% of Cr doping, thereafter, these parameters decrease with further doping of Cr which is clear from **Table 1**. It shows the values of dielectric parameters for different compositions of the Mg-Cr ferrite system at particular frequencies. The initial increase in dielectric constant up to x = 0.3 may be due to the formation of Fe^{3+}

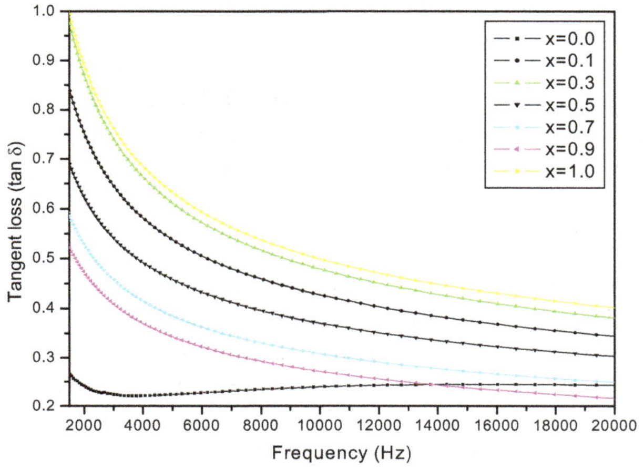

Figure 4. Variation of dielectric loss tangent (tan δ) with frequency.

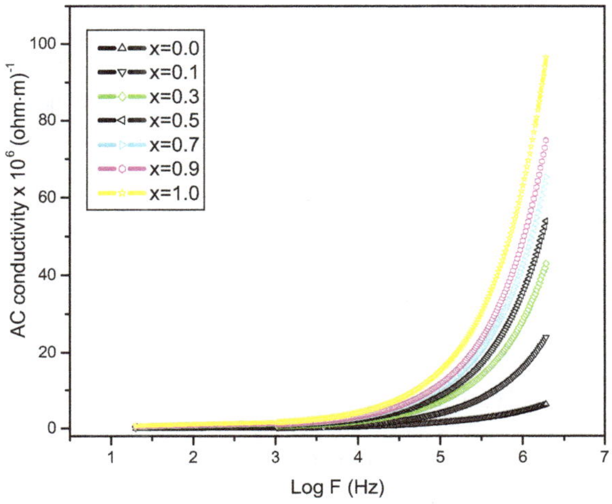

Figure 5. Variation of AC conductivity (σ_{AC}) with log F.

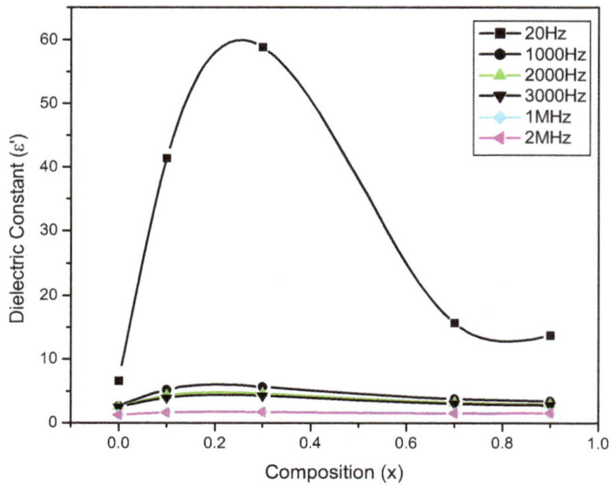

Figure 6. Variation of ε' with composition at selected frequencies.

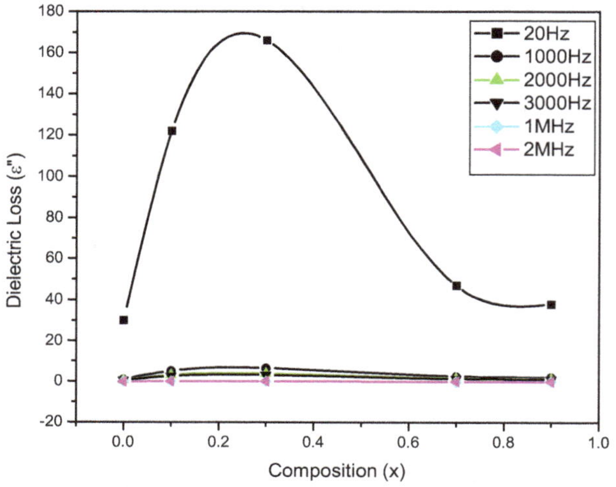

Figure 7. Variation of ε'' with composition at selected frequencies.

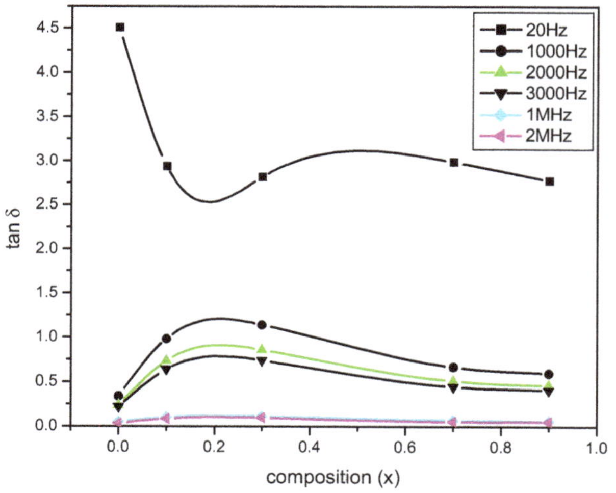

Figure 8. Variation of tan δ with composition at selected Frequencies.

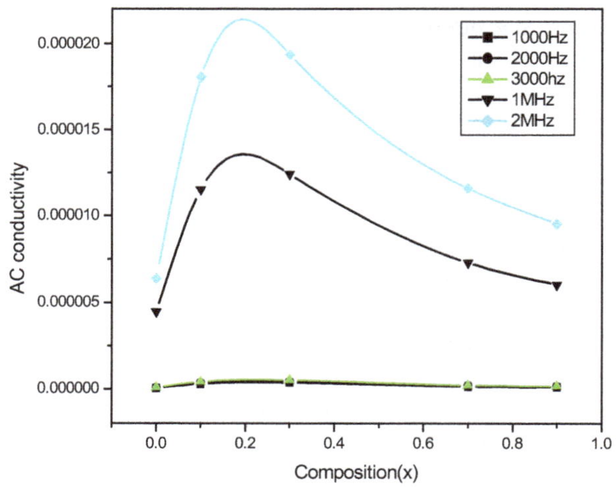

Figure 9. Variation of σ_{AC} with composition at selected frequencies.

ions on octahedral sites. This typical behavior can be explained on the basis that in Cr containing ferrites, Cr ions prefer to occupy the octahedral coordination until the ratio of Cr substitution becomes greater than 60%, whereafter, Cr ions may increase in tetrahedral sites causing migration of equal number ions to the octahedral sites [24]. The behavior can be explained by assuming that the mechanism of dielectric polarization is similar to that of the conduction in ferrites (Robinkin and Novikova, 1960). They observed that the electronic exchange interaction between $Fe^{2+}\leftrightarrow Fe^{3+}$ results in local displacement of the electrons in the direction of an electric field which determines the polarization of ferrites. The presence of Fe^{2+} ions in excess amount favors the polarization effect [25]. Thus more dispersion was observed in the sample with low Cr^{3+} ion substitution. This is because at low Cr^{3+} concentration the presence of Fe^{2+} ions is in excess amount. As the Cr^{3+} ion substitution increased it occupies the octahedral site in the ferrite system, thereby decreasing the number of Fe^{3+} ions and there is a least possibility of electronic exchange interaction between $Fe^{2+}\leftrightarrow Fe^{3+}$, which results in decrease in dielectric parameters with increasing Cr content in the present system. **Table 1** shows the values of dielectric parameters for different compositions of the Mg-Cr ferrite system at selected frequencies.

4. Conclusions

We have successfully synthesized single phase $MgCr_xFe_{2-x}O_4$ nano ferrites with cubic spinel structure through Citrate-gel auto combustion method with very fine crystallite size of the particles ranging from 7 - 23 nm.

- The dielectric constant of all the ferrites is very low at low frequency which may be due to low content of polarizable Fe^{2+} ions on the octahedral site.

Table 1. Dielectric parameters of different compositions of $MgCr_xFe_{2-x}O_4$ at selected frequencies (20 Hz, 3000 Hz and 2 MHz).

Composition	At frequency 20 Hz				At frequency 3000 Hz				At frequency 2 MHz			
	ε'	tan δ	ε''	σ_{AC}	ε'	tan δ	ε''	σ_{AC}	ε'	tan δ	ε''	σ_{AC}
0.0	6.65	4.51	30	3.34×10^{-8}	2.54	0.223	0.57	9.50×10^{-8}	1.35	0.04	0.06	6.43×10^{-6}
0.1	41.4	2.94	122	1.35×10^{-7}	3.99	0.64	2.58	4.31×10^{-7}	1.72	0.09	1.06	1.81×10^{-5}
0.3	58.8	2.85	166	1.84×10^{-7}	4.28	0.739	3.17	5.3×10^{-7}	1.76	0.1	0.17	1.94×10^{-5}
0.7	15.7	2.97	47	5.22×10^{-8}	3.06	0.45	1.40	2.34×10^{-7}	1.61	0.06	0.11	1.16×10^{-5}
0.9	13.7	2.78	38	4.25×10^{-8}	2.81	0.41	1.15	1.92×10^{-7}	1.61	0.05	0.09	9.56×10^{-6}

- A normal dispersion in dielectric parameters (ε', ε'', tan δ) with frequency was observed for all samples and this has been explained on the basis of space charge polarization mechanism as discussed in Maxwell-Wagner model.
- AC conductivity measurement indicates that with increase in Cr^{3+} substitution Mg-Cr nano ferrites, AC conductivity increases linearly with frequency which suggest that the conduction in the present system may be due to the polaron hopping mechanism.
- The values of dielectric loss tangent decrease from 0.09 (x = 0.1) to 0.054 (x = 1.0) at 2 MHz. This shows that with increase in Cr concentration in Mg-Cr nano ferrites the energy losses decrease at high frequencies. The low loss values at higher frequencies show the potential applications of these materials in high frequency micro wave devices.

5. Acknowledgements

One of the authors (MRS) is grateful to K. S. Ravikumar, Chairman, Jayaprakash Narayan College of Engineering, Mahabunagar (Dist) for his support and continuous encouragement in carrying out research work. One of the authors (D.R) is grateful to Prof. T. L. N. Swamy, Principal Nizam College for his encouragement to carry out this research work. The authors are thankful to Prof. C. Gyana Kumari, Head, Department of Chemistry, Osmania University, Hyderbad for her encouragement in carrying out the research activities.

REFERENCES

[1] Y. Yamamoto and J. Makino, "Core Losses and Magnetic Properties of Mn-Zn Ferrites with Fine Grain Sizes," *Journal of Magnetism and Magnetic Materials*, Vol. 133, No. 1-3, 1994, pp. 500-503. doi.10.1016/0304-8853(94)90607-6.

[2] J. Smit and H. P. G. Wijn, "Ferrites," John and Wiley & Sons, Inc., Hoboken, 1959, p. 136.

[3] M. Sugimoto, "The Past, Present, and Future of Ferrites," *Journal of Ceramic American Society*, Vol. 82, No. 2, 1999, p. 269.

[4] K. Kondo, T. Chiba and S. Yamada, "Effect of Microstructure on Magnetic Properties of Ni-Zn Ferrites," *Journal of Magnetism and Magnetic Materials*, Vol. 541, 2003, p. 254.

[5] M. P. Reddy, G. Balakrishnaiah, W. Madhuri, M. Venkata Ramana, N. Rammanohar Reddy, K. V. Siva Kumar, V. R. K. Murthy and R. Ramakrishna Reddy, "Structural, Magnetic and Electrical Properties of NiCuZn Ferrites Prepared by Microwave Sintering Method Suitable for MLCI Applications," *Journal of Physis and Chemistry of Solids*, Vol. 71, 2010, p. 1373.

[6] L. B. Kong, Z. W. Li, G. Q. Lin and Y. B. Gan, "Magneto-Dielectric Properties of Mg-Cu-Co Ferrite Ceramics: II. Electrical, Dielectric, and Magnetic Properties," *Journal of American Ceramic Society*, Vol. 90, 2007, p. 2014. doi:10.1111/j.1551-2916.2007.01691.x

[7] Y. Konseoglu, H. Kavas and B. Aktas, "Surface Effects on Magnetic Properties of Superparamagnetic Magnetite Nanoparticles," *Physics Status Solidi A*, Vol. 203, 2006, p. 1595. doi:10.1002/pssa.200563104

[8] S. L. Gupta and S. Gupta, "Electricity, Magnetism and Electronics," Vol. 1, Jaiprakash Nath and Company, 1991, p. 60.

[9] E. Pervaiz and I. H. Gul, "Enhancement of Electrical Properties Due to Cr^{+3} Substitution in Co-Ferrite Nano Particles Synthesized by Two Chemical Techniques," *Journal of Magnetism and Magnetic Materials*, Vol. 324, No. 22, 2012, pp. 3695-3703. doi.10.1016/j.jmmm.2012.05.050

[10] B. D. Cullity, "Elements of XRD," Addison Weseley Publishing, Redading, 1959, p. 132.

[11] M. J. Iqbal, M. N. Ashiq, P. Hernandez-Gomezb and J. M. Munoz, "Synthesis, Physical, Magnetic and Electrical Properties of Al-Ga Substituted Co-Precipitated Nanocrystalline Strontium Hexaferrite," *Journal of Magnetism and Magnetic Materials*, Vol. 320, 2007, No. 6, pp. 881-886. doi.10.1016/j.jmmm.2007.09.005

[12] B. Chandra Babu, V. Naresh, B. Jayaprakash and S. Buddhudu, "Structural, Thermal and Dielectric Properties of Lithium Zinc Silicate Ceramic Powders by Sol-Gel Method," *Ferro Electrics Letter*, Vol. 38, 2011, p. 124.

[13] S. Sindhu, M. R. Antharaman, B. P. Thampi, K. A. Malini and Ph. Kurian, "Tailoring Magnetic and Dielectric

Properties of Rubber Ferrite Composites Containing Mixed Ferrites," *Bulletin Materials Science*, Vol. 24, 2001, pp. 623-629. doi.10.1007/BF02704011

[14] A. A. Zaky and R. Hawley, "Dielectric Solids," Routledge and Kegan Paul Ltd., London, 1990, p. 21

[15] N. Rezlescu and E. Rezlescu, *Physics Status Solidi A*, Vol. 23, 1974, p. 575

[16] I. H. Gul, A. Maqsood, M. Naeem and M. Naeem Ashiq, "Optical, Magnetic and Electrical Investigation of Cobalt Ferrite Nanoparticles Synthesized by Co-Precipitation Route," *Journal of Alloys and Compounds*, Vol. 507, No. 1, 2010, pp. 201-206. doi.10.1016/j.jallcom.2010.07.155

[17] J. C. Maxwell, "Electricity and Magnetism," Vol. 1, Oxford University Press, Oxford, 1929.

[18] K. W. Wagner, "Zur Theorie der Unvolkommenen Di-Elektrika," *American Physics*, *American Physics*, Vol. 40, 1913, p. 817.

[19] I. T. Rabinkin and Z. I. Novikova, "Ferrites," IZV Acad. Nauk USSR Minsk, 1960.

[20] B. Baruwati, K. M. Reddy, V. Sunkara, R. K. Manorama, O. Singh and J. Prakash, "Tailored Conductivity Behavior in Nanocrystalline Nickel Ferrite," *Applied Physics Letters*, Vol. 85, No. 14, 2004. doi.10.1063/1.1801685

[21] R. S. Devan and B. K. Chougule, "Effect of Composition on Coupled Electric Magnetic and Dielectric Properties of Two Phase Particulate Magnetoelectric Composite," *Journal of Applied Physics*, Vol. 101, No. 1, 2007, Article ID: 014109

[22] C. G. Koops, *Physics Revision*, "On the Dispersion of Resistivity and Dielectric Constant of Some Semiconductors at Audiofrequencies," Vol. 83, No. 1, 1951, pp. 121-124. doi.10.1103/PhysRev.83.121

[23] D. Alder and J. Fienleib, "Electrical and Optical Properties of Narrow-Band Materials," *Physics Revision B*, Vol. 2, 1970, p. 3112. doi.10.1103/PhysRevB.2.3112

[24] D. Elkony, "Egypt," *Journal of Solids*, Vol. 27, No. 2, 2004, p. 285.

[25] S. E. Shirsath, B. G. Toksh, M. K. L. Mane, V. N. Dhage, D. R. Shengule and K. M. Jadhav, "Frequency, Temperature and In^{+3} Dependent Electrical Conduction in $NiFe_2O_4$ Powder," *Powder Technology*, Vol. 212, No. 1, 2011, pp. 218-223. doi.10.1016/j.powtec.2011.05.019

Dielectric Property Studies of Biologically Compatible Brushite Single Crystals Used as Bone Graft Substitute

M. P. Binitha, P. P. Pradyumnan*

Department of Physics, University of Calicut, Kerala, India.
Email: *drpradyumnan@gmail.com

ABSTRACT

The electrical characterization of bone is essential for the better understanding of the role of electrical stimulation in bone remodeling. Calcium Hydrogen Phosphate Dihydrate or brushite ($CaHPO_4$ $2H_2O$) has been used in bone substitution owing to their fast resorption under physiological condition. Brushite is a suitable matrix for osteoconductive bone grafts. In this work, Calcium Hydrogen Phosphate single crystals have been grown by single diffusion gel growth technique. The powder XRD studies revealed the monoclinic structure of the grown crystals. The vibrational analysis of the crystals is done with FTIR spectroscopy and the major functional groups and their assigned vibrations are discussed. The frequency dependence of dielectric constant and ac conductivity at different temperatures have been studied in detail. This study shows decrease in the dielectric constant with the increase in frequency and temperature. The variation of ac conductivity is found to be increasing with frequency and decreasing with temperature.

Keywords: Bone Grafts; Brushite; Gel Growth; Dielectric Properties; AC Conductivity

1. Introduction

Bone is one of the natural materials made up of organic as well as inorganic components. Although bone has the ability to repair itself following damage, in the case of complex bone fractures which fails to heal properly, causing a significant health risk to the patients, bone grafting is needed. Bone grafts may be autograft, allograft or synthetic with similar mechanical properties to bone. Most bone grafts are expected to be reabsorbed and replaced as the natural bone heals over a few months time. The ideal bone substitute should have osteoconductive and osteoinductive properties [1].

Approximately 60% of the bone graft substitutes currently available involve ceramics, either alone or in combination with another material. These include calcium sulfate, bioactive glass, and calcium phosphate [1]. In the past 20 years the most frequently used synthetic bone graft has been the calcium orthophosphates ceramics. One class of calcium orthophosphates used in bone grafting applications are the calcium phosphate cements, which are the combination of a solid component containing one or more calcium orthophosphate powder with an aqueous solution [2]. The calcium phosphate cements can overcome many problems associated with autologous grafting such as scarcity of grafting material and secondary morbidity and the risk of biological contamination associated with allograft [3]. In addition, calcium phosphates are osteoconductive, osteointegrative and in some cases, osteoinductive. Depending on the pH value of the cement paste the end product of the cement setting reaction may be either brushite or hydroxyapatite. The advantage of hydroxyapatite over brushite is that it is usually set at a pH value closer to the physiological environment. Hence it has been more frequently used. However, brushite is more soluble than hydroxyapatite in physiological condition and hence is more completely resorbed after its implantation in animal bodies [2]. Thus brushite is a suitable matrix as osteoconductive bone grafts [4,5]. Several clinical studies have shown that injectable brushite cements are capable of regenerating oral and maxillofacial bone in atrophic areas, buccal dehiscence defects and maxillary sinuses. Also brushite cements have been used to fill bone defects created by pathological tumors [6]. In the newly formed natural bone itself, the occurrence of brushite is reported [7].

While electrical stimulation is used for healing frac-

*Corresponding author.

tures, the electrical characterization of bone has much clinical importance. Furthermore, earlier studies have suggested that electrical and dielectric properties of natural bone depend on the bone density and could, therefore, be used to predict bone strength. Information on the inter relationship between electrical and dielectric properties with bone mineral density would enable the estimation of field distribution during electrical stimulation. In this context the electrical characterization of the bone grafting material is worthwhile and hence we have investigated the electric as well as dielectric properties of brushite. The electrical measurements were conducted in a wide frequency range, as motivated by the fact that electromagnetic pulses used in the stimulation of bone healing contain a wide range of frequencies as well [8].

2. Experimental

The growth of calcium phosphate has been carried out using single diffusion of calcium chloride through the hydrosilica gel medium. Because of the viscous nature of gel, this growth technique provides an in vitro model for crystallization of biomolecules [9]. The silica gel was set by acidification of sodium meta silicate solution of specific gravity 1.03 with 0.5 M orthophosphoric acid to get a pH of 6 [10,11]. This gel is found to set after 24 hours. After assuring proper gel setting, 7 ml of 0.5 M calcium chloride solution was slowly poured over it, without disturbing the gel surface. Small crystals appeared in the gel, which were mostly platelets or needle shaped. After a growth period of 6 weeks the crystals were harvested and characterization studies were carried out. The structural analyses were done by X-ray diffraction method using Cu-Kα monochromator of wavelength 1.541 Å (XRD). The vibrational analysis was done by Fourier Transform Infra Red (FTIR) spectroscopy on the powdered sample by the standard KBr pellet method in the region 400 - 4000 cm^{-1}. Differential Scanning Calorimeter (DSC) was used to measure the thermal characteristics of grown crystals. Dielectric studies were conducted by LCR Hi TESTER 3532-50.

3. Result and Discussion

XRD pattern of the grown crystal is given in **Figure 1**. XRD studies confirmed that the crystal structure is that of brushite having monoclinic structure. The values of cell parameters are a = 5.812 Å, b = 15.18 Å, c = 6.239 Å, $\alpha = 90^\circ$, $\beta = 116.43^\circ$, $\gamma = 90.00^\circ$ (JCPDS 72-0713).

FTIR spectrum of brushite crystals is shown in **Figure 2**. The broad absorption peak between 3500 - 2400 cm^{-1} is due to O-H stretching vibration. The H-O-H bending gives rise to absorption at 1648.27 cm^{-1}. The absorptions at 1214.35 and 1132.03 cm^{-1} are due to P=O associated stretching vibrations. The absorption at 1065.32 cm^{-1} is

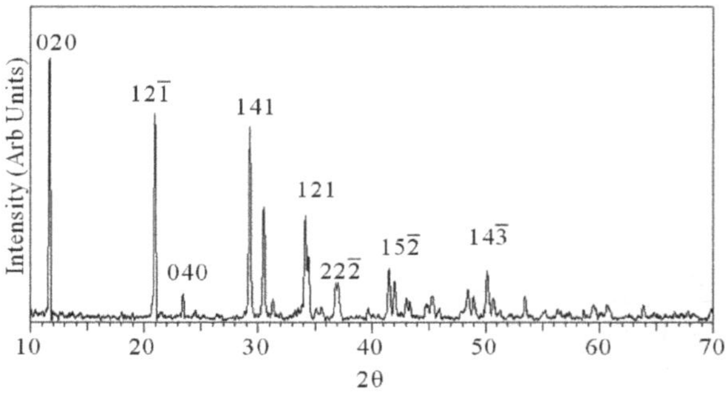

Figure 1. XRD pattern of grown brushite.

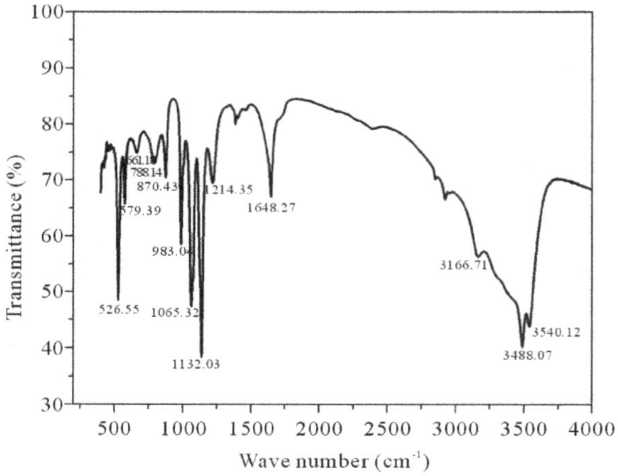

Figure 2. FT-IR spectrum of grown brushite.

due to P=O stretching vibrations. The P-O-P asymmetric stretching vibrations give rise to absorptions at 983.04, 870.43 and 788.14 cm^{-1}. The absorptions at 661.13, 579.39, 526.55 cm^{-1} are due to (H-O-)P=O for acid phosphates [12,13].

The differential thermal analysis was done with a heat flow from 50.00°C to 375.00°C at 10.00°C/min using the instrument Perkin Elmer DSC 4000. The thermograms of brushite crystals (**Figure 3**) show two endothermic peaks, a broad one at around 149.38°C corresponding to the onset of evaporation of water of crystallisation and therefore the partial dehydration of brushite and a well defined peak at 190.85°C associated with the phase transition of the crystal due to the total dehydration [13]. At this temparature brushite is completely transformed to monatite according to the relation

$$CaHPO_4\ 2H_2O \rightarrow CaHPO_4 + 2H_2O \quad (1)$$

4. Conductivity Studies

The variation of dielectric constant (ε_r) and ac conductivity (σ_{ac}) of the crystal with frequency of applied field ranging from 100 Hz to 5 MHz is studied at different temperatures. The samples were finely ground and made in the form of pellets. The dielectric constant (ε_r) is calculated using the relation

$$\varepsilon_r = \frac{Cd}{\varepsilon_0 A} \quad (2)$$

and the ac conductivity is calculated by the relation

$$\sigma_{ac} = \varepsilon_0 \varepsilon_r \omega \tan\delta \quad (3)$$

where C is the capacitance, d is the thickness, A is the area of cross section of pellet and $\tan\delta$ is the dielectric relaxation of the sample. It is observed that the dielectric constant and ac conductivity decrease with increasing temperature as shown in **Figures 4** and **5** respectively. Also the values of dielectric constant decreases with increasing frequency, while the ac conductivity value increases with increasing frequency. The dielectric constant is high at lower frequency region and decreases with increasing frequencies. The high value of dielectric constant at low frequencies is attributed to space charge polarization. At lower range of frequencies dielectric constant decreases drastically with frequency becoming a constant at higher frequencies. The polarization occurs due to the local displacement of electrons, which is the effect of the electronic exchange of the number of ions in the crystal. As frequency increases the electron exchange cannot follow the electric field and the polarization becomes independent of frequency.

Variation of dielectric constant with temperature is generally attributed to the orientational polarization, crystal expansion, the presence of impurities and crystal defects. When temperature increases the dielectric constant decreases, as evident from the **Figure 4**. The thermal energy disrupts the ion dipole interaction which is responsible for polarization at higher temperatures, causing the relaxation of polarization. The ac conductivity increases with frequency and decreases with temperature [14,15].

During open surgery for bone grafting or total hip replacement, there is no method for direct measurement of

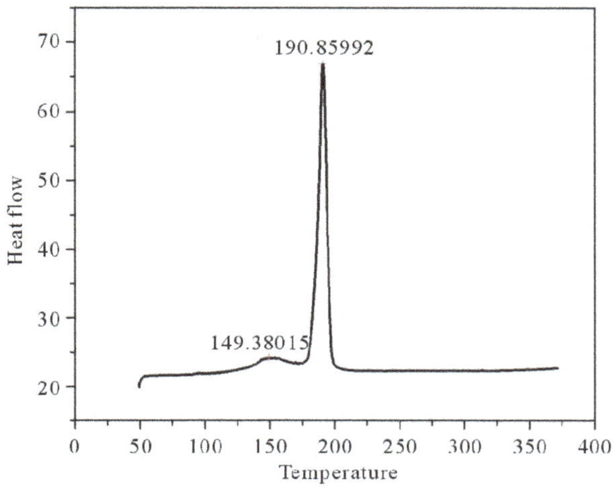

Figure 3. DSC plot of grown brushite.

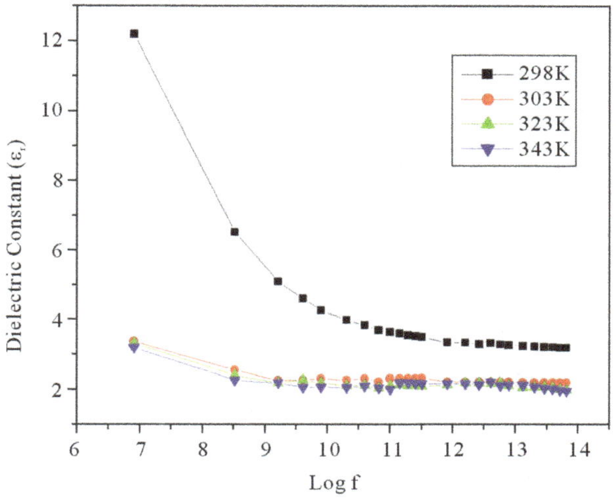

Figure 4. Variation of dielectric constant with frequency.

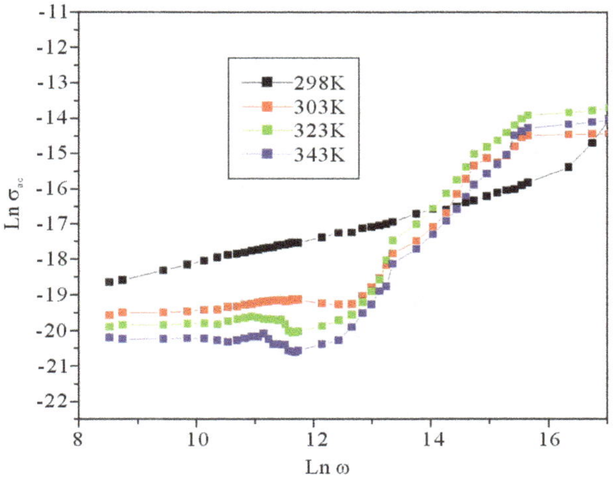

Figure 5. Variation of ac conductivity with frequency.

bone quality. Electrical measurements utilizing open-ended probe technology are potentially feasible in this kind of application. In addition, better understanding of the relationships between bone density, mechanical, electrical and dielectric properties may help us to improve techniques related to electrical stimulation and follow-up of healing processes in fractured bone. Several studies were made to study the variation of dielectric constant of bone as a function of frequency and found that the dielectric constant decreases with frequency [16,17].

5. Conclusion

Brushite crystals are grown successfully by gel method and crystal structure is determined to be monoclinic. The FTIR studies confirmed the major functional groups in the crystal. The dielectric constant shows decrease with frequency, attaining a constant value at higher frequencies. Also the dielectric constant decreases with temperature. The ac conductivity is found to be increasing with frequency and decreasing with temperature.

6. Acknowledgements

M.P.B would like to acknowledge UGC, Govt. of India, for the award of FDP (KLCA 062 TF 01). She is also thankful to University of Calicut and Govt. College, Kodanchery, for their support.

REFERENCES

[1] W. R. Moore, S. E. Graves and G. I. Bain, "Synthetic Bone Graft Substitutes," *ANZ Journal of Surgery*, Vol. 71, No. 6, 2001, pp. 354-361.

[2] Z. D. Xia, L. M. Grover, Y. Z. Huang, I. E. Adamopoulos, U. Gbureck, J. T. Triffitt, R. M. Shelton and J. E. Barralet, "*In Vitro* Biodegradation of Three Brushite Calcium Phosphate Cements by a Macrophage Cell-Line," *Biomaterials*, Vol. 27, No. 26, 2006, pp. 5457-4565. doi:10.1016/j.biomaterials.2006.04.030

[3] G. Penel, N. Leroy, P. Van Landuyt, B. Flautre, P. Hardouin, J. lemaitre and G. Leroy, "Raman Microspectrometry Studies of Brushite Cement: *In Vivo* Evolution in a Sheep Model," *Bone*, Vol. 25, No. 2, 1999, pp. 81S-84S. doi:10.1016/S8756-3282(99)00139-8

[4] H. N. Lim, A. Kassim, N. M. Huang, M. A. Armo, P. S. Khiew and W. S. Chiu, "Preparation and Characterization of Brushite Crystals Using High Internal Phase Emulsion," *Colloid Journal*, Vol. 71, No. 6, 2009, pp. 793-802. doi:10.1134/S1061933X09060088

[5] S. V. Dorozhkin, "Calcium Orthophosphates as Bioceramics: State of the Art," *Journal of Functional Biomaterials*, Vol. 1, No. 1, 2010, pp. 22-107

[6] F. Tamimi, Z. Sheikh and J. Barralet, "Dicalcium Phosphate Cements: Brushite and Monetite," *Acta Biomaterialia*, Vol. 8, No. 2, 2012, pp. 474-487. doi:10.1016/j.actbio.2011.08.005

[7] W. F. Neuman, M. W. Neuman, A. G. Diamond, J. Menanteau and W. Gibbons, "Blood: Bone Disequilibrium. VI. Studies of the Solubility Characteristics of Brushite: Apatite Mixtures and Their Stabilization by Noncollagenous Proteins of Bone," *Calcified Tissue International*, Vol. 34, No. 1, 1982, pp. 149-157

[8] J. Sierpowska, J. Toyras, M. A. Hakulinen, S. Saarakkala, J. S. Jurvelin and R. Lappalainen, "Electrical and Dielectric Properties of Bovine Trabecular Bone—Relationships with Mechanical Properties and Mineral Density," *Physics in Medicine and Biology*, Vol. 48, No. 6, 2003, pp. 775-786. doi:10.1088/0031-9155/48/6/306

[9] K. Suguna and C. Sekar, "Role of Strontium on the Crystallization of Calcium Hydrogen Phosphate Dihydrate (CHPD)," *Journal of Minerals and Materials Characterization and Engineering*, Vol. 10, No. 7, 2011, pp. 625-636. doi:10.1002/crat.200610826

[10] C. Justin Raj, G. Mangalam, S. Mary Navis Priya, J. Mary Linet, C. Vesta, S. Dinakaran, B. Milton Boaz and S. Jerome Das, "Growth and Characterization of Nonlinear Optical Zinc Hydrogen Phosphate Single Crystal Grown in Silica Gel," *Crystal Research and Technology*, Vol. 42, No. 4, 2007, pp. 344-348. doi:10.1002/crat.200610826

[11] V. Mahalakshmi, A. Lincy, J. Thomas and K. V. Saban, "Crystal Growth and Characterization of a New Co-Ordination Complex—Barium Tetrakismaleate Dihydrate," *Journal of Physics and Chemistry of Solids*, Vol. 73, No. 4, 2012, pp. 584-588. doi:10.1016/j.jpcs.2011.12.012

[12] G. Madhurambal, R. Subha and S. C. Mojumdar, "Crystallizaion and Thermal Characterization of Calcium Hydrogen Phosphate Dihydrate Crystals," *Journal of Thermal Analysis and Calorimetry*, Vol. 96, No. 1, 2009, pp. 73-76. doi:10.1007/s10973-008-9841-1

[13] K Rajendran and C. Dale Keefe, "Growth and Characterization of Calcium Hydrogen Phosphate Dihydrate Crystals from Single Diffusion Gel Technique," *Crystal Research and Technology*, Vol. 45, No. 9, 2010, pp. 939-945. doi:10.1002/crat.200900700

[14] B. Parekh, M. Joshi and A. Vaidya, "Characterisation and Inhibitive Study of Gel Grown Hydroxyapatite Crystals at Physiological Temperature," *Journal of Crystal Growth*, Vol. 310, No. 7, 2008, pp. 1749-1753. doi:10.1016/j.jcrysgro.2007.11.219

[15] K. Arora, V. Patel, B. Amin and A. Kothari, "Dielectric Behaviour of Strontium Tartrate Single Crystals," *Bulletin of Material Science*, Vol. 27, No. 2, 2004, pp. 141-147. doi:10.1007/BF02708496

[16] G. N. Reddy and S. Saha, "Electrical and Dielectric Properties of Wet Bone as a Function of Frequency," *IEEE Transactions on Biomedical Engineering*, Vol. BME-31, No. 3, 1984, pp. 296-303.

[17] A. A. Narino, R. O. Becker and C. H. Bachman, "Dielectric Determination of Bound Water of Bone," *Physics in Medicine and Biology*, Vol. 12, No. 3, 1967, pp. 367-378. doi:10.1088/0031-9155/12/3/309

Analysis on the Influence Factors of Capacitor Voltage Transformer Dielectric Loss Measurement

Yongdong Li, Qing-da Meng, Po, Yang Zheyuan Zhao, Wei Zhang, Zhuo Pan
Jibei Electric Power Maintenance Company, Beijing, China
Email: yangpo0416@163.com

ABSTRACT

Capacitor voltage transformer (CVT), which is with simple structure, convenient maintenance, functional diversity and high impact pressure strength, is widely used. And its capacitance and dielectric loss Angle measurement is an important test on testing the insulation of the equipment. This paper is mainly to introduce and discuss one of the CVT test methods——the "self excitation method", combined with the actual situation encountered in the work, sums up and expounds the maintenance of CVT and preventive test of influence factors.

Keywords: Capacitive Voltage Transformer; Dielectric Loss; the Influencing Factors

1. Introduction

In recent years, the capacitive voltage transformer is widely used, because of insulation structure reasonable, high dielectric strength, can make full use of the carrier communication required for the coupling capacitor, in more than 110kV voltage grade [1].

2. The CVT Transmission Structure

Capacitive voltage transformer (CVT) is combined with the capacitive voltage divider, electromagnetic unit (including medium voltage transformer, reactor) and the terminal box. And the wiring principles are shown in **Figure 1**.

Capacitive voltage transformer, which is set as a converting device between the high tension line and ground, is to drop the system voltage into medium first, and after converting secondary voltage through intermediate transformer, the voltage will be output to the measuring instrument and relay protection device [2].

3. The Significance and Principle of CVT Dielectric Loss Measurement

For CVT in operation, its running situation overall is good, however, how to discover the CVT defects, such as the manufacturing quality, water be affected with damp be affected with damp, oil shortage, insulation aging factors and determine the operation state is of great significance to its safe and reliable operation. At present, each capacitance differential pressure unit and the dielectric loss value of CVT was tested by the power outage method as one of the preventive test project is also the principal means of judging its running status.

According to the structure it is divided into separate loading CVT and Superposition CVT, and the Superposition CVT which has no intermediate extraction voltage terminal is in widely used in China. When this kind of CVT is in field operation, we take the Self-excitation method to measure the partial pressure capacitor of capacitance and dielectric loss instead of the common testing methods [3]. Self-excitation method is based on the middle of the transformer as a testing transformer, excited voltage from the secondary side, Induction of high pressure as power supply for measurement. In recent years,

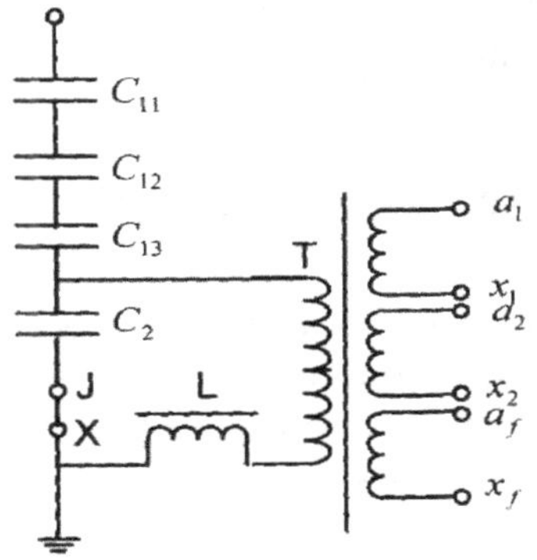

C_{11}, C_{12} - coupling capacitance; C_{13},C_2—partial-voltage capacitance ; L—compensation reactor; T—medium voltage transformer

Figure 1. The principle diagram of the capacitive voltage transformer.

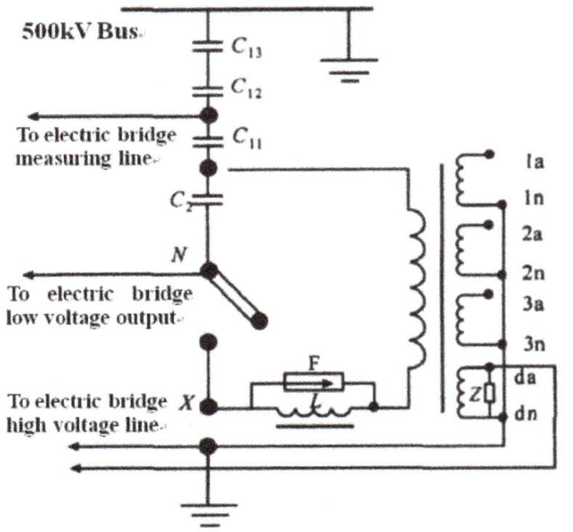

Figure 2. Self-excitation measurement schematic diagram.

the excitation method, which can achieve accurate measurement of the partial pressure of capacitance values and dielectric loss, has been used widely in the CVT routine test at the scene of the substation. Its measuring principle is shown in **Figure 2**.

In theory any secondary terminals can be used as a self-excited terminal test. But in fact, considering the capacitor voltage effect in the unit test process, we should choose the secondary terminal of large capacity as self-excited terminals to make the higher applied voltage and the test values are more accurate [4]. The capacity of da-dn in secondary terminal is the largest in most of the CVT, and with damping resistance winding, da-dn for the remainder of the experiment can make the process safer, so the self-excitation capacitance measuring pressure method generally choose da-dn terminal is pressurized.

Considering the insulation safety factors in low pressure end, the test voltage should not exceed the end of C_2 maximum allowable voltage, 2 kV is selected as a general method of self-excited test voltage in field experiment.

4. Analysis of Influence Factors of Measurement

In CVT dielectric loss measurement process, errors sometimes influence the accuracy of the measurement due to some of the factors. The followings analyzed some common influence factors.

4.1. The Influence of the Temperature

Temperature has a big influence on the measurement of dielectric loss Angle tanδ, the extent of the impact varies from the different materials, structures [5]. In general, tanδ increases with temperature rise. Some insulation material when temperature is below a certain threshold, the tanδ may be increased with the decreasing of temperature; And wet material under 0℃ when water freezes, tanδ will be lower. So, measured the insulation of the dielectric loss value in both too high and too low temperature cannot reflect the real situation, but easy to lead the wrong judgment. As a result, the measurement of dielectric loss should be not less than 5℃.

4.2. The Influence of the Test Voltage

In general, at its rated voltage range, tan delta value is almost unchanged. The tanδ of good insulation is not significantly increased with the rising of over voltage. But things are different if bubbles, delaminating, shell happens in the insulation [6]. When the test voltage is not enough to make the air bubbles or air gap of the insulation free, the tanδ and normal has no obvious difference; When test voltage makes the air in insulation free and produces corona or partial discharge, the tanδ will be increased as the test voltage increases. Several kinds of typical testing curve is shown in **Figure 3**.

4.3. Standard Capacitor

Capacitance and dielectric loss of the standard capacitor plays an important role in the measurement of CVT by applying self-excitation method. The accuracy of C_2 has certain relationship with these parameters since C_n is in series connection with C_1. For example, when a CVT (Model TYD110/√3—0.01 H, Nominal voltage divider C_1 = 12500 pF, C_2 = 50000 pF, dielectric loss tanδ = 0.1%) is tested with Cn = 100 pF, C_1 equivalent to series connection of a resister and a capacitor [7].

$$R_1 = \tan\frac{\delta}{\omega C_1} = \frac{0.001}{314} \times 12500 \times 10^{-12} = 255\Omega \quad (1)$$

$$C_n' = C_1 \times \frac{C_n}{(C_1 + C_n)} = 12500 \times 100 / \binom{12500}{+100} = 99.206\,\text{pF} \quad (2)$$

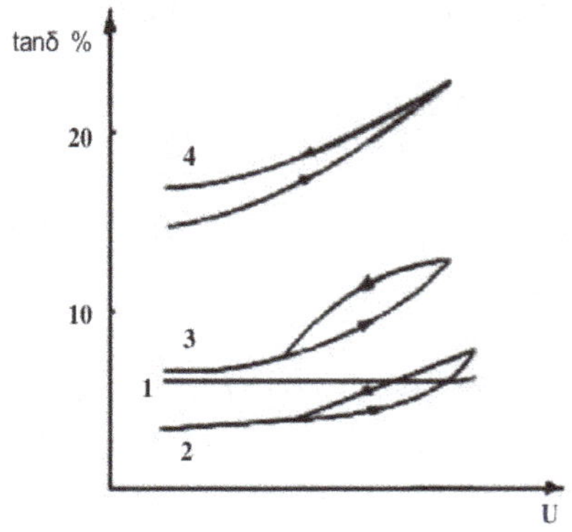

1- good insulation; 2 - when the insulation aging; 3 - there is air gap in insulation; 4 - insulation be affected with damp be affected with damp

Figure 3. Tan delta and the curve of voltage.

The error of the standard capacitor becomes:

$$(99.206 - 100) / 100 = -0.79\% \tag{3}$$

And the equivalent dielectric loss is:

$$\tan\delta' = R_0 C_n' = 255 \times 314 \times 99.206 \times 10^{-12} = 7.9 \times 10^{-6} \tag{4}$$

As is shown above, the measurement error is negligible as long as the standard capacitor is properly opted, i.e. capacitance high and dielectric loss low, or the accuracy will be affected.

4.4. Insulation of the Low Voltage End

As a result of insulation degradation of the terminal block on the low voltage end of the voltage divider, the measured dielectric loss of C_1 may exceed its true value by applying self-excitation method. Since terminal N has a relatively high potential about 2 kV in testing, leakage current upon the minor bushing and terminal block will lower down the accuracy due to the degradation of insulation, which is commonly caused by damp conditions of the secondary terminal box. In order to get the true value, measurements can be put out after drying process.

In the case of C_2 testing, the main influencing factor lies in the method of measurement, since terminal N has a relatively low potential and directly connects to the testing bridge. The error is negligible and the testing result makes approximately its true value.

Many factors will have influence on the testing result by applying self-excitation method, yet the measurement value of C_2 well indicates the real situation of the voltage divider, as it shares the same insulator and the corresponding dielectric loss are of the same, while the higher value of C_1 represents the damp conditions of secondary terminal and minor insulator.

4.5. Other Factors

The voltage of PT components interfere the voltage phase and amplitude of terminal N through the distributed capacitance coupling, thus affect the measurement accuracy [8].

Electric filed in the vicinity caused by other operating equipments also influence the testing result. To eliminate the interference, frequency conversion method is often applied in field testing.

5. Conclusions

As is discussed, the accuracy in the measurement of CVT dielectric loss is influenced by various factors, i.e. ambient temperature, testing voltage, standard capacitor and the insulation status of low voltage end, the last two of which have greater impact on the testing result and are commonly emerged in field. When the measurement value exceeds its true value, further analysis should be carried out before judgment so as to ensure the stable operation of power grid.

REFERENCES

[1] T. X. Chen, Y. Z. Wang and S. J. Hai, "Electrical Test," China Electric Power Press, Beijing, 2008.

[2] W. He, "Measurement of Tangent Loss in Capacitor Voltage Transformer," *Northwest China Electric Power*, Vol. 31, No. 5, 2003.

[3] W. L. Yang, H. Q. He and H. S. Wang, "Impacts of CVT Damped Secondary Outlet Block on Dielectric Loss Measurement by Self-Excitation Method," *East China Electric Power*, Vol. 39, No. 9, 2011.

[4] T. Li, X. P. Du and H. G. Liu, "Discussion about Self-excited Method Error on Capacitive Voltage Transformer," *Power System Protection and Control*, Vol. 37, No. 5, 2009.

[5] Q. Rao, "Discussion about Measurement on 110-220 kV Capacitive Potential Transformer," *Guangxi Electrical Engineering*, 2006.

[6] Z. M. Liang and Y. M. Tan, "Tanδ Test and Analysis for CVT's EM Unit," *High Voltage Engineering*, 2006.

[7] X.D. Jin and K. Y. Jing , "Self-excited Method for Superposition CVT," *Jiangsu Electrical Engineering*, 2005, Vol.11.

[8] Y. G. Yue, J. B. Yin and Y. P. Wang, "Analysis on 500 kV Capacitor Voltage Transformer Site Testing By Self-excitation Method," *Power Capacitor & Reactive Power Compensation*, Vol. 31, No. 4, 2010.

Synthesis and Dielectric Properties of $(0.80 - x)Pb(Cr_{1/5},Ni_{1/5},Sb_{3/5})O_3$-$xPbTiO_3$-$0.20PbZrO_3$ Ferroelectric Ceramics

Abdelhek Meklid[1*], Ahmed Boutarfaia[1,2], Zelikha Necira[1], Hayet Menasra[1], Malika Abba[1]

[1]Laboratory of Applied Chemistry, Materials Science Department, Mohamed Kheider University of Biskra, Biskra, Algeria; [2]University of Ouargla, Ouargla, Algeria.
Email: [*]abdelhek.meklid@yahoo.fr, aboutarfaia@yahoo.fr

ABSTRACT

Perovskite PZT variants were synthesized from stoichiometric oxide ratios of Pb, Zr, Ti, Cr, Ni and Sb. The oxide powders were mixed mechanically and calcinated, and then sintered to form the desired perovskite phase. The detailed structural and ferroelectric properties were carried out for sintered specimens. The results of X-ray diffraction showed that all the ceramics specimens have a perovskite phase. The multi-component ceramic system consists of the $(0.80 - x)Pb(Cr_{1/5},Ni_{1/5},Sb_{3/5})O_3$-$xPbTiO_3$-$0.20PbZrO_3$ (PZT-CNS), with $0.30 \leq x \leq 0.42$, and the ternary system near the rhombohedral/tetragonal morphotropic phase boundary(MPB) was investigated by X-ray diffraction and dielectric properties. In the present system, the MPB that coexists with the tetragonal and rhombohedral phases is a narrow composition region of x = 0.38 - 0.42. The scanning Electron Microscopy (SEM) showed an increase of the mean grain size when the sintering temperature was increased. A sintered density of 92.93% of the theoretical density was obtained for Ti = 42% after sintering at 1180˚C. Ceramics sintered at 1180˚C with Ti = 42% achieve excellent dielectric properties, which are as follows $\varepsilon_r = 4262.48$, and Tc = 340˚C.

Keywords: PZT; Calcination; Dielectric Properties; MPB; Ceramic; X-Ray Methods

1. Introduction

Lead zirconate titanate (PZT) with the perovskite structure is the most popular ferroelectric material, which plays a remarkable role in modern electroceramic industry [1]. Moreover, PZT has high dielectric constant, high electromechanical coupling and high piezoelectric coefficient and has been employed as sensors, actuators and transducers [2-6]. The PZT are often modified by the introduction of the doping agents into the sites A or/and in the sites B of perovskite ABO_3 structure. The principal role of the doping agents is generally the improvement of the physical and mechanical properties of these materials. Substitutions in the crystal lattice called doping are often led with the aim to improve the specific properties of the PZT or sometimes to adapt them to specific applications. These properties are generally improved by the additions of one or more cations which will replace Pb^{2+} in site A and/or couple (Zr^{4+}/Ti^{4+}) in site B of perovskite structure (ABO_3) [7]. The selection of dopants or substitutions to tailor some physical properties of PZT was based on many factors which are the following: 1) charge neutrality, 2) tolerance factors, 3) ionic radius, and 4) solubility/miscibility. However, the sintering of PZT at high temperatures gives rise to a lead loss, which drastically degrades the device performance. Generally, a lead loss at high temperatures can be prevented by atmosphere controlled sintering of PZT. However, such composition requires sintering at a high temperature (>1250˚C) in a controlled atmosphere to contain lead volatilization so as to avoid a shift in composition. To get around the problem, different sintering aids have been tried by various workers [8-10]. However, for practical applications, such sintering aids need proper selection so that the electrical and piezoelectric properties of the ceramics do not degrade. The width and the properties of the coexistence region are associated with the occurrence of the compositional fluctuation of Ti^{4+} and Zr^{4+}

[*]Corresponding author.

ions in the PZT materials [11]. The compositional fluctuation, which is due to a non-uniform distribution of Titanium and Zirconium ions, leads to a broad variation in the dielectric constant accompanied with a Titanium concentration in the MPB region [12]. The width of this coexistence region and the structure of the PZT ceramics were greatly affected by the firing time and temperature [13].

In this study, $(0.80 - x)Pb(Cr_{1/5},Ni_{1/5},Sb_{3/5})O_3$-$xPbTiO_3$-$0.20PbZrO_3$ piezoelectric ceramics were investigated near the MPB by varying the ratio of Zr/Ti. The purpose of this work was to study the phase structure, the dielectric, and the piezoelectric properties of these ceramics near the MPB in detail.

2. Experimental Procedure

The compositions used for the present study were $(0.80 - x)Pb(Cr_{1/5},Ni_{1/5},Sb_{3/5})O_3$-$xPbTiO_3$-$0.20PbZrO_3$ with x varying as 30, 33, 36, 39 and 42 wt% respectively. The samples were prepared by a conventional oxide mixing technique. The starting materials were Pb_3O_4 (99.90%), ZrO_2 (99.90%), TiO_2 (99.90%), Cr_2O_3 (99.6%), Sb_2O_3 (99.90%) and NiO (99.6%).

Raw materials were mixed in acetone medium by using a magnetic stirrer during two hours. The obtained paste is being dried at 80˚C in a drying oven for two hours, and then crushed in a mortar out of glass during six hours. After crushing, the obtained powder is compacted in a form of pastilles with a pressure of 300 kg/cm^2. Then, a preliminary calcination with 800˚C is carried out during two hours with a heating rate of 2˚C/mn. The calcined mixture is crushed for a second time during four hours, and then was quickly crushed in a form of pellets with a pressure of 1000 kg/cm^2. These pellets are agglomerated at various temperatures of sintering (1100˚C, 1150˚C, 1180˚C, and 1200˚C) during two hours. It is important to note that a lead loss is possible by evaporation of PbO which is very volatile in T ≥ 900˚C. To limit this effect; an atmosphere rich in PbO was maintained with the powder of $PbZrO_3$ to the minimum to reduce this loss during sintering. The pastilles are metalized by using a thin layer of silver paste on the two faces.

X-ray diffraction (XRD, Siemens D500) was used to determine the crystalline phases present in the powder. The compositions of the PZT phases were identified by the analysis of the peaks [(002)T, (200)R, (200)T] in the 2θ range 43˚ - 46˚. The tetragonal (T), rhombohedral (R) and tetragonal-rhombohedral phases were characterized and their lattice parameters were calculated. The rhombohedral lattice parameter was calculated on the assumption that the rhombohedral distortion was constant (unit cell angle α_R = 89.9˚) [14,15]. In order to ensure an accurate determination of the lattice parameters, the X-ray peaks were recorded gradually with 0.01˚ steps.

Electronic micrographs scanning (SEM) were taken from fractured as well as chemically etched surfaces. A section of the sintered sample was etched in a 5% HCl solution for 3 minutes. The fractured surfaces were used for grain size and morphology determination. The size distribution of the grains was measured and the results compared with each other. The size distribution of the pores and the total value of porosity were determined on a polished cross-section of the samples with an image analyzer. To investigate the electrical properties, the electrodes were made by applying a silver paste on the two major faces of the sintered disks followed by a heat treatment at 750˚C for thirty minutes. The dielectric constant ε was calculated from the capacitance at a frequency of one kHz. It was measured at temperatures ranging from 25˚C to 450˚C with a heating rate of one ˚C/minute by using an impedance analyzer (HP 4192A, Hewlett-Packard).

3. Results and Discussion

3.1. Phase Analysis and Microstructure

Sintered powders were examined by X-ray diffractometry to ensure phase purity, and to identify the crystal structure. The coexistence of tetragonal and rhombohedral phases near the morphotropic phase boundary implies the existence of compositional fluctuations which can be determined from the width of the X-ray diffraction peaks. However, determination of the compositional fluctuation for samples near the morphotropic phase boundary is difficult. XRD patterns of PZT powders were analyzed for detecting the characteristic rhombohedral and tetragonal splittings. The (2 0 0) reflections form a doublet in the tetragonal phase while (1 1 1) is a singlet. For the rhombohedral phase, (1 1 1) is a doublet while the (2 0 0) is a singlet. The powder X-ray patterns of $(0.80 - x)Pb(Cr_{1/5},Ni_{1/5},Sb_{3/5})O_3$-$xPbTiO_3$-$0.20PbZrO_3$ ceramics with different x values are shown in **Figure 1**. For x in the range of 0.39 - 0.42, the diagram indicates that a mixture of phases should be present, which is illustrated by the (0 0 2) and (2 0 0) tetragonal doublet enclosing the (2 0 0) rhombohedral line. For x in the range of 0.30 - 0.36 there is a little evidence of the (2 0 0) R peak, indicating a virtually single-phase tetragonal structure. It is evident from "**Figure 1**" that as the Ti content increases, the morphotropic phase becomes more prominent whereas the tetragonal decreases.

Figure 2(a) shows the variation of density with sintering temperature. The density increases in the initial period with sintering temperature and saturates beyond 1180˚C. From these results, the optimum firing temperature for the maximum density, ρ, of the ceramic is be-

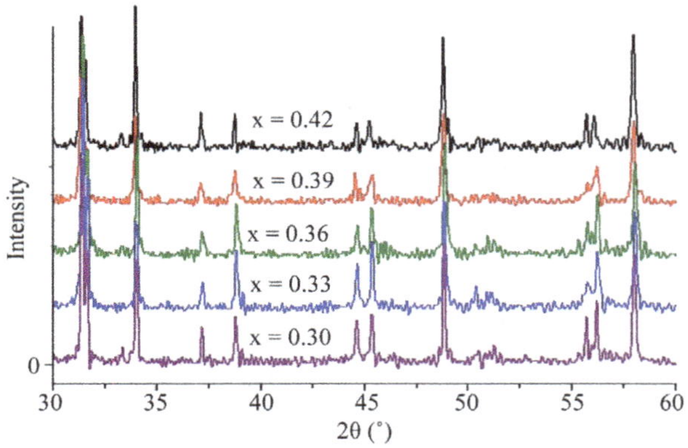

Figure 1. X-ray diffraction patterns of $(0.80 - x)Pb(Cr_{1/5},Ni_{1/5},Sb_{3/5})O_3$-$xPbTiO_3$-$0.20PbZrO_3$ ceramics sintered at 1180°C for 2 h with $0.30 \leq x \leq 0.42$.

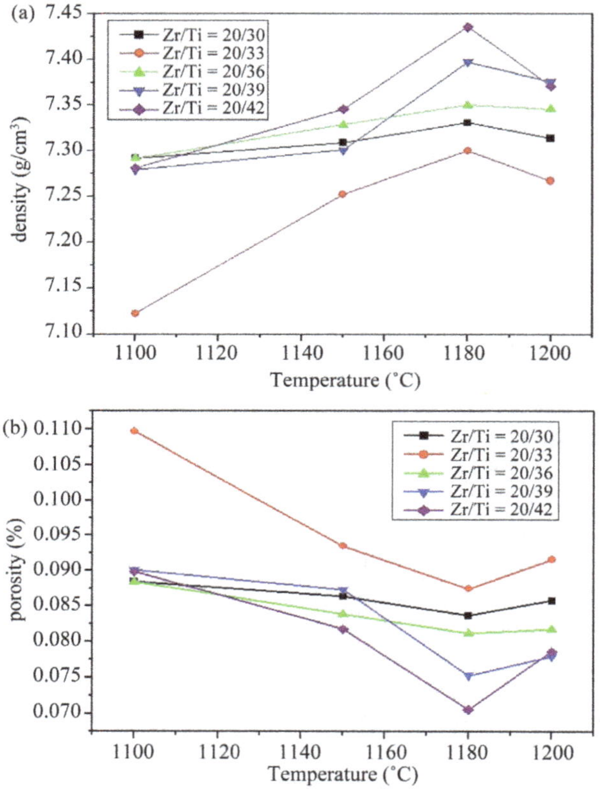

Figure 2. Density and porosity versus sintering temperature of PZT-CNS ceramics (sintering time 2 h).

tween 1150 and 1200°C. At 1180°C and Ti = 42%, 92.93% of the theoretical value was achieved. Sintering at 1200°C caused the density to decrease. The optimum value of the sintering temperature was affected by the additions of impurities and other processing parameters, such as the rate of heating, time of thermal treatment, and composition of the protecting atmosphere. The optimum sintering temperature was taken as the point when the PbO vapor pressure evaporation-recondensation equilibrium for the reaction: PbO-PbO (vapor)-Pb(vapor) + 1/2 O_2 was established [16]. Increase of the porosity for temperatures higher than the optimum can, therefore, be attributed to a greater rate of evaporation of PbO compared to that recondensed. Additions of different oxides to PZT-type ceramics influence the densification and the grain size. The process involves a decrease in the number and size of the pores together with an increase in the grain sizes. The porosity, determined by means of the image analyzer as a function of sintering temperature, is given in **Figure 2(b)**.

Figure 3 shows the lattice constant at room temperature as function of x. It can be seen that the tetragonal lattice parameter a_T increases linearly with increasing x, while the c_T parameter decreases linearly to a smaller extent. In all the composition range where the tetragonal phase is present, c_T and a_T are closing to gather a_T when Ti content increases, particularly inside the co-existence region, meaning that the structure is approaching the cubic geometry. The rhombohedral lattice parameter aR appears to oscillate between 4.023 and 4.024 Å. According to these results, we find that there is a region where the two phases tetragonal (T) and rhombohedral (R) co-exist. This region is detected for compositions: Ti = 39%, Ti = 42%. As against the compositions correspond to Ti ≤ 36%, show that the material obtained is of tetragonal structure. The influence of the substitution of Zr/Ti ratio on the structure of the parameters can be explained by the difference between the ionic rays of Zr and Ti (0.68 and 0.79 Å, respectively). This cannot provide a total homogeneity in the solid solutions containing both tetragonal and rhombohedral phases.

Figures 4(a)-(c) shows the SEM images of PZTCNS (20/36), PZTCNS (20/39) and PZTCNS (20/42) ceramics sintered at 1180°C. All the sintered ceramics appear to be very dense and of a homogeneous granular structure. At first sight, the three compositions seem homogeneous and there do not seem to be grains of the pyrochlore

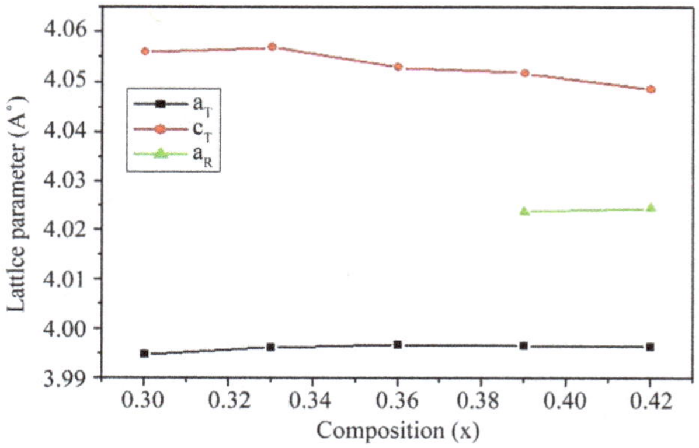

Figure 3. Variation of the unit cell dimensions as a function of composition (Ti%).

(a)

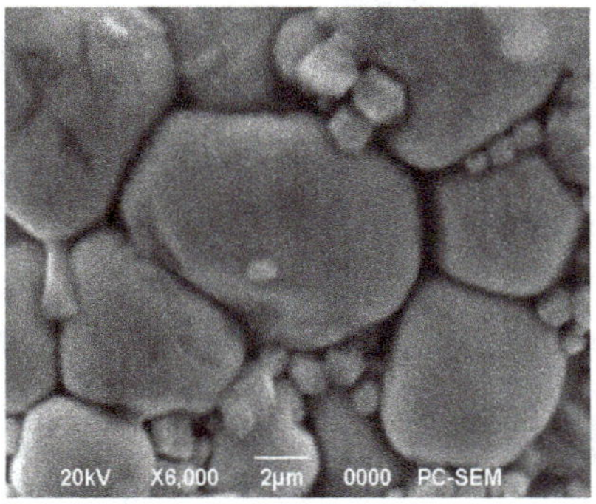

(b)

(c)

Figure 4. Microstructure of $(0.80 - x)Pb(Cr_{1/5},Ni_{1/5},Sb_{3/5})O_3$-$xPbTiO_3$-$0.20PbZrO_3$ ceramics sintered at 1180˚C for 2 h, (a) Ti = 36; (b) Ti = 39; (c) Ti = 42.

phase which are identifiable by their pyramidal form. The ruptures with the grain boundaries are synonymous with a good sintering. It is noticed that the average diameter of the grains increases significantly with increasing TiO_2. The intermediate size of the grains is 1.842 μm for the sample "**Figure 4(a)**" with Ti = 36%. For cons, the intermediate size of sample "**Figure 4(b)**" of the grains is larger (2.283 μm). In the case of ceramics "**Figure 4(c)**" with Ti = 42%, the intermediate size of the grains is larger than that of "**Figure 4(a)**" and "**Figure 4(b)**" (of the order 2.521 μm); and the broader the granulo-metric distribution "**Figure 4(c)**", the more the size of the grains gets bigger [17].

3.2. Dielectric Properties

Figure 5 shows the variation of the dielectric constant as a function: of composition and of temperature at sintering temperatures 1100˚C, 1150˚C and 1180˚C. For the three temperatures of sintering 1100˚C, 1150˚C and 1180˚C, observed that the permittivity increases gradually with the increase in the composition of x and takes a maximum of 290,15 for the sample with Ti = 42% included in the morphotropic phase boundary (MPB) at the temperature 1180˚C. This maximum of dielectric activity can be explained by the presence of several directions of spontaneous polarization relating to the existence of the two structures rhombohedral and tetragonal._Sample No. 4 (20/39/41) at the sintering temperature of 1180˚C has an exception in the evolution of ε_r (T). The dielectric constant increases continuously as a function of temperature, so this sample does not have a Curie temperature to a temperature between (0, 450˚C) [18,19].

4. Conclusion

The compounds of the solution solid zirconate-titanate lead, noted PZT, general formula $(0.80 - x)Pb(Cr_{1/5},Ni_{1/5},Sb_{3/5})O_3$-$xPbTiO_3$-$0.20PbZrO_3$ as x varies from 0.30 to 0.42 by setup of 0.03, and it has been prepared from a mixture of oxides by the method ceramics. The

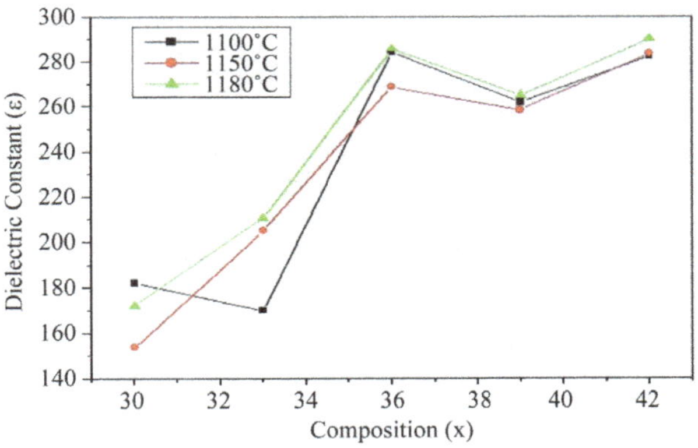

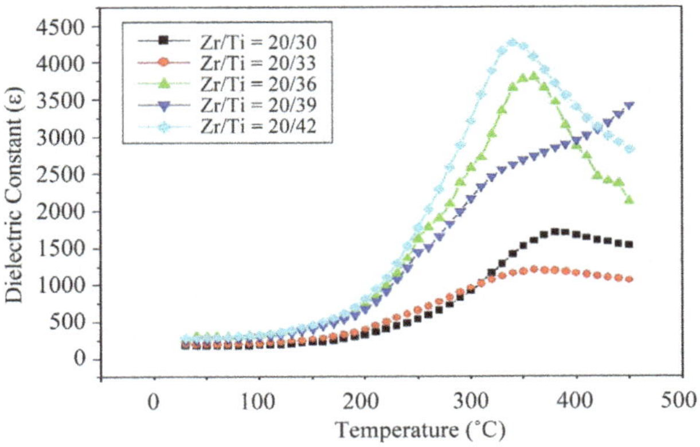

Figure 5. Dielectric constant (ε) according to the variation of x (Ti) in the composition and temperature.

effect of sintering temperature on density and porosity was studied to achieve the optimum sintering temperature corresponding to the maximum density and minimum value of porosity, because this temperature (1180°C) corresponds to a better quality product. The parameters of the lattice: a_T and c_T of the tetragonal structure and a_R of the rhombohedral structure were found to change when composition is modified. The $0.20PbZrO_3$-$0.42PbTiO_3$-$0.38Pb(Cr_{1/5},Ni_{1/5},Sb_{3/5})O_3$ ceramics sintered at 1180°C exhibit good Dielectric properties at the new MPB $\varepsilon_r = 4262.48$, and Tc = 340°C.

REFERENCES

[1] Z. He, J. Ma and R. Z. Hang, "Investigation on the Microstructure and Ferroelectric Properties of Porous PZT Ceramics," *Ceramics International*, Vol. 30, No. 7, 2004, pp. 1353-1356. http://dx.doi.org/10.1016/j.ceramint.2003.12.108

[2] R. E. Newnham and A. Amin, "Smart Systems: Microphones, Fish Farming, and Beyond-Smart Materials, Acting as Both Sensors and Actuators, Can Mimic Biological Behavior," *Chemtech*, Vol. 29, No. 12, 1999, pp. 38-47.

[3] K. Uchino, "Materials Issues in Design and Performance of Piezoelectric Actuators: An Overview," *Acta Materialia*, Vol. 46, No. 11, 1998, pp. 3745-3753. http://dx.doi.org/10.1016/S1359-6454(98)00102-5

[4] R. Ranjan, R. Kumar, B. Behera and R. N. P. Choudhary, "Effect of Sm on Structural, Dielectric and Conductivity Properties of PZT Ceramics," *Materials Chemistry and Physics*, Vol. 115, No. 1, 2009, pp. 473-477. http://dx.doi.org/10.1016/j.matchemphys.2009.01.017

[5] S. T. Lau, K. W. Kwok, H. L. W. Chan and C. L. Choy, "Piezoelectric Composite Hydrophone Array," *Sensors and Actuators A: Physical*, Vol. 96, No. 1 ,2002, pp. 14-20. http://dx.doi.org/10.1016/S0924-4247(01)00757-9

[6] T. Zeng, X. L. Dong, S. T. Chen and H. Yang, "Processing and Piezoelectric Properties of Porous PZT Ceramics," *Ceramics International*, Vol. 33, No. 3, 2007, pp. 395-399. http://dx.doi.org/10.1016/j.ceramint.2005.09.022

[7] S. S. Chandratreya, R. M. Fulrath and J. A. Y. Pask, "Reaction Mechanisms in the Formation of PZT Solid Solutions," *Journal American Ceramic Society*, Vol. 64, No. 7, 1981, pp. 422-425. http://dx.doi.org/10.1111/j.1151-2916.1981.tb09883.x

[8] S. Y. Cheng, S. L. Fu, C. C. Wei and G. M. Ke, "The Properties Low-Temperature Fixed Piezoelectric Ceramics," *Journal of Materials Science*, Vol. 21, No. 2, 1986, pp. 571-576. http://dx.doi.org/10.1007/BF01145525

[9] H. G. Lee, J. H. Choi and E. S. Kim, " Low-Temperature Sintering and Electrical Properties of $(1-x)Pb(Zr_{0.5}Ti_{0.5})O_3$-$xPb(Cu_{0.33}Nb_{0.67})O_3$ Ceramics," *Journal of Electroceramics*, Vol. 17, No. 2-4, 2006, pp. 1035-1040. http://dx.doi.org/10.1007/s10832-006-0384-1

[10] R. Mazumder, A. Sen and H. S. Maiti, "Impedance and Piezoelectric Constants of Phosphorous-Incorporated $Pb(Zr_{0.52}Ti_{0.48})O_3$ Ceramics," *Materials Letters*, Vol. 58, No. 25, 2004, pp. 3201-3205. http://dx.doi.org/10.1016/j.matlet.2004.06.011

[11] A. V. Turik, M. F. Kupriyanov, E. N. Sidorenko and S. M. Zaitsev, "Behavior of Piezoceramics of Type $Pb(Zr,Ti)O_3$, near the Region of the Morphotropic Transition," *Soviet Physics-Technical Physics*, Vol. 25, No. 10, 1980, pp. 1251-1254.

[12] K. Kakegawa, J. Mohri, T. Takahashi, H. Yamamura and S. Shirasaki, "A Compositional Fluctuation and Properties of $Pb(Zr, Ti)O_3$," *Solid State Communications*, Vol. 24, No. 11, 1977, pp. 769-772. http://dx.doi.org/10.1016/0038-1098(77)91186-3

[13] S. A. Mabud, "The Morphotropic Phase Boundary in PZT Solid Solution," *Journal of Applied Crystallographic*, Vol. 13, 1980, pp. 211-216. http://dx.doi.org/10.1107/S0021889880011958

[14] P. Ari-Gur and L. Benguigui, "X-Ray Study of the PZT Solid Solutions near the Morphotropic Phase Transition," *Solid State Communications*, Vol. 15, No. 6, 1974, pp. 1077-1079. http://dx.doi.org/10.1016/0038-1098(74)90535-3

[15] A. Boutarfaia and S. E. Bouaoud, "Tetragonal and Rhombohedral Phase Co-Existence in the System: $PbZrO_3$-$PbTiO_3$-$Pb(Fe_{1/5}, Ni_{1/5},Sb_{3/5})O_3$," *Ceramics International*, Vol. 22, No. 4, 1996, pp. 281-286. http://dx.doi.org/10.1016/0272-8842(95)00102-6

[16] R. B. Atkin and R. M. Fulrath, "Point Defects and Sintering of Lead Zirconate-Titanate," *Journal of American Ceramic Society*, Vol. 54, No. 5, 1971, pp. 265-270. http://dx.doi.org/10.1111/j.1151-2916.1971.tb12286.x

[17] O. Ohtaka, R. Von Der Mühll and J. Ravez, "Low-Temperature Sintering of $Pb(Zr,Ti)O_3$ Ceramics with the Aid of Oxyfluoride Additive: X-Ray Diffraction and Dielectric Studies," *Journal American Ceramic Society*, Vol. 78, No. 3, 1995, pp. 805-808. http://dx.doi.org/10.1111/j.1151-2916.1995.tb08251.x

[18] H. R. Rukmini, R. N. P. Choudhary and D. L. Prabhakara, "Sintering Temperature Dependent Ferroelectric Phase Transition of $Pb_{0.91}(La_{1-z/3}Liz)_{0.09}(Zr_{0.65}Ti_{0.35})_{0.9775}O_3$," *Journal of Physics and Chemistry of Solids*, Vol. 61, No. 11, 2000, pp. 1735-1743. http://dx.doi.org/10.1016/S0022-3697(00)00040-8

[19] W. Chaisan, R. Yimnirun, S. Ananta and D. P. Cann, "Phase Development and Dielectric Properties of $(1-x)Pb(Zr_{0.52}Ti_{0.48})O_3-xBaTiO_3$ Ceramics," *Materials Science and Engineering B*, Vol. 132, No. 3, 2006, pp. 300-306. http://dx.doi.org/10.1016/j.mseb.2006.04.033

22

The Growth and the Optical, Mechanical, Dielectric and Photoconductivity Properties of a New Nonlinear Optical Crystal—L-Phenylalanine-4-nitrophenol NLO Single Crystal

Sagadevan Suresh

Crystal Growth Centre, Anna University, Chennai, India.
Email: sureshsagadevan@yahoo.co.in

ABSTRACT

Nonlinear optical single crystals of L-phenylalanine-4-nitrophenol have been grown by the slow evaporation method. The grown crystal was subjected to the single crystal X-ray diffraction analysis, to confirm that it belongs to the monoclinic crystal structure, with space group $P2_1$. The optical transmission study reveals the transparency of the crystal in the entire visible region and the cut off wave length has been found to be 320 nm. The optical band gap is found to be 3.87 eV. The transmittance of the L-phenylalanine-4-nitrophenol crystal has been used to calculate the refractive index (n), the extinction coefficient (K) and the real (ε_r) and imaginary (ε_i) components of the dielectric constant. The mechanical behaviour of the grown crystals was studied using Vicker's microhardness tester. The dielectric constant and dielectric loss of L-phenylalanine-4-nitrophenol are measured in the frequency range of 50 Hz to 5 MHz at different temperatures. The photoconductivity study confirms the negative photoconductive nature of the sample.

Keywords: Crystal Growth; Slow Evaporation Method; NLO; Single Crystal XRD; Dielectric Studies; Photoconductivity Studies

1. Introduction

Recently, a very novel approach has been adopted for obtaining new materials for non linear optics. The fast development in the field of optoelectronics necessitates the search for new and efficient nonlinear optical (NLO) materials, which can be utilized for many applications such as optical communications, optical data storage, optical computing and electro-optic shutters [1]. Among nonlinear optical crystals, organic salts occupy an intermediate position between molecular organic compounds with covalent bonds and inorganic compounds with mainly ionic bonds [2]. The design and synthesis of organic molecules exhibiting nonlinear optical properties have been motivated by their potential for applications in optical communications, optical computing, data storage, dynamic holography, harmonic generators, frequency mixing, and optical switching [3,4]. In the present work, the L-phenylalanine-4-nitrophenol single crystal has been grown by slow evaporation technique and the cell parameters calculated using the single crystal XRD. The mechanical behaviour was analyzed using Vicker's microhardness test. The dielectric constant and dielectric loss measurements were taken at different temperatures and frequencies. The photoconductivity studies reveal that the crystal shows negative the photoconductive nature of the L-phenylalanine-4-nitrophenol single crystal.

2. Experimental Details

L-phenylalanine-4-nitrophenol single crystals were synthesized by dissolving L-phenylalanine and 4-nitrophenol in the molar ratio of 1:1 in distilled water. The solution was stirred continuously using a magnetic stirrer. The prepared solution was filtered and kept undisturbed at room temperature. Tiny seed crystals with good transparency were obtained, due to spontaneous nucleation. Among them, defect free seed crystal was suspended in the mother solution, which was allowed to evaporate at room temperature. Large size single crystals were obtained, due to the collection of monomers at the seed crystal sites from the mother solution.

3. Results and Discussion

3.1. Single Crystal XRD

The single crystal XRD analysis for the grown crystals has been carried out to identify the lattice parameters. The calculated lattice parameters are; a = 5.84 Å, b = 7.01 Å, c = 17.87 Å and volume of the crystal V = 728.32 $Å^3$. The crystal has a monoclinic crystal structure with space group $P2_1$. These values have good agreement with the reported values [5].

3.2. UV-Visible Spectroscopy

The optical absorption of the crystal is an important factor for analyzing the second harmonic generation property. Hence, the grown crystal was subjected to UV-VIS-NIR absorption studies. The absorption of the present L-phenylalanine-4-nitrophenol crystal was in the range of 250 nm to 2000 nm; the corresponding spectrum is shown in **Figure 1**. From the absorption spectrum, the lower cut-off wavelength is found to be 320 nm and the lower percentage absorption indicates that the crystal readily allows the transmission of the laser beam in 250 nm and 2000 nm. It shows that the grown crystal has a good transparency in UV, visible and near IR region (*i.e.*, for the entire wavelength range studied) indicating that it can be used for NLO applications. The measured transmittance (T) was used to calculate the absorption coefficient (α) using the formula

$$\alpha = \frac{2.3026 \log(1/T)}{t} \tag{1}$$

where t is the thickness of the sample. The optical band gap (E_g) was evaluated from the transmission spectrum and optical absorption coefficient (α) near the absorption edge was calculated using the formula [6].

$$\alpha h\nu = A\left(h\nu - E_g\right)^{1/2} \tag{2}$$

where A is a constant, Eg the optical band gap, h the Planck constant and n the frequency of the incident photons. The band gap of the L-phenylalanine-4-nitrophenol crystal was estimated by plotting $(ahv)^{1/2}$ versus hv as shown in **Figure 2** and extrapolating the linear portion near the onset of absorption edge to the energy axis. From the figure, the value of the band gap was found to be 3.87 eV. This high band gap value indicates that the grown crystal possesses dielectric behavior to induce polarization when powerful radiation is incident on the material. The extinction coefficient (K) can be obtained from the equation,

$$K = \frac{\lambda \alpha}{4\pi} \tag{3}$$

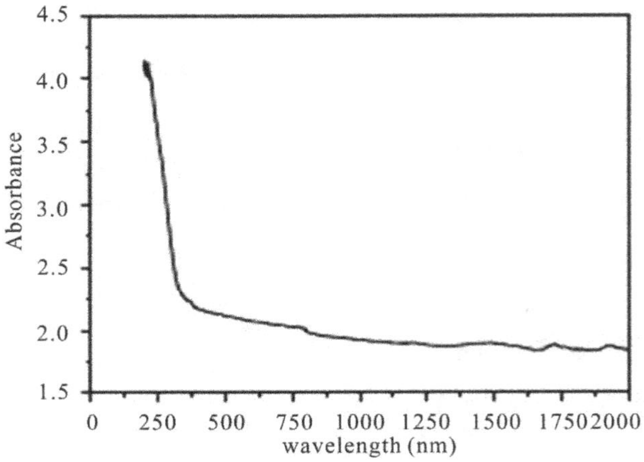

Figure 1. Transmission spectrum of L-phenylalanine-4-nitrophenol crystal.

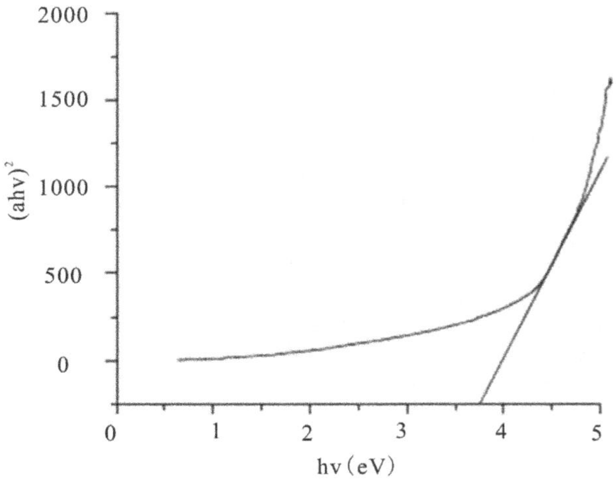

Figure 2. Plot of (αhυ)2 versus hυ for L-phenylalanine-4-nitrophenol single crystal.

The transmittance (T) is given by

$$T = \frac{(1-R)^2 \exp(-\alpha t)}{1-R^2 \exp(-2\alpha t)} \tag{4}$$

The reflectance (R) in terms of the absorption coefficient can be obtained from the above equation.

Hence,

$$R = \frac{\exp(-\alpha t) \pm \sqrt{\exp(-\alpha t)T - \exp(-3\alpha t)T + \exp(-2\alpha t)T^2}}{\exp(-\alpha t) + \exp(-2\alpha t)T} \tag{5}$$

Refractive index (n) can be determined from the reflectance data, using the following equation,

$$n = -(R+1) \pm 2\frac{\sqrt{R}}{(R-1)} \tag{6}$$

The refractive index (n) is 1.36 at λ = 2000 nm. From the optical constants, the electric susceptibility (χc) can be calculated according to the following relation [7].

$$\varepsilon_r = \varepsilon_0 + 4\pi\chi_C = n^2 - k^2 \quad (7)$$

Hence,

$$\chi_C = \frac{n^2 - k^2 - \varepsilon_0}{4\pi} \quad (8)$$

The value of electric susceptibility χ_c is 0.152 at λ = 2000 nm. The real part dielectric constant ε_r and the imaginary part dielectric constant ε_i can be calculated from the following relations [8],

$$\varepsilon_r = n^2 - k^2 \quad \& \quad \varepsilon_i = 2nk \quad (9)$$

The values of the real ε_r and ε_i imaginary dielectric constants at λ = 2000 nm are 1.26 and 5.204×10^{-5} respectively.

3.3. Microhardness Test

The microhardness studies were carried out on the as grown crystal of L-phenylalanine-4-nitrophenol using the Shimadzu HMV-2000 Microhardness tester. To evaluate the Vickers microhardness, several indentations were made on the smooth surface of the crystal. The Vicker's microhardness number was evaluated using the relation,

$$H_V = 1.8544P / \mathrm{d}^2 \mathrm{kg/mm}^2 \quad (10)$$

where H_V is the Vicker's micro hardness number, P is the applied load and d the average diagonal length of the indentation impression. The dependence of the micro hardness on the load for the L-phenylalanine-4-nitrophenol crystal has been calculated and shown in **Figure** 3. The indentation was made on the sample with the load ranging from 10 - 50 g and the indentation time was kept constant as 10 s. The **Figure 3** indicates that the microhardness number increases with increasing load. The work hardening coefficient value (n) of the grown crystal would be n > 2. Hence, the L-phenylalanine-4-nitrophenol crystal is a soft material.

3.4. Dielectric Properties

The study of the dielectric constant of a material gives an outline about the nature of atoms, ions and their bonding in the material. From the analysis of the dielectric constant and dielectric loss as a function of frequency and temperature, the different polarization mechanism in solids can be understood. Variations in the dielectric constant and dielectric loss as functions of temperatures and frequencies are shown in **Figures 4** & **5**. The dielectric constant and the dielectric loss of the L-phenyla lanine-4-nitrophenol crystal were studied at different

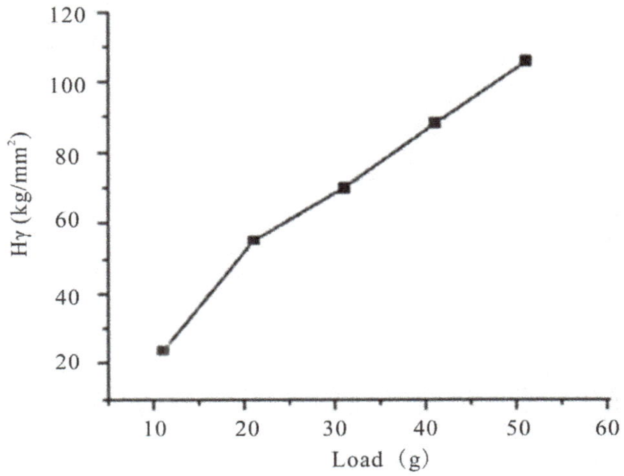

Figure 3. Vickers hardness profile of L-phenylalanine-4-nitrophenol as a function of applied load.

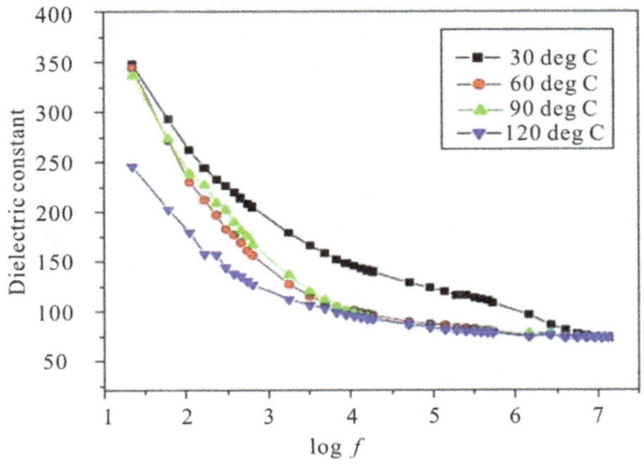

Figure 4. Variation of dielectric constant of L-phenylalanine-4-nitrophenol with log frequency.

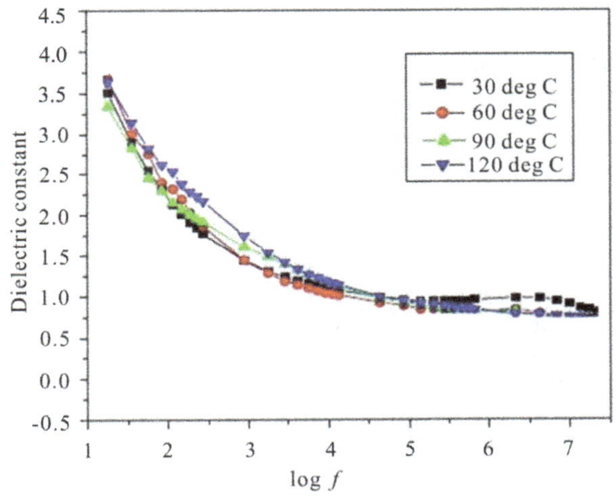

Figure 5. Variation of dielectric loss of L-phenylalanine-4-nitrophenol with log frequency.

temperatures using the HIOKI 3532 LCR HITES-TER in the frequency region of 50 Hz to 5 MHz.The dielectric constant was measured as a function of frequency at different temperatures of 30˚C, 60˚C, 90˚C, and 120˚C and is shown in **Figure 4**, while the corresponding dielectric losses are depicted in **Figure 5**. The dielectric constant is evaluated using the relation,

$$\varepsilon_0 = \frac{Cd}{\varepsilon_0 A} \tag{11}$$

where d is the thickness of the sample, A, the area of the sample. **Figure 4** shows the plot of the dielectric constant (ε) versus log frequency for 30˚C, 60˚C, 90˚C, and 120˚C. It is seen that the value of the dielectric constant is high in the lower frequency region for all the temperatures and then it decreases with an increase in the frequency. The high value of the dielectric constant in the low frequency region is attributed to space charge polarization, due to the charged lattice defects [9]. The large values of the dielectric constant at low frequency suggest that there are contributions from the electronic, ionic, dipolar, and space charge polarizations. Space charge polarization is generally active at lower frequencies and high temperatures and indicates the perfection of the crystals. Further space charge polarization will depend on the purity and perfection of the material. Its influence is large at higher temperatures and is noticeable in the low frequency [10]. A graph is drawn between the dielectric loss and log frequency for different temperatures (30˚C, 60˚C, 90˚C and 120˚C) and is shown in **Figure 5**. The low value of dielectric loss at high frequency suggests that the grown crystals possess good optical quality. This parameter is of vital importance for nonlinear optical materials in their applications [10].

3.5. Photoconductivity Studies

Photoconductivity studies have been carried out using the Keithley 485 picoammeter instrument. The dark current was recorded by keeping the crystal unexposed to any radiation. **Figure 6** shows the variation of both the dark current and photocurrent with applied field at different levels of illumination. The required current is noted for varying applied fields. It is seen from the plots that both dark current and photocurrent of the sample increase linearly with the applied field. The dark current is seen to be higher than the photocurrent for the same applied field, which is termed as negative photoconductivity. The negative photoconductivity exhibited by the sample may be due to the reduction in the number of charge carriers in the presence of radiation. The decrease in the mobile charge carriers during negative photoconductivity can be explained by the Stockmann model [11].The negative photoconductivity in a solid is due to the decrease in the

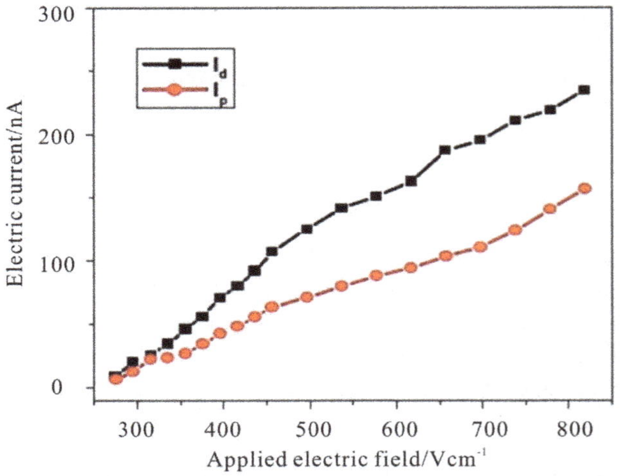

Figure 6. Field dependence of dark and photocurrents.

number of charge carriers or their lifetime, in the presence of radiation [12]. For a negative photoconductor, the forbidden gap holds two energy levels in which one is placed between the Fermi level and the conduction band while the other is located close to the valence band. The second state has a higher capture cross-section for electrons and holes. As it captures electrons from the conduction band and holes from the valence band, the number of charge carriers in the conduction band gets reduced and the current decreases in the presence of radiation.

4. Conclusion

Single crystals of L-phenylalanine-4-nitrophenol have been grown from an aqueous solution by the slow evaporation technique. The grown single crystals have been characterized by the single crystal XRD and it is confirmed that the crystal belongs to the monoclinic system, with space group $P2_1$. The optical band gap (E_g), absorption coefficient (α), extinction coefficient (K), refractive index (n), electric susceptibility χ_c and dielectric constants were calculated. The mechanical strength of the grown crystal was found from the Vicker's microhardness measurement. Dielectric studies have been carried out to examine the dielectric constant and dielectric loss at different frequencies and different temperatures. The photocurrent was less than the dark current, signifying a negative photoconducting nature.

REFERENCES

[1] J. Zyss, "Molecular Nonlinear Optics: Materials Physics and Devices," Academic Press, Boston, 1994.

[2] J. F. Nicoud, R. J. Twieg, D. S. Chemla and J. Zyss, "In Nonlinear Optical Properties of Organic Molecules and

Crystals," Academic Press, New York, 1987, pp. 227-296.

[3] D. S. Chemla and J. Zyss, "Nonlinear Optical Properties of Organic Molecules and Crystals," Academic Press, New York, 1987.

[4] P. N. Prasad and D. J. Williams, "Introduction to Nonlinear Optical Effects in Organic Molecules and Polymers," Wiley, New York, 1991.

[5] V. H. Rodrigues, M. M. R. R. Costa, E. M. Gomes, E. Nogueirab and M. S. Belslsey, "L-phenylalanine-4-nitrophenol," *Acta Crystallographica*, Vol. 62, 2006, pp. O699-O701.

[6] A. Ashour, N. El-Kadry and S. A. Mahmoud, "On the Electrical and Optical Properties of CdS Films Thermally Deposited by a Modified Source," *Thin Solid Films*, Vol. 269, No. 1-2, 1995, 117-120. doi:10.1016/0040-6090(95)06868-6

[7] V. Gupta and A. Mansingh, "Influence of Postdeposition Annealing on the Structural and Optical Properties of Sputtered Zinc Oxide Film," *Journal of Applied Physics*, Vol. 80, No. 2, 1996. p. 1063. doi:10.1063/1.362842

[8] M. A. Gaffar, A. A. El-Fadl and S. B. Anooz, "Influence of Strontium Doping on the Indirect Band Gap and Optical Constants of Ammonium Zinc Chloride Crystals," *Physica B: Condensed Matter*, Vol. 327, No. 1, 2003, pp. 43-54. doi:10.1016/S0921-4526(02)01700-3

[9] C. P. Smyth, "Dielectric Behaviour and Structure," Megraw Hill, New York, 1965.

[10] C. Balarew and R. Dehlew, "Application of the Hard and Soft Acids and Bases Concept to Explain Ligand Coordination in Double Salt Structures," *Journal of Solid State Chemistry*, Vol. 55, No. 1, 1984, pp. 1-6. doi:10.1016/0022-4596(84)90240-8

[11] V. N. Joshi, "Photoconductivity," Marcel Dekker, New York, 1990.

[12] R. H. Bube, "Photoconductivity of solids," Wiley, New York, 1981.

23

Dielectric Properties of Ni-Zn Ferrites Synthesized by Citrate Gel Method

K. Rama Krishna[1], Dachepalli Ravinder[2*], K. Vijaya Kumar[3], Utpal S. Joshi[4], V. A. Rana[4], Abrham Lincon[5]

[1]Department of Physics, Malla Reddy College of Engineering & Technology, Secunderabad, India; [2]Department of Physics, P.G. College of Science, Osmania University, Hyderabad, India; [3]Department of Physics, Jawaharlal Nehru Technological University, Hyderabad, India; [4]Department of Physics, Gujarat University, Ahmedabad, India; [5]Department of Chemsitrys, P.G. College of Science, Osmania University, Hyderabad, India.
Email: *ravindergupta28@rediffmail.com

ABSTRACT

Ni-Zn ferrite with composition of $Ni_{1-x}Zn_xFe_2O_4$ (x = 0.0, 0.2, 0.4, 0.6, 0.8, 0.9, 1.0) were prepared by citrate gel method. The Dielectric Properties for all the samples were investigated at room temperature as a function of frequency. The dielectric constant shows dispersion in the lower frequency region and remains almost constant at higher frequencies. The frequency dependence of dielectric loss tangent (tanδ) is found to be abnormal, giving a peak at certain frequency for mixed Ni-Zn ferrites. A qualitative explanation is given for the composition and frequency dependence of the dielectric loss tangent.

Keywords: Ferrites; Citrate Method; Lattice Parameter; Dielectric Constant; Dielectric Loss

1. Introduction

Ni-Zn ferrite is a well-known spinel magnetic material. In the inverse spinel structure of $NiFe_2O_4$, the tetrahedral sites are occupied by ferric ions and octahedral sites by ferric and nickel ions. $Ni_{1-x}Zn_xFe_2O_4$ ferrites are ferrimagnetic materials with a large number of technological applications in telecommunications and entertainment electronics. Ni-Zn ferrites are among the most widely used soft magnetic materials because of high frequency applications as they posses high electrical resistivity and low eddy current losses [1-3]. It is known that magnetic properties of ferrites are sensitive to preparation technique and their microstructures [4]. The electrical and magnetic properties of such ferrites depend strongly on distribution of cations at the tetrahedral (A) and octahedral (B) sites in the lattice [5-7]. It is well known that zinc ions can be used to alter the saturation magnetization. It is believed that the addition of zinc ions also affects the lattice parameter and it would therefore be expected to change the Curie temperature of the material [8]. The substitution of divalent ions in pure ferrites leads to the modification of the structural, electrical and magnetic properties [9]. The conventional solid-state reaction route is widely used for the production of ferrite because of its low cost and suitability for large scale production. The citrate method is used to speed up the synthesis of complex materials. It is a simple process, which offers a significant saving in time and energy consumption over the traditional methods. Several researchers have reported the synthesis of Ni-Zn ferrites using different techniques like, refluxing process [10], ceramic [11], hydrothermal [12], combustion [13], co-precipitation [14], reverse micelle process [15], spark plasma sintering [16], micro emulsion [17] and ball milling etc.

In this work, we present the results of systematic doping of non-magnetic Zn content on the dielectric properties of Ni-Zn ferrite synthesized by citrate method.

2. Experimental

The starting materials were nickel nitrate, zinc nitrate, iron nitrate, citric acid and ammonia all of analytical grade. The solution of nickel nitrate ($Ni(NO_3)_2\cdot 6H_2O$, ferric nitrate ($Fe(NO_3)_3\cdot 9H_2O$) and zinc nitrate ($Zn(NO_3)_2\cdot 6H_2O$) in their stoichiometry were dissolved in a de ionized water. Citric acid was then added to the prepared aqueous solution to chelate Ni^{2+}, Zn^{2+} and Fe^{3+} in the solution. The molar ratio of citric acid to total moles of nitrate ions was adjusted at 1:3. The mixed solution was neutralized to pH 7 by adding ammonia (NH_3) solution. The neutralized solution was evaporated to dry ness by heating at 100˚C on a hot plate with continuous stirring, until it becomes

*Corresponding author.

viscous and finally formed a very viscous gel. Increasing the temperature up to 200°C leads to ignition of gel. The dried gel burnt completely in a self propagating combustion manner to form a loose powder. Finally the burnt powder was calcined in air at temperature of 1000°C for one hour to obtain spinel phase. Afterwards the powder was pressed into pellets of thickness 3 mm and a diameter of 10 mm with press by applying a pressure of 2 tons/in^2. The final sintering was done at 1000°C, afterwards the pellets were coated with silver paint for better electrical contact to measure the dielectric properties.

The structural characterizations of all samples were carried out by X-ray diffraction (XRD) and conforms the well defined single phase spinel structure. XRD data were taken at room temperature using Cu$K\alpha$ radiation. The dielectric data are measured by LCR meter at room temperature in the frequency range 2 Hz to 2 MHz.

3. Results and Discussions

All the zinc substitute nickel ferrites of the various compositions show the crystalline cubic spinel structure. The sharp peaks showed all-crystalline nature of single phase ferrite. The lattice parameter of individual composition was calculated by using the formula

$$a = d\left(h^2 + k^2 + 1^2\right)^{1/2}$$

where, a = lattice constant;

d = inter planar distance;

(h, k, l) = the Miller indices.

The variation of lattice parameter with zinc composition is shown in **Figure 1**. The lattice parameter is found vary linearly with increasing zinc concentration, there by indicating that the Ni-Zn ferrite system obeys Vegard's law [18]. A similar behavior of lattice constant with dopant concentration was observed by several investigators in various ferrite systems [19-21]. The variation in lattice constant with zinc content can be explained on the basis of the ionic radii of Zn^{2+} (0.82 Å) ions is higher than that of Ni^{2+} (0.78 Å) [22].

3.1. Dielecric Constant with Frequency

The effect of frequency on the real dielectric constant (ε') can be seen from **Figure 2** and that the value of dielectric constant decreases continuously with increasing frequency. The decrease in the values of dielectric constant as the frequency increases can be due to electron exchange interaction between Fe^{2+} and Fe^{3+} ions, which cannot follow the alternating electric field. The decrease of dielectric constant with increase of frequency as observed in the case of Ni-Zn ferrite. A similar behavior was also observed in [23-26] of various ferrite systems. The explanation for the decrease in the values of ε' as the frequency increases can be related to electron exchange interaction between Fe^{2+} and Fe^{3+} ions which cannot follow the alternation of the electric field beyond a certain frequency.

3.2. Dielectric Loss Tangent (tanδ) with Frequency

The variation of tanδ with frequency can be seen from **Figure 3**, it can be seen from the figure that in the case of $NiFe_2O_4$, $Ni_{0.8}Zn_{0.2}Fe_2O_4$, $Ni_{0.6}Zn_{0.4}Fe_2O_4$ and $Ni_{0.4}Zn_{0.6}Fe_2O_4$ tanδ shows maximum at a frequency of 21 Hz, in the case of $Ni_{0.1}Zn_{0.9}Fe_2O_4$, $ZnFe_2O_4$ tanδ shows maximum at a frequency of 22 Hz and $Ni_{0.2}Zn_{0.8}Fe_2O_4$ tanδ shows maximum at 25 Hz. A qualitative explanation can be given for the occurrence of the maxima in tanδ verses frequency curves in $NiFe_2O_4$, $Ni_{0.8}Zn_{0.2}Fe_2O_4$, $Ni_{0.6}Zn_{0.4}Fe_2O_4$, $Ni_{0.4}Zn_{0.6}Fe_2O_4$, $Ni_{0.1}Zn_{0.9}Fe_2O_4$, $ZnFe_2O_4$ and $Ni_{02}Zn_{0.8}Fe_2O_4$, as pointed out by Iwauchi [27], there is a strong correlation between the conduction mechanism and the dielectric behavior of ferrites. The conduction mechanism in n-type ferrites is considered as due

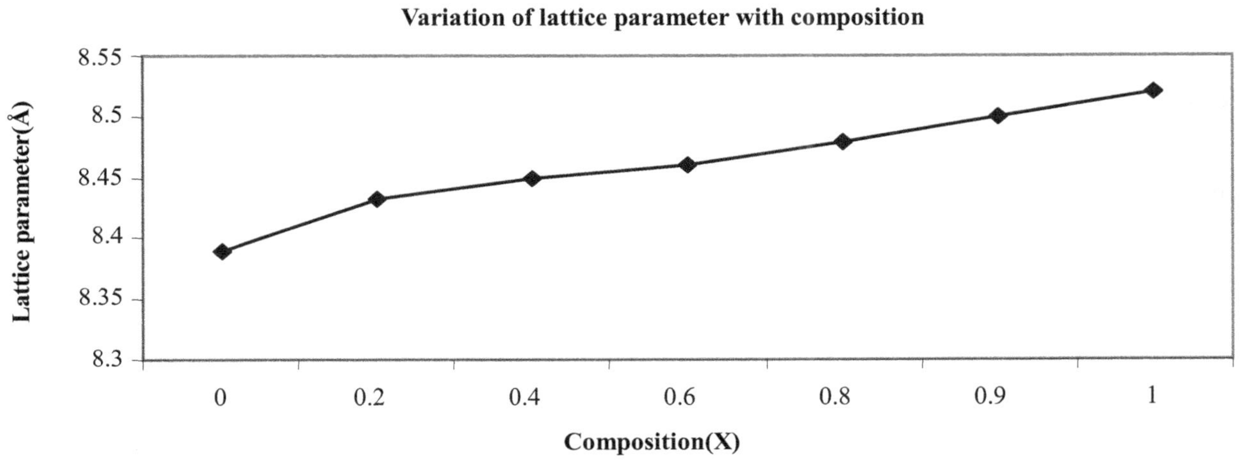

Figure 1. Variation of lattice parameter with composition.

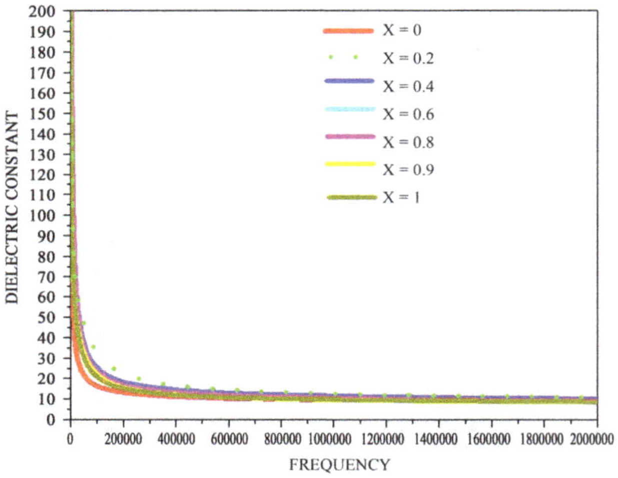

Figure 2. Variation of dielectric constant with composition.

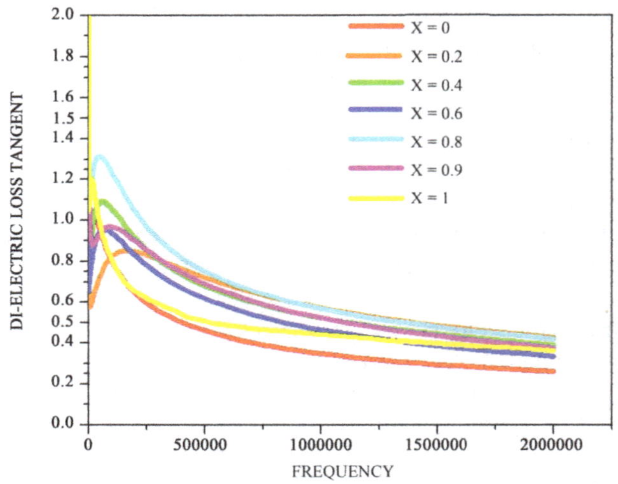

Figure 3. variation of dielectric loss tangent with composition.

to hopping of electrons between Fe^{2+} and Fe^{3+} situated on the octahedral sites. As seen, when the hopping frequency is nearly equal to that of external applied electric field a maximum of loss tangent may be observed [28]. As such it is possible that in the case of in $NiFe_2O_4$, $Ni_{0.8}Zn_{0.2}Fe_2O_4$, $Ni_{0.6}Zn_{0.4}Fe_2O_4$, $Ni_{0.4}Zn_{0.6}Fe_2O_4$, $Ni_{0.1}Zn_{0.9}Fe_2O_4$, $ZnFe_2O_4$ and $Ni_{02}Zn_{0.8}Fe_2O_4$, the hoping frequencies are of the approximate magnitude to observe a loss maximum at 21, 22, 25 Hz respectively.

3.3. Conclusion

It may be concluded that a series of Ni-Zn ferrite with compositional formula $Ni_{1-x}Zn_xFe_2O_4$ where $x = 0.0$, 0.2, 0.4, 0.6, 0.8, 0.9 & 1.0 are prepared by citrate gel method. The lattice constant was found to be increases with zinc composition. The variation of dielectric constant and dielectric loss tangent was explained on the basis of electronic exchange between the Fe^{2+} and Fe^{3+} ions.

4. Acknowledgements

The authors are grateful Dr. Partha sarthy Principal, P.G. College of Science, Osmania University, Hyderabad for his encouragement in research work. One of the author K. Rama Krisna is grateful to V. S. K. Reddy, Principal, Malla Reddy College Of Engineering & Technology, Hyderabad. And the other author K. Vijaya Kumar is grateful to Dr. A. Govardhan, Principal JNTUH College of Engineering, Nachupally, Karim Nagar (Dist).

REFERENCES

[1] X. He, G. Song and J. Zhu, "Non-Stoichiometric Ni-Zn Ferrite by Sol-Gel Processing," *Materials Letters*, Vol. 59, No. 14-15, 2005, pp. 1941-1944. doi:10.1016/j.matlet.2005.02.031

[2] Y. Matsuo, M. Inagaki, T. Tomozawa and F. Nakao, "High Performance Ni-Zn Ferrite," *IEEE Transactions on Magnetics*, Vol. 37, No. 4, 2001, pp. 2359-2361. doi:10.1109/20.951172

[3] P. S. A. Kumar, J. J. Shrotri, S. D. Kulkarni, C. E. Deshpande and S. K. Date, "Low Temperature Synthesis of $Ni_{0.8}Zn_{0.2}Fe_2O_4$ Powder and Its Characterization," *Materials Letters*, Vol. 27, No. 6, 1996, pp. 293-296. doi:10.1016/0167-577X(96)00010-9

[4] H. Su, H. Zhang, X. Tang, Y. Jing and Y. Liu, "Effects of Composition and Sintering Temperature on Properties of Ni-Zn and NiCuZn Ferrites," *Journal of Magnetism and Magnetic Materials*, Vol. 310, No. 1, 2007, pp. 17-21. doi:10.1016/j.jmmm.2006.07.022

[5] J. L. Dormann and M. Nogues, "Magnetic Structures in Substituted Ferrites," *Journal of Physics: Condensed Matter*, Vol. 2, No. 5, 1990, pp. 1223-1237. doi:10.1088/0953-8984/2/5/014

[6] N. Rezlescu, E. Rezlescu, C. Pasnicu and M. L. Craus, "Effects of the Rare-Earth Ions on Some Properties of a Nickel-Zinc Ferrite," *Journal of Physics: Condensed Matter*, Vol. 6, No. 29, 1994, pp. 5707-5716. doi:10.1088/0953-8984/6/29/013

[7] A. E. Virden and K. O'Grady, "Structure and Magnetic Properties of Ni-Zn Ferrite Nanoparticles," *Journal of Magnetism and Magnetic Materials*, Vol. 290-291, No. 2, 2005, pp. 868-870. doi:10.1016/j.jmmm.2004.11.398

[8] A. M. El-Sayed, "Influence of Zinc Content on Some Properties of Ni-Zn Ferrites," *Ceramics International*, Vol. 28, No. 4, 2002, pp. 363-367. doi:10.1016/S0272-8842(01)00103-1

[9] Z. Zhong , Q. Li, Y. Zhang, H. Zhong, M. Cheng and Y. Zhang, "Synthesis of Nanocrystalline Ni-Zn Ferrite Powders by Refluxing Method," *Powder Technology*, Vol. 155, No. 3, 2005, pp. 193-195. doi:10.1016/j.powtec.2005.05.060

[10] A. Dias and R. L. Moreira, "Chemical, Mechanical and Dielectric Properties after Sintering of Hydrothermal Nickel-Zinc Ferrites," *Materials Letters*, Vol. 39, No. 1, 1999, pp. 69-76. doi:10.1016/S0167-577X(98)00219-5

[11] S. E. Jacobo, S. Duhalde and H. R. Bertorello, "Rare Earth Influence on the Structural and Magnetic Properties of NiZn Ferrites," *Journal of Magnetism and Magnetic Materials*, Vol. 272-276, No. 3, 2004, pp. 2253-2254. doi:10.1016/j.jmmm.2003.12.564

[12] S. D. Shenoy, P. A. Joy and M. R. Anantharaman, "Effect of Mechanical Milling on the Structural, Magnetic and Dielectric Properties of Coprecipitated Ultrafine Zinc Ferrite," *Journal of Magnetism and Magnetic Materials*, Vol. 269, No. 2, 2004, pp. 217-226. doi:10.1016/S0304-8853(03)00596-1

[13] S. A. Morrison, C. L. Cahill, E. E. Carpenter, S. Calvin, R. Swaminathan, M. E. McHenry and V. G. Harris, "Magnetic and Structural Properties of Nickel Zinc Ferrite Nanoparticles Synthesized at Room Temperature," *Journal of Applied Physics*, Vol. 95, No, 11, 2004, pp. 6392-6395. doi:10.1063/1.1715132

[14] J. Sun, J. Li, G. Sun and W. Qu, "Synthesis of Dense NiZn Ferrites by Spark PLasma Sintering," *Ceramics International*, Vol. 28, No. 8, 2002, pp. 855-858. doi:10.1016/S0272-8842(02)00064-0

[15] A. Verma, T. C. Goel, R. G. Mendiratta and M. I. Alam, "Dielectric Properties of NiZn Ferrites Prepared by the Citrate Precursor Method," *Materials Science and Engineering: B*, Vol. 60, No. 2, 1999, pp. 156-162. doi:10.1016/S0921-5107(99)00019-7

[16] G. P. Lopez, S. P. Silvetti, S. E. Urreta and E. D. Cabanillas, "Magnetic Interactions in High-Energy Ball-Milled $NiZnFe_2O_4/SiO_2$ Composites," *Physica B*, Vol. 398, No. 2, 2007, pp. 241-244. doi:10.1016/j.physb.2007.04.024

[17] C. Upadhyay, D. Mishra, H. C. Verma, S. Anand and R. P. Das, "Effect of Preparation Conditions on Formation of Nanophase Ni-Zn Ferrites through Hydrothermal Technique," *Journal of Magnetism and Magnetic Materials*, Vol. 260, No. 1-2, 2003, pp. 188-194. doi:10.1016/S0304-8853(02)01320-3

[18] L. Vegard, "The Constitution of Mixed Crystals and the Space Occupied by Atoms," *Zeitschrift für Physik*, Vol. 5, No. 17, 1921, pp. 17-26.

[19] Y.-P. Fu and S.-H. Hu, "Electrical and Magnetic Properties of Magnesium-Substituted Lithium Ferrite," *Ceramics International*, Vol. 36, No. 4, 2010, pp. 1311-1317.

[20] R. G. Kharake, R. S. Devan and B. K. Chougalu, "Structural and Electrical Properties of Cd-Substituted Li-Ni Ferrites," *Journal of Alloys and Compounds*, Vol. 463, No. 1-2, 2008, pp. 67-72.

[21] B. R. Karache, B. V. Khasbardar and A. S. Vanigam, "X-Ray, SEM and Magnetic Properties of Mg Cd Ferrites," *Journal of Magnetism and Magnetic Materials*, Vol. 168, No. 3, 1997, pp. 292-298. doi:10.1016/S0304-8853(96)00705-6

[22] R. D. Shanoss and C. T. Prewitt, "Revised Values of Effective Ionic Radii," *Acta Crystallographica*, Vol. 1326, No. 7, 1970, pp. 1046-1048.

[23] S. Mahalakshmi and K. Srinivasa Manja, "ac Electrical Conductivity and Dielectric Behavior of Nanophase Nickel Ferrites," *Journal of Alloys and Compounds*, Vol. 457, No. 1-2, 2008, pp. 522-525.

[24] D. Ravinder and P. V. B. Redddy, "High-Frequency Dielectric Behaviour of Li-Mg Ferrites," *Materials Letters*, Vol. 57, No. 26-27, 2003, pp. 4344-4350. doi:10.1016/S0167-577X(03)00093-4

[25] N. Rezlescu and E. Rezlescu, "Abnormal Dielectric Behaviour of Copper Containing Ferrites," *Solid State Communications*, Vol. 14, No. 1, 1974, pp. 69-72.

[26] V. R. Murthy and J. Sobbhandari, "Dielectric Properties of Some Nickel-Zinc Ferrites at Radio Frequency," *State Solid Physics A*, Vol. 36, No. 2, 1976, pp. K133-K135.

[27] K. Iwauchi, "Dielectric Properties of Fine Particles of Fe_3O_4 and Some Ferrites," *Japanese Journal of Applied Physics*, Vol. 10, No. 11, 1971, pp. 1520-1528. doi:10.1143/JJAP.10.1520

[28] A. M. Abdeen, O. M. Hemeda, E. E. Assem and M. M. El-sehly, "Structural, Electrical and Transport Phenomena of Co Ferrite Substituted by Cd," *Journal of Magnetism and Magnetic Materials*, Vol. 238, No. 1, 2008, pp. 75-83. doi:10.1016/S0304-8853(01)00465-6

Microstructure and Dielectric Properties of PZS-PLZT Ceramics System

Sakri Adel*, Boutarfaia Ahmed

Laboratory of Applied Chemistry, University of Biskra, Biskra, Algeria.
Email: *adelsak@yahoo.fr

ABSTRACT

Piezoelectric ceramics are an important class of solid-state materials as they exhibit a wide range of applications. This article investigates the sintering temperature effect on microstructure and dielectric properties of a new ceramics material Pb (Zn, Sb) O_3-PbLa (Zr, Ti) O_3 (PZS-PLZT), which was prepared by a solid mixed-oxide solution method. Scanning electron micrograph (SEM) and X-ray diffraction (XRD) techniques was employed to examine the crystallization of the ceramics. The results of XRD show that the phase structure of the samples is tetragonal. The lattice parameters and the density increased with increase of the sintering temperature. The dielectric constant and the dielectric loss of the investigated samples decreased with increase in the frequency.

Keywords: Ferroelectric; Microstructure; Piezoelectric Ceramics; Dielectric Properties

1. Introduction

The piezoelectric property plays an important role for electric and electronic materials. The most widely used piezoelectric materials are based on the $PbTiO_3$-$PbZrO_3$ system (PZT) [1-3]. The application of the PZT component is wide, including nonvolatile memory elements, pyroelectric detectors, piezoelectric transducers, and photoelectric devices [4,5], but the electrical and thermal properties of PZT ceramics depend to a large extent on their surface microstructures [6]. Many researchers have considered modified PZT compositions with suitable substitution at the A and B sites [7-9] with some dopants to improve their electrical and mechanical property of piezoelectric materials in recent years, this dopants can be classified into two categories: donor dopants which induce Pb vacancies by changing with a higher valence ion for Pb^{2+} or $(Ti, Zr)^{4+}$ like Sb^{5+}, La^{3+}. And the acceptor dopants induce O vacancies by changing with a lower valance ion for Pb^{2+} or $(Ti, Zr)^{4+}$ example K^{+}, Fe^{3+} [10, 11]. So the properties of a PZT samples are very sensitive to additives, composition and synthesis methods.

Whatmore *et al.* [12] and Chaipanich *et al.* [13] have reported that the electric properties of PZT ceramics could be improved by the doping of Sb_2O_3. [14]. It was found that sintering behavior had obvious effects on the microstructure and piezoelectric properties. With the increase of sintering temperature, the microstructure of the ceramic samples changed accordingly and the porosity decreased [15].

There have been several reports on La substituted PZT. It has been reported that substitution of Pb^{2+} by La^{3+} ions created vacancies in the A site of perovskite ABO_3 structured PZT ceramics. The 7% La doped PZT ceramics PLZT (7/60/40) composition is one of the most important candidates for piezoelectric applications due to its extremely high piezoelectric and electromechanical coupling coefficients reported in literature [16-18].

In this work, a new ceramics material based PZT is prepared by a solid mixed-oxide solution method, and we investigate the structural, dielectrical properties of a doped PZT ceramic with La, Sb and Zn for different sintered temperature.

2. Experimental

The composition of PZS-PLZT system was $0.3Pb(Zn_{1/3}, Sb_{2/3})O_3$-$0.7Pb_{0.98}La_{0.02}(Zr_{0.48},Ti_{0.52})O_3$ were synthesized from high purity oxide powders PbO (99.90%), ZrO_2 (99.90%), TiO_2 (99.80%), Sb_2O_3 (99.90%), ZnO (99.90%), and La_2O_3 (99.90%), the samples were prepared by the conventional ceramic procedure. Stoichiometric amount of metal for the designated PZS-PLZT

*Corresponding author.

composition was mixed for 2 hours as the grinding media and ethanol as the solvent. After milling for 6 hours, the resultant slurry was dried in an oven and then the powders were calcined in a high-temperature furnace at 850˚C for 120 min. Then, the calcined powder was ball-milled again to ensure a fine particle size for 6 hours. After drying, the powder was pressed as disks and sintered at two different temperatures 1050˚C and 1200˚C in a closed aluminum crucible, the atmosphere was enriched in PbO vapor using $PbZrO_3$ powder, the density of the sintered component was measured from its mass and dimensions. The sintered discs were next mechanically processed by abrasion, dried and electrified with a silver paste burned out at 700˚C for 30 min. The polling was carried out in a silicon oil bath at 100˚C under an electric field of 3 kV/mm and cooled down to room temperature still being under the influence of the electric field. They were aged for 24 h prior to testing. The temperature dependence of dielectric properties was measured at temperatures ranging from room temperature to 650˚C with a heating using an impedance analyzer. An X-ray diffract meter BRUKER-AXE, D8 with CuKα radiation (λ = 1.5406 Å). Using the JCPDS database we can identify the presence of the perovskite structure and phases of the sintered samples. Surface microstructures of the ceramics were observed using a scanning electron microscopy (SEM) (JEOL JSM-6390LV) at room temperature.

3. Results and Discussion

3.1. X-ray Diffraction (XRD)

The XRD patterns of PZS-PLZT ceramics shown in **Figure 1** were identified as a material with perovskite structure having tetragonal symmetry according to JCPDS (Joint Committee of Powder Diffraction Standards) card n 00-046-0504. The tetragonal lattice parameters were determined from the evolution of the tetragonal peaks (200) and (002) by using *CELREF* software.

The results given in **Table 1** revealed that the tetragonal cell parameters are associated with some physical properties of the sintered samples. The lattice parameters and the density increased with increase the sintering temperature.

3.2. Scanning Electron Micrograph (SEM)

Figure 2 shows the scanning electron micrograph (SEM) microstructure of the porous PZS-PLZT prepared samples sintered at 1050˚C and 1200˚C by (SEM: JEOL JSM-6390LV). The increase of grain size from ≈1.83 μm to ≈3.12 μm is observed significantly with the increasing of sintering temperature from 1050˚C to 1200˚C respectively, and it can clearly be seen that the prepared PZS-PLZT ceramics is porous and has a not uniform distributions of grain shape and size.

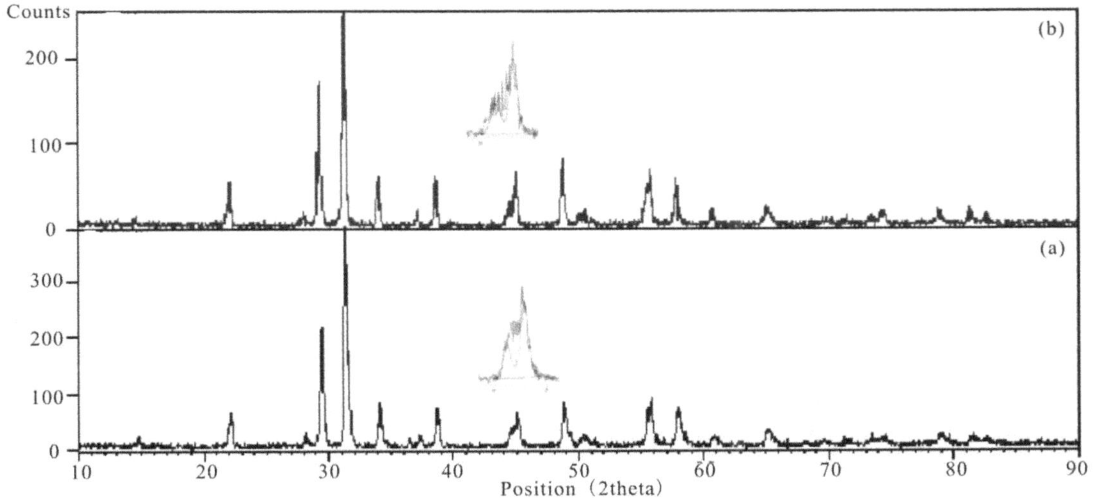

Figure 1. XRD patterns of PZS-PLZT ceramics sintered at (a) 1050˚C and (b) 1200˚C.

Table 1. Some physical properties of the sintered PZS-PLZT ceramics.

Sintering temperature (˚C)	Thickness [cm]	Diameter [cm]	specific weight [g/cm^3]	phase	lattice parameters		
					a [Å]	*b* [Å]	*c* [Å]
1050˚C	0.18	1.20	6.635	T	4.0155	4.0155	4.0517
1200˚C	0.18	1.15	7.598	T	4.0346	4.0346	4.0714

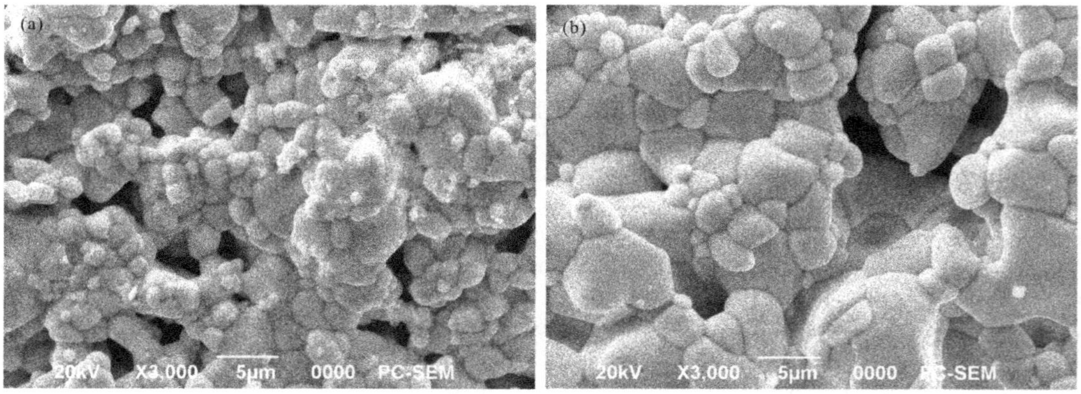

Figure 2. SEM micrographs of PZS-PLZT ceramics sintered at (a) 1050˚C; (b) 1200˚C for 2 hours.

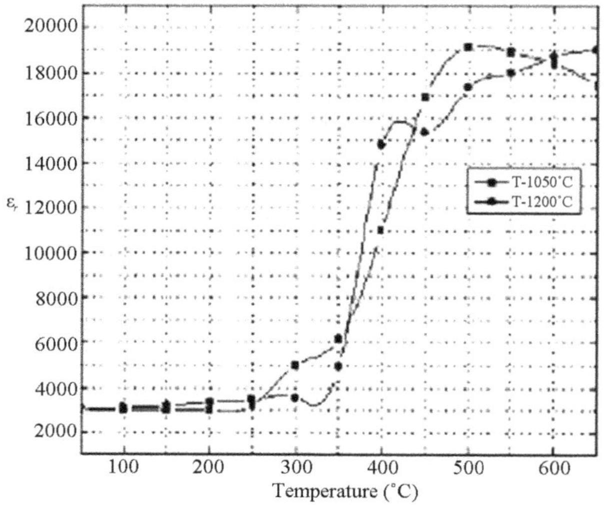

Figure 3. Temperature dependence of ε_r for PZS-PLZT samples sintered at 1050˚C and 1200˚C.

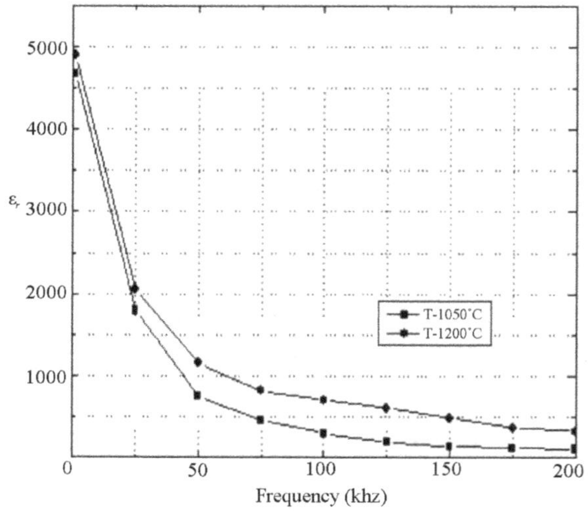

Figure 4. Frequency dependence of ε_r for PZS-PLZT samples sintered at 1050˚C and 1200˚C.

3.3. Dielectric Properties and the Loss Factor

Figure 3 shows the variation of the dielectric constant at 5 kHz with increment of temperature for the two samples witch sintered at 1050˚C and 1200˚C. Similar to normal ferroelectrics, the dielectric constant increases gradually with increasing temperature up to the transition temperature, then it decreases. Here we can note the most important characteristic of a disordered perovskite structure. The dielectric constant reaches a maximum value of 19,130 (T_c = 500˚C) for sample sintered at 1050˚C, but for the sintering temperature 1200˚C, the dielectric constant continues increasing (out of range temperature T_c > 650˚C). There, may be that this result is due to the change in the bulk density and the grain morphology with the change of sintering temperature [19]. This constant decreases gradually with increasing frequency (**Figure 4**) it reaches a maximum value at small values of frequency which explained by existence of different polarization models [20,21]. The dielectric constant of sample sintered at 1200˚C decreases slowly than of

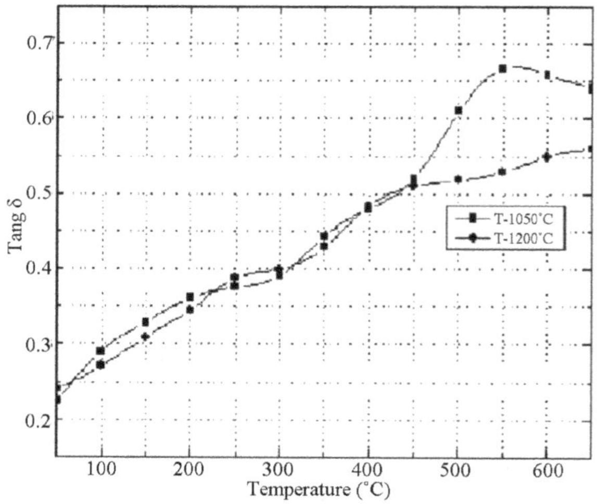

Figure 5. Temperature dependence of Tangδ for PZS-PLZT samples sintered at 1050˚C and 1200˚C.

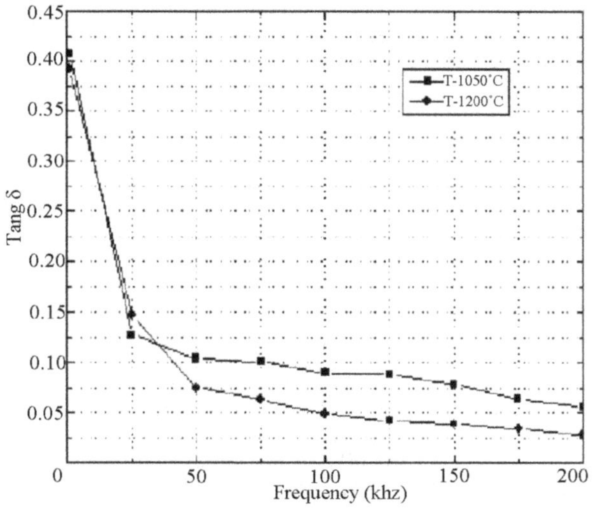

Figure 6. Frequency dependence of Tangδ for PZS-PLZT samples sintered at 1050˚C and 1200˚C.

1050˚C sintered sample.

Figure 5 shows the variation of the loss factor *Tang*δ with temperature. Tangδ increases gradually with increasing temperature. For sintered temperature 1050˚C the loss factor reaches a maximum value of 0.65 at (T_c = 550˚C), but for the sintering temperature 1200˚C, the dielectric constant continues increasing (out of range temperature T_c > 650˚C). This may be caused by the losses due to the electrical conduction. This factor decreases with increasing frequency (**Figure 6**), which is a characteristic of ferroelectrics [22,23].

4. Conclusion

PZS-PLZT was successfully prepared by a solid mixed-oxide solution method; The X-ray diffraction measurements for all $0.3Pb(Zn_{1/3},Sb_{2/3})O_3$-$0.7Pb_{0.98}La_{0.02}(Zr_{0.48},Ti_{0.52})O_3$ sintered samples indicated the presence of tetragonal phase and the scanning electron micrograph shows a porous nature of ceramics and the grain size increases with increasing the temperature. The influence of the sintering temperature on the properties of PZS-PLZT ceramics was studied. The lattice parameters and the density increased with increase of the sintering temperature. The dielectric constant and the dielectric loss of the sintered samples decreased with increase in the frequency. The effects of the sintering temperature on the dielectric properties can be attributed to the change in the bulk density and the grain morphology with the change of the sintering temperature.

REFERENCES

[1] E. Sawaguchi, "Ferroelectricity versus Antiferroelectricity in the Solid Solutions of $PbZrO_3$ and $PbTiO_3$," *Journal of the Physical Society of Japan*, Vol. 8, 1953, pp. 615-629. doi:10.1143/JPSJ.8.615

[2] T. Yamamoto, "Ferroelectric Properties of the $PbZrO_3$-Pb TiO_3 System," *Japanese Journal of Applied Physics*, Vol. 35, 1996, pp. 5104-5108.

[3] T. Takenaka, H. Nagata, Y. Hiruma, Y. Yoshii and K. Matumoto, "Lead-Free Piezoelectric Ceramics Based on Perovskite Structures," *Journal of Electroceramics*, Vol. 19, No. 4, 2007, pp. 259-265. doi:10.1007/s10832-007-9035-4

[4] R. E. Newnham and G. R. Ruschau, "Smart Electroceramics," *Journal of the American Ceramic Society*, Vol. 74 No. 3, 1991, pp. 463-480. doi:10.1111/j.1151-2916.1991.tb04047.x

[5] T. Chen, T. Zhang, J. F. Zhou, J. W. Zhang, Y. H. Liu and G. C. Wang, "Piezoelectric Properties of $[(K_{1-x}Na_x)_{0.95}Li_{0.05}]_{0.985}Ca_{0.015}(Nb_{0.95}Sb_{0.05})_{0.985}Ti_{0.015}O_3$ Lead-Free Ceramics," *Indian Journal of Physics*, Vol. 86, No. 6, 2012, pp. 443-446. doi:10.1007/s12648-012-0087-1

[6] T. K. Mandal, "X-Ray Diffraction and Scanning Electron Microscopic Studies on the Microstructure of Laser Induced Surface Alloyed PZT Ceramics," *Indian Journal of Physics*, Vol. 86, No. 10, 2012, pp. 881-888. doi:10.1007/s12648-012-0163-6

[7] B. Jaffe, W. R. Cook and H. Jaffe, "Piezoelectric Ceramics," Academic Press, New York, 1971. doi:10.1016/0022-460X(72)90684-0

[8] R. La, S. C. Sharma and R. Dayal, "Effects of Doping Pairs on the Preparation and Dielectricity of PLZT Ceramics," *Ferroelectrics*, Vol. 67, No. 1, 1986, pp. 93-102. doi:10.1080/00150198608245011

[9] R. Rai and S. Sharma, "Structural and Dielectric Properties of Sb-Doped PLZT Ceramics," *Ceramics International*, Vol. 30, No. 7, 2004, pp. 1295-1299. doi:10.1109/ISAF.2009.5307560

[10] W. Qiu and H. H. Hng, "Effects of Dopants on the Microstructure and Properties of PZT Ceramics," *Materials Chemistry and Physics*, Vol. 75, No. 1-3, 2002, pp. 151-156. doi:10.1016/S0254-0584(02)00045-7

[11] A. Govindan, A. Sharma, A. K. Pandey and S. K. Gaor, "Piezoelectric and Pyroelectric Properties of Lead Lanthanum Zirconate Titanate (PLZT) Ceramics Prepared by Sol Gel Derived Nano Powders," *Indian Journal of Physics*, Vol. 85, No. 12, 2011, pp. 1829-1832. doi:10.1007/s12648-011-0187-3

[12] R. W. Whatmore, O. Molter and C. P. Shaw, "Electrical Properties of Sb and Cr-Doped $PbZrO_3$-$PbTiO_3$-$PbMg_{1/3}$-$Nb_{2/3}O_3$ Ceramics," *Journal of the European Ceramic Society*, Vol. 23, No. 5, 2003, pp. 721-728. doi:10.1016/S0955-2219(02)00162-0

[13] A. Chaipanich, G. Rujijanagul and T. Tunkasiri, "Properties of Sr- and Sb-doped PZT-Portland Cement Composites," *Applied Physics A*, Vol. 94, No. 2, 2009, pp. 329-337. doi:10.1007/s00339-008-4798-2

[14] Y. D. Hou, M. K. Zhou, J. L. Tang, X. M. Song, C. S. Tian and H. Yan, "Effects of Sintering Process and Mn-Doping on Microstructure and Piezoelectric Properties of $Pb((Zn_{1/3}Nb_{2/3})_{0.20}(Zr_{0.47}Ti_{0.53})_{0.80})O_3$ System," *Materials*

Chemistry and Physics, Vol. 99, No. 1, 2006, pp. 66-70. doi:10.1016/j.matchemphys.2005.09.076

[15] Z. He, J. Ma and R. Z. T. Li, "PZT-Based Materials with Bilayered Structure: Preparation and Ferroelectric Properties," *Journal of the European Ceramic Society*, Vol. 23, No. 11, 2003, pp. 1943-1947. doi:10.1016/S0955-2219(02)00423-5

[16] Y. Xu, "Ferroelectric Materials and Their Applications," North-Holland, Amsterdam, 1991.

[17] M.-S. Cao, D.-W. Wang, J. Yuan, H.-B. Lin, Q.-L. Zhao, W.-L. Song and D.-Q. Zhang, "Enhanced Piezoelectric and Mechanical Properties of ZnO Whiskers and Sb_2O_3 Co-Modified Lead Zirconate Titanate Composites," *Materials Letters*, Vol. 64, No. 16, 2010, pp. 1798-1801. doi:10.1016/j.matlet.2010.05.037

[18] A. Yang, C. Wang, R. Guo, Y. Huang and C. Nan, "EfFects of Sintering Behavior on Microstructure and Piezoelectric Properties of Porous PZT Ceramics," *Ceramics International*, Vol. 36, No. 2, 2010, pp. 549-554. doi:10.1016/j.ceramint.2009.09.022

[19] J. Zhao, Q. M. Zhang, *et al.*, "Electromechanical Properties of Relaxor Ferroelectric Lead Magnesium Niobate-Lead Titanate Ceramics," *Japanese Journal of Applied Physics*, Vol. 34, No. 10, 1995, pp. 5658-5663. doi:10.1143/JJAP.34.5658

[20] G. H. Haertling and C. E. Land, "Hot-Pressed (Pb,La)$(Zr,Ti)O_3$ Ferroelectric Ceramics for Electrooptic Applications," *Journal of the American Ceramic Society*, Vol. 54, No. 1, 1971, pp. 1-11. doi:10.1111/j.1151-2916.1970.tb12105.x-i1

[21] A. Fawzi, A. D. Sheikh and V. L. Mathe, "Effect of Magnetostrictive Phase on Structural, Dielectric and Electrical Properties of $NiFe_2O_4$+$Pb_{0.93}La_{0.07}(Zr_{0.6}Ti_{0.4})O_3$ Composites," *Solid State Sciences*, Vol. 11, No. 11, 2009, pp. 1979-1984 doi:10.1016/j.solidstatesciences.2009.07.006

[22] Z. He, J. M. R. Zhang and T. Li, "Fabrication and Characterization of Bilayered $Pb(Zr,Ti)O_3$-Based Ceramics," *Materials Letters*, Vol. 56, No. 6, 2002, pp. 1084-1088. doi:10.1016/S0167-577X(02)00683-3

[23] Z. He, J. Ma and R. Zhang, "Investigation on the Microstructure and Ferroelectric Properties of Porous PZT Ceramics," *Ceramics International*, Vol. 30, No. 7, 2004, pp. 1353-1356. doi:10.1016/j.ceramint.2003.12.108

Dielectric Properties and the Phase Transition of Pure and Cerium Doped Calcium-Barium-Niobate

Alexander Niemer[1], Rainer Pankrath[1], Klaus Betzler[1], Manfred Burianek[2], Manfred Muehlberg[2]

[1]Department of Physics, University of Osnabrück, Osnabrück, Germany; [2]Institute of Crystallography, University of Cologne, Cologne, Germany.
Email: Klaus.Betzler@uos.de

ABSTRACT

The complex dielectric constant of pure and cerium doped calcium-barium-niobate (CBN) was studied at frequencies 20 Hz $\leq f \leq$ 1 MHz in the temperature range 300 K $\leq T \leq$ 650 K and compared with the results for the well known ferroelectric relaxor strontium-barium-niobate (SBN). By the analysis of the systematically taken temperature and frequency dependent measurements of the dielectric constant the phase transition characteristic of the investigated materials was evaluated. From the results it must be assumed that CBN shows a slightly diffuse phase transition without relaxor behavior. Doping with cerium yields a definitely different phase transition characteristic with some indications for a relaxor type ferroelectric material, which are common from SBN.

Keywords: 77.80.-e Ferroelectricity and Antiferroelectricity; 77.80.Jk Relaxor Ferroelectrics; 77.84.Ek Niobates and Tantalates; 77.22.Ch Permittivity (Dielectric Function)

1. Introduction

The ferroelectric material strontium-barium-niobate ($Sr_xBa_{1-x}Nb_2O_6$, SBN) is well known for its excellent pyroelectric, piezoelectric, electrooptic, acoustooptic, photorefractive and non-linear optical properties [1-4]. But SBN exhibits a low phase transition temperature of about 353 K for the congruently melting composition with 61 mole% strontium content (SBN:61) which hampers its applicability.

By the search for an alternative for SBN the isostructural material calcium-barium-niobate ($Ca_xBa_{1-x}Nb_2O_6$, CBN) was found. CBN is a novel tungsten bronze which was first grown as single crystals a few years ago [5]. CBN offers similar physical properties like SBN, so it is also well suited for a wide range of applications. One difference is that CBN has an about 200 K higher phase transition temperature for the congruently melting composition with 28 mole% of calcium (CBN:28) [5]. This is one of the main advantages of the material compared to SBN. The existence region of CBN with $0.2 \leq x \leq 0.4$ of calcium content is smaller than the region for SBN with $0.32 \leq x \leq 0.82$ for experimentally realized compositions [6,7].

In literature it was found that CBN does not show a clear relaxor like behavior as SBN [8]. To investigate the phase transition behavior in more detail, in this work the dielectric properties of pure and cerium doped CBN are studied. The measurements of the dielectric constant and a comparison to the results of SBN are used to study the character of the phase transition of CBN. For SBN it is well known, that the material shows a relaxor type phase transition and that there is an influence of the concentration and type of rare earth dopants on the phase transition temperature and characteristic [9-12]. Previous investigations on CBN were mainly done on undoped crystals without methodical investigation of the dielectric constant. This paper presents the systematical measurement and analysis of the temperature and frequency dependency of the dielectric constant of CBN, cerium doped CBN (CBN:Ce) and a comparison to SBN to get a characterization of the phase transition of CBN for the doped case, too.

2. Experimental Details

All investigated crystals, pure CBN, cerium doped CBN, and SBN were grown by the Czochralski technique from the congruently melting composition with x = 0.28 and x = 0.61, respectively. The cerium doped samples were grown from a melt with the congruently melting composition and a cerium content of 1%. For the measurements c-cut samples were used. The thickness of the samples was in the range of 1 to 4 mm and the c-faces were polished to optical quality. To electrically contact the samples the c-faces were covered with silver paste and then the samples with the silver paste electrodes were heated up to 670 K to get reproducible results. Otherwise one would measure side effects from the electrodes. The complex dielectric constant $\varepsilon = \varepsilon' + i\varepsilon''$ was calculated from the values of the capacity and dissipation factor measured with an Hewlett-Packard 4284A LCR-meter in the temperature range from 300 to 650 K for the CBN samples and from 300 to 460 K for SBN. The measurements were performed in a frequency range from 20 Hz to 1 MHz and taken under isothermal conditions. To ensure that, the CBN samples were heated with a low rate of 20 K/h with respect to their larger thickness of 3 - 4 mm. The SBN samples were only about 1 mm thick and were therefore heated with a higher rate of 60 K/h by a Peltier element. To achieve the higher temperatures, a high precision laboratory oven was used. As electrical field for the measurements about 3 V/mm were used to minimize the influence on the sample and to obtain minimal noise.

3. Results and Discussion

The experimental results for the complex dielectric constant of CBN, CBN:Ce and SBN are shown in **Figures 1-3**. Both materials CBN and CBN:Ce show a ferroelectric phase transition indicated by the peak in the real part of the dielectric constant, as it is known for SBN.

The phase transition temperature was estimated as the peak temperature at a frequency of 20 Hz. For CBN and CBN:Ce phase transition temperatures of 552 K and 462 K, respectively, were measured, which are considerably higher than the value of 354 K for SBN. The cerium doping causes a lowering of the phase transition temperature of about 76 K compared to undoped CBN. Such a lowering of the phase transition temperature by doping with rare earth elements is known from SBN [12,13].

Regarding the absolute value of ε' it has to be pointed out that the values for CBN are about forty times larger than the values of SBN and the values of CBN:Ce are still by a factor of four greater. The absolute values for ε'' show nearly the same behavior.

The frequency dependency of ε' and ε'' is shown as an example for the case of CBN:Ce in the inset of **Figure 2**. From low to high frequencies the value of ε'_m decreases and starts to increase at very high frequencies again. ε''_m shows qualitatively the same behavior but there is a more dramatic increase in the high frequency range.

All samples show a relatively broad peak in the dielectric constant around the phase transition temperature (**Figures 1-3**) which is typical for relaxor ferroelectrics or ferroelectrics with a diffuse phase transition [3,14]. But there are also significant differences in the behavior of CBN compared to the reference material SBN. For a better comparison **Figure 4** shows the normalized dielectric

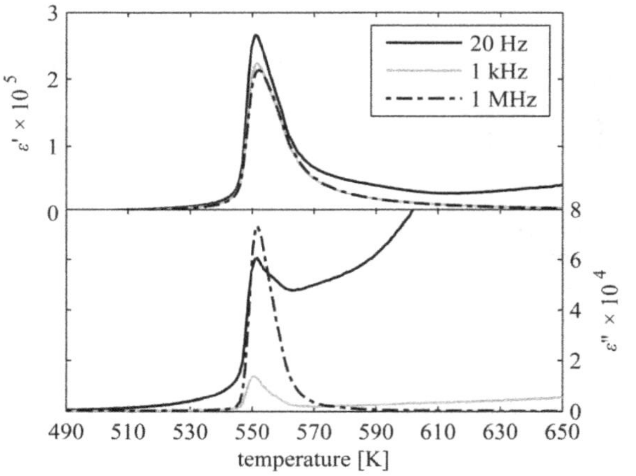

Figure 1. Temperature dependency of the real (ε') and imaginary (ε'') part of the dielectric constant for CBN:28.

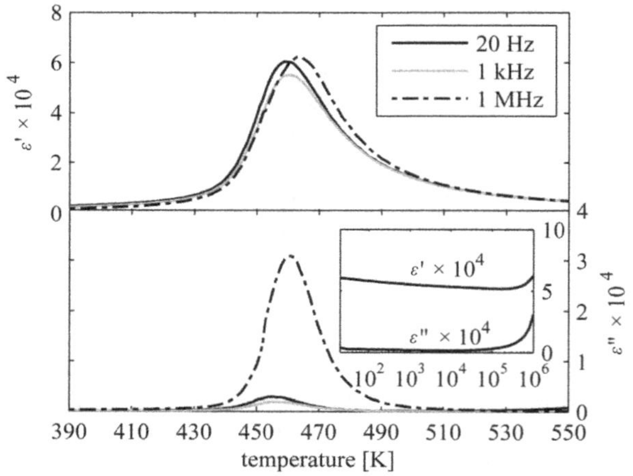

Figure 2. Temperature dependency of the real (ε') and imaginary (ε'') part of the dielectric constant for CBN:Ce. The inset shows the values of ε' and ε'' as a function of frequency.

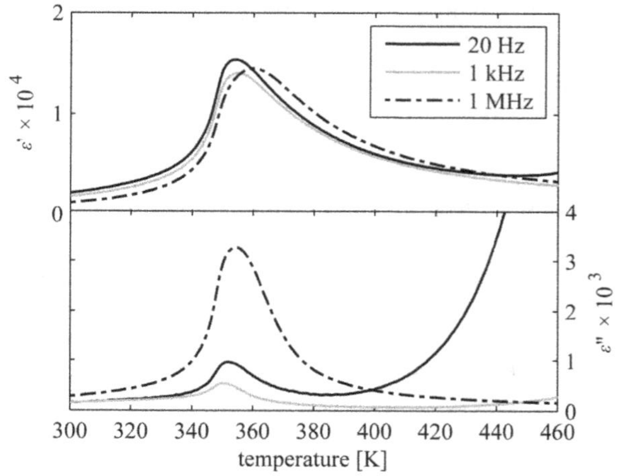

Figure 3. Temperature dependency of the real (ε') and imaginary (ε'') part of the dielectric constant for SBN:61.

constant $\varepsilon'/\varepsilon'_m$ for all investigated samples. The peak in the real part of the dielectric constant of undoped CBN exhibits a smaller value of the full width at half maximum (FWHM) than the one for SBN. By comparison with the cerium doped sample it is evidenced that the doping induces a broadening of the peak in the dielectric constant which indicates a more diffuse phase transition characteristic of the material [10,15]. It can be concluded that the dopant has a strong influence on the characteristic of the ferroelectric phase transition of CBN.

The FWHM of the normalized dielectric constant can be used to determine the degree of diffuseness of the phase transition, higher values of FWHM mean a higher degree of diffuseness [15]. The frequency dependence of the FWHM values is shown in **Figure 5**. The values of the FWHM for pure CBN are about four times smaller than for SBN and indicate a phase transition with a very low degree of diffuseness, whereas for SBN a high degree of diffuseness of the phase transition is known. CBN:Ce shows a diffuse phase transition where the values of the FWHM are only about 1.5 times smaller than the ones for SBN. For CBN and CBN:Ce no frequency dependency of the FWHM can be observed, for SBN, however, an increase of the FWHM with the frequency is found.

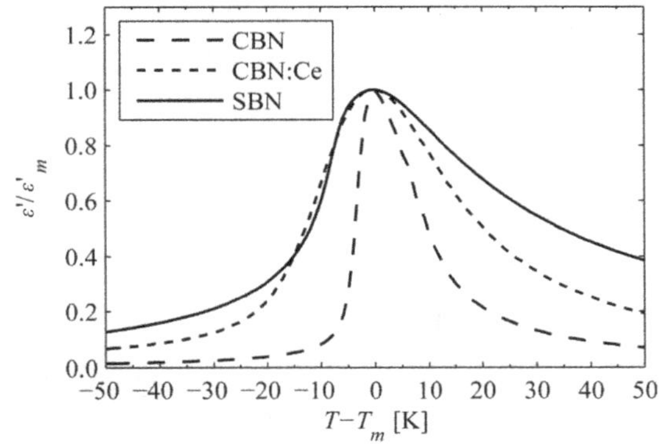

Figure 4. The normalized dieelectric constant $\varepsilon'/\varepsilon'_m$ around the peak-temperature T_m at a frequency of 5 kHz.

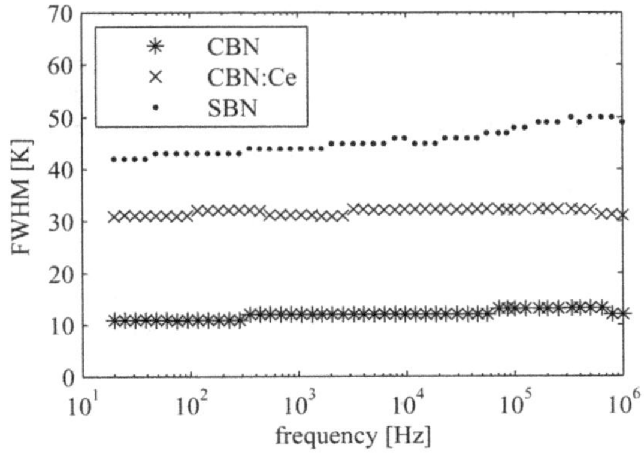

Figure 5. Frequency dependent plot of the FWHM for the investigated samples.

In ferroelectric relaxors a dependency of the peak temperature T_m on the measurement frequency is expected [16], which can be described by the Vogel-Fulcher law [17-19]:

$$f = f_0 \exp\left(\frac{T_0}{T_m - T_f}\right) \tag{1}$$

The best fitting parameters are listed in **Table 1** and the measurement data and results of the fit are presented in **Figure 6**. For CBN:Ce and SBN the Vogel-Fulcher law fits excellent to the experimental results. For CBN a fit of the Vogel-Fulcher equation is not reasonable because there is no frequency dependence of T_m in the investigated region from 20 Hz to 1 MHz. This evidences that CBN shows no or only very weak relaxor behavior, but nevertheless a diffuse phase transition because of the broadening of the peak in the dielectric constant shown above. For comparison, also the phase transition temperatures T_c are listed in **Table 1**—defined as the peak temperatures T_m of ε' measured at lowest frequencies. For both samples CBN:Ce and SBN, extrapolated freezing temperature T_f is lower than T_c which indicates that the region of glassy freezing is never reached but replaced by a quasi critical behavior. Such a behavior is known from literature for rare earth doped SBN and is considered to be typical for an Ising-type relaxor [10].

Table 1. Best fit parameters of the T_m vs. f data shown in Figure 6 to the Vogel-Fulcher law. For comparison, also the values of T_c are shown.

sample	f_0 [Hz]	T_0 [K]	T_f [K]	T_m [K]
CBN	-	-	-	552
CBN:Ce	4.59×10^6	5.9	459.5	462
SBN	1.22×10^8	40.5	351.3	354

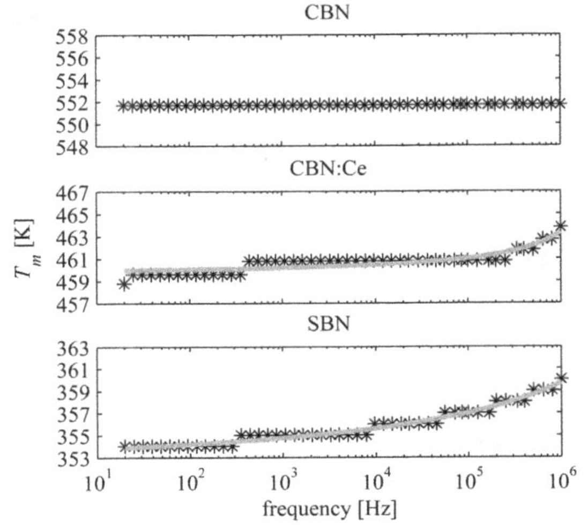

Figure 6. Frequency dependency of the peak-temperature of $\varepsilon'(T_m)$ and fit according to Vogel-Fulcher's law described by the solid lines.

The parameter $T_0 = E_a/k_B$ where k_B is Boltzmann's constant, describes the average activation energy. The activation energy for CBN:Ce is nearly ten times smaller than the one for SBN. Higher values of this energy are an indication for a rougher landscape of free energy of the system. Doping with cerium in SBN leads to higher values of T_0 [20] thus it can be concluded that CBN:Ce has a mainly flat landscape of free energy.

The values of the calculated frequency f_0 are far below the soft mode frequency, which is in the order of 10^{10} - 10^{11} Hz. Following the argumentation in [20] random field correlated polar clusters are assumed to be responsible for this untypical behavior.

For ferroelectric materials with a first or second order phase transition described by Landau's theory the behavior of the reciprocal dielectric constant above the Curie-Temperature (T_c) can be described by the Curie-Weiss Law. But for ferroelectrics with a diffuse or relaxor like phase transition the normal Curie-Weiss Law can only be used in an adequate distance from T_m [21]. Near T_m the reciprocal dielectric constant can be better described by a modified Curie-Weiss Law [22,23]:

$$\frac{1}{\varepsilon'} - \frac{1}{\varepsilon'_m} = \frac{(T - T_m)^\gamma}{C_C} \tag{2}$$

This modified version transforms to the normal Curie-Weiss Law for conventional ferroelectrics because for this materials one gets $\gamma = 1$; for materials with a complete diffuse phase transition γ has values near 2. The best fitting of Equation (2) to the experimental results is shown in **Figure 7**. From the calculated values of γ it can be seen that all investigated materials show a partially diffuse phase transition because γ has a value around 1.5. The value of γ for CBN is slightly smaller than for CBN:Ce which is nearly the same evaluated for SBN, but there is not such a clear difference like in the results for the FWHM. Thence the FWHM should not be used as the sole property for the characterization of the phase transition.

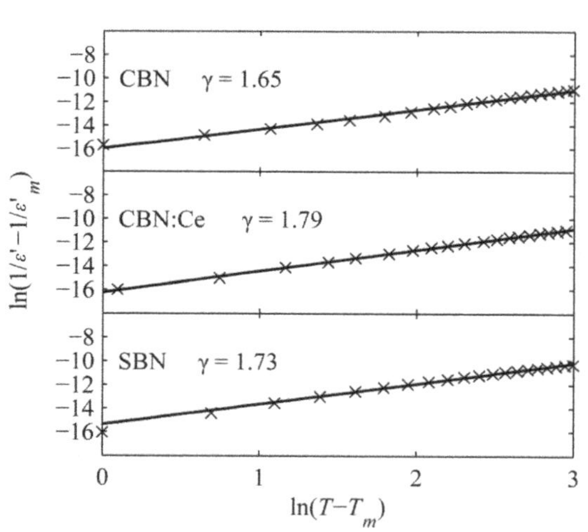

Figure 7. Frequency dependent plot of the FWHM for the investigated samples.

4. Conclusions

The dielectric constant data of CBN, CBN:Ce, SBN show, that the phase transition characteristic of the three materials is considerably different. The analysis of the full-width at half maximum yields a slightly diffuse phase transition characteristic for CBN and a distinct diffuseness of the phase transition for CBN:Ce which is in the range of the phase transition of SBN. From the results of the Vogel-Fulcher law can be obtained that CBN shows no relaxor behavior whereas CBN:Ce shows like SBN a clear frequency dependency of the peak temperature which follows Vogel-Fulcher's law. Such a deviation is a typical sign of ferroelectric relaxors [10]. Further it was shown, that CBN:Ce has a very flat free energy landscape compared to the well-known ferroelectric relaxor SBN. By the reciprocal dielectric constant and the critical exponent calculated from it was shown that all investigated materials show a partially diffuse phase transition, as the critical exponent of CBN is marginally smaller than the one for CBN:Ce or SBN.

It can be concluded that CBN shows a slightly diffuse phase transition without the typical relaxor behavior whereas CBN:Ce shows a distinctly diffuse phase transition with a typical behavior for ferroelectric relaxors like SBN. Thus, as in the case of SBN, doping with rare earth elements has a strong influence on the phase transition characteristic of CBN.

5. Acknowledgements

Financial support from the Deutsche Forschungsgemeinschaft and the federal state of Niedersachsen within the graduate college GRK 695 "Nonlinearities of Optical Materials" is gratefully acknowledged.

REFERENCES

[1] A. S. Kewitsch, M. Segev, A. Yariv, G. J. Salamo, T. W. Towe, E. J. Sharp and R. R. Neurgaonkar, "Tunable Quasi-Phase Matching Using Dynamic Ferroelectric Domain Gratings Induced by Photorefractive Space— Charge Fields," *Applied Physics Letters*, Vol. 64, No. 23, 1994, pp. 3068-3070. doi:10.1063/1.111349

[2] R. R. Neurgaonkar, W. F. Halla, J. R. Olivera, W. W. Hoa and W. K. Corya, "Tungsten Bronze $Sr_{1-x}Ba_xNb_2O_6$: A Case History of Versatility," *Ferroelectrics*, Vol. 87, No. 1, 1988, pp. 167-179. doi:10.1080/00150198808201379

[3] L. E. Cross, "Relaxor Ferroelectrics," *Ferroelectrics*, Vol. 76, No. 1, 1987, pp. 241-267. doi:10.1080/00150198708016945

[4] M. Horowitz, A. Bekker and B. Fischer, "Broadband

Second-Harmonic Generation in $Sr_xBa_{1-x}Nb_2O_6$ by Spread Spectrum Phase Matching with Controllable Domain Gratings," *Applied Physics Letters*, Vol. 62, No. 21, 1993, pp. 2619-2621. doi:10.1063/1.109264

[5] M. Eßer, M. Burianek, D. Klimm and M. Mühlberg, "Single Crystal Growth of the Tetragonal Tungsten Bronze $Ca_xBa_{1-x}Nb_2O_6$ (x = 0.28; CBN-28)," *Journal of Crystal Growth*, Vol. 240, No. 1-2, 2002, pp.1-5. doi:10.1016/S0022-0248(02)00868-0

[6] M. Eßer, M. Burianek, P. Held, J. Stade, S. Bulut, C. Wickleder and M. Mühlberg, "Optical Characterization and Crystal Structure of the Novel Bronzetype $Ca_xBa_{1-x}Nb_2O_6$ (x = 0.28; CBN-28)," *Crystal Research and Technology*, Vol. 38, No. 6, 2003, pp. 457-464. doi:10.1002/crat.200310057

[7] M. Ulex, R. Pankrath and K. Betzler, "Growth of Strontium Barium Niobate: The LIquidus-Solidus Phase Diagram," *Journal of Crystal Growth*, Vol. 271, No. 1-2, 2004, pp. 128-133. doi:10.1016/j.jcrysgro.2004.07.039

[8] Y. J. Qi, C. J. Lu, J. Zhu, X. B. Chen, H. L. Song, H. J. Zhang and X. G. Xu, "Ferroelectric and Dielectric Properties of $Ca_xBa_{1-x}Nb_2O_6$ Single Crystals of Tungsten Bronzes Structure," *Applied Physics Letters*, Vol. 87, No. 8, 2005, Article ID 082904. doi:10.1063/1.2010614

[9] O. Kersten, A. Rost and G. Schmidt, "Dielectric Dispersion of Relaxor Ferroelectrics (SBN 75 and PLZT 8/65/35)," *Physica Status Solidi A*, Vol. 75, No. 2, 1983, pp. 495-500. doi:10.1002/pssa.2210750220

[10] I. A. Santos, D. U. Spinola, D. Garcia and J. A. Eiras, "Dielectric Behavior and Diffuse Phase Transition Features of Rare Earth Doped $Sr_{0.61}Ba_{0.39}Nb_2O_6$ Ferroelectric Ceramic," *Journal of Applied Physics*, Vol. 92, No. 6, 2002, pp. 3251-3256. doi:10.1063/1.1481210

[11] S. Kuroda and K. Kubota, "Diffuse Phase Transition in Rare Earth Ion Doped SBN," *Journal of Physics and Chemistry of Solids*, Vol. 42, No. 7, 1981, pp. 573-577. doi:10.1016/0022-3697(81)90104-9

[12] T. Volk, L. Ivleva, P. Lykov, N. Polozkov, V. Salobutin, R. Pankrath and M. Wöhlecke, "Effects of Rare-Earth Impurity Doping on the Ferroelectric and Photorefractive Properties of Strontium-Barium Niobate Crystals," *Optical Materials*, Vol. 18, No. 1, 2001, pp. 179-182. doi:10.1016/S0925-3467(01)00162-8

[13] I.-I. Oprea, U. Voelker, A. Niemer, R. Pankrath, S. Podlozhenov and K. Betzler, "Influence of Erbium Doping on Phase Transition and Optical Properties of Strontium Barium Niobate," *Optical Materials*, Vol. 32, No. 1, 2009, pp. 30-34. doi:10.1016/j.optmat.2009.05.015

[14] L. E. Cross, "Relaxor Ferroelectrics: An Overview," *Ferroelectrics*, Vol. 151, No. 1, 1994, pp. 305-320. doi:10.1080/00150199408244755

[15] J. Kim, M. Jang, I. Kim and K. Lee, "Niobium Doping Effects and Ferroelectric Relaxor Behavior of Bismuth Lantanium Titanate," *Journal of Electroceramics*, Vol. 17, No. 2-4, 2006, pp. 129-133. doi:10.1007/s10832-006-5410-9

[16] A. K. Tagantsev, "Vogel-Fulcher Relationship for the Dielectric Permittivity of Relaxor Ferroelectrics," *Physical Review Letters*, Vol. 72, No. 7, 1994, pp. 1100-1103. doi:10.1103/PhysRevLett.72.1100

[17] G. Fulcher, "Analysis of Recent Measurements of the Viscosity of Glasses," *Journal of the American Ceramic Society*, Vol. 8, No. 6, 1925, pp. 339-355. doi:10.1111/j.1151-2916.1925.tb16731.x

[18] F. Chu, I. M. Reaney and N. Setter, "Spontaneous (Zero-Field) Relaxor-to-Ferroelectric-Phase Transition in Disordered $Pb(Sc_{1/2}Nb_{1/2})O_3$," *Journal of Applied Physics*, Vol. 77, No. 4, 1995, pp. 1671-1676. doi:10.1063/1.358856

[19] D. H. Vogel, "Das Temperaturabhaengigkeitsgesetz der Viskositaet von Fluessigkeiten," *Physikalische Zeitschrift*, Vol. 22, 1921, p. 645.

[20] J. Dec, W. Kleemann, T. Woike and R. Pankrath, "Phase Transitions in $Sr_{0.61}Ba_{0.39}Nb_2O_6{:}Ce^{3+}$: I. Susceptibility of Clusters and Domains," *European Physical Journal B*, Vol. 14, No. 4, 2000, pp. 627-632. doi:10.1007/s100510051071

[21] W. H. Huang, D. Viehland and R. R. Neurgaonkar, "Anisotropic Glasslike Characteristics of Strontium Barium Niobat Relaxors," *Journal of Applied Physics*, Vol. 76, No. 1, 1994, pp. 490-496. doi:10.1063/1.357100

[22] K. Uchino and S. Nomura, "Critical Exponents of the Dielectric Constants in Diffused-Phase-Transition Crystals," *Ferroelectrics Letters*, Vol. 44, 1982, pp. 55-61. doi:10.1080/07315178208201875

[23] Y. Guo, K. Kakimoto and H. Ohsato, "Dielectric and Piezoelectric Properties of Lead-Free $(Na_{0.5}K_{0.5})$ NbO_3-$SrTiO_3$ Ceramics," *Solid State Communications*, Vol. 129, No. 5, 2004, pp. 279-284. doi:10.1016/j.ssc.2003.10.026

Dielectric and Optical Characterization of Boron Doped Ammonium Dihydrogen Phosphate

Delci Zion[1], Shyamala Devarajan[1], Thayumanavan Arunachalam[2]

[1]Department of Physics, D.G.Vaishnav College, Chennai, India; [2]Department of Physics, A.V.V.M. Sri Pushpam College, Thanjavur, India.
Email: delcidgvc@gmail.com

ABSTRACT

Single crystals of pure and boron doped ammonium dihydrogen phosphate were grown from aqueous solution by slow solvent evaporation process. ICP studies were done to confirm the presence of the dopant boron in the parent crystal. The values of the lattice parameters were determined by single crystal X-ray diffraction. The pure and doped ADP crystals were found to have tetragonal structure. Complete optical characterizations of the crystals were done using the FTIR, UV-Vis and NLO studies. The presences of the various functional groups in the crystals were identified by FTIR spectrum. The band gap energies of the pure and doped crystals have been calculated at their cut off frequencies using the UV-Vis spectrum. The second harmonic generation efficiency of the crystals was determined. The electric properties of the grown crystal have been analyzed by studying the variation of dielectric constant and dielectric loss with frequency.

Keywords: Solution Growth; ADP; Boric Acid; Optical Transparency; NLO; Dielectric Constant

1. Introduction

Growth and studies of ammonium dihydrogen phosphate (ADP) crystal is a favourite topic to researchers because of its unique properties and wide applications. Single crystals of ADP are used for frequency doubling and frequency tripling of laser systems, optical switches in inertial confinement fusion and acoustic-optical devices [1,2]. ADP has been the subject of a wide variety of investigations over the past decades. Reasonable studies have been done on the growth and properties of pure ADP [3-5]. ADP belongs to scalenohedral class of tetragonal crystal systems. ADP has unit cell parameters of a = b = 7.510 Å and c = 7.564 Å [6,7]. It is known that, very little amount of additives can strongly suppress the metal ion impurities and promote the crystal quality. Oxalic acid and amino acids as additives in ADP crystals give appreciable change in optical, thermal, dielectric and mechanical behaviours [8,9]. In a crystal metastable zone width is an essential parameter for the growth of good crystals from solution. Organic additives urea and thiourea increase the metastable zone width. In ADP crystal, using urea as additive, meta stable zone width increased by 3.7°C, saturated at 30°C [10]. The effect of additives depends on the additive concentration, super saturation, temperature and PH. The growth promoting effect is observed in the presence of organic additives. In recent years, efforts have been taken to improve the quality, growth rate and properties of ADP, by employing new growth techniques, and also by the addition of organic, inorganic and semiorganic impurities [11-14]. In the light of the research work being done on ADP crystals, to improve the properties, the present work focuses on boron as a dopant in ADP and this is expected to enhance the nonlinearity of the crystal. We have studied and reported the optical and dielectric behaviour of boron doped ADP and pure ADP crystals.

2. Materials and Method

Experimental Procedure

All starting materials were of AR grade and the growth process was carried out in aqueous solution. The starting materials namely ADP and boric acid (H_3BO_3) were taken in the ratio 1:0.1. The calculated amount of salts was dissolved in double dissolved water. This solution was then stirred well for more than six hours using a magnetic stirrer and filtered using Whatman filter paper. The solution was then poured into a Petri dish and allowed to evaporate at room temperature. Optically good quality single crystals of dimension 12.6 × 5.2 × 2.8 mm^3 were harvested in a period of 20 - 25 days. The photograph of the as grown crystal of pure ADP and boron doped ADP are shown in **Figures 1(a)** and **1(b)**.

3. Structural Confirmation

3.1. ICP Studies

In ICP technique, the wavelength at which emission occurs identifies the element and quantifies its concentration. So in boron doped ADP, ICP studies were conducted to identify the boron content in the grown crystal. The concentration of the boron was found to be 210 ppm. So the presence of dopant was confirmed by the ICP technique.

3.2. Single Crystal X-Ray Diffraction Studies

The single crystal X-ray diffraction analysis on the grown crystals was done using Bruker Kappa APEX II single crystal X-ray diffractometer at IIT, Chennai, to determine the unit cell dimensions. The lattice parameters determined are presented in **Table1**. It was seen that there is only a slight variation in the values of the lattice parameters. Both of them belong to tetragonal I4-2d space group. This implies that the structure is not changed due to the presence of the dopant but the influence of the dopant with respect to the properties was further investigated.

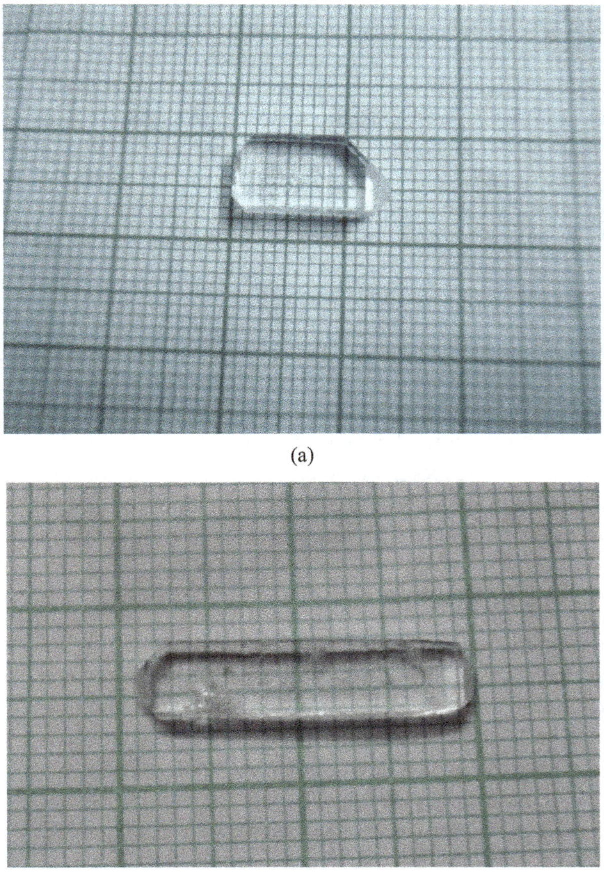

(a)

(b)

Figure 1. (a) Photograph of the as grown pure ADP crystals; (b) Photograph of the as grown boron doped ADP crystals.

4. Optical Characterization

4.1. FTIR

The Fourier transform infrared spectrum of pure ADP and boron doped ammonium dihydrogen phosphate crystals were recorded using Perkin Elmer model RXI Spectrometer in the range 400 - 4000 cm^{-1}. The recorded spectrum is given in **Figures 2** and **3**. The strong band centred at about 3933 cm^{-1} is generally O-H due to OH stretching and water bending. The strong and broad band at 3420 cm^{-1} is due to KBr null absorption, which appears in the spectrum of any compound [15]. The band at 2353 cm^{-1} can be assigned to a second overtone of P-O deformation vibration appearing near 910 cm^{-1}. The PO stretching vibration leads to a distinct band at 1288 cm^{-1}. At room temperature PO_4^{-3} is a symmetric complex anion. The P=O deformation vibrations are found between 840 and 1120 cm^{-1} [16]. At 552 and 459 cm^{-1} P=O-H wagging and rocking vibrational bandsare found. The spectrum of boron doped ADP is almost the same as pure ADP except that a few shifting of peaks is noticed. The B=O stretching which usually occurs at 1651 cm^{-1} is found to be shifted to 1715 cm^{-1}. Also the P=O deformation at 459 cm^{-1} has disappeared in the FTIR spectrum of boron doped ADP. These effects can be due to the presence of boron in the crystal influencing the P=O functional group. A detailed assignment of the frequencies observed in the FTIR spectrum is given **Table 2**.

4.2. UV-Vis Spectral Studies

Optical absorption data collected for the pure and doped ADP polished sample is shown in **Figure 4**. The measurements were done using Perkin Elmer Model-Lambda 35 spectrometer between 200 - 1200 nm. The spectrum indicates that the pure and doped ADP crystals have minimum absorption in the entire region between 200 - 1200 nm. It is seen that the doped ADP crystal has better lower cut off wavelength than that of pure ADP crystal. It is interesting to note that the boron doped ADP crystal has reduced absorption when compared to pure ADP. It is already well established that if the UV-Vis spectrum of

Table 1. Unit cell parameters of pure ADP and boron doped ADP crystals.

Sample	Lattice parameter A = b, c (Å)		Cell volume (Å^3)	$\alpha=\beta=\gamma$	Structure
Pure ADP	7.491	7.546	423	90°	Tetragonal
Boron doped ADP	7.512	7.564	426	90°	Tetragonal

Table 2. Observed FTIR frequencies (cm^{-1}) of pure ADP and boron doped ADP crystals.

Observed FTIR Frequencies (cm^{-1})		Assignments
Pure ADP	Boron Doped ADP	
3933 M	3935 VS	O-H Stretching and Water Bending
3125 VW	3139 VW	B-OH Stretching
2890 M	-	N-H Stretching of Ammonia
2353 W	2428 W	Combination Band of Vibrations
1651 VW	1715 VW	Asymmetric B-O Stretching
1409 VW	1411 VW	Bending Vibration of Ammonia
1288 VW	1289 VW	P-O Stretching Vibration
1098 VW	1099 VW	P-O-H Vibration
910 VW	910 VW	P-O-H Vibrations
552 W	555 W	PO_4 Vibrations
459 M	-	PO_4 Vibration

VS: very strong; S: strong; M: medium; W: weak; VW: very weak.

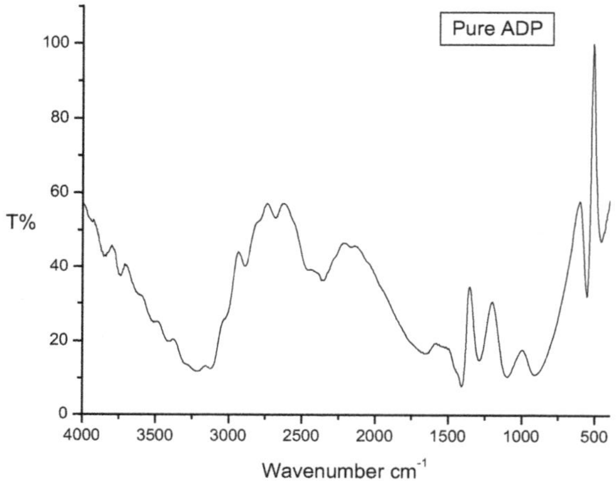

Figure 2. FTIR spectrum of pure ADP crystal.

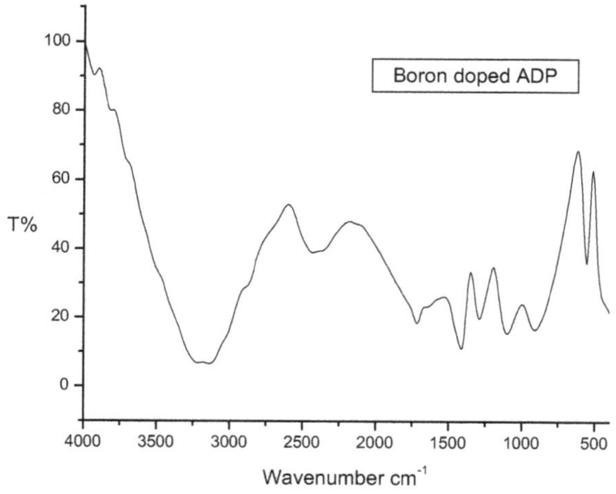

Figure 3. FTIR spectrum of boron doped ADP crystal.

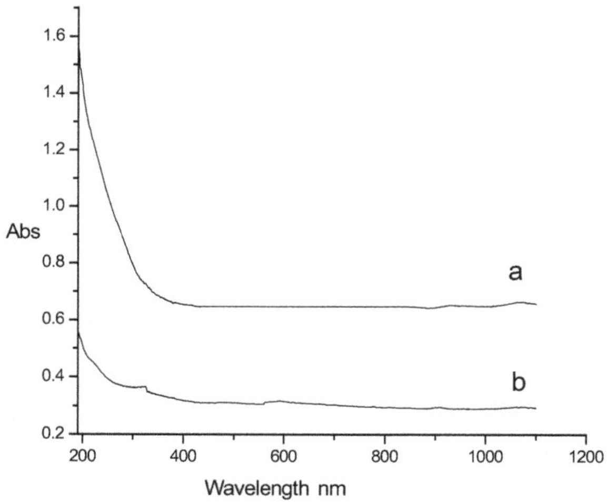

Figure 4. UV-Vis spectrum of (a) pure ADP (b) boron doped ADP crystal.

a crystal shows minimum absorption and lower cut off wavelength then the probability that the crystal is NLO active is more. It is observed that in boron doped ADP the above stated condition is satisfied hence it can be concluded that the crystal is NLO active. Absorption in the near ultraviolet region arises from electronic transitions associated within the sample. Using the formula $Eg = hc/\lambda$, the band gap energies were found where h is the Planck's constant and c is the velocity of light. It was seen that the band gap energy is 4.1 eV for pure ADP and 4.7 eV for doped ADP. Therefore the improved band gap energy of 4.7 eV and better crystal transparency in doped ADP can be attributed to the presence of boron.

4.3. SHG Studies

To conform the NLO nature, the grown crystals of pure and doped ADP were subjected to Kurtz-Perry Second Harmonic Generation (SHG) test using the Nd: YAG Q switched laser beam. The second harmonic signal of 3.6 mJ and 4.7 mJ, respectively were obtained for pure and boron doped ADP. Thus, the SHG efficiency of boron doped crystal is nearly 1.31 times greater than pure ADP.

5. Dielectric Studies

The dielectric analysis is an important characteristic that can be used to fetch knowledge based on the electrical properties of a material medium as a function of temperature and frequency. Based on this analysis, the capability of storing electric charges by the material and capability of transferring the electric charge can be assessed. Dielectric properties are correlated with electro optic property of the crystals: particularly when they are non conducting materials [17]. The dielectric constant and dielectric loss of pure and doped ADP crystals were determined using Multi-frequency LCR meter (LCR-800

SERIES). Samples were cut to a proper thickness and polished. Each sample was electroded on both sides with high purity silver paste so that it behaved like a parallel plate capacitor. The dielectric constant is calculated using the formula

$$\varepsilon_r = Ct/\varepsilon_0 A \tag{1}$$

where C is capacitance (F), t the thickness (m), A the area (m^2), and ε_0 the absolute permittivity in the free space having a value of 8.854×10^{-12} $F \cdot m^{-1}$. Repeated trials were performed to ascertain the correctness of the observed results.

Figures 5-8 show the variation of dielectric constant and dielectric loss with respect to frequency for temperatures (40˚C, 80˚C, 120˚C and 160˚C) for both pure and boron doped ADP crystals. From **Figures 5** and **7**, it is seen that the dielectric constant increases with the increase in temperature for both pure and boron doped ADP crystals. Compared to the pure ADP, the dielectric constant (ε_r) of doped ADP is lesser in value. This is found to be similar to the data as reported in pure and urea-doped KDP single crystals [18]. The magnitude of dielectric constant depends on the degree of polarization in the crystals. It was seen that the dielectric constant has high value in the lower frequency region and then it decreases with the increasing frequency. In **Figures 6** and **8**, that dielectric loss is high at low frequency and decreases with higher frequencies.

A low dielectric loss at high frequency implies that the optical quality of the crystal is higher because of lesser defects, which is a desirable property for NLO applica-

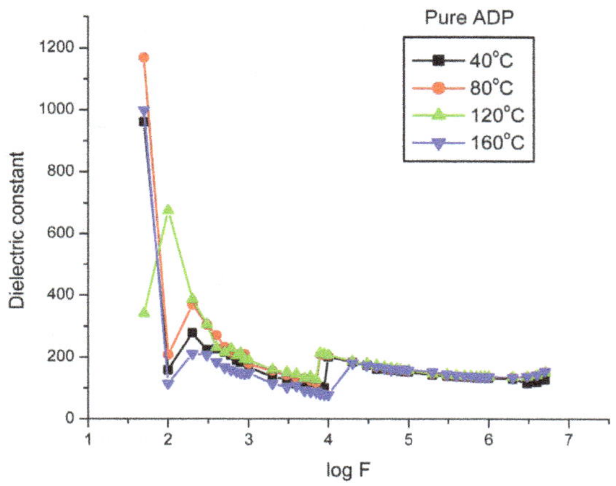

Figure 5. Variation of dielectric constant with frequency at different temperature.

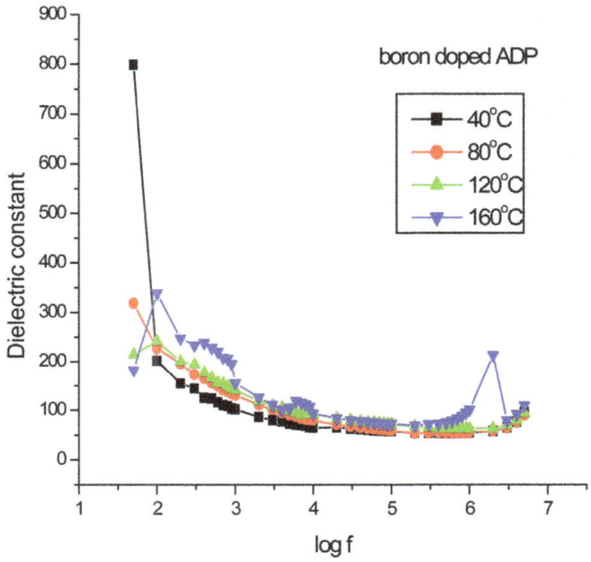

Figure 7. Variation of dielectric constant with frequency at different temperature.

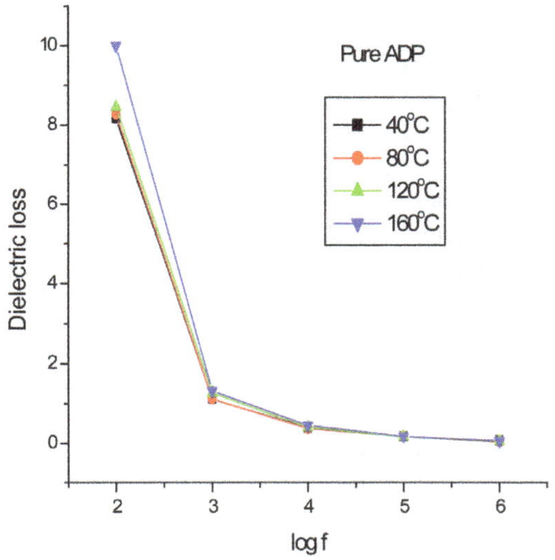

Figure 6. Variation of dielectric loss with frequency at different temperature.

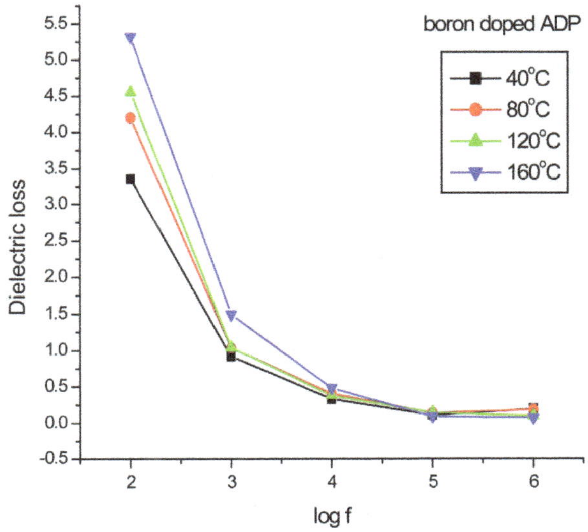

Figure 8. Variation of dielectric loss with frequency at different temperature.

tions [19]. The variation of dielectric constant with temperatures at frequencies 100 Hz, 1 KHz, 10 KHz, 100 KHz and 1 MHz was determined. The results are shown in **Figures 9-12**. It was found that the values of dielectric constant and dielectric loss increase with an increase in temperature and decrease with the increasing frequency. At low frequencies the dipoles can easily switch alignment with the changing field [20]. As the frequency increases the dipoles can rotate less and maintain phase with the field; thus they reduce their contribution to the polarization field, and hence the observed reduction in dielectric constant and dielectric loss. According to Miller rule, the lower value of dielectric constant at higher frequencies is a suitable parameter for the enhancement of SHG coefficient [21,22]. The low dielectric loss with

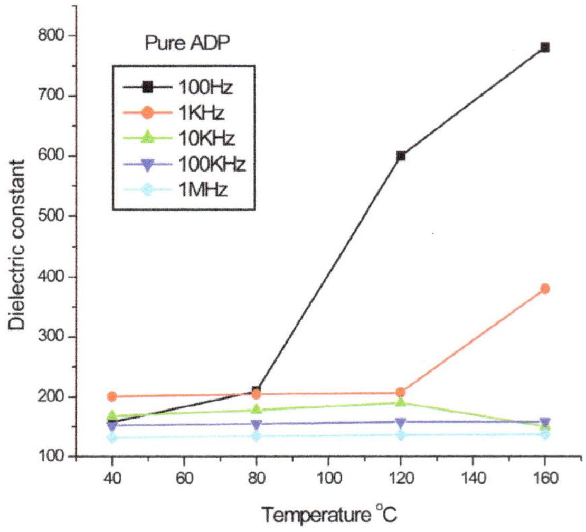

Figure 9. Variation of dielectric constant with temperature at different frequencies.

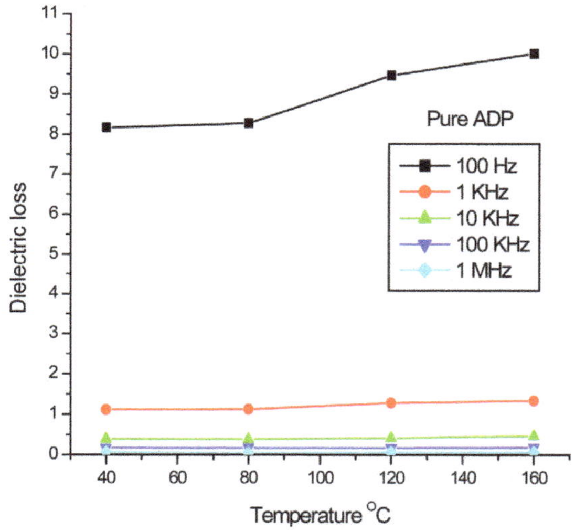

Figure 10. Variation of dielectric loss with temperature at different frequencies.

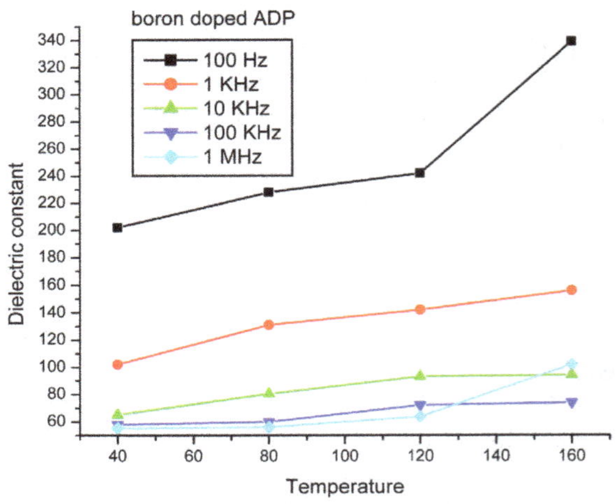

Figure 11. Variation of dielectric constant with temperature with temperature at different frequencies.

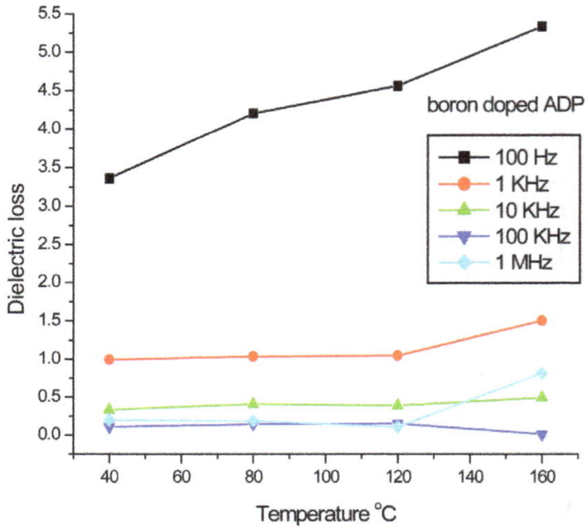

Figure 12. Variation of dielectric loss with temperature at different frequencies.

high frequency for a given sample suggests that the sample possesses enhanced optical quality with lesser defects and this parameter is of vital importance for nonlinear optical materials [23,24]. In the case of boron doped ADP there is an appreciable decrease in dielectric constant value compared to pure ADP suggesting the possibility of realizing ADP crystals with low ε_r value.

6. Conclusion

Pure and boron doped ADP single crystals were grown by slow evaporation technique at room temperature. Colourless and transparent crystals were obtained. The presence of boron in ADP was confirmed by ICP. Both the crystals are found to belong to the same tetragonal I4-2d space group. The values of lattice parameter of both pure and doped samples show no appreciable difference. In the Fourier transform infrared spectrum the dopant boron was found to effectively influence the P-O stretching and deformation. From the UV-Vis spectrum the band energy of the doped ADP was calculated to be 4.7 eV suggesting a positive NLO activity. The crystal has a wide transmission window ranging from 300 to 1100 nm. The SHG efficiency is 1.3 times greater than pure ADP and this suggests that boron doped ADP is a promising candidate for NLO applications. The lower value of dielectric constant and dielectric loss on doping with boron is observed. From the above characterization studies it is concluded that the addition of boron in pure ADP has enhanced the transparency and NLO efficiency of ADP crystals. So it is suggested that the crystal is a potential candidate for the fabrication of NLO devices.

REFERENCES

[1] D. Xu and D. Xue, "Chemical Bond Simulation of KADP Single-Crystal Growth," *Journal of Crystal Growth*, Vol. 310, No. 7-9, 2008, pp. 1385-1390. doi:10.1016/j.jcrysgro.2007.12.008

[2] R. J. Davey and J. W. Mullin, "Growth of the {100} Faces of Ammonium Dihydrogen Phosphate Crystals in the Presence of Ionic Species," *Journal of Crystal Growth*, Vol. 26, 1974, p. 45.

[3] K. Sethuraman, R. R. Babu, R. Gopalakrishnan and P. Ramasamy, "Unidirectional Growth of (1 1 0) Ammonium Dihydrogen Orthophosphate Single Crystal by Sankaranarayanan-Ramasamy Method," *Journal of Crystal Growth*, Vol. 294, No. 2, 2006, pp. 349-352. doi:10.1016/j.jcrysgro.2006.06.033

[4] S. Nagalingam, S. Vasudevan and P. Ramasamy, "Effect of Impurities on the Nucleation of ADP from Aqueous Solution," *Crystal Research and Technology*, Vol. 16, No. 6, 1981, p. 647-650.

[5] R. Ramesh and C. Mahadevan, "Nucleation Studies in Supersaturated Aqueous Solutions of $(NH_4)H_2PO_4$ Doped with $(NH_4)_2C_2O_4{\cdot}H_2O$," *Bulletin of Materials Science*, Vol. 21, No. 4, 1998, pp. 287-290. doi:10.1007/BF02744954

[6] M. E. Lines and A. M. Glass, "Principles and Applications of Ferroelectrics and Related Materials," Clarendon Press, Oxford, 1977.

[7] N. P. Rajesh, K. Meera, K. Srinivasan, S. Ragavan and P. Ramasamy, "Effect of EDTA on the Metastable Zone width of ADP," *Journal of Crystal Growth*, Vol. 213, No. 3-4, 2003, pp. 389-394. doi:10.1016/S0022-0248(00)00374-2

[8] K. Srinivasan, P. Ramasamy, A. Cantoni and G. Bocelli, "Mixed Crystals of $NH_4H_2PO_4$-KH_2PO_4: Compositional Dependence of Morphology, Microhardness and Optical Transmittance," *Materials Science and Engineering: B*, Vol. 52, No. 2-3, 1998, pp. 129-133. doi:10.1016/S0921-5107(97)00286-9

[9] M. Yoshimatsu, "Some Observations of Imperfections in ADP Single Crystals by X-Ray Diffraction Micrography," *Japanese Journal of Applied Physics*, Vol. 5, 1966, pp. 29-35. doi:10.1143/JJAP.5.29

[10] R. Reintjes and E. C. Eckardt, "Evaporated Inhomogeneous Thin Films," *Applied Optics*, Vol. 5, No. 1, 1966, pp. 29-34. doi:10.1364/AO.5.000029

[11] P. Kumaresan, S. M. Babu and P. M. Anbasan, "Thermal, Dielectric Studies on Pure and Amino Acid (l-Glutamic Acid, l-Histidine, l-Valine) Doped KDP Single Crystals," *Optical Materials*, Vol. 30, No. 9, 2008, pp. 1361-1368. doi:10.1016/j.optmat.2007.07.002

[12] P. Rajesh and P. Ramasamy, "Growth of DL-Malic Acid-Doped Ammonium Dihydrogen Phosphate Crystal and Its Characterization," *Journal of Crystal Growth*, Vol. 311, No. 13, 2009, pp. 3491-3497. doi:10.1016/j.jcrysgro.2009.04.020

[13] J. Podder, "The Study of Impurities Effect on the Growth and Nucleation Kinetics of Potassium Dihydrogen Phosphate," *Journal of Crystal Growth*, Vol. 237-239, No. 1, 2002, pp. 70-75. doi:10.1016/S0022-0248(01)01854-1

[14] P. V. Dhanaraj, G. Bhagavanarayana and N. P. Rajesh, "Effect of Amino Acid Additives on Crystal Growth Parameters and Properties of Ammonium Dihydrogen Orthophosphate Crystals," *Materials Chemistry and Physics*, Vol. 112, No. 2, 2008, pp. 490-495.

[15] T. Ananthi, S. M. Delphine, M. M. Freeda, R. K. Priya and A. Wahabmusallam, "Growth and Characterization of Doped ADP Crystal," *Recent Research in Science and Technology*, Vol. 3, No. 1, 2011, pp. 32-40.

[16] Z. Delci, D. Shyamala, S. Karuna, A. Senthil and A. Thayumanavan, "Enhancement of Optical, Thermal and Hardness in KDP Crystals by Boron Doping," *International Journal of ChemTech Research*, Vol. 4, No. 2, 2012, pp. 816-826.

[17] S. Boomadevi, H. P. Mittal and R. Dhansekaran, "Synthesis, Crystal Growth and Characterization of 3-Methyl 4-Nitropyridine 1-Oxide (POM) Single Crystals," *Journal of Crystal Growth*, Vol. 261, No. 1, 2004, pp. 55-62. doi:10.1016/j.jcrysgro.2003.09.005

[18] B. T. Hatton, K. Landskron, W. J. Hunks, *et al.*, "Materials Chemistry for Low-K Materials," *Materials Today*, Vol. 9, No. 3, 2006, pp. 22-31. doi:10.1016/S1369-7021(06)71387-6

[19] S. Goma, C. M. Padma and C. K. Mahadevan, "Dielectric Parameters of KDP Single Crystals Added with Urea," *Materials Letters*, Vol. 60, No. 29-30, 2006, pp. 3701-3705. doi:10.1016/j.matlet.2006.03.092

[20] S. Suresh, A. Ramanand, D. Jayaraman and P. Mani, "Growth, Photoconductivity and Dielectric Properties of Triglycine Sulfate (TGS) Single Crystals," *Optoelectronics and Advanced Materials*, Vol. 4, No. 11, 2010, pp. 1763-1765.

[21] C. Balarew and R. Duhlew, "Application of the Hard and Soft Acids and Bases Concept to Explain to Ligand Coordination in Double Salt Structures," *Journal of Solid State Chemistry*, Vol. 55, No. 1, 1984, pp. 1-6. doi:10.1016/0022-4596(84)90240-8

[22] M. Meena and C. K. Mahadevan, "Growth and Electrical Characterization of L-Arginine Added KDP and ADP Single Crystals," *Crystal Research and Technology*, Vol. 43, No. 2, 2008, pp. 166-172.

doi:10.1002/crat.200711064

[23] S. Aruna, G. Bhagavannarayana, M. Palanisamy, P. C. Thomas, B. Varghese and P. Sagayaraj, "Growth, Morphological, Mechanical and Dielectric Studies of Semi Organic NLO Single Crystal: l-Argininium Perchlorate," *Journal of Crystal Growth*, Vol. 300, No. 2, 2007, pp. 403-408. doi:10.1016/j.jcrysgro.2006.11.296

[24] N. Pattanaboonmee, P. Ramasamy, R. Yimnirun and P. Manyum, "A Comparative Study on Pure, l-Arginine and Glycine Doped Ammonium Dihydrogen Orthophosphate Single Crystals Grown by Slow Solvent Evaporation and Temperature Gradient Method," *Journal of Crystal Growth*, Vol. 314, No. 1, 2010, pp. 196-201.

27

Study of Ionic Conduction and Dielectric Behavior of Pure and K^+ Doped Ag_2CdI_4

Khalid Siraj*, Rafiuddin

Physical Chemistry Division, Department of Chemistry, Faculty of Sciences, Aligarh Muslim University, Aligarh, India.
Email: *chemdocprof@gmail.com

ABSTRACT

The ionic conductivity of 8, 9 and 10 mol% K^+ doped Ag_2CdI_4 showed slight decrease whereas the phase transition was observed almost at the same temperature as it reported for pure Ag_2CdI_4. This decrease in conductivity likely results from decrease in free volume because of the larger K^+ ions (r_K^+ = 133 pm and r_{Ag}^+ = 129 pm) entering Ag_2CdI_4 lattice which is unchanging in size. The dielectric constant of Ag_2CdI_4 was found to increase with increasing temperature as the orientation of dipoles is facilitated in rising temperature.

Keywords: Electrical Conductivity; Dielectric Constant; Ionic Conductor; Nanocrystalline

1. Introduction

Ag_2CdI_4 compound belongs to the so called superionic materials of A_2BI_4 type (A = Cu, Ag, In, Tl; B = Hg, Cd, Zn), that undergo phase transition into superionic state. Structural investigation of superionic materials Ag_2HgI_4, Cu_2HgI_4 and others show the existence of low temperature β-phase, as a rule, tetragonal syngony, and high temperature α-phase [1,2]. It is considered that during $\beta \rightarrow \alpha$ phase transition iodine sub-lattice remains, while there in α phase A^+, B^{2+} ions and stoichiometric vacancies, V_{Ag}^+, which are distributed randomly in crystal lattice determine high conductivity of the material.

Ag_2CdI_4 compounds possess smeared phase transitions into the superionic state in the temperature region 330 - 380 K. The solid fast ion conductor Ag_2CdI_4 exhibits a number of solid state phase transitions upon heating. Room temperature, covalent phase Ag_2CdI_4 crystallizes in a well-defined structure. Ag_2CdI_4 at T < 330 K belongs predominantly to hexagonal, with space group P_6/mmm and unit cell dimensions a = 4.578 Å and C = 7.529 Å and changes to cubic at T > 380 K with space group Pm3m and unit cell dimension a = 5.05 Å [3].

Sudharsanan *et al.* [4,5] studied the IR and Raman spectra of this compound. Bolesta *et al.*, [6] calculated the band structure of Ag_2CdI_4 by semi-empirical method using strong band approximation. Yunakova *et al.* [7], studied the temperature dependence of the spectral position and half-width of the *A* band in the range 90 - 430 K in Ag_2CdI_4. Nair *et al.* [8] studied the behaviour of Ag^+ in Ag_2CdI_4 by cation substituted smaller Li^+ and the larger Tl^+ ions. Beeken *et al.* [9] tried to determine the effect of Cu-substitution on the conductivity and phase transition temperature of Ag_2CdI_4. Inspired by the results we have tried to prepare Ag_2CdI_4 and study the effect of K^+ substitution on the conductivity and phase transition temperature of Ag_2CdI_4.

*Corresponding author.

2. Experimental

2.1. Material Preparation

Silver tetraiodocadmiate (Ag_2CdI_4) was prepared from AgI and CdI_2 obtained from BDH (India), with stated purity 99.2% and 99.6% respectively, by the solid state reaction method. AgI and CdI_2 were mixed in the requisite composition in an agate mortar and were heated at 400°C for 48 hours in a silica crucible with intermittent grinding. X-ray diffraction studies were carried out for the material after the reaction was completed. X-ray diffraction pattern suggest the formation of the product Ag_2CdI_4.

Potassium ion was mixed with Ag_2CdI_4 in requisite amount in an agate mortar by solid state reaction and the powder is kept for 24 hours at 473 K with intermittent grinding so that the reaction takes place properly.

2.2. Measurements

2.2.1. Electrical Conductivity Measurements

Electrical conductivity was measured by preparing pellet of the compound by pouring its fine powder into a stainless steel die and applying a pressure of about 4 tonnes/cm^2

with the help of hydraulic press, SPECTRA LAB, model LB-89. The pellet so formed has area of 4.524 cm^2 and thickness of 0.1 cm. The pellet is annealed at 400 K for 12 hours in order to eliminate any grain boundary effect. The electrical conductivity measurements were performed by means of a two probe method. The pellet was mounted on a stainless steel samples holder assembly between copper leads using two polished platinum electrodes. The copper leads were electrically insulated from the samples holder by Teflon sheets. The electrical conductivity of the samples was measured in the temperature range of 298 - 473 K by using a GenRad 1659 RLC Digibridge at a fixed frequency.

2.2.2. Dielectric Measurement

The dielectric constant of Ag_2CdI_4 was calculated over the frequency range 100Hz to 10 KHz and in the temperature range of 298 - 473 K using the relation given as,

$$\varepsilon' = Cd/\varepsilon_0 A$$

where ε_0 is the constant of permittivity for free space, C is the capacitance which is measured by the RLC Digibridge (mentioned above), d the thickness of the pellet and A is the cross-sectional area of the flat surface of the pellet.

3. Result and Discussion

The temperature dependence of conductivity is given by the Arrhenius expression

$$\sigma = \left(ne2\lambda 2\nu\gamma/k_b T\right) \exp\left(-\Delta G^*/k_b T\right) \quad (1)$$

$$= \left(ne2\lambda 2\nu\gamma/k_b T)\right) \exp\left(-\Delta S^*/k_b - \Delta H^*/k_b T\right) \quad (2)$$

where n is the number of ions per unit volume, e the ionic charge, λ the distance between two jumps positions, ν the jump frequency, γ the intersite geometric constant, k the Boltzmann constant and ΔG^*, ΔS^* and ΔH^* are activation free energy, entropy and enthalpy terms. The equation can be written in a simpler form as

$$\sigma T = \sigma_o \exp\left(-Ea/k_b T\right) \quad (3)$$

$$\log\sigma T = \log\sigma_o - Ea/2.303 k_b T \quad (4)$$

where $\sigma_o = ne2\lambda 2\nu\gamma/k \exp(-\Delta S^*/k)$ and $\Delta H^* = Ea$, *i.e.*, the activation enthalpy equals experimental activation energy for ionic motion, which may include a defect formation enthalpy contribution [10].

Figure 1 shows a typical heat and cool mode plots of log σT vs 1/T for undoped Ag_2CdI_4, using Equation (4). It observed that the conductivity data were reproducible within experimental error (≤ ±2%) for heating and cooling cycles. The conductivity values of undoped Ag_2CdI_4 were found to have increased slowly initially but a sharp increase in conductivity was noted above 380 K. The presence of hysterisis loop in the heating and cooling phase of conductivity in the unannealed sample of Ag_2CdI_4 confirm the fact that abrupt change in conductivity is due to the phase transition.

Electrical conductivity (σ), measurements were performed on several potassium doped Ag_2CdI_4 sample throughout the temperature range 298 - 473 K. Arrhenius plot of pure Ag_2CdI_4 shows gradual increase in conductivity up to 380 K and above this temperature a sharp change in conductivity indicates the occurrence of phase transition at 380 K (**Figure 2**). The conductivity data of doped Ag_2CdI_4 samples with K^+ ion found to exceed in conductivity as compare to the Li^+ and Na^+ ions [2,8].

The conductivity increase observed in the present investigation can be explained on the basis of the space charge model reported by Maier *et al.* [11]. In thermal equilibrium, the surface and grain boundaries of an ionic crystal may carry an electric charge resulting from the presence of excess ions of one sign. This charge is just compensated by a space charge cloud of the opposite sign

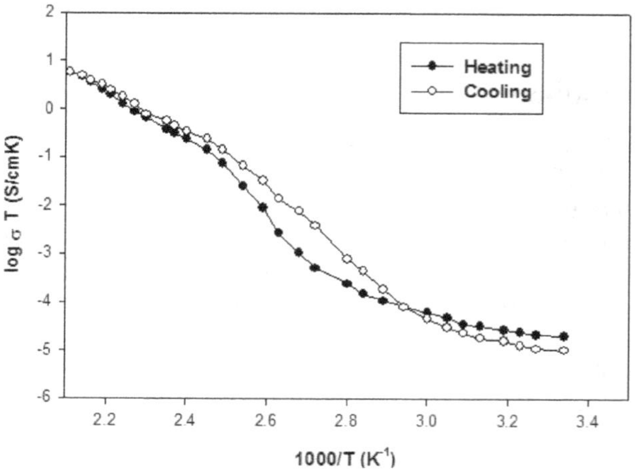

Figure 1. Temperature dependence electrical conductivity of Ag_2CdI_4.

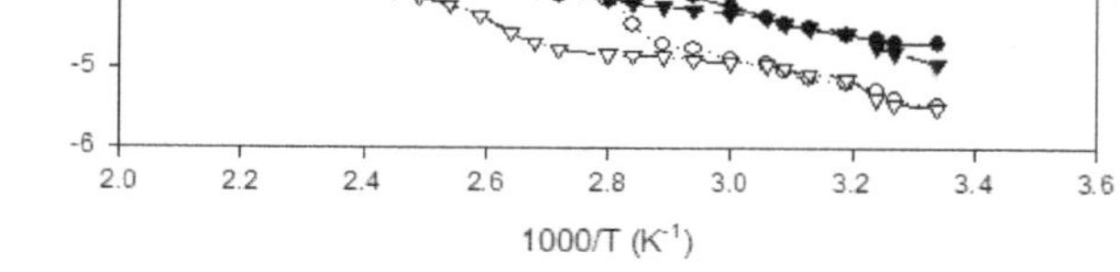

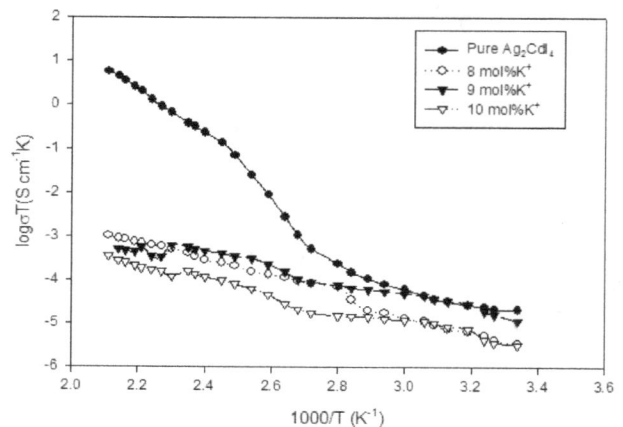

Figure 2. Effect of the cation dopant on the conductivity of Ag_2CdI_4.

adjacent to the boundary. For a pure material, this charge arises if the energies to form anion and cation vacancies or interstitials at the boundary are different. The magnitude and sign of the boundary charge changes if there are aliovalent solutes present in the matrix which alter the concentration of the lattice defects in the crystal. Space charge effects will be predominantly important in very small crystals since the grain boundaries contain a large number of defects compared to the coarser grained polycrystalline materials.

K^+ ion was chosen as the dopant with the expectation that it would enhance the conductivity of Ag_2CdI_4 on the basis of lattice expansion due to the larger K^+ radius, (r_K^+ = 133 pm and r_{Ag}^+ = 129 pm). But during the course of reaction the conductivity of Ag_2CdI_4 doped with different mol % of K^+ ions was found to decrease. This observed decrease likely results from decrease in free volume as larger potassium ion entering Ag_2CdI_4 lattice which is unchanging in size. Another important feature observed in the K^+ doped Ag_2CdI_4 is erratic conductivity behaveiour in the post transition region of 9 and 10 mol% K^+ doped Ag_2CdI_4. After 433 K the conductivity further decreases which are due to the collapse of CdI_4^{2-} framework and subsequently recovered on cooling which implies the restructuring of that sub-lattice [12].

The dielectric constant, ε' of Ag_2CdI_4 was calculated over the frequency range 100 Hz to 10 KHz and in the temperature range of 298 - 473 K using relation mentioned in experimental section.

The plot between log ε' and temperature (**Figure 3**), shows almost linear increase in ε' with temperature. There is a small upward shift in the values of ε' with decrease of frequencies near the transition temperature of Ag_2CdI_4, the slope of the graph before and after the transition remains the same.

This increase in dielectric constant of Ag_2CdI_4 with temperature may be attributed due to the reason that the molecules cannot orient themselves in dielectrics. As temperature rises, the orientation of dipoles is facilitated and ionic polarization increases resulting in the increase of dielectric constant (ε') [13,14]. Also at higher temperatures, the conductivity due to the hopping of mobile ions becomes important and increased dielectric constant results.

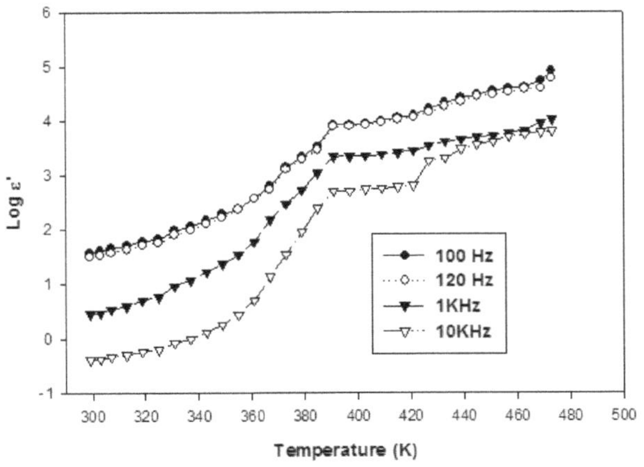

Figure 3. Temperature dependence dielectric constant of Ag_2CdI_4 on different frequency.

According to Samara [15], higher the dielectric constant of an ionic crystal, lower the energy of formation of lattice defects. Since the grain boundary of nanoparticles contain a large density of defects [16,17], it can be argued that the dielectric constant of nanoparticles should have higher values.

Nano-crystalline materials consist of grain or interphase boundaries. These boundaries contain defects such as dangling bond, vacancies, vacancy clusters etc. [18]. These defects can cause a positive or negative space charge distribution at interfaces. The space charges can move under the application of an external field and when they are trapped by the defects, lots of dipole moments are formed (space charge polarization). Hence the space charge effect will be an important factor which decides the dielectric properties in materials with small particles sizes [11]. In addition, ion jump polarization may also be greater in nanocrystalline materials since there will be a number of position in the grain boundaries for the ions to occupy. Thus the high values of the dielectric constant in Ag_2CdI_4 may be attributed to the increased ion jump orientation effect and the increased space charge effect exhibited by nanoparticles.

3. Conclusion

Electrical conductivity (σ) plots of several K^+ doped Ag_2CdI_4 samples show decrease in conductivity. This observed decrease in conductivity was due to decrease in free volume which results from the larger potassium ion entering Ag_2CdI_4 lattice. The erratic conductivity behaveiour observed in the post transition region of K^+ doped Ag_2CdI_4 was due to the collapse of CdI_4 framework. The dielectric constant of Ag_2CdI_4 was found to increase with temperature which facilitates the orientation of dipoles in Ag_2CdI_4.

4. Acknowledgements

We thankfully acknowledge Department of Chemistry, Aligarh Muslim University, Aligarh, India, for providing necessary facilities to carryout this work.

REFERENCES

[1] H. G. LeDuc and L. B. Celeman, "Far-Infrared Studies of the Phase Transition and Conduction Mechanism in the Fast-Ion Conductors Ag_2HgI_4 and Cu_2HgI_4," *Physical*

Review B, Vol. 31, No. 2, 1985, pp. 933-941. doi:10.1103/PhysRevB.31.933

[2] S. M. Nair, A. I. Yahiya and A. Ahmad, "Ion Conduction in the Ag_2HgI_4-Cu_2HgI_4 Systems Doped with Cd^{2+}, K^+, and Na^+," *Journal of Solid State Chemistry*, Vol. 122, No. 2, 1996, pp. 349-352. doi:10.1006/jssc.1996.0125

[3] I. M. Bolesta, O. V. Futey and O. G. Syrbu, "Optical Investigations of Superionic Phase Transition in Ag_2CdI_4 Thin Films," *Solid State Ionics*, Vol. 119, No. 1-4, 1999, pp. 103-107. doi:10.1016/S0167-2738(98)00489-5

[4] R. Sudharsanan, S. Radhakrishna and K. Hariharan, "Electrical Conductivity and Electronic Absorption Studies on Ag_2CdI_4 Thin Films," *Solid State Ionics*, Vol. 9-10, No. 2, 1983, pp. 1473-1476. doi:10.1016/0167-2738(83)90198-4

[5] R. Sudharsanan, T. K. K. Srinivasan and S. Radhakrishna, "Raman and Far IR Studies on Ag_2CdI_4 and Cu_2CdI_4 Superionic Compounds," *Solid State Ionics*, Vol. 13, No. 4, 1984, pp. 277-283. doi:10.1016/0167-2738(84)90069-9

[6] I. Bolesta, O. Futey and S. Velgosh, "Crystalline and Band Energy Structure of Ag_2CdI_4," *Ukrainian Journal of Physical Optics*, Vol. 1, No. 1, 2000, pp. 13-15. doi:10.3116/16091833/1/1/13/2000

[7] O. N. Yunakova, V. K. Miloslavskii and E. N. Kovalenko, "The Absorption Spectrum and Excitons in an Ag_2CdI_4 Ionic Conductor," *Physics of the Solid State*, Vol. 43, No. 6, 2001, pp. 1072-1076. doi:10.1134/1.1378146

[8] S. M. Nair and A. Ahmad, "Effect of Cation Substitution on Fast Ag^+ Ion Conductivity in Ag_2CdI_4," *Journal of Physics and Chemistry of Solids*, Vol. 58 No. 2, 1997, pp. 331-333. doi:10.1016/S0022-3697(96)00116-3

[9] R. B. Becken, J. C. Falud, W. M. Schreier and J. M. Tritz, "Ionic Conductivity in Cu-Substituted Ag_2CdI_4," *Solid State Ionics*, Vol. 154-155, 2002, pp. 719-722. doi:10.1016/S0167-2738(02)00433-2

[10] K. Siraj and Rafiuddin, "Electrical Conductivity Behavior of $CdHgI_4$-CuI Mixed System," *International Journal of Chemical*, Vol. 3, No. 2, 2011, pp. 174-179.

[11] J. Maier, S. Prill and B. Reichert, "Space Charge Effects in Polycrystalline, Micropolycrystalline and Thin Film Samples: Application to AgCl and AgBr," *Solid State Ionics*, Vol. 28-30, 1988, pp. 1465-1469. doi:10.1016/0167-2738(88)90405-5

[12] M. S. Kumari and E. A. Secco, "IV. Order-Disorder Transitions: Solid State Kinetics, Thermal Analyses, X-Ray Diffraction and Electrical Conductivity Studies in the Ag_2SO_4-K_2SO_4 System," *Canadian Journal of Chemistry*, Vol. 63, No. 2, 1985, pp. 324-328. doi:10.1139/v85-054

[13] B. Tareev, "Physics of Dielectric Materials," Mir Publication, Moscow, 1979.

[14] W. D. Kingery, "Introduction to Ceramics," John Wiley, New York, 1976.

[15] G. A. Samara, "High-Pressure Studies of Ionic Conductivity in Solids," *Solid State Physics*, Vol. 38, 1984, pp. 1-80. doi:10.1016/S0081-1947(08)60311-2

[16] W. P. Halperin, "Quantum Size Effects in Metal Particles," *Reviews of Modern Physics*, Vol. 58, No. 3, 1986, pp. 533-607. doi:10.1103/RevModPhys.58.533

[17] H. Gleiter, "Nanocrystalline Materials," *Progress in Materials Science*, Vol. 33, No. 4, 1989, pp. 223-315. doi:10.1016/0079-6425(89)90001-7

[18] C.-M. Mo, L. D. Zhang and G. Z. Wang, "Characteristics of Dielectric Behavior in Nanostructured Materials," *Nanostructured Materials*, Vol. 6, No. 5-8, 1995, pp. 823-826. doi:10.1016/0965-9773(95)00186-7

Dielectric Spectroscopy of PVAc at Different Isobaric-Isothermal Paths

Soheil Sharifi

Department of Physics, University of Sistan and Baluchestan, Zahedan, Iran.
Email: soheil.sharifi@gmail.com, sharifi@df.unipi.it

ABSTRACT

We studied broadband dielectric spectroscopy of a glass from systems that the dynamics of the primary α- and the Johari-Goldstein (JG) β-processes are strongly correlated in Poly(vinyl acetate) over a wide temperature T and pressure P range. Analysing the temperature and pressure behaviour of the α- and (Non-JG) β-processes, a correlation hasnot been found between the structural relaxation time, the (Non-JG) β-processes relaxation time and the dispersion of the structural relaxation. These results support the idea that the (JG)-processes relaxation acts as a precursor of the structural relaxation and therefore of the glass transition phenomenon but it isnot clear relation in the (Non-JG) β-processes and structural relaxation at PVAc .

Keywords: Dielectric Relaxation; Nan-Size Motion; Pressure and Temperature

1. Introduction

Usually in the Glass former materials we can find complex relaxation pattern, which evolves over several time decades. A technique that it is useful for study the dynamic of relaxation inside the liquid and glass state is Dielectric spectroscopy [1-3]. It has turned out that in such a broad dynamic range several molecular processes take place, and usually most of them are characterized by non-exponential relaxation functions. In polymeric materials the slowest of these processes is called normal mode: simplifying, if we consider a vector connecting the two ends of a polymeric chain, the normal mode reflects the motion of such vector. In non polymeric materials the slowest process is usually called main, structural or α-relaxation [2-6]. It reflects the cooperative motion of the molecules and its characteristic time can be related to the overall viscosity of the material. The origin of the structural α-relaxation is ascribed to cooperative motions that involve an increasing number of molecules and slow down dramatically when the glass transition is approached, for example either by decreasing temperature T or increasing pressure P (*i.e.*, density) [1-3]. So, it is challenging to distinguish between intermolecular and intramolecular secondary processes also in systems with a complex molecular structure, even showing more than one secondary process.

In this work, we contribute to such discussion by investigating the relation between the structural and the secondary processes in the poly (vinyl acetate) (PVAc) in an interval of pressure ranging from 0.1 to 600 MPa and temperature from 100 to 350 K. The main goal is to find relation between alpha and secondary relaxation in PVAc and compare this system with relation of alpha and secondary relaxation in the mixture of the rigid polar molecule quinaldine (QN) and tristyrene.

2. Experiment and Materials

The complex dielectric constant $\varepsilon = \varepsilon' - i\varepsilon''$ was measured in the frequency range from 10^{-2} Hz up to 10^{7} Hz at different isothermal and isobaric conditions using Novocontrol Alpha analyser. The temperature at atmospheric pressure was varied from 100 K and 320 K by means of a conditioned nitrogen flow cryostat. The high pressure experiment was carried out by means of an hydrostatic press and silicon oil as a pressure transmitting medium. A Teflon membrane prevented the oil to contact the dielectric cell. The temperature of the whole pressure chamber was controlled by a thermal jacket connected to a liquid circulator.

Poly (vinyl acetate) (PVAc), with M.W. = 167 Kg/mol, and T_g = 310.7 K is purchased from Aldrich. PVAc has interesting structural variations and can be obtained in the atactic, and thus amorphous form, which is crucial for investigations of the glass transition. PVAc has been often chosen to test the current aspects of various theories related with glass transition phenomena. An ample dipole moment makes PVAc a good candidate of dielectric spectroscopy [7-10].

3. Results

Dielectric loss spectra, $\varepsilon(\omega) = \varepsilon' - i\varepsilon''$ of PVAc was measured from above to below the glass transition. We applied the different thermodynamic paths to studied the relation between the structural and the secondary relaxation close to T_g (P).

We acquired dielectric spectra along isothermal paths by varying pressure from 0.1 MPa up to the maximum value of 600 MPa, with step of 10 MPa in the supercooled liquid and step of 50 MPa in the PVAC, **Figure 2**. We also measured dielectric spectra by varying temperature at high pressure from 100 to 300 K in supercooled liquid and in the glass. The increase of pressure slows down the structural relaxation similarly to the decrease of temperature: the structural relaxation peak shifts to lower frequencies until the structure of the liquid is eventually arrested in a glassy state. At ambient pressure in PVAc only the α-structural process is visible above T_g, when spectra are collected in the frequency interval 10^{-2} - 10^6 Hz. At temperatures below T_g two further relaxation processes, β-, is visible in the frequency interval, **Figure 1**. All the observed relaxation processes move towards lower frequencies on decreasing temperature or increaseing pressure, the α-process being the most sensitive and the β - the less. The β-process occurs only at very high frequency and we can measure only at very low temperature at ambient pressure, but never at high pressure due to limitation of the temperature interval, **Figure 1**. In the PVAc below T_g we observe only one secondary relaxation.

Dielectric spectra were fitted with a superposition of different HN and CC functions [1-3], one for each relaxation process. Since for any value of temperature not more than two relaxation processes appear, in the fitting procedure we never used more than two relaxation functions. Moreover, the β-process was usually described by the CC function. The fitting program, based on a least square minimizing procedure, contemporary fits the real and the imaginary part of the dielectric spectrum and furnishes the values of the parameters and the corresponding errors. **Table 1**, report some representative values of the parameters α and β for the considered systems at different pressure and temperature.

Finally, since in all the cases, near T_g, the structural and the secondary relaxations are well separated in the frequency scale we reproduced the structural relaxation peak by the Fourier transform of the KWW function. This further analysis allowed a direct estimation of the stretching parameter characterizing the broadness of the structural peak.

Table 1. Parameters of α-(structural), β-(secondary) relaxation at different pressure and temperature for PVAc.

		α-relaxation		β-relaxation	
P (MPa)	T (K)	α	β	α	β
250	323	0	0	0.55	1
300	323	0	0	0.56	1
350	323	0	0	0.57	1
400	323	0	0	0.58	1
51	342	0	0	0.56	1
101	342	0	0	0.56	1
0.1	244	0	0	0.66	1
0.1	244	0	0	0.68	1

As model independent parameter for the characteristic time scale of the process we considered the maxima of frequency, $\nu^{i}_{max} = 1/(2\pi\tau^{i}_{max})$ corresponding to the loss peak maximum frequency, which was calculated by

$$\nu^{i}_{max} = 1/2\pi\tau_i \times \{\sin[((1-\alpha_i)\times\pi)/(2+2\times\beta_i)]\}^{1/(1-\alpha_i)} \times \{\sin[((1-\alpha_i)\times\beta_i\times\pi)/(2+2\times\beta_i)]\}^{-1/(1-\alpha_i)} \quad (1)$$

The temperature dependence of ν^{α}_{max} can be well fitted by a Vogel-Fulcher-Tammann, VFT, equation, $\nu^{\alpha}_{max} = \nu^{\alpha}_{0}\exp(DT_0/(T-T_0))$ over the entire temperature interval. The pressure dependence of ν^{α}_{max} can be well fitted by a Vogel-Fulcher-Tammann like, PVFT, equation, $\log(\nu_{max}) = \log(\nu_{0max}) + [A\times P/(B-P)]$ over the entire pressure interval. In the isothermal paths, pressure dependence of ν^{α}_{max} below T_g are well reproduced by the Arrhenius equation, $\nu_{max} = \nu_0\exp(-P\cdot\Delta V_\beta/k_B T)$ (where ΔV_β is the activetion volume of the secondary β-relaxation), and in isobaric condition the temperature dependence of ν^{α}_{max} below T_g is well reproduced by the Arrhenius equation, $\nu^{\beta}_{max} = \nu^{\beta}_{0}\exp(-E_a/(k_B T))$. The different VFT and Arrhenius parameters for the isobaric dielectric relaxation spectra of the different are listed in **Table 2**.

The value of the glass transition temperature, T_g, determined by $\tau_\alpha(T_g) = 10$ s at ambient temperature is (310 K $\pm$ 2 K) for PVAc. From the **Figure 2**, it is clear that the position of secondary relaxation at T_g for isobaric paths, **Figure 2(a)**, isn't the same of isothermal path, **Figure 2(b)**.

4. Discussion

The combined variation of both temperature and pressure allows to reach dynamic states characterized by the same value of structural relaxation time, but different thermal

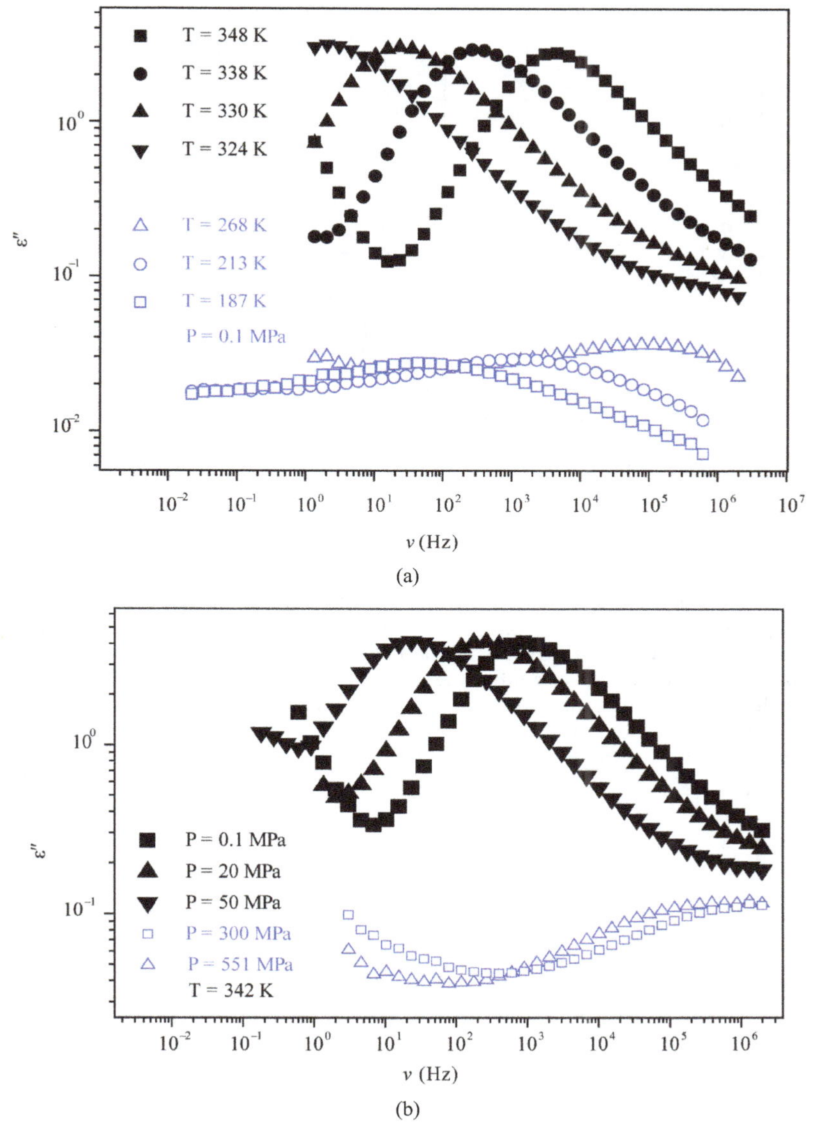

Figure 1. (a) Isobaric (0.1 MPa) dielectric loss spectra of PVAc measured at different temperatures (values reported in the figure); (b) Isothermal (342 K) dielectric loss spectra of PVAc measured at different pressure (values reported in the figure). The black spectra are measured in the supercooled liquid and are dominated by the α-peak and the blue open symbols spectra are measured in the glassy state and are dominated by the β-peak.

Table 2. The glass transition temperature and the relevant VFT parameter for three different systems, as determined by dielectric relaxation measurement [parameter of A and B come from the equation, $\log(\nu_{max}) = \log(\nu_{0max}) + A \cdot P/(B - P)$].

P (MPa)	T (K)	log($\nu_{0\alpha}$)	DT$_0$ (K)	T$_g$(P) (K)	log(ν_{0max})	A	B (MPa)	P$_g$ (T) (MPa)
#	342	2.99	#	342 ± 1	16.6	16.6	618.3	132.9 ± 0.1
#	323	0.49	#	322 ± 1	98.8	98.8	1777.5	57.3 ± 0.1
0.1	318.4	13.7	2.95	318 ± 1	#	#	#	0.1 ± 0.1

energies and densities. The relative role of density and thermal energy on the slowing the structural relaxation on approaching the glass transition is a matter of study. Even if a similar behavior can be found for materials belonging to the same class, the detail of their relative role varies for each system [11]. From the dynamic point of view the glass transition is traditionally defined by considering the structural relaxation time being a fixed long value. In other words, at different of temperature and of pressure the ratio $\tau_\alpha(P,T)/\tau_\beta(P,T)$ isn't con-

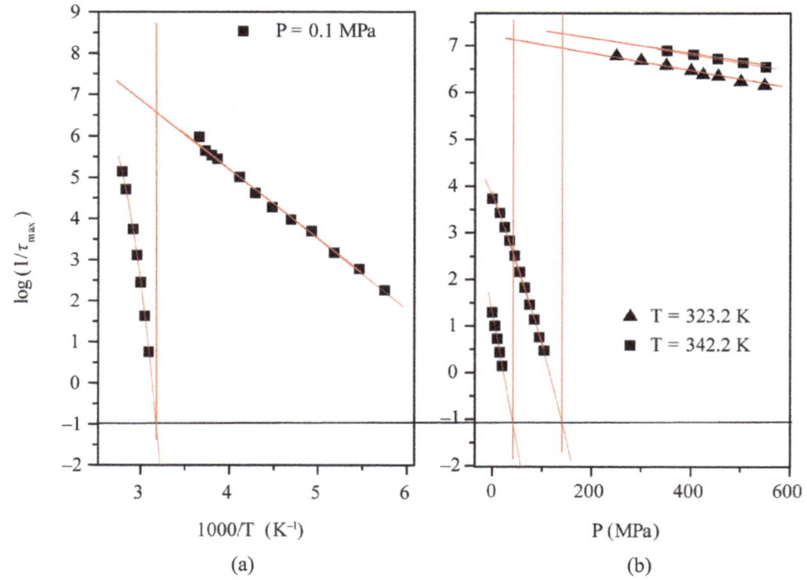

Figure 2. (a) Pressure dependence of the logarithmic of τ_{max} for the α-(full symbols) and the β-(secondary) process, at two different values of T: 342K (squares), and 323.2 K (triangles) of PVAc (b) Temperature dependence of the logarithmic of ν^{α}_{max} for the α-(full symbols) and the β-(secondary) process, at ambient pressure: 0.1 MPa (squares) of PVAc. In both panels the continuous lines represent fit with the VFT (α-relaxation) and Arrhenius (β-relaxation) equations. The horizontal dotted line show the relaxation time value used to define the glass transition. Crosses represent the values of τ_{max} at $(T, P)_g$.

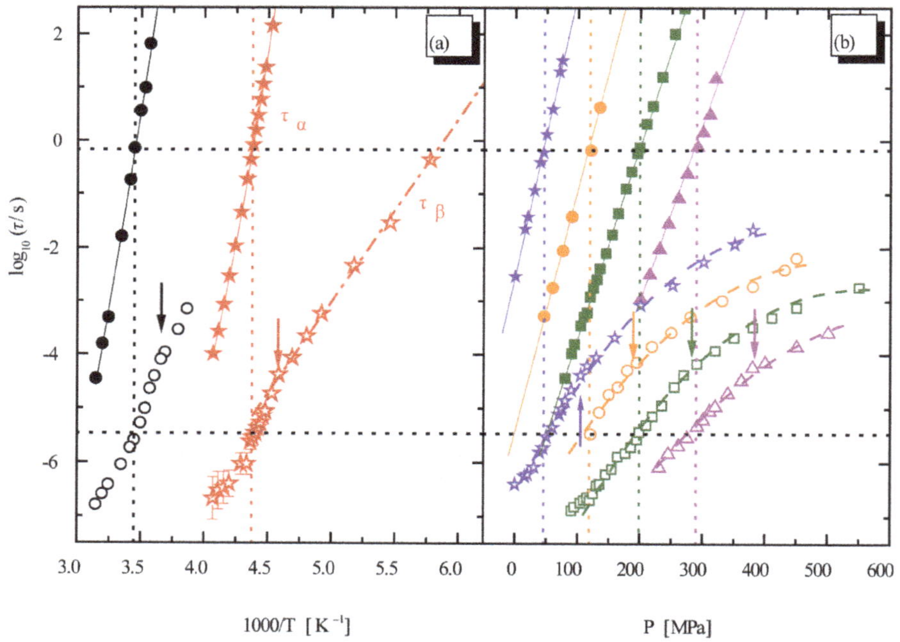

Figure 3. Relaxation map for the mixture 10% QN in tristyrene. (a) Isobaric data: 0.1 MPa (stars), 380 MPa (circles). (b) Isothermal data: 238 K (stars), 253 K (circles), 263 K (squares), 278 K (triangles). The ratio $\tau_{\alpha}(P,T)/\tau_{\beta}(P,T)$ is constant [12].

stant for PVAC for all thermodynamic paths. It is likely to suppose that the connection between α- and β-dynamoics isn't a universal feature of all β-processes, but only of those local processes intimately connected to the cooperative structural dynamics. For comparison of β-process at JG and Non-JG relaxation we compare the Non-JG relaxation at PVAc with JG relaxation at mixture of quinaldine (QN) with tristyrene.

At mixtures of quinaldine (QN) with tristyrene, **Figure 3**, ref. [12], the $\tau_{\alpha}(P,T)/\tau_{\beta}(P,T)$, T is constant for all the isobaric-isothermal paths but at PVAc the $\tau_\alpha(\mathrm{P,T})/\tau_\beta(\mathrm{P,T})$ isn't constant. So, it is likely to suppose that the connection between α- and β-dynamics isn't a universal feature of all β-processes, but only at the JG relaxation.

5. Conclusion

We investigated the relation between secondary and structural dynamics of PVAc under variations of temperature and pressure. We found that the Non-JG relaxations and α-relaxation inside of PVAc is not relation to each other that it is different behaviour than the JG relaxation inside the mixture systems. The analysis consists in investigating the ratio τ_α/τ_β for different values of pressure and temperature (density and thermal energy), but the same value of structural relaxation time. According to such analysis we evidenced that the β-process in PVAc isnot related with the structural dynamics at the glass transition.

REFERENCES

[1] S. Sharifi, "Temperature Dependence of the Activation Volume of Secondary Relaxation in Glass Formers," *ISRN Materials Science*, 2011, Article ID: 460751. doi:10.5402/2011/460751

[2] S. Sharifi and J. M. Asl, "Secondary Relaxation inside the Glass," *ISRN Materials Science*, 2011, Article ID: 764874. doi:10.5402/2011/764874

[3] S. Sharifi, "Activation Volume of Secondary Relaxation," *Materials Sciences and Applications*, Vol. 2, No. 6, 2011, pp. 624-628. doi:10.4236/msa.2011.26084

[4] H. Jansson and J. Swenson, "The Slow Dielectric Debye Relaxation of Monoalcohols in Confined Geometries," *Journal of Chemical Physics*, Vol. 134, No. 10, 2011, Article ID: 104504. doi:10.1063/1.3563630

[5] R. Bergman, H. Jansson and J. Swenson, "Slow Debye-Type Peak Observed in the Dielectric Response of Polyalcohols," *Journal of Chemical Physics*, Vol. 132, No. 4, 2010, Article ID: 044504. doi:10.1063/1.3294703

[6] J. E. McKinney and R. Simha, "Configurational Thermodynamics Properties of Polymer Liquids and Glasses," *Macromolecules*, Vol. 7, No. 6, 1974, pp. 894-901. doi:10.1021/ma60042a037

[7] G. Dlubek, D. Kilburn and M. A. Alam, "Temperature and Pressure Dependence of α-Relaxation and Free Volume in Poly(vinyl acetate)," *Macromolecular Chemistry and Physics*, Vol. 206, No. 8, 2005, pp. 818-826. doi:10.1002/macp.200400495

[8] M. Tyagi, A. Aleg and J. Colmenero, "Broadband Dielectric Study of Oligomer of Poly(vinyl acetate): A Detailed Comparison of Dynamics with Its Polymer Analog," *Physical Review E*, Vol. 75, No. 6, 2007. doi:10.1103/PhysRevE.75.061805

[9] A. Alegr, L. Goitiand, I. Teller and J. Colmenero, "α-Relaxation in the Glass-Transition Range of Amorphous Polymers. 2. Influence of Physical Aging on the Dielectric Relaxation," *Macromolecules*, Vol. 30, No. 13, 1997, pp. 3881-3887. doi:10.1021/ma961266m

[10] L. Goitiandi and A. Alegría, "Physical Aging of Poly (vinyl acetate). A Thermally Stimulated Depolarization Current Investigation," *Journal of Non-Crystalline Solids*, Vol. 287, No. 1-3, 2001, pp. 237-241. doi:10.1016/S0022-3093(01)00578-6

[11] G. Floudas, "Effects of Pressure on Systems with Intrinsic Orientational Order," *Progress in Polymer Science*, Vol. 29, No. 11, 2004, pp. 1143-1171. doi:10.1016/j.progpolymsci.2004.08.004

[12] K. Kessairi, S. Capaccioli, D. Prevosto, M. Lucchesi, S. Sharifi and P. A. Rolla, "Interdependence of Primary and Johari-Goldstein Secondary Relaxations in Glass-Forming Systems," *The Journal of Physical Chemistry B*, Vol. 112, No. 15, 2008, pp. 4470-4473. doi:10.1021/jp800764w

The Effects of Annealing Process on Dielectric and Piezoelectric Properties of BMT-Base Lead-Free Ceramics

Mahdi Ghasemifard[1], Meisam Daneshvar[2], Misagh Ghamari[1]
[1]Nanotechnology Laboratory, Esfarayen University, Esfarayen, Iran
[2]Department of Physisc (Electroceramics Laboratory), Shahid Beheshti University, Tehran, Iran
Email: mahdi.ghasemifard@gmail.com

ABSTRACT

By using nitric acid as the fuel, the lead-free ceramic of $Ba(Ti_{1-x},Mg_x)O_3$ (x = 0.31) was prepared by auto combustion method (ACM). To make a comparison, this ceramic was also prepared using mixed oxide method (MOM). By X-ray diffraction, the phase structures of two samples were studied, and the results showed that rising temperatures would reduce unwanted phases. The piezoelectric and electrical properties as a function of calcination and sintering temperatures were investigated. The results showed that the outstanding electrical properties were obtained for nanoceramic with this composition. The SEM image of the grain size was estimated around 2 micrometers, and the grain size increased with the increasing of sintering temperature for two samples. The curie temperature of the BMT-ACM was 126°C and it's significantly larger than the curie temperature of BMT-MOM which was 118°C. The results of electrical properties emphasized that the synthesis optimum temperature for two samples was about 1200°C and it was the best temperature that led to improved properties such as dielectric constant, polarization and piezoelectric coefficients.

Keywords: Dielectric Properties; $Batio_3$ and Titanates; Capacitors

1. Introduction

The microstructure of ceramics strongly depends on annealing process and it determines the ceramic's physical behavior. Density, homogeneity and grains size in the annealing steps are important to achieve the desired electrical properties [1]. Crystal formation and complete perovskite phase are the most important events that occur during annealing. The grain size growth and the optimum electrical properties of the samples are done at this stage. With the rise of annealing temperature, the grain size becomes larger and ceramic to be denser. But if the temperature is too high, it can cause cracks in the sample because of thermal expansion. Then this reduces the density and increases porosity of the ceramic and the resulting reduction in mechanical strength [2]. Barium titanate (BT) can be developed as the first lead-free piezoelectric ceramic with perovskite structure (ABO_3) which below the Curie temperature (130°C) has the tetragonal phase and above the Curie temperature has the cubic phase [3]. Nowadays, the BT-based ceramics such as barium zirconium titanate (BZT) for applications in multi-layer capacitors (MLCs), piezoelectric transducers and electro optic devices are further studied. The BT-based ceramics have a high dielectric constant and a relatively large strain [4-7]. The dielectric constant of capacitors in the multilayer form is higher than that in the disc form [8]. The dielectric constant can be improved significantly by adding metals such as Ag [9] and Ni [10]. BT-based nano-powder has been prepared by a few methods, including coprecipitation, sol-gel process, combustion, hydrothermal method, etc. However, finding an efficient way is a challenge to prepare BT-based nano-powder with particle size in about several nanometers. Therefore, we made an attempt to prepare Mg doping BT by two methods, namely auto combustion method (ACM), and mixed oxide method (MOM) for comparative investigation of their properties. The effect of calcination and sintering temperature on the $Ba(Ti_{1-x}, Mg_x)O_3$ lead-free ceramics prepared by two methods were studied. The dielectric, piezoelectric and electrical properties of samples were measured and discussed with respect to the sintering temperature.

2. Experimental Procedure

The nano-powder of $0.94Ba(Mg_{0.33},Ti_{0.67})O_3$-$0.06BaTiO_3$ was synthesized by using salt's precursors as starting materials. Raw materials used in auto combustion method consist of barium nitrate [$Ba(NO_3)_2$], titanium isopropoxide Ti[OCH(CH3)2]$_4$ and magnesium acetate [$(CH_3COO)_2Mg\cdot 4H_2O$]. Aqueous solution of each single cation (*i.e.* Ba^{+2}, Mg^{+2} and Ti^{+4}) was prepared by dissolving barium nitrate and magnesium acetate in distilled water and titanium isopropoxide was dissolved in the mixture of nitric acid, citric acid and hydrogen peroxide for preparation of Ti^{+4}. The solutions of barium, magnesium and titanium were added to the aqueous solution of citric acid under continuous stirring at 65°C - 75°C and finally at the end the pH of the sol was maintained at 6.5 by the addition of ammonium hydroxide. In order to obtain the gel the peroxo-citrato-nitrate sol of BMT was heated at about 80°C. After auto combustion of the gels by addition nitric acid as a fuel the resultant powders were calcinated at different temperatures to obtain the desired single-phase powders. BMT nano-powders were produced by the auto combustion technique was labeled as BMT-ACM (obtained from ACM). Raw materials used in mixed oxide method consist of barium nitrate [$Ba(NO_3)_2$], magnesium acetate [$(CH_3COO)_2Mg.4H_2O$] and titanium dioxide [TiO_2]. First, for improvement mixed oxides we added the BT and BMT powders together which prepared separately. With using the pure ethanol and stirring them for two hours at room temperature, a homogenous BMT-BT milky precursor was prepared. After drying a white powder was obtained. The heating rate of calcination step powder was 2°C/min from room temperature to various temperatures ranging from 700°C to 850°C for 2 h, and it was labeled as BMT-MOM (obtained from MOM). The flow diagram of the samples processing method employed in this study are shown in **Figure 1**.

X-ray diffraction patterns used to study the structure of phase and we found that the suitable calcination temperature is 1000°C. The XRD patterns of BMT powders calcinated at 1000°C are shows in **Figure 2**. Pellets made under 3 Mpa pressure after aggregation and shaping of the powders and sintered at different temperatures from 1100°C to 1250°C.

3. Results and discussion

3.1. X-Ray Analysis

According to **Figure 2**, the XRD results reveal the existence of a tetragonal type phase for MOM and ACM method at 1000°C. According to **Figure 3**, there is a percentage of the unwanted phase at 900°C temperature.

As can be seen from **Figure 3**, the percentage unwanted phase at BMT-MOM is more than BMT-ACM and this result indicates that for synthesis of BMT in nano scale, we require less energy than micro scale. Because the maximum phase purity was obtained only for powder at 1000°C, this powder was selected as starting powder for electrical measurement. **Table 1** shows the X-ray diffraction analysis data.

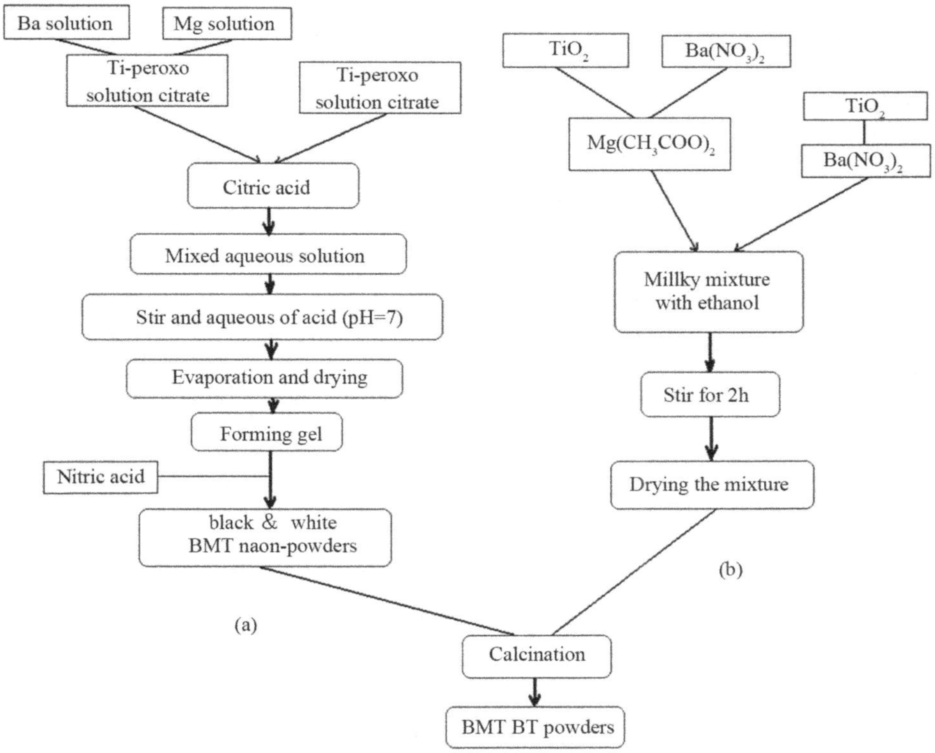

Figure 1. The flow diagrams of BMT. (a) ACM; (b) MOM.

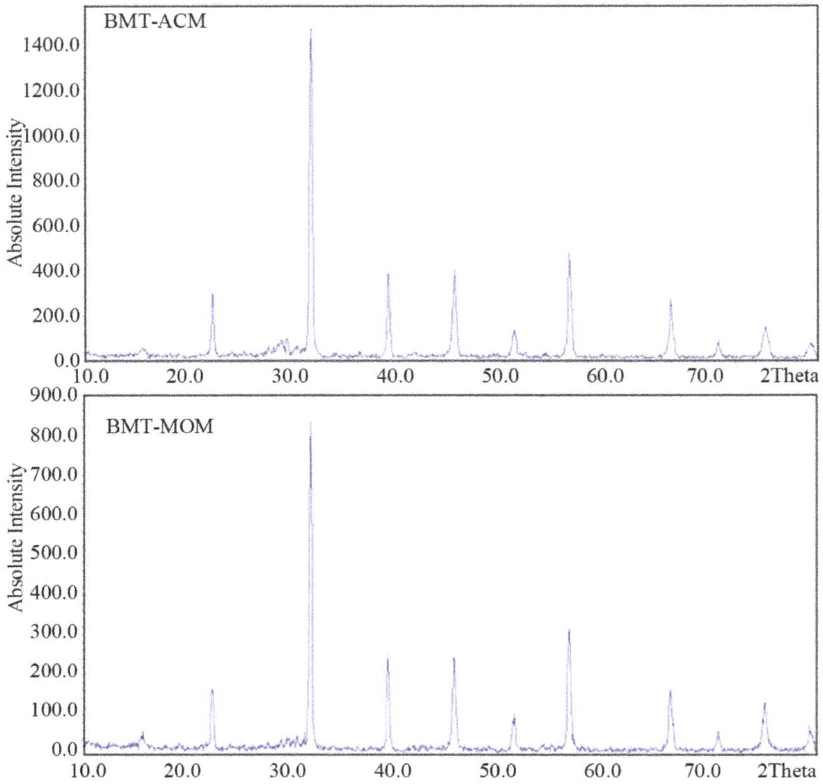

Figure 2. XRD spectra of samples of the BMT calcinated at 1000°C temperatures.

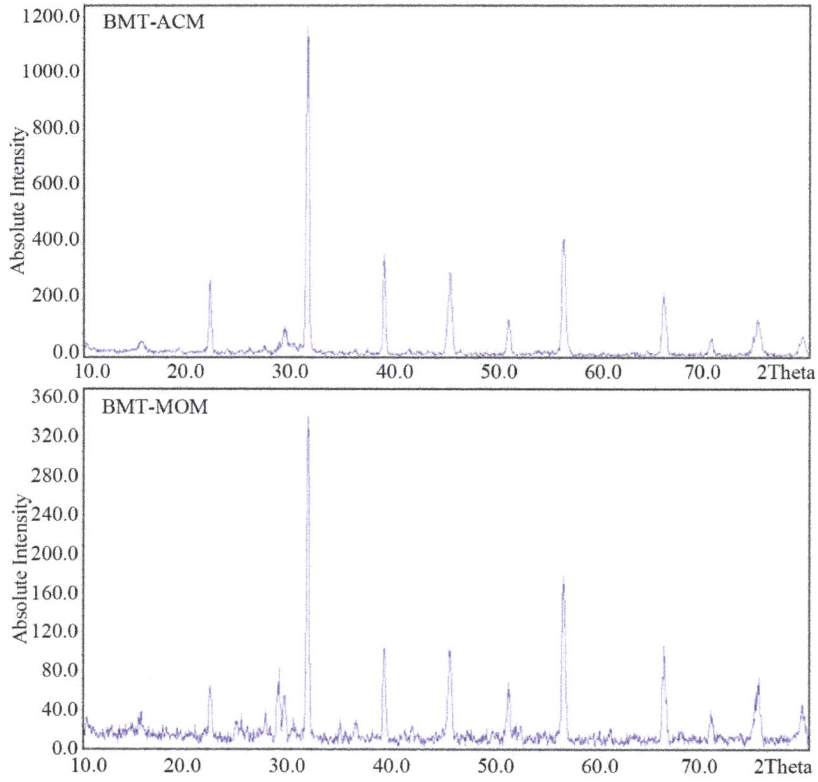

Figure 3. XRD spectra of samples of the BMT calcinated at 900°C.

Table 1. Parameters of X-ray diffraction analysis.

Temperature (°C)	Main peak 2θ	hkl	Phase structure	Lattice parameters (Å)	Volume (Å)^3	Main Phase percent
900-MOM	14.11	110	Tetragonal	a = b = 4.09 c = 4.40	73.604	88
900-ACM	14.02	110	Tetragonal	a = b = 4.04 c = 4.38	71.49	83
1000-MOM	14.58	110	Tetragonal	a = b = 3.95 c = 4.01	62.746	98
1000-ACM	14.44	110	Tetragonal	a = b = 3.89 c = 3.99	60.38	100
Ref. [11]	BT		Tetragonal	a = b = 3.99 c = 4.03	64.366	100
Ref. [12]	$BaTiO_3$-Nb_2O_5-MgO		Tetragonal	a = b = 4.004 c = 4.015	64.37	93

3.2. TEM Image

The typical TEM image of the BMT nano-powder calcinated at temperatures of 1000°C prepared by the auto combustion method is shown in **Figure 4**. From TEM analysis the primary particle size of the nano-powders can be determined. The particle size of the BMT-ACM powder was found to be approximately 24 nm in diameter.

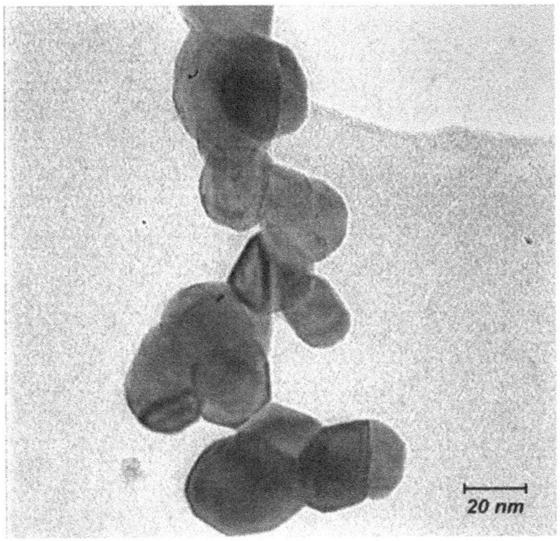

Figure 4. TEM micrograph of the prepared BMT-ACM powders at 1000°C.

3.3. Microstructure Investigation

Figures 5 and **6** show the microstructure and histogram of the BMT-MOM and BMT-ACM lead free ceramics that have been sintered at different temperatures. According to these figures it can be observed that increasing of annealing temperature leads to the grain size increase. For example, at sintering temperature of 1100°C, the grain size of the BMT-MOM ceramic is about 2.02 μm, when sintering temperature will increase to 1300°C the grain size grew unusually and reaches the size of about 2.69 μm. The grain size has strong effects on dielectric properties and polarization of piezoelectric materials [11]. The relationships between the grain size and the ceramic dielectric and piezoelectric properties are discussed in the next section.

3.4. Density

Generally, synthesis single phase of $Ba(Mg_{1/3},Ti_{2/3})O_3$ ceramic is difficult. However, the synthesis of $0.94Ba(Mg_{0.33},Ti_{0.67})O_3$-$0.06BaTiO_3$ composition is more difficulties and sintering stage becomes much easier. All the compositions were sintered in air at 1100°C, 1150°C, 1200°C and 1250°C for 2 h and the temperature gradient for annealing were 5°C/min and 1°C/min for BMT-MOM and BMT-ACM, respectively. Generally, as shown in **Figure 7**, density increases with increasing temperature until the density reaches to its maximum at 1200°C, then at 1250°C the density decreases. The pattern of the graph may be interpreted according to the sintering model of Coble [12].

3.5. Hysteresis Loop

The hysteresis loop is one of the important characteristics that shows ceramic is considered to be ferroelectric and gives comprehensive information about the polarization. **Figure 8** shows the P-E hysteresis loops of the BMT lead free piezoelectric ceramics measured at room temperature. As can be seen in **Figure 8**, with increasing temperature hysteresis loops are thinner and it reaches to zero at Curie temperature and P_r has a maximum value at 1200°C. The reason may be the presence of nano-sized domains that have been polarized in the phase boundary polarization are leading a lot of quality [13,14].

3.6. Piezoelectric Constant d_{33}

The piezoelectric coefficient (d_{33}) for two BMT lead-free ceramic at room temperature (Pennebaker, model:8000) are shown in **Figure 9**. As can be seen from **Figure 9**, the amount of d_{33} reached a peak to 87.6 pC/N in 1215°C, which led to a drop to 72.3 pC/N at 1300°C. According to **Figure 9**, the d_{33} amount of BMT-MOM is more than BMT-ACM in each of the sintering temperatures. This could be due to the insufficient grain growth in the sintering step.

3.7. Dielectric Constant

The polarization is active by thermal processes and thus

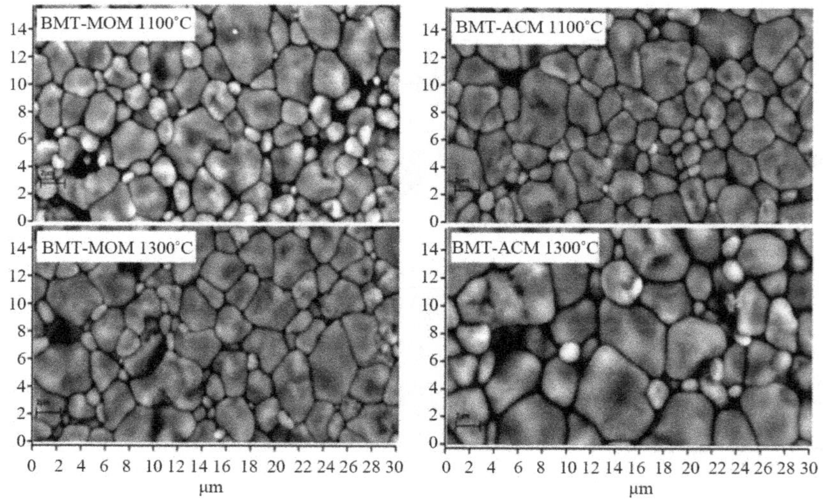

Figure 5. SEM graphs of surface of BMT-MOM and BMT-ACM ceramics.

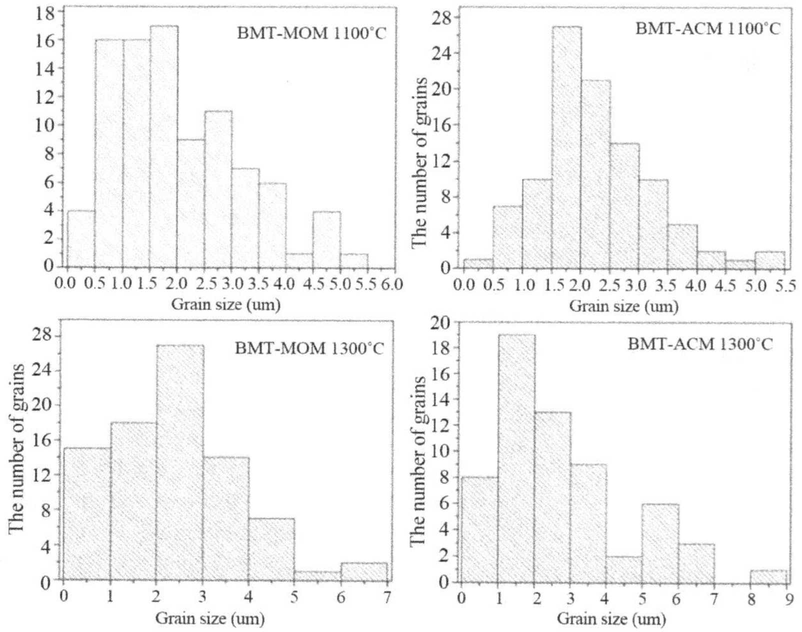

Figure 6. The grain histogram of BMT-MOM and BMT-ACM ceramics.

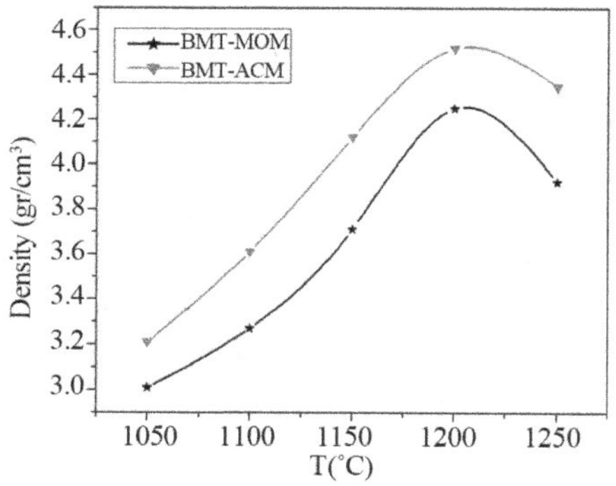

Figure 7. The density of the BMT-MOM and BMT-ACM ceramics.

the dielectric constant is dependent on temperature. Dielectric constant may be increased or decreased through polarizetion. At Curie temperature the BMT-MOM and BMT-ACM have a very large spontaneous polarization that leads to a large dielectric constant. At below Curie temperature, the structure of samples changes from cubic to tetrahedral and the position of magnesium and titanium ions becomes an off-center position corresponding to a permanent electrical dipole. To the investigation of dielectric properties, the temperature dependence of the dielectric constant for ceramics polarized, were measured at 100 kHz frequency. **Figure 10** shows the variation of dielectric constant of the BMT-MOM and BMT-ACM

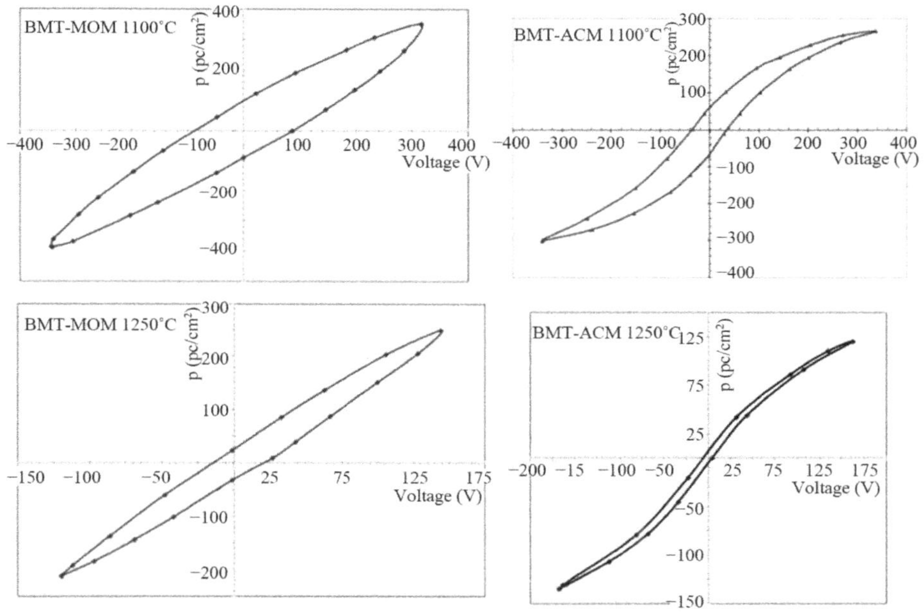

Figure 8. The P-E hysteresis loops of the BMT-MOM and BMT-ACM ceramics sintered at 1100˚C and 1200˚C.

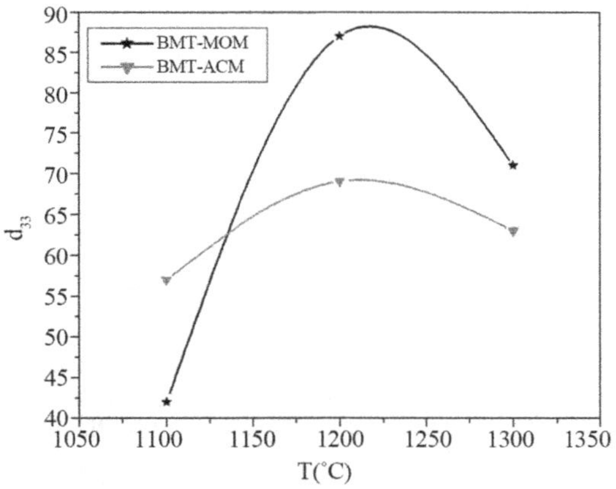

Figure 9. Piezoelectric coefficient d_{33} as a function of sintering temperature.

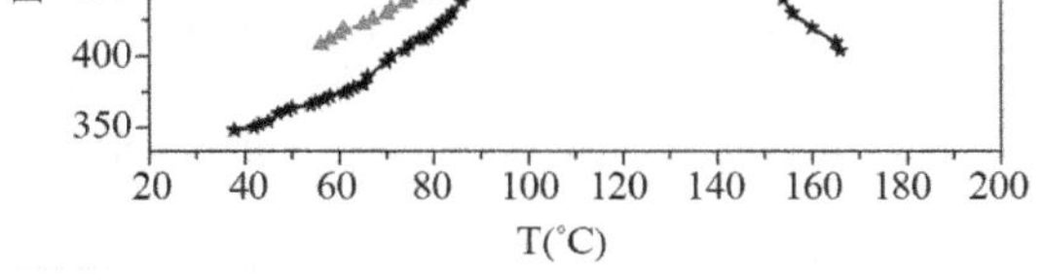

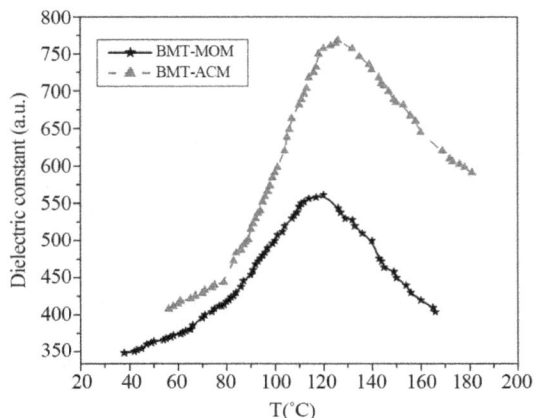

Figure 10. Dielectric constants of BMT-MOM and BMT-ACM as a function of temperature.

ceramics which sintered at 1200˚C. It was found that the dielectric constants for the BMT-ACM are higher than 750. According to **Figure 10**, the dielectric constant increases steeply from 412 to 773 by 55˚C and 126˚C, respectively, then its value is reduced to 418 at 180˚C. Moreover, the variations of the dielectric constant through polarize also depend on the domain alignment, and this leads to a rise of dielectric constant.

To the investigation of relaxor properties, the temperature dependence of the dielectric constant for ceramics polarized was measured at different frequencies (100, 500, 1000 kHz).

According to **Figure 11**, the curve peak with increaseing frequency, has not an outstanding displacement along the vertical axis and this represents which the BMT-BT lead free ceramic is not ferroelectric relaxor. **Table 2** listed the electrical parameters of the BMT-MOM and BMT-ACM ceramics that sintered at different temperatures. According to **Table 2**, the grain size has strong effects on dielectric properties and polarization of piezoelectric materials [15]. On the other hand, all electrical coefficients at optimum temperatures (1200˚C) were increased.

4. Conclusion

The ferroelectric of $0.94Ba(Mg_{0.33},Ti_{0.67})O_3$-$0.06BaTiO_3$ was prepared by the mixed oxide method and auto combustion method using oxide and non-oxide precursor. The powders with maximum tetragonal phase and ceramic's sample uniform microstructure were obtained at the optimum temperature of 1000˚C and 1200˚C, respec-

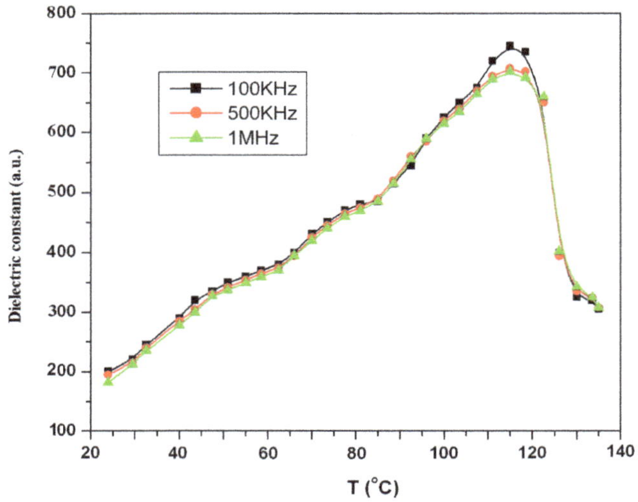

Figure 11. Temperature dependence of dielectric constant for BMT-MOM ceramics at different frequencies.

Table 2. Various parameters on some electrical properties.

Methods	S. T. (°C)	Grain size (μm)	Density (gr/cm^3)	d_{33} (pC/N)	Q_m	k_p	Curie T. (°C)	Dielectric constant (a.u.)
	1100	2.02	3.27	41.9	34.6	0.38	-	-
BMT-MOM	1200	2.26	4.25	87.1	41.9	0.64	120	396.60
	1300	2.69	3.92	72.3	64.7	0.93	208	427.13
	1100	1.75	3.61	57.4	37.11	0.41	-	-
BMT-ACM	1200	1.97	4.52	67.4	40.33	0.63	126	520.56
	1300	2.39	4.35	63.1	58.92	0.87	221	687.71

tively. Electrical property investigations indicated that excellent ferroelectric and piezoelectric properties were obtained at this composition. This result may be due to grain growth. In particular, the BMT ceramics have a relatively high Curie temperature and a better temperature stability of permittivity. Dielectric investigations show that a peculiar relaxor behavior cannot be observed in this system.

REFERENCES

[1] W. Maison, S. Ananta, T. Tunkasiri, P. Thavornyutikarn and S. Phanichphant, "Effect of Calcination Temperature on Phase Transformation and Particle size of Barium Titanate Fine Powders Synthesized by the Catecholate Process," *Science Asia*, Vol. 27, No. 4, 2001, pp. 239-243. doi:10.2306/scienceasia1513-1874.2001.27.239

[2] M. Ghasemifard, "Dielectric, Piezoelectric and Electrical Study of 0.65PMN-0.20PZT-0.15PT Relaxor Ceramic," *The European Physical Journal Applied Physics*, Vol. 54, No. 2, 2011, pp. 20701-20707. doi:10.1051/epjap/2011100495

[3] B. Jaffe and W. R. Cook, "Piezoelectric Ceramics," Academic Press, London and New York, 1971.

[4] W. Maison, R. Kleeberg, R. B. Heimann and S. Phanichphant, "Phase Content, Tetragonality, and Crystallite Size of Nanoscaled Barium Titanate Synthesized by the Catecholate Process: Effect of Calcinations Temperature," *Journal of the European Ceramic Society*, Vol. 23, No. 1, 2003, pp. 127-132. doi:10.1016/S0955-2219(02)00071-7

[5] W. Heywang, "Barium Titanate as a PTC Thermistor," *Solid-State Electronics*, Vol. 3, No. 1, 1961, pp. 51-58. doi:10.1016/0038-1101(61)90080-6

[6] W. Heywang, "Resistivity Anomaly in Doped Bariumtitanate," *Journal of the American Ceramic Society*, Vol. 47, No. 10, 1964, pp. 484-490. doi:10.1111/j.1151-2916.1964.tb13795.x

[7] B. Huybrechts, K. Ishizaki and M. Takata, "The Positive-temperature Coefficient of Resistivity in Barium Titanate," *Journal of Materials Science*, Vol. 30, No. 10, 1995, pp. 2463-2474. doi:10.1007/BF00362121

[8] M. S. H. Chu and C. E. Hodgkins, "Multilayer Ceramic Devices," Advanced Ceramic, Vol. 19, 1986, pp. 203-207.

[9] C. Y. Chen and W. H. Tuan, "Mechanical and Dielectric Properties of $BaTiO_3$/Ag Composites," *Journal of Materials Science Letters*, Vol. 18, No. 5, 1999, pp. 353-354. doi:10.1023/A:1006612129503

[10] C. Pecharroman and J. S. Moya, "Experimental Evidence

of a Giant Capacitance in Insulator-Conductor Composites at the Percolation Threshold," *Advanced Materials*, Vol. 12, No. 4, 2000, pp. 294-297.

[11] M. Alguero, A. Moure, L. Pardo, J. Holc and M. Kosec, "Processing by Mechano Synthesis and Properties of Piezoelectric $Pb(Mg_{1/3}Nb_{2/3})O_3$-$PbTiO_3$ with Different Compositions," *Acta Materialia*, Vol. 54, No. 3, 2006, pp. 501-511. doi:10.1016/j.actamat.2005.09.020

[12] R. L. Coble, "Grain Growth in Sintered ZnO and ZnO-Bi_2O_3 Ceramics," *Journal of Applied Physics*, Vol. 56, 1985, pp. 131-141.

[13] K. Kakegawa and J. Mohri, "A Compositional Fluctuation and Proporties of $Pb(Zr,Ti)O_3$," *Solid State Communications*, Vol. 24, No. 11, 1977, pp. 769-772. doi:10.1016/0038-1098(77)91186-3

[14] K. Carl and K. H. Hardtl, "Composition Dependences in Solid Solution on the Basis of Lead-Zirconate-Titanate and Sodium Niobate," *Physica Status Solidi* (*a*), Vol. 8, No. 1, 1971, pp. 87-91. doi:10.1002/pssa.2210080108

[15] K. Okazaki and K. Nagata, "Effects of Grain Size and Porosity on Electrical and Optical Properties of PLZT Ceramics," *Journal of the American Ceramic Society*, Vol. 56, No. 2, 1973, pp. 82-86. doi:10.1111/j.1151-2916.1973.tb12363.x

Study of Dielectric and Piezoelectric Properties in the Ternary System $Pb_{0.98}Ca_{0.02}[\{(Zr_{0.52}Ti_{0.48})_{0.98}(Cr^{3+}_{0.5}, Ta^{5+}_{0.5})_{0.02}\}_{1-z}P_z]O_3$ Doping Effects

Hamzioui Louanes[1,2*], Kahoul Fares[1,2], Abdessalem Nora[1], Boutarfaia Ahmed[1,2]

[1]Laboratoire de Chimie Appliquée, Université de Biskra, Biskra, Algérie; [2]Département de Science de la Matière, Université de Ouargla, Ouargla, Algérie.
Email: {hamzioui_louanes, aboutarfaia}@yahoo.fr

ABSTRACT

The effects of P_2O_5 oxide on microstructure, dielectric and piezoelectric properties of $Pb_{0.98}Ca_{0.02}[\{(Zr_{0.52}Ti_{0.48})_{0.98}(Cr^{3+}_{1/2}, Ta^{5+}_{1/2})_{0.02}\}_{1-z}P_z]O_3$ ternary ceramics were investigated. Specimens with various contents of P_2O_5 from 0 to 12 wt% were prepared by a conventional oxide mixing technique. The effect of P_2O_5 doping with regard to the development of the crystalline phase, density, microstructure, dielectric, ferroelectric and piezoelectric characteristics has been investigated. It has been found that the sintering temperature of piezoelectric $Pb_{0.98}Ca_{0.02}[\{(Zr_{0.52}Ti_{0.48})_{0.98}(Cr^{3+}_{1/2}, Ta^{5+}_{1/2})_{0.02}\}_{1-z}P_z]O_3$ can be reduced by phosphorus addition without compromising the dielectric properties. A sintered density of 94 % of the theoretical density was obtained for 4 wt% P_2O_5 addition after sintering at 1050˚C for 4 h. Ceramics sintered at 1050˚C with 4 wt% P_2O_5 achieve excellent properties, which are as follows: kp = 0.73, $\rho = 0.09 \times 10^{+4}$ (Ω·cm), ε_r = 18800, $\tan\delta = 0.0094$ and Tc = 390˚C.

Keywords: PZT; Piezoelectricity; Electronic Materials; Dielectric Properties; Methods Physico-Chemical of Analysis

1. Introduction

Lead-based perovskite-type solid solutions consisting of the ferroelectric and relaxor materials have attracted a growing fundamental and practical interest because of their excellent dielectric, piezoelectric and electrostrictive properties which are useful in actuating and sensing applications [1,2]. However, the sintering of PZT at high temperatures gives rise to a lead loss, which drastically degrades the device performance. Generally, a lead loss at high temperatures can be prevented by atmosphere-controlled sintering of PZT. However, such composition requires sintering at a high temperature (>1250˚C) in a controlled atmosphere to contain lead volatilization so as to avoid a shift in composition. To get around the problem, different sintering aids have been tried by various workers [3-5]. However, for practical applications, such sintering aids need proper selection so that the electrical and piezoelectric properties of the ceramics do not degrade.

The dielectric constants increased with the addition of NiO, Fe_2O_3, Gd_2O_3, Nb_2O_5 or WO_3 and decreased with Cr_2O_3 or MnO_2 addition [6-12]. Duran *et al.* studied the effect of MnO addition on the sintering and piezoelectric properties of Sm-modified lead titanate ceramics. The maximum density observed was 96.8% of the theoretical densit for 1% MnO addition at a sintering temperature of 1150˚C [13]. The main role of dopants is generally improved physical and mechanical properties of these materials. This work aims at, to study the influence of P_2O_5 on the properties dielectric and piezoelectric of a ceramics material of general formula: $Pb_{0.98}Ca_{0.02}[(Zr_{0.52}Ti_{0.48})_{0.98}(Cr^{3+}_{0.5}, Ta^{5+}_{0.5})_{0.02}]O_3$ and of structure perovskite.

2. Experimental Procedure

The compositions used for the present study were $Pb_{0.98}Ca_{0.02}[\{(Zr_{0.52}Ti_{0.48})_{0.98}(Cr^{3+}_{0.5}, Ta^{5+}_{0.5})_{0.02}\}_{1-z}P_z]O_3$ with z varying as 0, 2, 4, 6, 8, 10 and 12 wt% respectively. The samples were prepared by a conventional oxide mixing technique. The appropriate amounts of PbO (99.9%), TiO_2 (99.9%), ZrO_2 (99.0%), Ta_2O_5 (99.9%), CaO (99.9%), Cr_2O_3 (99.9%) and P_2O_5 (99.9%) powders were weighed and mixed by ball milling with partially stabilized zirconia balls as media in isopropyl alcohol for 6 h.

*Corresponding author.

After drying, the mixture was calcined in a covered alumina crucible at 800°C for 4 h. The calcined powders were again ball milled for 24 h. The resulting powders were uniaxially compacted into pellets of 10 mm in diameter at a pressure of 5 MPa, followed by isostatically pressing at 150 MPa. To investigate their sintering behavior, the specimens were sintered in a sealed alumina crucible at temperatures ranging from 1000°C to 1180°C for 2 h. To limit PbO loss from the pellets, a PbO-rich atmosphere was maintained by placing an equimolar mix ture of PbO and ZrO_2 inside the crucible. The weight loss of a well-sintered specimen was less than 0.5 wt%, thus a 0.5 wt% excess PbO was added to compensate for the lead loss during sintering. The bulk density was measured using the Archimedean method. The sintered compounds are carefully ground, then analyzed by the scanning electron microscopy (SEM) is a technical for estimating the size distribution, the average size of grains after sintering and qualitatively assess the presence of porosity. The micrographics are made using a Microscope JMS 6400. To investigate the electrical properties, the sintered disks were lapped on their major faces, and then sliver electrodes were deposited with a low temperature paste at 700°C for 30 min. The piezoelectric samples were poled in a silicone oil bath at 100°C by applying 20 kV/cm for 20 min. then cooling them under the same electric field. They were aged for 24 h prior to testing. The temperature dependence of dielectric properties was measured at temperatures ranging from room temperature to 420°C with a heating rate of 2 °C/min using an impedance analyzer—HP4192A, Hewlett-Packard, Palo Alto, CA. The electromechanical coupling factor, *kp*, was determined by the resonance and anti-resonance technique using another impendence analyzer (SI1260 Impedance/Gain-Phase Analyzer, Solartron, UK). ($kp = [2.51(fa - fr)/fr)]^{1/2}$, where *fr* and *fa* are the resonance and anti-resonance frequencies, respectively [14]. Variation of the dielectric constant ε_r, resistivity and also the angle of the losses were examined by using a measuring bridge type RLC (bridge Schering) depending on temperature, concentration, the frequency.

3. Results and Discussion

3.1. Sintered Density

Figure 1(a) shows the variation of density with sintering temperature and the amount of P_2O_5 addition. This curves show the similar variation trend with increasing sintering temperature. The density of specimens sintered at 1050°C showed the maximum value of 7.52 $g \cdot cm^{-3}$ at 4 wt% P_2O_5 and then was decreased after the maximum value. This variation is mainly attributed to the formation of liquid phase of excess PbO that improves densification of the ceramics. However, a large amount of this liquid phase leads to low density which may result from the formation of voids [15]. The densities of the sintered pellets are shown in **Figure 1(b)**. From this figure, it is evident that initially the bulk density (as a percentage of theoretical density) of the pellets decreases with 1 wt% P_2O_5 addition, then also with an increasing amount of P_2O_5 addition up to 4.0 wt%, the bulk density increased and then greater than 4.0 wt% P_2O_5 addition, the density again decreased. This behavior can be explained by decomposition of P_2O_5 (melting point around 570°C) can aid in liquid-phase sintering. Initially, the density of the pellets decreased on addition of 4 wt% P_2O_5, because the amount of liquid formed was probably too low to get an appreciable densification. With increasing P_2O_5, the liquid-phase sintering dominated and was optimum for 4 wt% P_2O_5 addition leading to maximum densification. With a higher amount of P_2O_5 addition, the densification again decreased because there should be an optimum amount of liquid above which densification is inhibited

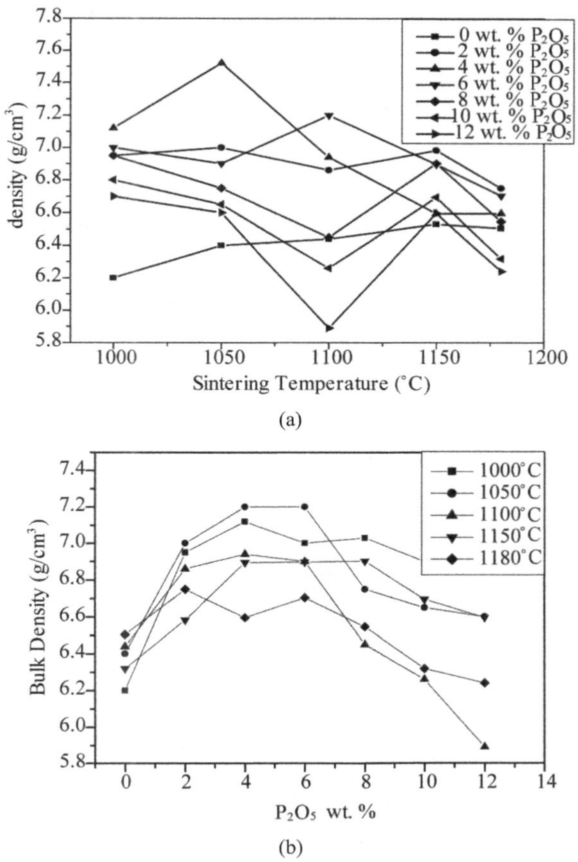

Figure 1. (a) Variation of the density of PZT-CCT samples with sintering temperature at different P_2O_5 addition; (b) Variation of bulk density of sintered PZT-CCT samples with P_2O_5 addition at different sintering temperature.

due to the formation of a thick coating of the liquid around the grains [16]. Recently, Saha *et al.* [17] reported that a small amount of phosphorous addition (as P_2O_5) can improve the sintering of the PZT ceramics.

3.2. Study of Morphological

Figure 2 shows SEM microstructures of the fracture surfaces of samples sintered at various temperatures. The

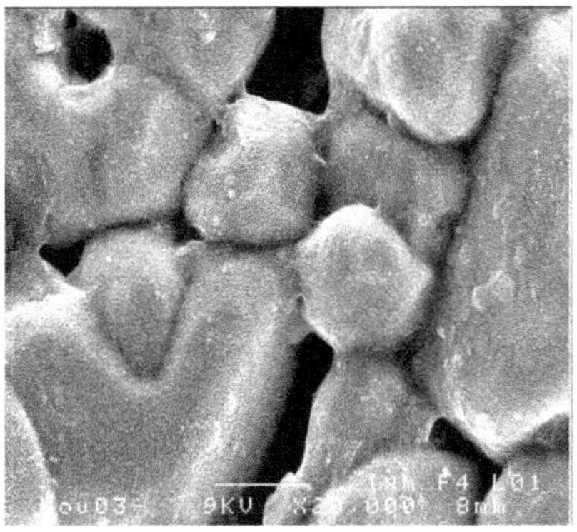

(a)

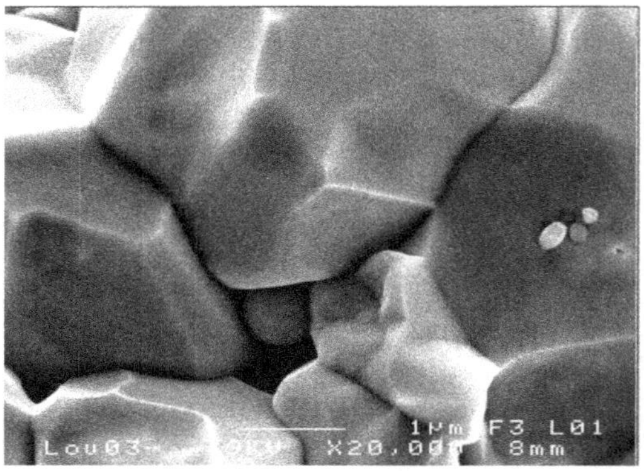

(b)

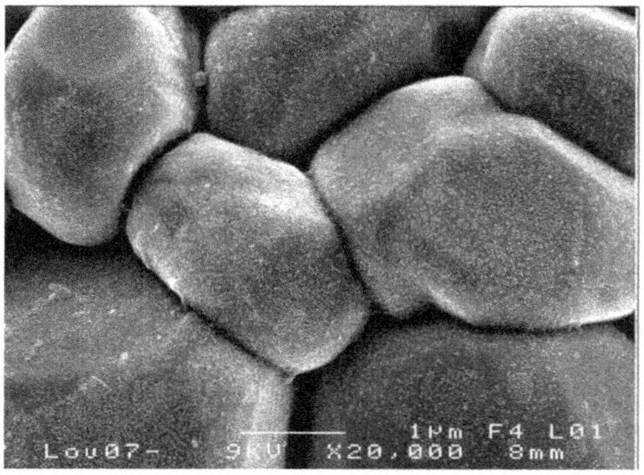

(c)

Figure 2. SEM micrographs of $Pb_{0.98}Ca_{0.02}[(Zr_{0.52}Ti_{0.48})_{0.98}(Cr^{3+}_{0.5}, Ta^{5+}_{0.5})_{0.02}]O_3$ specimens doped with 4 wt% P_2O_5 and sintered at (a) 1000˚C, (b) 1050˚C and (c) 1100˚C.

distributions in the grain shape and size of the samples are rather uniform. All the samples showed an intergranular fracture mechanism indicating that the grain boundaries are mechanically weaker than the grains. These samples appear very dense and of a homogeneous granular structure, the three samples seem homogeneous and there do not seem to be grains of the pyrochlore phase which are identifiable by their pyramidal form. It is noted that the average grain size increases with increasing sintering temperature for 2 μm at 1000˚C up to 2.671 μm at 1050˚C then this size decreases slightly to 1100˚C to reach the value of 2.51 μm; and the broader the granulo-metric distribution "**Figure 2(b)**", the more the size of the grains gets bigger. With increasing P_2O_5, the liquid-phase sintering dominated and was optimum for 4 wt% P_2O_5 addition leading to maximum densification. The increase in grain size may have led to the decrease of oxygen vacancies in PZT [18]. **Figure 2** also shows that the porosity decreases with the increase of sintering temperature (it reaches a minimum at 1050˚C), which is consistent with the increase of the densification of specimens.

3.3. Phase Structure

Sintered powders were examined by X-ray diffractometry to ensure phase purity and to identify the phases of the materials. The results of X-rays on the samples sintered at 1050˚C are illustrated in **Figure 3**. The ceramics with various P_2O_5 contents all exist as pure perovskite phase. The ceramics exist as tetragonal phase which is indicated by the single $(002)_T$ and $(200)_T$ peak at z = 0.00 and z = 0.02. As P_2O_5 content increases from 4 to 8 wt%, the ceramics coexist as tetragonal and rhombohedral phase revealed by the coexistence of $(002)_T$ and $(200)_R$ peaks in the 2θ range from 43.8˚ to 45.3˚. The ceramics with z = 0.08 and z = 0.10 exist as tetragonal phase revealed by the splitting of (002)T and (200)T peaks in the 2θ range from 43.5˚ to 45.4˚.

3.4. Dielectric Properties

The electrical properties were strongly dependent on the phase of the specimens. **Figure 4** shows the dielectric constant ε_r and dielectric loss tanδ of the ceramics sintered at 1050˚C as a function of P_2O_5 content measured at 1 kHz. The temperature of the maximum dielectric constant (*Tm*) increased and the dielectric constant peak sharpened with increasing P_2O_5 concentration. Dielectric loss tanδ slowly deceases at first (4 wt%), and begins to increase when P_2O_5 content is up to 8 wt%.

Figure 5 shows dielectric loss (tanδ) of ceramics sintered at 1050˚C as a function of P_2O_5 content it is noticeable that there is a variation of the dielectric losses

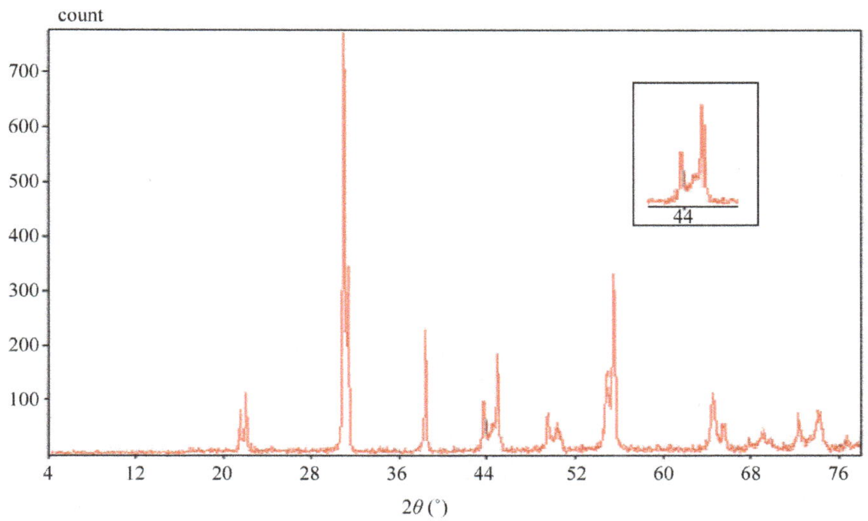
count
700
600
500
400
300
200
100
44
4
12
20
28
36
44
52
60
68
76
2θ (°)

(a)

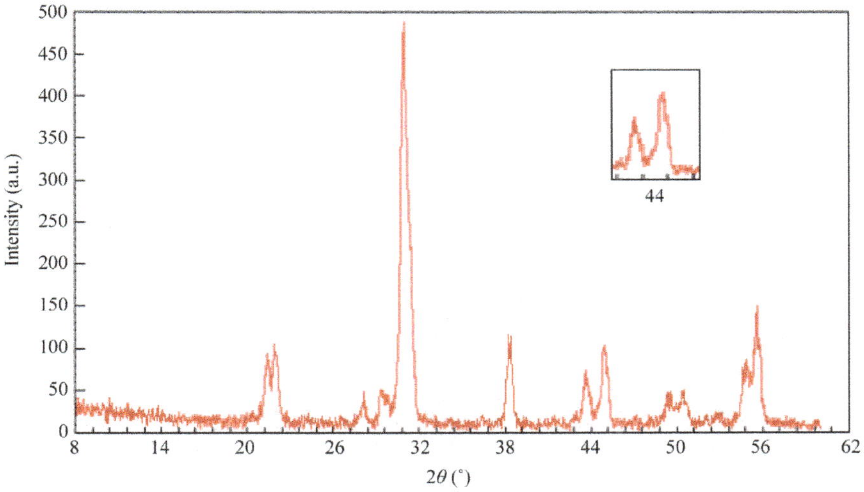
Intensity (a.u.)
500
450
400
350
300
250
200
150
100
50
0
44
8
14
20
26
32
38
44
50
56
62
2θ (°)

(b)

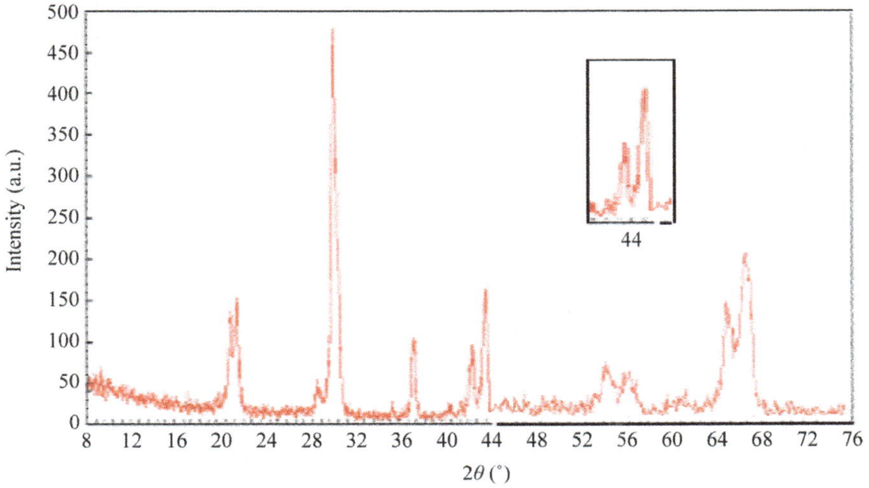
Intensity (a.u.)
500
450
400
350
300
250
200
150
100
50
0
44
8
12
16
20
24
28
32
36
40
44
48
52
56
60
64
68
72
76
2θ (°)

(c)

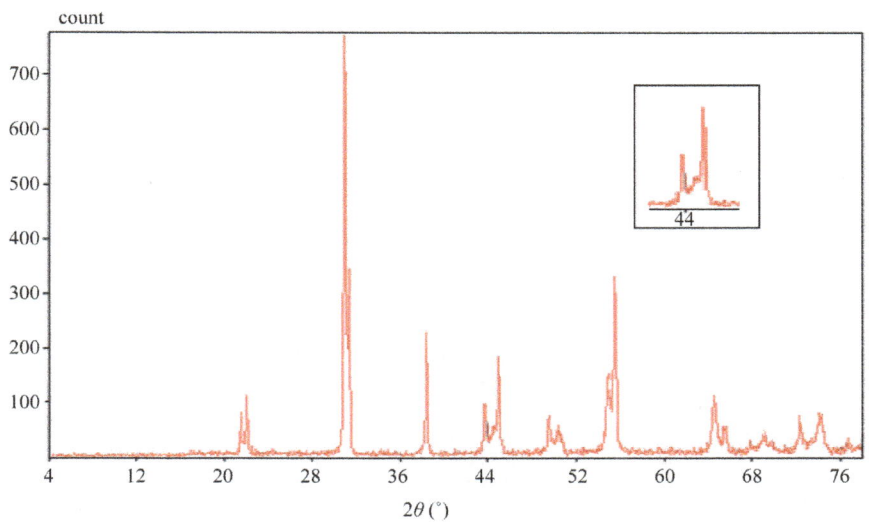

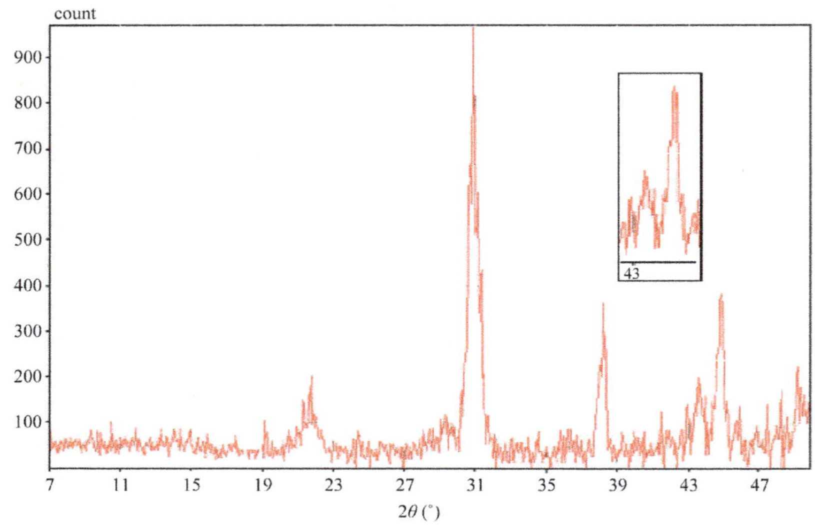
count
900
800
700
600
500
400
300
200
100
43
7
11
15
19
23
27
31
35
39
43
47
2θ (°)

(d)

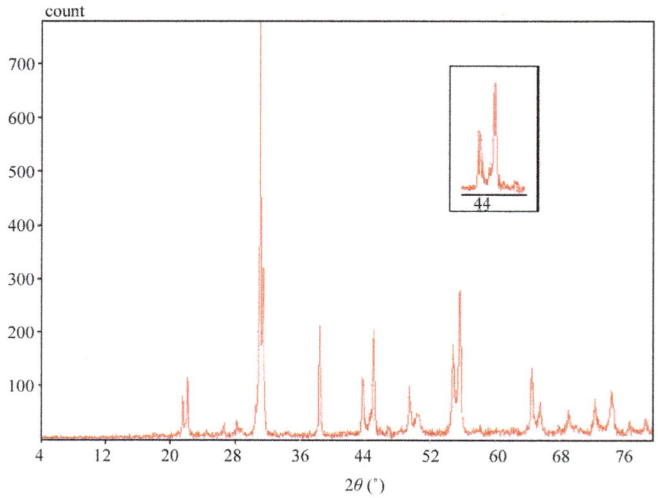
count
700
600
500
400
300
200
100
44
4
12
20
28
36
44
52
60
68
76
2θ (°)

(e)

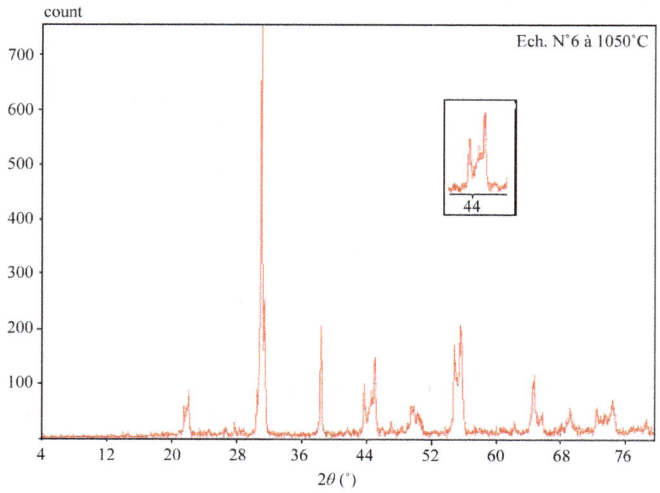
count
Ech. N°6 à 1050°C
700
600
500
400
300
200
100
44
4
12
20
28
36
44
52
60
68
76
2θ (°)

(f)

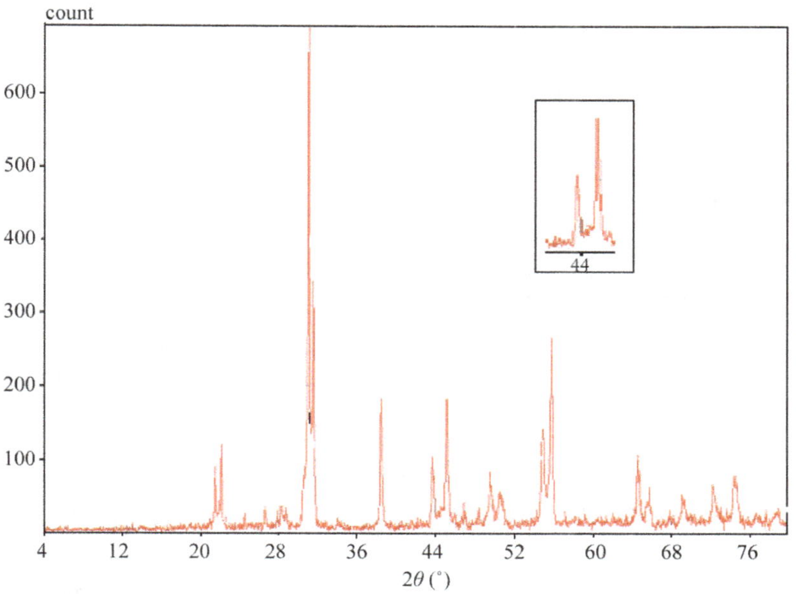

(g)

Figure 3. XRD patterns of sintered PZT-CCT ceramics with varying P_2O_5 addition: (a) 0 wt%; (b) 2 wt%; (c) 4 wt%; (d) 6 wt%; (e) 8 wt%; (f) 10 wt% and (g) 12 wt% sintered at 1050°C.

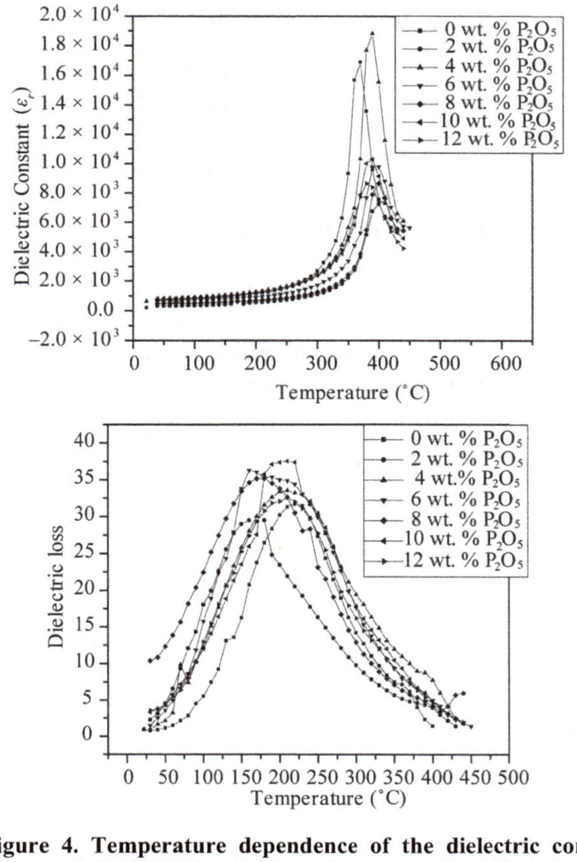

Figure 4. Temperature dependence of the dielectric constant ε_r and dielectric loss tanδ for perovskite $Pb_{0.98}Ca_{0.02}[\{(Zr_{0.52}Ti_{0.48})_{0.98}(Cr^{3+}_{0.5}, Ta^{5+}_{0.5})_{0.02}\}_{1-x}P_x]O_3$ ceramics.

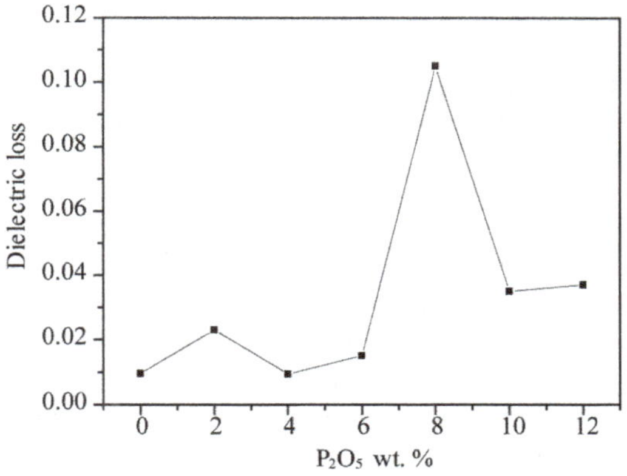

Figure 5. tanδ of ceramics sintered at 1050°C as a function of concentration of P_2O_5.

with increasing concentration of P_2O_5. Dielectric loss (tanδ) slowly deceases at first, and begins to increase when P_2O_5 content is up to 6 wt%. This indicates that the compound doped with 4 wt% of P_2O_5 is denser than other doped.

The effects of P_2O_5 on the dielectric constant and dissipation factor measured at 1 kHz are shown in **Figure 6**. The position of anomalies in the ε_r and tanδ curves, corresponding to $F_{R(HT)}$-P_C phase transition, is dependent on the P_2O_5 content. The dielectric peak temperatures (Tc) shift slightly to lower temperatures with increasing P_2O_5 content. It is well known that Tc varies with substitution. Therefore, the decrease in Tc can be explained by assumption that P_2O_5 substituted in the perovskite structure.

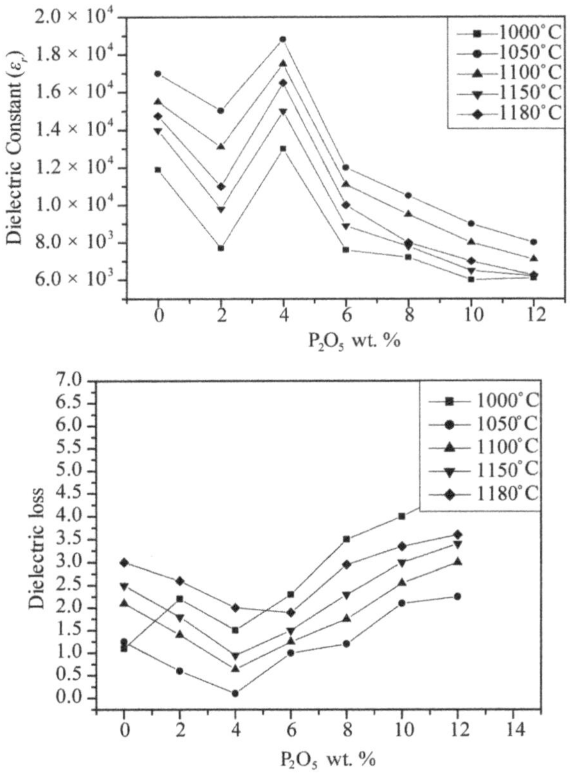

Figure 6. Evolution of the dielectric constant and dielectric loss tanδ as a function of P_2O_5 additive at different sintering temperature.

The maxima in ε_r curves become broadened when the P_2O_5 content exceeds 4.0 wt%. The broadening or diffuseness of dielectric peaks occurs mainly due to the statistical composition fluctuations because large amount of the second phases appear in PZT ceramics. A statistical treatment based on a postulated Gaussian distribution of local Curie temperature is associated with the composition fluctuations [19].

The variation of the dielectric constant as a function of frequenc from 1 kHz to 100 kHz for $Pb_{0.98}Ca_{0.02}[\{(Zr_{0.52}Ti_{0.48})_{0.98}(Cr^{3+}_{0.5}, Ta^{5+}_{0.5})_{0.02}\}_{1-x}P_x]O_3$ is illustrated in **Figure 7**. With the increase of frequency, dielectric constant varies moderately before 30 kHz, and then goes up abruptly. According to these results, we can consider the compound doped by 4 wt% P_2O_5 as a ceramic soft (depolarize easily) and can be used at high frequencies in the transmission technology. Some authors explain the maximum value of the dielectric constant at room temperature and values of low frequency by the existence of different types of polarization [20,21].

3.5. Electromechanical Properties

The electrical properties were strongly dependent on the

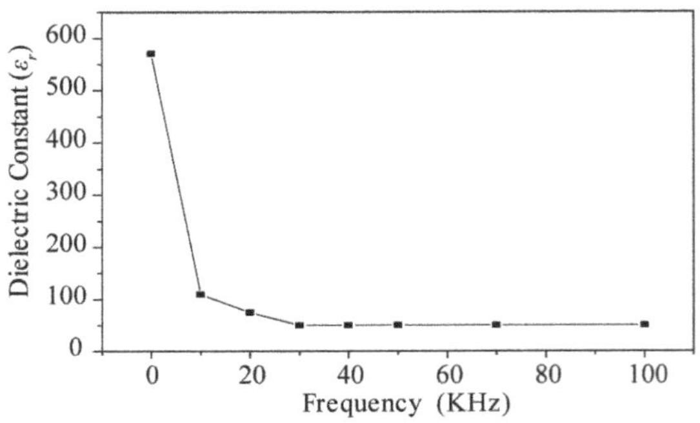

Figure 7. Evolution of the dielectric constant depending on the frequency for the sample doped by 4 wt% of P_2O_5.

phase of the specimens. The room-temperature electromechanical coupling factor (*kp*) of $Pb_{0.98}Ca_{0.02}[\{(Zr_{0.52}Ti_{0.48})_{0.98}(Cr^{3+}_{0.5}, Ta^{5+}_{0.5})_{0.02}\}_{1-z}P_z]O_3$ ceramics sintered at 1050°C as a function of P_2O_5 content is plotted in **Figure 8**. It can be observed that both of the *kp* curve possess a peak with increasing P_2O_5 content. The *kp* of P_2O_5 ceramics at z = 0 is 59.0%. With increasing P_2O_5 content (4 wt%), the *kp* of PZTMN ceramics reach their maximum values of 72.3%.

3.6. Study of Resistivity and Conductivity

The resistivity and conductivity of PZTMN sintered at 1050°C as a function of temperature are plotted in **Figures 9** and **10**, independently. It can be seen that both resistivity decrease monotonically with increasing temperature. The above variation of conductivity is different from resistivity shown in **Figure 9**. The curves of this figure show that there is a relationship between the relative change in temperature and the two electrical factors (conductivity and resistivity). More temperature increases, the resistivity of each sample decreases more and more (**Figure 9(a)**). It decreases for the sample of doped 4 wt%.

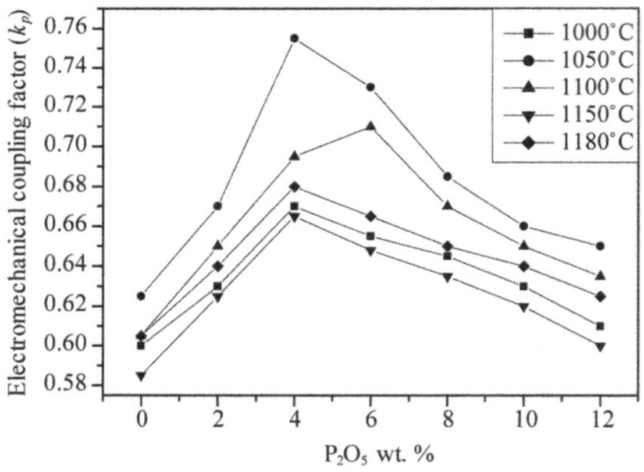

Figure 8. Evolution of k_p as a function of P_2O_5 additive at different sintering temperature.

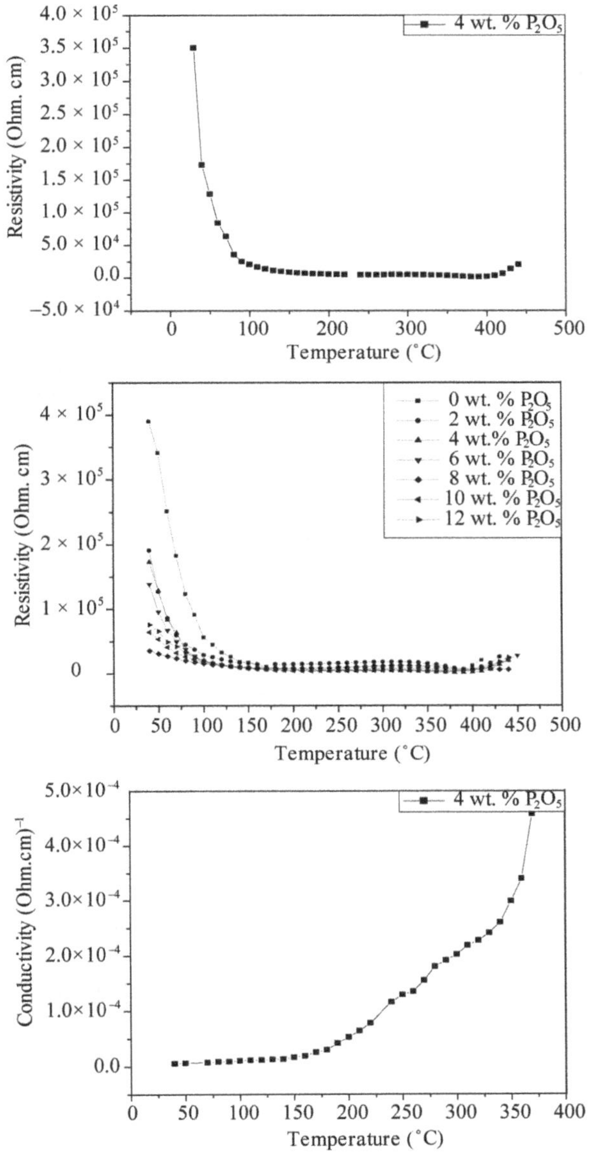

Figure 9. Variation in resistivity and conductivity sintered at 1050˚C as a function of temperature.

P_2O_5 of 35 × 10^{+4} (Ω·cm) when T = 50˚C until it reaches the value 0.09 × 10^{+4} (Ω·cm) to 350˚C (**Figure 9 (b)**). This is due to the fact that high-temperature thermal energy may be sufficient to break some connections or ionic and covalent causes some mobility of ions. However, the electrical conductivity varies in the opposite direction of resistivity; it grows with the increase of temperature (**Figure 9(c)**). It can reach a maximum value of 7.5 × 10^{-4} $(\Omega \cdot cm)^{-1}$ at a temperature of 350˚C.

4. Conclusions

The compounds of the solution solid zirconate-titanate lead, noted PZT, general formula $Pb_{0.98}Ca_{0.02}[\{(Zr_{0.52}Ti_{0.48})_{0.98}(Cr^{3+}_{0.5}, Ta^{5+}_{0.5})_{0.02}\}_{1-z}P_z]O_3$ as z vary from 0.00 to 0.12 by setup of 0.02, it has been prepared from a mixture of oxides by the method ceramics. The effect of sintering temperature on density and porosity was studied to achieve the optimum sintering temperature corresponding to the maximum density and minimum value of porosity, because this temperature (1050˚C) corresponds to a better quality product. Low-temperature densification of PZT can be achieved by the incorporation of a small amount of P_2O_5 as a sintering aid without sacrificing the dielectric properties.

The preferable sintering temperature was about 1050˚C and presented the maximum bulk density of 7.62 g/cm^3. The study of dielectric properties and piezoelectric of this compound in the solid as a function of temperature allows us to have high values of dielectric constant and planar electromechanical coupling factor at z = 0.04. The results of measurement of dielectric and piezoelectric properties of this material are reported in the **Table 1**.

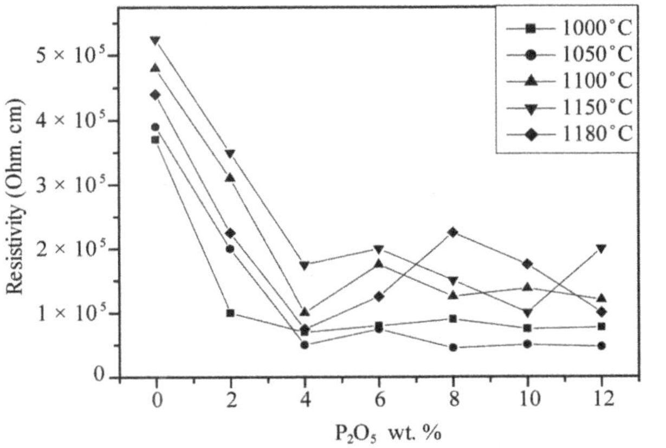

Figure 10. Evolution of the resistivity as a function of P_2O_5 additive at different sintering temperature.

Table 1. Dielectric and piezoelectric properties of PZTMN doped with 4 wt% of P_2O_5 and sintered at 1050˚C.

properties	d (g/cm^3)	ε_r	tanδ	ρ (Ω·cm)	k_p (%)
values	7.6	18800	0.0094	0.09 × 10^4	0.73

REFERENCES

[1] G. H. Haertling, "Ferroelectric Ceramics: History and Technology," *Journal American Ceramic Society*, Vol. 82, No. 4, 1999, pp. 797-818. doi:10.1111/j.1151-2916.1999.tb01840.x

[2] K. Uchino, "Ferroelectric Device," Marcel Dekker, New York, 2000.

[3] S. Y. Cheng, S. L. Fu, C. C. Wei and G. M. Ke, "The Pro-

perties Low-Temperature Fixed Piezoelectric Ceramics," *Journal of Materials Science*, Vol. 21, No. 2, 1986, pp. 571-576. doi:10.1007/BF01145525

[4] H. G. Lee, J. H. Choi and E. S. Kim, " Low-Temperature Sintering and Electrical Properties of $(1-x)Pb(Zr_{0.5}Ti_{0.5})O_3$-$xPb(Cu_{0.33}Nb_{0.67})O_3$ Ceramics," *Journal of Electroceramics*, Vol. 17, No. 2-4, 2006, pp. 1035-1040. doi:10.1007/s10832-006-0384-1

[5] R. Mazumder, A. Sen and H. S. Maiti, "Impedance and Piezoelectric Constants of Phosphorous-Incorporated $Pb(Zr_{0.52}Ti_{0.48})O_3$ Ceramics," *Materials Letters*, Vol. 58, No. 25, 2004, pp. 3201-3205. doi:10.1016/j.matlet.2004.06.011

[6] G. Robert, M. D. Maeder, D. Damjanovic and N. Setter, "Synthesis of Lead Nickel-Niobate Zirconate Titanate Solid Solutions by a B-Site Precursor," *Journal American Ceramic Society*," Vol. 84, No. 12, 2001, pp. 2863-2868. doi:10.1111/j.1151-2916.2001.tb01107.x

[7] L. Pdungsap, S. Boonyeun, P. Winotai, N. Udomkan and P. Limsuwan, "Effects of Gd^{3+} Doping on Structural and Dielectric Properties of PZT (Zr:Ti = 52:48) Piezoceramics," *The European Physical Journal B*, Vol. 48, No. 3, 2005, pp. 367-372. doi:10.1140/epjb/e2005-00407-9

[8] S. J. Yoon, A. Joshi and K. Uchino, "Effect of Additives on the Electromechanical Properties of $Pb(Zr,Ti)O_3$-$Pb(Y_{2/3}W_{1/3})O_3$ Ceramics," *Journal of the American Ceramic Society*, Vol. 80, No. 4, 2005, pp. 1035-1039. doi:10.1111/j.1151-2916.1997.tb02942.x

[9] G. A. Smolenskii and A. I. Agranovskaya, "Dielectric Polarization of a Number of Complex Compounds," *Soviet Physics Solid State*, Vol. 1, No. 10, 1960, pp. 1429-1437.

[10] F. Kulcsar, "Electromechanical Properties of Lead Titanate Zirconate Ceramics Modified with Tungsten and Thorium," *Journal American Ceramic Society*, Vol. 48, No. 1, 1965, pp. 48-54. doi:10.1111/j.1151-2916.1965.tb11796.x

[11] N. Abdessalem and A. Boutarfaia, "Effect of Composition on the Electromechanical Properties of $Pb[Zr_xTi_{(0.9-x)}$-$(Cr_{1/5}, Zn_{1/5}, Sb_{3/5})_{0.1}]O_3$ Ceramics," *Ceramics International*, Vol. 33, No. 2, 2007, pp. 293-296. doi:10.1016/j.ceramint.2005.08.008

[12] J. S. Kim and K. H. Yoon, "Physical and Electrical Properties of MnO_2-Doped $Pb(Zr_xTi_{1-x})O_3$ Ceramics," *Journal of Materials Science*, Vol. 29, No. 3, 1994, pp. 809-815. doi:10.1007/BF00445997

[13] P. Duran, J. F. Fernandez and C. Moure, "Effect of MnO Additions on the Sintering and Piezoelectric Properties of Samarium-Modified Lead Titanate Ceramics," *Journal of Materials Science Letters*, Vol. 10, No. 15, 1991, pp. 917-919. doi:10.1007/BF00724781

[14] Z. He, J. Ma, R. Z. Hang, "Investigation on the Microstructure and Ferroelectric Properties of Porous PZT Ceramics," *Ceramics International*, Vol. 30, No. 7, 2004, pp. 1353-1356. doi:10.1016/j.ceramint.2003.12.108

[15] R. Sumang and T. Bongkarn, "The Effect of Excess PbO on Crystal Structure, Microstructure, Phase Transition and Dielectric Properties of $(Pb_{0.75}\,Sr_{0.25})TiO_3$ Ceramics," *Taylor & Francis Group LLC*, Vol. 403, No. 1, 2010 , pp. 82-90. doi:10.1080/00150191003748949

[16] P. Goel, S. Sharma, K. L. Yadav and A. R. James, "Structural and Dielectric Properties of Phosphorous-Doped PLZT Ceramics," *Pramanas*, Vol. 65, No. 6, 2005, pp. 1127-1132. doi:10.1007/BF02705288

[17] A. K. Saha, D. Kumar, O. Parkash, A. Sen and H. S. Maiti, "Effect of Phosphorus Addition on the Sintering and Dielectric Properties of $Pb(Zr_{0.52}Ti_{0.48})O_3$," *Materials Research Bulletin*, Vol. 38, No. 7, 2003, pp. 1165-1174. doi:10.1016/S0025-5408(03)00112-0

[18] O. Ohtaka, R. Von Der Mühll and J. Ravez, "Low-Temperature Sintering of $Pb(Zr,Ti)O_3$ Ceramics with the Aid of Oxyfluoride Additive: X-Ray Diffraction and Dielectric Studies," *Journal American Ceramic Society*, Vol. 78, No. 3, 1995, pp. 805-808. doi:10.1111/j.1151-2916.1995.tb08251.x

[19] W. Heywang, "Ferroelektrizität in Perowskitischen Systemen und Ihre Technischen Anwendungen," *Zeitschrif Angewandte Physik*, Vol. 19, 1965, pp. 473-481.

[20] S. Babu, D. Singh and A. Govindan, "Electrical Properties of Calcium Modified PZT System," *International Journal of Computer Science et Technologie*, Vol. 2, No. 1, 2011, pp. 128-131.

[21] IEEE Standard on Piezoelectricity, IEEE Standard 176-1978, Institute of Electrical and Electronic Engineers, New York, 1978.

Optical and Dielectric Studies on L-Valinium Picrate Single Crystal

P. Koteeswari[1], P. Mani[1], S. Suresh[2*]

[1]Department of Physics, Hindustan Institute of Technology, Padur, India; [2]Department of Physics, Loyola College, Chennai, India.
Email: *sureshsagadevan@yahoo.co.in

ABSTRACT

Single crystals of L-Valinium picrate were grown from aqueous solution by slow evaporation technique. Single crystal X-ray diffraction analysis reveals that the crystal belongs to monoclinic system. The optical transmission study reveals the transparency of the crystal in the entire visible region and the cut off wave length has been found to be 470 nm. The optical band gap is found to be 2.55 eV. The transmittance of L-Valinium picrate crystal has been used to calculate the refractive index (n), the extinction coefficient (K) and both the real (ε_r) and imaginary (ε_i) components of the dielectric constant as functions of wavelength. Low dielectric loss at high frequency region is indicative of enhanced optical quality with lesser defects. Photoconductivity measurements carried out on the grown crystal reveal the negative photoconducting nature.

Keywords: Solution Growth; Single Crystal XRD; Optical Transmission; Dielectric Studies; Photoconductivity Studies

1. Introduction

Crystal growth is a frontier area of science and technology, which plays a major role in the technology of photonics. The field of nonlinear optics has been in the hands of materials scientists for the past five decades for which organic materials are attracting a great deal of attention, as they have large optical susceptibilities, inherent ultrafast response time and good optical properties as compared to that of inorganic crystals. Research in organic and inorganic functionalized nonlinear optical materials plays a crucial role because of their molecular interactions, bond strength, high molecular polarizability, easy incorporation of ions in the lattice, etc. [1] and [2]. In the recent years there has been a growing interest in nonlinear optical materials due to their effective usage in the field of electro-optical devices, data storage technology and optical signal processing [3]. However, semi-organic single crystals are attracting great attention in the field of nonlinear optics because of their high optical nonlinearity, chemical flexibility of ions, thermal stability and excellent transmittance in the UV—visible region [4,5]. In the present investigation, we report optical, dielectric, and Photoconductivity properties of L-Valinium picrate single crystals.

*Corresponding author.

2. Experimental

2.1. Crystal Growth

Single crystals of L-Valinium picrate were grown, from aqueous solution by slow evaporation technique. The solution was prepared by dissolving equimolar amounts of picric acid and L-Valine in deionized water and stirred well to yield a homogenous mixture of solution. A saturated solution was prepared and the solution was filtered. The filtered solution was taken in a beaker which was hermetically sealed to avoid the evaporation of the solvent.

3. Results and Discussion

3.1. Single Crystal X-Ray Diffraction Studies

Single crystal X-ray diffraction analysis for the grown crystals has been carried out to identify the cell parameters using an ENRAF NONIUS CAD 4 automatic X-ray Diffractometer. The title crystal belongs to monoclinic crystal system and the lattice parameters are a = 9.96 Å; b = 6.23 Å; c = 12.64 Å, β = 110.40° and agree well with the reported literature [6].

3.2. Optical Studies

The optical transmittance spectrum of L-Valinium picrate was recorded in the range 300 - 1100 nm with a crystal of thickness 2 mm. The **Figure 1** shows that the crystal has a wide transmission of above 70% in the entire range without any absorption peak. The lower cutoff wavelength of L-Valinium picrate is 470 nm. The crystal has good optical transmission in the visible region. The transparency in the visible region for this crystal suggests its suitability for second harmonic generation.

The measured transmittance (*T*) was used to calculate the absorption coefficient (α) using the formula

$$\alpha = \frac{2.3026\log\left(\frac{1}{T}\right)}{t} \quad (1)$$

where *t* is the thickness of the sample. Optical band gap (E_g) was evaluated from the transmission spectrum and optical absorption coefficient (α) near the absorption edge is given by [7].

$$\alpha = \frac{A\left(h\nu - E_g\right)^{\frac{1}{2}}}{h\nu} \quad (2)$$

where *A* is a constant, E_g the optical band gap, *h* the Planck constant and n the frequency of the incident photons. The band gap of L-Valinium picrate crystal was estimated by plotting $(\alpha h\nu)^2$ versus $h\nu$ as shown in **Figure 2**. From the figure, the value of band gap was found to be 2.55 eV.

Extinction coefficient (*K*) can be obtained from the following equation:

$$K = \frac{\lambda\alpha}{4\pi} \quad (3)$$

The transmittance *(T)* is given by

$$T = \frac{(1-R)^2\exp(-\alpha t)}{1-R^2\exp(-2\alpha t)} \quad (4)$$

Reflectance (*R*) in terms of absorption coefficient can be obtained from the above equation. Hence, Equation (5)

Refractive index (*n*) can be determined from reflectance data using the following equation;

$$n = -(R+1) \pm 2\frac{\sqrt{R}}{(R-1)} \quad (6)$$

The refractive index (*n*) is 1.41 at λ = 1100 nm. From the optical constants, electric susceptibility (χ_C) can be calculated according to the following relation [8]

$$\varepsilon_r = \varepsilon_0 + 4\pi\chi_C = n^2 - k^2 \quad (7)$$

Hence,

$$\chi_C = \frac{n^2 - k^2 - \varepsilon_0}{4\pi} \quad (8)$$

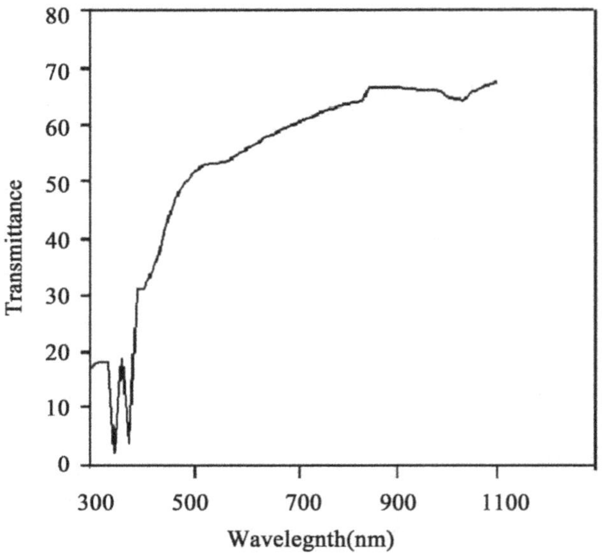

Figure 1. Transmittance spectrum of LVP.

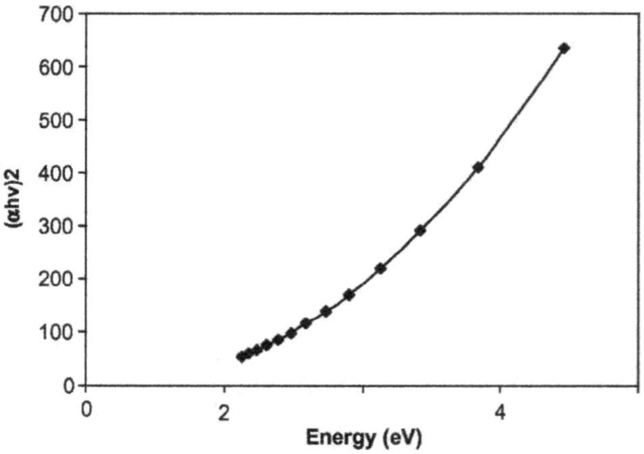

Figure 2. $(\alpha h\nu)^2$ **vs photon energy (*hv*).**

where ε_0 is the dielectric constant. The value of electric susceptibility χ_C is 0.163 at λ = 1100 nm. The real part dielectric constant ε_r and imaginary part dielectric constant ε_i can be calculated from the following relations [9]

$$\varepsilon_r = n^2 - k^2 \;\&\; \varepsilon_i = 2nk \quad (9)$$

The value of real ε_r and ε_i imaginary dielectric constants at λ = 1100 nm are 1.352 and 5.612 × 10^{-5}, respectively.

3.3. Dielectric Studies

The dielectric constant and the dielectric loss of the L-

$$R = \frac{\exp(-\alpha t) \pm \sqrt{\exp(-\alpha t)T - \exp(-3\alpha t)T + \exp(-2\alpha t)T^2}}{\exp(-\alpha t) + \exp(-2\alpha t)T} \quad (5)$$

Valinium picrate crystals were studied at different temperatures using a HIOKI 3532 LCR HITESTER in the frequency region from 50 Hz to 5 MHz. The dielectric constant and dielectric loss have been calculate using Equations (10) and (11)

$$\varepsilon = \frac{Cd}{\varepsilon_0 A} \tag{10}$$

$$\varepsilon = \varepsilon \tan \delta \tag{11}$$

where d is the thickness of the sample; A is the area of the sample. **Figure 3** shows the plot of dielectric constant versus log frequency. The high value of dielectric constant at low frequencies may be due to the presence of all the four polarizations, namely, space charge, orientation, electronic and ionic polarization and its low value at higher frequencies may be due to the loss of significance of these polarizations gradually [10]. From the plot, it is also observed that dielectric constant increases with an increase in temperature. The variation of dielectric loss with frequency is shown in **Figure 4**. The characteristics

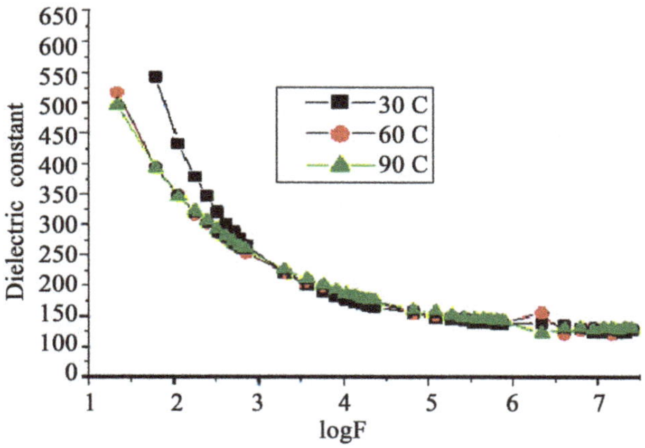

Figure 3. Plot of dielectric constant vs. log *f*.

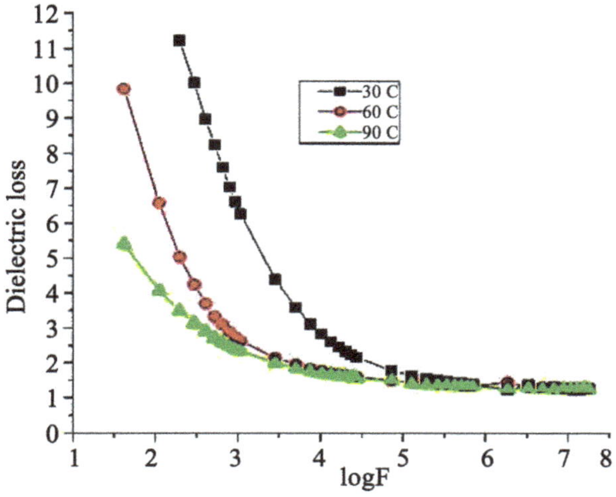

Figure 4. Plot of dielectric loss vs log *f*.

of low dielectric loss with high frequency for the sample suggest that it possesses enhanced optical quality with lesser defects and this parameter is imperative for nonlinear optical applications [11].

3.4. Photoconductivity Studies

Photoconductivity measurements are carried out on a cut and polished sample of the grown single crystal by fixing it onto a microscope slide. The sample is connected in series with a DC power supply and KEITHLEY 485 Picoammeter. The sample is covered with a black cloth and the voltage applied is increased from 0 to 300 V in steps of 20 V and the dark current is recorded. The photocurrentis recorded for the same values of the applied voltage. Field dependence of dark and photo currents of grown crystal is shown in **Figure 5**. The photocurrent is found to be lessthan the dark current at every applied electric field. This phenomenon is known as negative photoconductivity. Generally, this may be attributed to the loss of water molecules in the crystal [12]. However, the negative photoconductivity in this case may be due to the reduction in the number of charge carriers or their lifetime in the presence of radiation [13]. Decrease in lifetime with illumination could be due to the trapping process and increase in carrier velocity according to the relation

$$\tau = \left(\nu s N\right)^{-1} \tag{12}$$

where ν is the thermal velocity of the carriers, s is the capture cross-section of the recombination centers and N is the carrier concentration. As intense light falls on the sample, the lifetime decreases. In the Stockmann model, a two level scheme is proposed to explain negative pho-

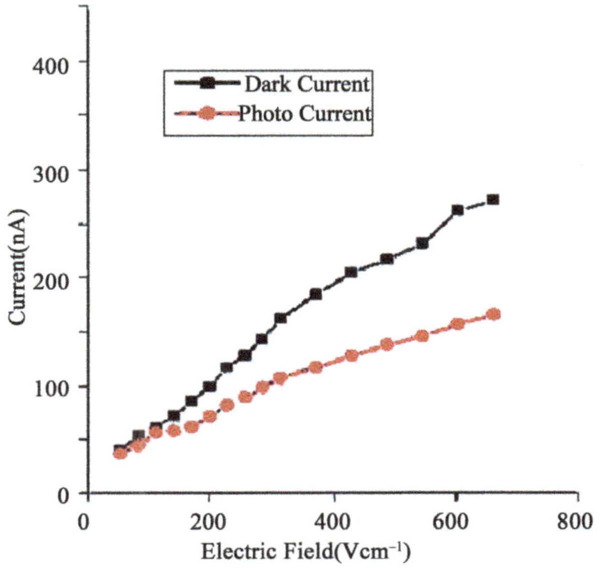

Figure 5. Field dependent photoconductivity of grown single crystal.

toconductivity [14]. As a result, the recombination of electrons and holes take place resulting in decrease in the number of mobile charge carriers, giving rise to negative photoconductivity.

4. Conclusion

Single crystals of L-Valinium picrate were grown by slow evaporation technique. Single-crystal XRD analysis confirmed that the crystals belong to monoclinic system. Optical band gap (E_g), absorption coefficient (α), extinction coefficient (K), refractive index (n), electric susceptibility χ_C and dielectric constants were calculated as a function of wavelength. The characteristics of low dielectric loss for the sample suggest that it possesses enhanced optical quality with lesser defects and this parameter is of vital significance for nonlinear optical applications. Photoconductivity investigations reveal the negative photoconducting nature of the title material.

REFERENCES

[1] C. Zhang, Z. Li, H. J. Cong, J. Y. Wang, H. J. Zhang and R. I. Boughton, "Crystal Growth and Thermal Properties of Single Crystal Monoclinic NdCOB ($NdCa_4O(BO_3)_3$)," *Journal of Alloys and Compounds*, Vol. 507, No. 2, 2010, pp. 335-340. doi:10.1016/j.jallcom.2010.07.174

[2] M. J. Rosker, P. Cunningham, M. D. Ewbank, H. O. Marcy, F. R. Vachss, L. F. Warren, R. Gappinger and R. Borwick, "Salt-Based Approach for Frequency Conversion Materials," *Pure and Applied Optics*, Vol. 5, No. 5, 1996, p. 667. doi:10.1088/0963-9659/5/5/020

[3] S. K. Gao, W. J. Chen, G. M. Wang and J. Z. Chen, "Synthesis, Crystal Growth and Characterization of Organic NLO Material: *N*-(4-Nitrophenyl)-N-Methyl-2-Aminoacetonitrile (NPAN)," *Journal of Crystal Growth*, Vol. 297, No. 2, 2006, pp. 361-365. doi:10.1016/j.jcrysgro.2006.09.047

[4] A. P. Jeyakumari, J. Ramajothi and S. Dhanuskodi, "Structural and Microhardness Studies of a NLO Material—Bisthiourea Cadmium Chloride," *Journal of Crystal Growth*, Vol. 269, No. 2-4, 2004, pp. 558-564. doi:10.1016/j.jcrysgro.2004.05.059

[5] H. Q. Sun, D. R. Yuan, X. Q. Wang, X. F. Cheng, C. R. Gong, M. Zhou, H. Y. Xu, X. C. Wei, C. N. Luan, D. Y. Pan, Z. F. Li and X. Z. Shi, "A Novel Metal-Organic Coordination Complex Crystal: Tri-Allylthiourea Zinc Chloride (ATZC)," *Crystal Research and Technology*, Vol. 40, No. 9, 2005, pp. 882-886.

[6] K. Anitha, B. Sridhar and R. K. Rajaram, "L-Valinium Picrate," *Acta Crystallographica*, Vol. E60, 2004, pp. o1530-o1532. doi:10.1107/S160053680401949X

[7] A. Ashour, N. El-Kadry and S. A. Mahmoud, "On the Electrical and Optical Properties of CdS Films Thermally Deposited by a Modified Source," *Thin Solid Films*, Vol. 269, No. 1-2, 1995, pp. 117-120. doi:10.1016/0040-6090(95)06868-6

[8] V. Gupta and A. Mansingh, "Influence of Postdeposition Annealing on the Structural and Optical Properties of Sputtered Zinc Oxide Film," *Journal of Applied Physics*, Vol. 80, No. 2, 1996, pp. 1063-1073. doi:10.1063/1.362842

[9] M. A. Gaffar, A. Abu El-Fadl and S. Bin Anooz, "Influence of Strontiumdoping on the Indirectbandgap and Opticalconstants of Ammoniumzincchloridecrystals," *Physica B: Condensed Matter*, Vol. 327, No. 1, 2003, pp. 43-54. doi:10.1016/S0921-4526(02)01700-3

[10] C. P. Smyth, "Dielectric Behavior and Structure," McGraw-Hill, New York, 1965.

[11] C. Balarew and R. Duhlew, "Application of the Hard and Soft Acids and Bases Concept to Explain Ligand Coordination in Double Salt Structures," *Journal of Solid State Chemistry*, Vol. 55, No. 1, 1984, pp. 1-6. doi:10.1016/0022-4596(84)90240-8

[12] R. H. Bube, "Photoconductivity of Solids," Wiley, New York, 1981.

[13] I. M. Ashraf, H. A. Elshaik and A. M. Badr, "Photoconductivity in Tl4S3 Layered Single Crystals," *Crystal Research and Technology*, Vol. 39, No. 1, 2004, pp. 63-70. doi:10.1002/crat.200310150

[14] V. N. Joshi, "Photoconductivity," Marcel Dekker, New York, 1990.

32

Comparative Study of the One Dimensional Dielectric and Metallic Photonic Crystals

Arafa H. Aly[1,2], Mohamed Ismaeel[3], Ehab Abdel-Rahman[2,3]
[1]Department of Physics, Faculty of Sciences, Beni-Suef University, Beni-Suef, Egypt
[2]YJ-STRC, The American University in Cairo, New Cairo, Egypt
[3]Department of Physics, The American University in Cairo, New Cairo, Egypt
Email: arafaaly@aucegypt.edu

ABSTRACT

The optical transmission properties of two types of photonic crystals have been analyzed by using the transfer matrix method. The first one is the dielectric photonic crystal (DPC), and the second is the metallic photonic crystal (MPC). We found the dielectric and metallic photonic crystals have different transmission spectra. The effect of the most parameters on the transmission spectra of the dielectric and metallic photonic crystals has been studied.

Keywords: Transmission; Metallic Photonic Crystals; Dielectric Photonic Crystals; Photonic Band Gap

1. Introduction

Photonic crystals (PCs) are macroscopic media which arranged periodically with different refractive indices and their periodicities are in the range of the incident light [1]. In such structures the permittivity is a periodic function in space. In this case, the dielectric permittivity function repeats itself in one dimension (1D) the structure called one dimensional photonic crystal (1D-PC), if it repeats itself in 2D or 3D the structure called 2D or 3D PC. The one dimensional photonic crystal (**Figure 1**) is a multilayered media. It is worthy to mention that, the propagation of photons in the PCs is similar to the propagation of electrons in the semiconductor crystals, where the effect of the periodic dielectric function on the propagating photon in PCs is much like the effect of the periodic potential function on the propagating electron in semiconductor crystal. Consequently, a photonic band is created in PCs similar to the electronic bad gap in semiconductor crystal [2].

On the other hand, when electromagnetic waves (EM) incident on the PCs Bloch states create in the crystal, if the Bloch wave falls in the so called forbidden bands (photonic band gap) such a wave is evanescent and can't propagate in the crystal. Thus the light energy is expected to be totally reflected, and the crystal acts as a high reflectance reflector for the incident wave. The photonic band gap of the photonic crystal makes us able to control the light even the spontaneous emission [3]. In this paper, we are going to do comparative study between the 1D-DPCs and -MPCs pointing to the general applications of each kind according to its characteristics.

2. Analysis

2.1. Dielectric Photonic Crystals (DPCs)

In the last decades dielectric photonic crystals have attracted much research interest due to their various applications for example, optical filters, waveguides, and optical fibres [4-7]. In this section, we restrict our communications on the characteristics of the 1D-DPC showing its various applications. The reflection of the EM waves through DPCs exhibit resonance reflection very much like the diffraction of x-rays by crystal lattice planes, therefore it's called Bragg reflector.

We have designed 1D-DPC composed of a low index material (Cryolite = 1.34) and a high index material (Silicon = 3.4) stacked alternatively on a glass substrate. The number of periods, lattice constant, effective refractive index, and the filling factors of the low and the high index materials are taken to be 10, 250, 2.389, 0.6, and 0.4 nm, respectively. The filling factor (*f*) of a material in a 1D-PC can be given by [8];

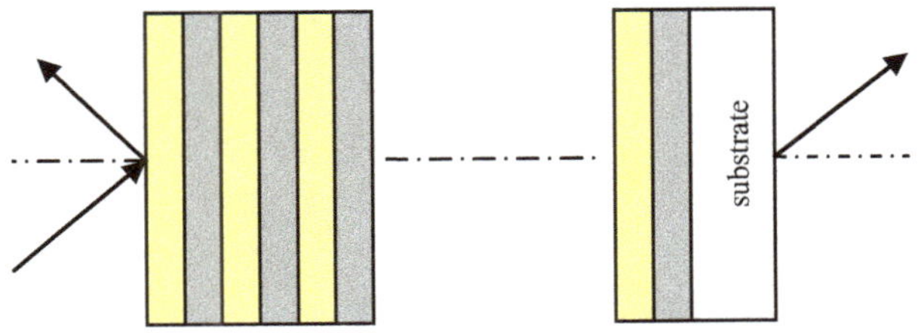

Figure 1. Schematic diagram shows a one dimensional photonic crystal.

$$f = d/\Lambda \quad (1)$$

where d and Λ are the material layer thickness in the unit cell of the PC and the lattice constant (spatially periodic constant), respectively. The transmission spectra of the DPC are displayed in **Figure 2**. The figure shows that the DPC presents a transmission band for the low frequencies (long wavelengths), and the first band gap is associated with the Bragg condition [2];

$$\lambda = 2n\Lambda \quad (2)$$

where λ is the centre wavelength of the first band gap, Λ is the lattice constant, and n is the effective refractive index. The effective refractive index can be given by [2];

$$n = \sqrt{\varepsilon_{eff}} \text{ and } \varepsilon_{eff} = f_l\varepsilon_l + f_h\varepsilon_h \quad (3)$$

The term $f_l\varepsilon_l$ represents the filling factor of the low index material multiplied by its permittivity, and the term $f_h\varepsilon_h$ represents the filling factor of the high index material multiplied by its permittivity. The centre wavelength of the first stop band determined by Bragg condition is approximately equal to 1194 nm which is consistent with the value deduced from the transfer matrix method that shown in **Figure 2**. The photonic band gap of the DPC can be tuneable by varying n or Λ, and the band gap can be shift to longer (shorter) wavelengths with increasing (decreasing) the lattice constant or the effective index as Bragg condition predict.

It's known that, in a specified frequency range the band gap width depends on the difference in the refractive indices of the two constituent materials (Δn). So to show effect of Δn we have designed DPC from Cryolite/Silicon dioxide ($n = 1.46$). The transmission spectra are displayed in **Figure 3**, it is obvious that, when the number of periods is equal to ten, Δn of Cryolite/Silicon dioxide is small not enough to open deep gap, but when the number of periods become equal to fifty, a narrow deep gap can be open. We have observed that the number of periods doesn't effect on the position or the width of the band gap. But increasing number of periods enhances the reflectivity of the bad gap and makes the band gap edges steeper.

In **Figure 3**, the number of the resonance transmission peaks (RTPs) for the PC of fifty periods is larger than number of RTPs for the PC of ten periods. We have noticed that, the RTPs of the DPC is directly proportional to the number of periods, and the RTPs have become closer to each other and sharper as the wavelength decreases and vice versa.

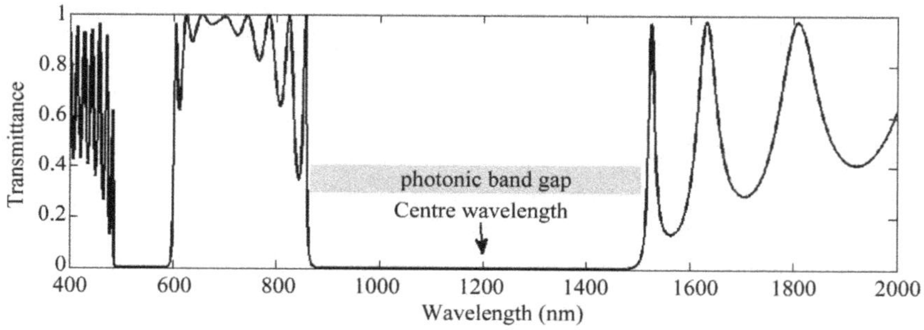

Figure 2. Calculated transmission spectra of a dielectric photonic crystal with $n_1 = 1.34$, $n_2 = 3.4$, $d_1 = 150$ nm, $d_2 = 100$ nm, number of periods = 10, and $\theta = 0°$.

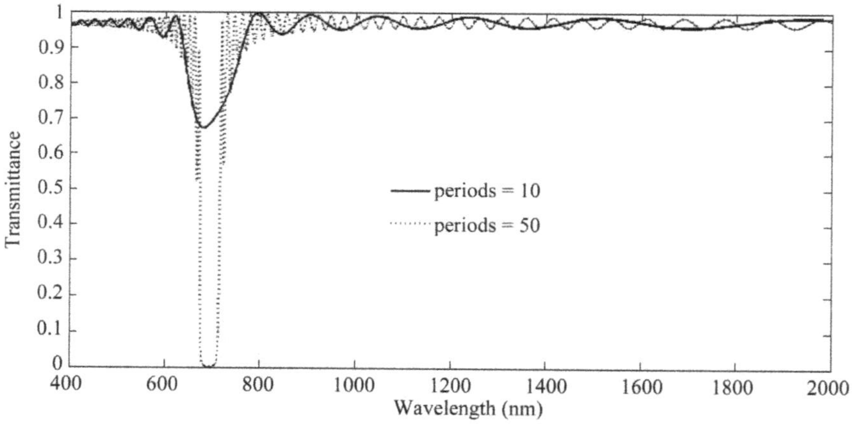

Figure 3. Calculated transmission spectra of a dielectric PC with $n_1 = 1.34$, $n_2 = 1.46$, $d_1 = 150$ nm, $d_2 = 100$ nm, period = 10, and $\theta = 0°$.

On the other hand, in order to study the filling factor effect, we have designed two DPCs both composed of Cryolite/Silicon but it is differ in the filling factors of the two constituent materials. The transmission spectra of the two PCs are shown in **Figure 4** by increasing the filling factor of the high index material (HIMF) red shift of the band gap occurs, this due to the increase of the effective refractive index of the dielectric stack. It is observed that, the band gap width slightly decreases with increasing the HIMF.

The incidence angle effect on the transmittance of the DPC for *S*- and *P*-polarized waves is displayed in the **Figures 5** and **6**, respectively. When the incidence angle of the electromagnetic waves increase blue shift of the band gaps of the *S*- and *P*-polarized waves occur. The band gap of the *P*-polarized wave shrinks due to Brewster effect at the interface between low and high index layers [9]. But the band gap of the *S*-polarized wave increases slightly. The forbidden gaps for the two polarizations not coincide due to the loss of the degeneracy. From the **Figures 5** and **6**, we have observed that, the *P*-Polarized wave more sensitive to the change of angle than *S*-polarized wave. 1D-DPCs structure has many applications such as filters [10], omnidirectional reflectors [11-19], polarisers [20-25], antireflection coatings, distributed Bragg reflectors for vertical-Cavity surface emitting lasers (VCSEL), and wavelength division multiplexers/demultiplexers on the basis of fibre Bragg

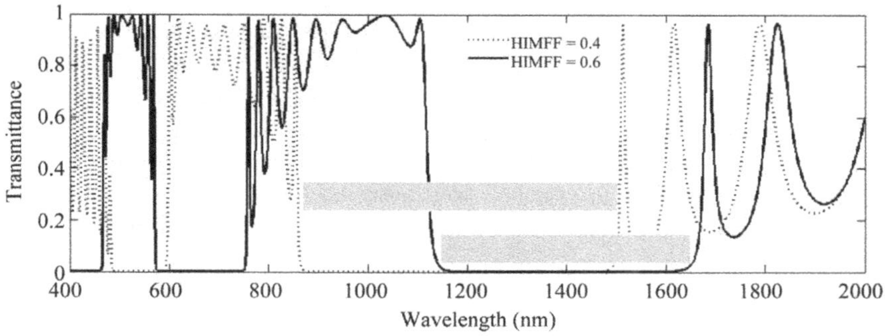

Figure 4. Calculated transmission spectra of two DPCs with n_1 = 1.34, n_2 =3.4, Λ = 250 nm, θ = 0° and different only in the filling factors of the constituent materials.

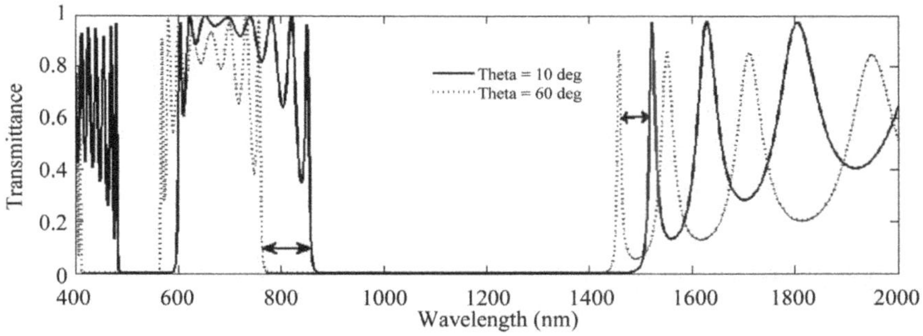

Figure 5. Calculated transmission spectra of DPC for *S*-polarized wave at different incident angles; the DPC with n_1 = 1.34, n_2 = 3.3, d_1 = 150 nm, d_2 = 100 nm, and number of periods = 10.

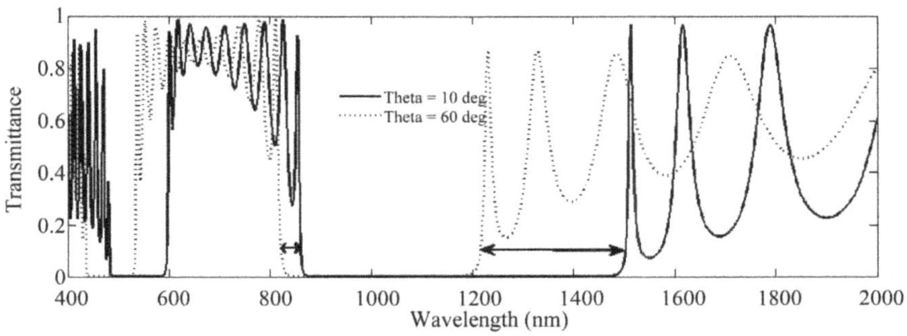

Figure 6. Calculated transmission spectra of DPC for *P*-polarized wave at different incident angles; the DPC with n_1 = 1.34, n_2 = 3.3, d_1 = 150 nm, d_2 = 100 nm, and number of periods = 10.

grating (FBG) [3].

2.2. Metallic Photonic Crystals (MPCs)

We have shown in the previous section that, in order to achieve photonic band gap, the system must has high contrast in the refractive index with negligible the absorption of light. These conditions have restricted the set of dielectrics that exhibit a photonic band gap. One suggestion is to use metals which have large value of dielectric permittivity rather than dielectrics. Accordingly a fewer numbers of periods would be enough to achieve photonic band gap [26,27]. We have designed 1D-MPC composed of Cryolite/Silver with 10 periods, lattice constant = 210 nm, and the filling factors of Sillver and Cryolite are 0.0476, and 0.9634, respectively. The dispersion has been taken into account by using Drude model and then we can alculate the refractive index of metals. The transmission spectra of the MPC are displayed in **Figure 7**. As shown in the figure, the MPC present like the DPC alternation of transmission bands and band gaps with the same progressive decrease of the transmission contrast. However for low frequency region starting from zero frequency of the spectrum, MPC exhibit plasmonic band gap. This plasmonic gap extends from 309.3 THz (970 nm) to zero frequency. This band gap not originated from the structure but from the bulk silver properties. In addition to the plasmonic band gap, the MPC exhibits structural band gap extends from 420 to 570 nm. The structural band gap follows the first transmission band that extends from 570 to 970 nm. The Plasmonic band gap is followed by a first transmission band whose centre wavelength corresponds to Bragg condition. The situation here turns out to be reversed compared to the case of the DPC, where the same exact relation corresponds to the first band gap. The value of the centre wavelength of the first transmission band determined from Bragg condition (750 nm) nearly consistent with the value deduced from the transfer matrix method shown in **Figure 7**. The first transmission band or the band gaps of the MPC can be tuned by varying n or Λ as in the DPC.

Figure 8 shows the transmission spectra of the previous designed MPC at the number of periods equal to five periods. By decreasing number of periods, no change in

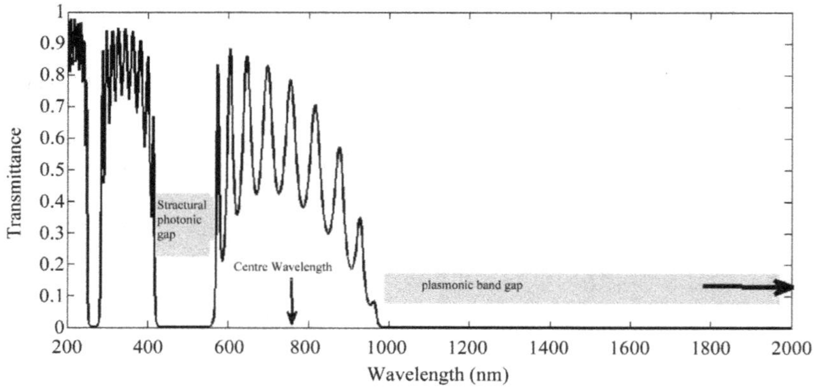

Figure 7. Calculate transmission spectra of MPC composed of Cryolite/Silver with d_1 = 200 nm, d_2 = 10 nm, number of periods = 10, and $\theta = 0°$.

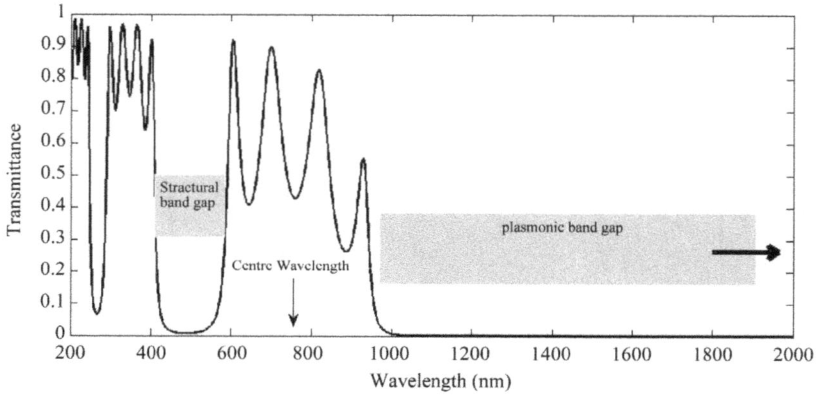

Figure 8. Calculated transmission spectra of MPC composed of Cryolite/Silver with d_1 = 200 nm, d_2 = 10 nm, $\theta = 0°$, and periods = 5.

the width of both structural and plasmonic gaps has been noticed, also there is no shift in the transmission spectra has been recorded. But the resonance transmission peaks became less (four), this due to the MPC behaves as ensemble of Fabry-Periot cavities coupled to one another along the propagation direction. So the MPC that composed of ten (five) periods can be regarded as nine (four) Fabry-Periot cavities.

The effect of the filling factor of the dielectric and metal on the transmittance of the MPC is displayed in **Figure 9**, the figure shows, when the metal filling factor (M.F) is doubled to be 0.096 keeping the lattice constant without change, the low energy edge of the first transmission band move to shorter wavelength without moving the high energy edge causing shrinking the first transmission band width and increasing the plasmonic band gap width. Moreover, the width of the structural band gap increases and the transmittance of the Ag-PC decreases. This behaviour of the MPC can be understood if we regarded the MPC as composite structure which can be described by the effective plasma frequency. The effective plasma frequency proportional to the metallic plasma frequency times the square root of the metal filling factor [28]. So with increasing the metal filling factor the effective plasma frequency increases and the transparent region (transmission band) shift to shorter wavelength. The incidence angle effect on the transmission spectra of the MPC for S- and P- polarized is displayed in the **Figures 10**, and **11**, respectively. For S- and P- polarized waves, an increase in the incidence angle cause a shift in the structural band gaps and the first transmission band to shorter wavelengths. However,

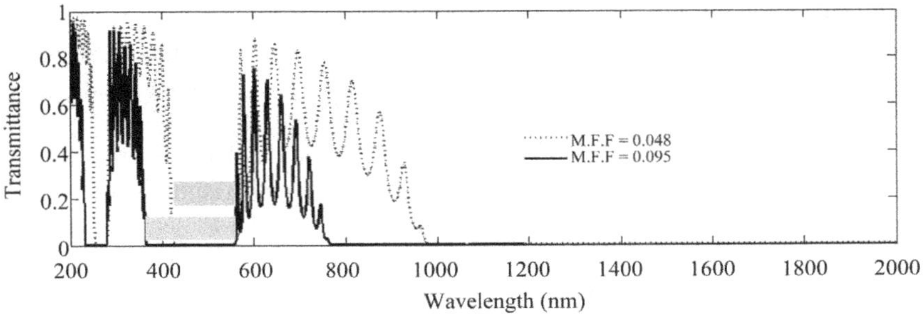

Figure 9. Calculated transmission spectra of two MPCs composed of Cryolite/Silver with periods = 10, $\theta = 0°$, $\Lambda = 210$ nm and different in the Silver filling factor.

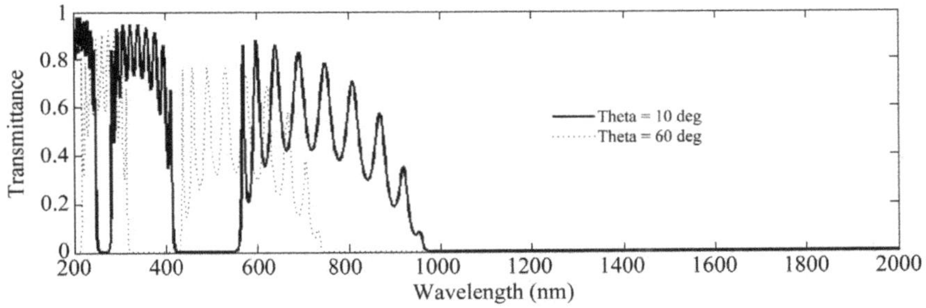

Figure 10. Calculated transmission spectra of MPC of TE waves, where the MPC composed of Cryolite/Silver with $d_1 = 200$ nm, $d_2 = 10$ nm, and periods = 10.

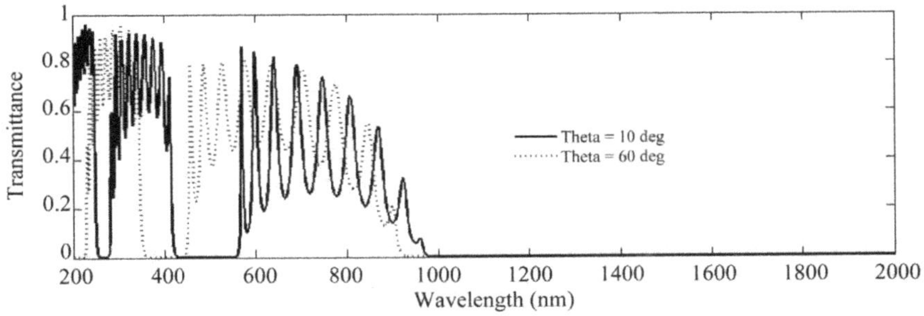

Figure 11. Calculated transmission spectra of MPC of TM waves, where the MPC composed of Cryolite/Silver with $d_1 = 200$ nm, $d_2 = 10$ nm, and periods = 10.

the S-polarized wave has shown blue shift larger than the P-polarized wave. In addition, the low energy edge of the first transmission band for S-polarized wave shifts from 970 to 770 nm nearly 200 nm with the angle change from 10° to 60°. But for P-polarized wave the shift is nearly 50 nm. This behaviour can be understood if we regarded the MPC as a system of several Fabry-Perot cavities coupled. Each metal/dielectric/metal structure in the MPC acts as a Fabry-Perot Cavity and the finite thickness of metal layers makes the cavity modes overlaps. Away from the normal incidence, Fabry-Perot modes satisfy the condition [29];

$$\left(\frac{2\pi}{\lambda}\right) n_d d \cos\theta_d + \delta_{S,P} = m\pi \text{, for } m = 1, 2, 3, \ldots \quad (4)$$

where θ_d is the angle of propagation inside a dielectric layer, and $\delta_{S,P}$ is the induced phase shift by the reflection at the dielectric metallic interface for S- and P-polarized waves and given by [29];

$$\delta_{S,P} = \cos^{-1}\left(\frac{n_d^2 - n_i^2}{n_d^2 + n_i^2}\right) \quad (5)$$

where n_i is the imaginary part of the metal refractive index. The induced phase shift of the S- and P-polarized waves is different, the Fabry-Perot modes for S- and P-polarized waves occur at different frequencies. Away from the normal incidence, the refractive index of the metal layers can be regarded as $n_m \cos(\theta_m)$ and $n_m / \cos(\theta_m)$ for S- and P-polarized waves, respectively. So the effective refractive index for the S-polarized waves is smaller than that for P-polarized waves. According to Equation (5), δ_s is smaller than δ_P, and according to Equation (4), the wavelength of the Fabry-Perot modes for S-polarized wave is smaller than that for P-polarized wave. Therefore, the transmission band in MPCs shifts more toward the shorter wavelength region for S-polarized wave than for P-polarized wave with increasing the angle of incidence.

In order to study the effect of the Plasmon frequency and the damping coefficient of metals on the transmittance of the MPC (**Figures 12** and **13**), we have designed MPCs composed of different metals (silver, gold, and Aluminium). The magnitudes of Plasmon frequencies and damping coefficients are 2175/4.35 THz (Ag), 2175/6.5 THz (Au), and 3750/19.4 THz (Al). In **Figure 12**, the transmission spectra of the Cryolite/Ag- and Cryolite/Au-PCs are displayed. The resonance transmission peaks of the AU-PC occurs at the same wavelengths of the Ag-PC. The plasmonic and the structure band gaps coincide, and have the same width, this due to Ag and Au have the same Plasmon frequency. We can see in **Figure 12**, Ag-PC has higher resonance transmission peaks than Au-PC, this due to Ag has lower damping coefficient. In **Figure 13**, the transmittance of the Al-PC is compared to the Ag-PC. The transmittance of the Al-PC is very small compared to the Ag- or Au-PC

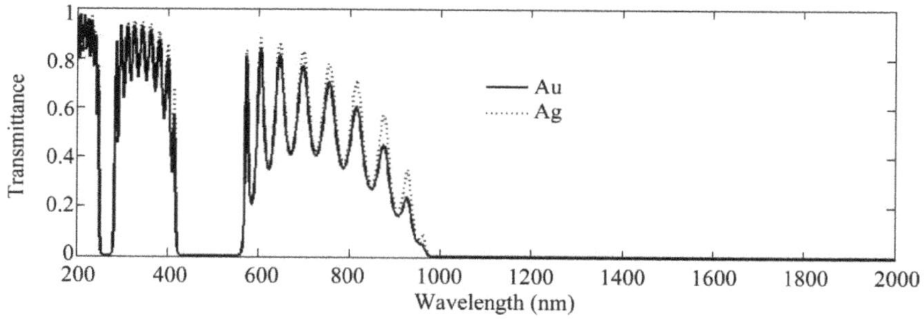

Figure 12. Calculated transmission spectra of Cryolite/Silver and Cryolite/Gold PCs with d_1 = 200 nm, d_2 = 10 nm, periods = 10, and θ = 0°.

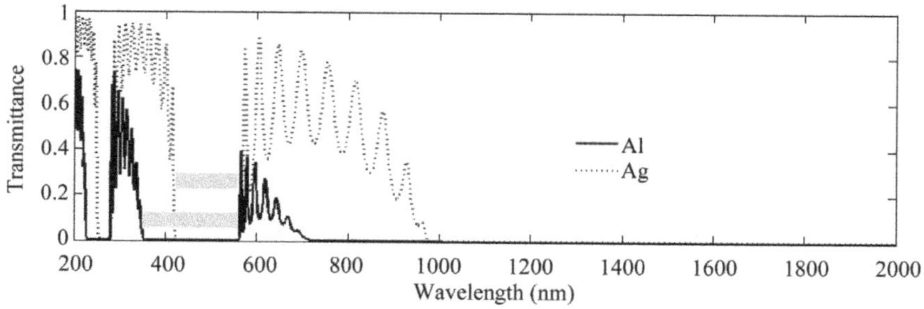

Figure 13. Calculated transmission spectra of Cryolite/Silver and Cryolite/Aluminium PCs with d_1 = 200 nm, d_2 =10 nm, periods = 10, and θ = 0°.

this due to the Al has very large damping coefficient. The resonance transmission peaks of the Al-PC occur at shorter wavelengths this because Al has larger value of Plasmon frequency. **Figure 13** shows the Al-PC has wider structural and plasmonic band gaps.

The 1D-MPCs can work as mirrors better than the DPCs because it needs fewer numbers of periods to give high reflectance gap. Plasmonic gap of the MPC block the longest wavelengths (microwaves and radiofrequency) and this candidate it to work as radiofrequency shield, microwave ovens doors. It also can work as UV protective.

3. Conclusion

We have designed one dimensional-dielectric and -metallic Photonic crystals showing the difference in the transmission spectra of both. The MPC has both structural and bulk metal band gaps, but the DPC has only Structural band gaps that produce from EM waves interferences. The spectra of dielectric photonic crystal begins with transmission band at lower frequencies, on contrary to the metallic photonic crystal that begins with stop band. The first band gap of the DPC and the first transmission band of the MPC occur at Bragg condition. The resonance transmission peaks of the MPC and the DPC are comparable to the number of periods. In the DPC the *P*-Polarized waves is more sensitive to the incident angle than the *S*-Polarized waves, but in the MPC the matter is turned out to be reversed. MPCs of larger damping coefficient metal have lower transmittance, and MPCs of larger Plasmon frequency metal have resonance transmission peaks at shorter wavelengths. Increase the metal filling factor of the MPC do as the Plasmon frequency of the metal increase.

REFERENCES

[1] A. H. Aly, "Metallic and Superconducting Photonic Crystal," *Journal of Superconductivity and Novel Magnetism*, Vol. 21, No. 7, 2008, pp. 421-425.

[2] J.-M. Lourtioz, H. Benisty, V. Berger, J.-M. Gerard, D. Maystre and A. Tchelnokov, "Photonic Crystals: Towards Nanoscale Photonic Devices," 2*nd* Edition, Springer-Verlag Berlin Heidelberg, New York, 2008.

[3] J. D. Joannopoulos., S. G. Johnson, J. N. Winn and R. D. Meade, "Photonic Crystals: Molding the Flow of Light," 2nd Edition, Princeton University Press, Princeton, 2008.

[4] M. Scalora, M. J. Bloemer, A. S. Pethel, J. P. Dowling, C. M. Bowden and A. S. Manka, "Transparent, Metallo-Dielectric, One-Dimensional, Photonic Band-Gap Structures," *Journal of Applied Physics* , Vol. 83, No. 5, 1998, p. 2377.

[5] Q.-H. Gong and X.-Y. Hu, "Ultrafast Photonic Crystal Optical Switching," *Frontiers of Physics in China*, Vol. 1, No. 2, 2006, pp. 171-177.

[6] Z. Jaksic, M. Maksimovie and M. Sarajlic, "Silver-Silica Transparent Metal Structures as Bandpass Filters for the Ultraviolet Range," *Journal of Optics A*: *Pure and Applied Optics*, Vol. 7, No.1, 2005, pp. 51-55

[7] S. K. Awasthi and S. P. Ojha, "Design of a Tunable Optical Filter by Using a One-Dimensional Ternary Photonic Band Gap Material," *Progress in Electromagnetics Research M*, Vol. 4, 2008, pp. 117-132. doi:10.2528/PIERM08061302

[8] S. K. Awasthi and S. P. Ojha, "Wide-Angle Broadband Plate Polarizer with 1D Photonic Crystal," *Progress in Electromagnetics Research*, Vol. 88, 2008, pp. 321-335.

[9] D. N. Chigrin, A. V. Lavrinenko, D. A. Yarotsky and S. V. Gaponenko, "Observation of Total Omnidirection Reflection from a One-Dimensional-Dielectric Lattice," *Applied Physics A*, Vol. 68, No. 1, 1999, pp. 25-28. doi:10.1007/s003390050849

[10] P. K. Choudhury, P. Khastgir, S. P. Ojha, D. K. Mahapatra and O. N. Singh, "Design of an Optical Filter as a Monochromatic Selector from Atomic Emission," *Journal of the Optical Society of America A*, Vol. 9, No. 6, 1992, pp. 1007-1010. doi:10.1364/JOSAA.9.001007

[11] J. N. Winn, S. Fan, C. Chen, J. Michel, J. D. Joannopoulos and E. L .Thomas, "A Dielectric Omnidirectional Reflector," *Science*, Vol. 282, No. 5394, 1998, pp. 1679-1682.

[12] O. Zandi, Z. Atlasbaf and K. Fabororaghi, "Flat Multilayer Dielectric Reflector Antennas," *Progress in Electromagnetics Research*, Vol. 72, 2007, pp. 1-19. doi:10.2528/PIER07022604

[13] M. Aissaoui, J. Zaghdoudi, M. Kanzari and B. Rezig, "Optical Properties of the Quasi-Periodic One-Dimensional Generalized Multilayer Fibnoacci Structures," *Progress in Electromagnetics Research*, Vol. 59, 2006, pp. 69-83. doi:10.2528/PIER05091701

[14] T. Maka, D. N. Chaigrin, S. G. Romanov and C. M. S. Torres, "Three Dimensional Photonic Crystals in the Visible Regime," *Progress in Electromagnetics Research*, Vol. 41, 2003, pp. 307-335. doi:10.2528/PIER0201089e

[15] C.-J. Wu, "Transmission and Reflection in a Period Superconductor/Dielectric Film Multilayer Structure," *Journal of Electromagnetic Waves and Applications*, Vol. 19, No. 15, 2005, pp. 1991-1996. doi:10.1163/156939305775570468

[16] L.-P. Zhao, X. Zhao, B. Wu, T. Su, W. Xue and C.-H. Liang, "Novel Design of Dual—Mode Bandpass Filter Using Rectangle Structure," *Progress in Electromagnetics Research B*, Vol. 3, 2008, pp. 131-141. doi:10.2528/PIERB07121003

[17] J. A. M. Rojas, J. Alpuente, J. Pineiro and R. Sanchez-Montero, "Regorous Full Vertical Analysis of Electromagnetic Wave Propagation in 1D," *Progress in Electromagnetics Research*, Vol. 63, 2006, pp. 89-105. doi:10.2528/PIER06042501

[18] Q.-R. Zheng, Y.-Q. Fu and N.-C. Yuan, "Characteristics of Planar PBG Structures with Cover Layer," *Journal of Electromagnetic Waves and Applications*, Vol. 120, No, 11, 2006, pp. 1439-1453. doi:10.1163/156939306779274264

[19] M. Deopua, C. K. Ullal, B. Temelkuran and Y. Fink, "Dielectric Omnidirectional-Visible Reflector," *Optics Letters*, Vol. 26, No. 15, 2001, pp. 1197-1199.

[20] M. Thomsen and Z. L. Wu, "Polarizing and Reflective Coatings Based on Half-Wave Layer Pairs," *Applied Optics*, Vol. 36, No. 1, 1997, pp. 307-313. doi:10.1364/AO.36.000307

[21] J. C. Monga, "Multilayer Thin-Film Polarisers with Reduced Electric-Field Intensity," *Journal of Modern Optics*, Vol. 36, No. 6, 1989, pp. 769-784. doi:10.1080/09500348914550841

[22] S. M. MacNeille, "Beam Splitter," US Patent No. 2403731, 1946.

[23] J. Mouchart, J. Begel and E. Duda, "Modified MacNeille Cube Polarizer for a Wide Angular Field," *Applied Optics*, Vol. 28, No. 14, 1989, pp. 2847-2853. doi:10.1364/AO.28.002847

[24] L. Li and J. A. Dobrowolski, "Visible Broadband, Wide-Angle, Thin-Film Multilayer Polarizing Beam Splitter," *Applied Optics*, Vol. 35, No. 13, 1996, pp. 2221-2225. doi:10.1364/AO.35.002221

[25] L. Li, and J. A. Dobrowolski, "Hig-Performance Thin-Film Polarizing Beam Splitter Operating at Angles Greater than the Critical Angle," *Applied Optics*, Vol. 39, No. 16, 2000, pp. 2754-2771 doi:10.1364/AO.39.002754

[26] A. H. Aly, S.-W. Ryu, H.-T. Hsu and C.-J. Wu, "THz Transmittance in One-Dimenssional Superconducting Nanomaterial-Dielectric Superlattice," *Materials Chemistry Physics*, Vol. 113, No. 16, 2009, pp. 382-384.

[27] M. M. Sigalas, C. T. Chan, K. M. Ho and C. M. Soukoulis, "Mettalic Photonic Band Gap Materials," *Physical Review B*, Vol. 52, No. 16, 1995, pp. 11744-11751.

[28] J. Manzanares-Martinez, "Analytical Expression for the Effective Plasma Frequency in One Dimentional Metallic-Dielectric Photonic Crystal," *Progress in Electromagnetic Research M*, Vol. 13, 2010, pp. 189-202.

[29] Y.-K. Choi, Y.-K. Ha, J.-E. Kim, H. Y. Park and K. Kim, "Antireflection Film in One-Dimensional Metallo-Dielectric Photonic Crystals," *Optics Communications*, Vol. 230, No. 4-6, 2004, pp. 239-243.

Electromagnetic Oscillations in a Spherical Conducting Cavity with Dielectric Layers. Application to Linear Accelerators

Wladyslaw Zakowicz[1], Andrzej A. Skorupski[2], Eryk Infeld[2]

[1]Institute of Physics, Polish Academy of Sciences, Warsaw, Poland; [2]Department of Theoretical Physics, National Centre for Nuclear Research, Warsaw, Poland.
Email: wladyslaw.zakowicz@ifpan.edu.pl, askor@fuw.edu.pl, einfeld@fuw.edu.pl

ABSTRACT

We present an analysis of electromagnetic oscillations in a spherical conducting cavity filled concentrically with either dielectric or vacuum layers. The fields are given analytically, and the resonant frequency is determined numerically. An important special case of a spherical conducting cavity with a smaller dielectric sphere at its center is treated in more detail. By numerically integrating the equations of motion we demonstrate that the transverse electric oscillations in such cavity can be used to accelerate strongly relativistic electrons. The electron's trajectory is assumed to be nearly tangential to the dielectric sphere. We demonstrate that the interaction of such electrons with the oscillating magnetic field deflects their trajectory from a straight line only slightly. The Q factor of such a resonator only depends on losses in the dielectric. For existing ultra low loss dielectrics, Q can be three orders of magnitude better than obtained in existing cylindrical cavities.

Keywords: Spherical Cavity; Spherical Dielectric Layer; TE Mode; TM Mode; Q Factor; Linear Accelerator

1. Introduction

It has been shown [1-3] that, if a plane electromagnetic wave is scattered on a finite dielectric object, structural resonances can be excited in the object (e.g., whispering gallery modes). They are associated with very high amplitudes of oscillating EM fields in the dielectric and its vicinity. Their maxima exceed values reached in resonant cavities of typical linear accelerators by several orders of magnitude. Therefore, one can think of applying these fields to accelerate charged particles [1-3]. Many other applications of the whispering gallery modes are described in [4-6].

As for the proposals given in [1-3], both light produced by lasers and microwaves are conceivable. However, it is difficult to achieve the required synchronization of wave particle in the optical frequency range. In the microwave frequency range, this mechanism would require excessive total excitation energy and so may not be practical.

In this paper we demonstrate that the last mentioned problem can be overcome by locating the dielectric object in a resonant cavity. This appeals to traditional accelerating structures used in SLAC, see **Figure 1**. In the latter case, maximum amplitudes of accelerating fields are restricted by Joule heating losses in conducting walls and electric breakdown. In this connection, in existing accelerators (e.g., in LHC) one avoids sharp edges of the walls and uses superconductive resonant

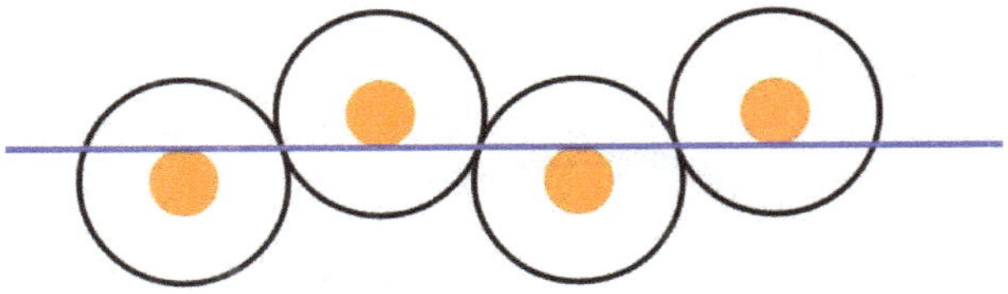

(a)

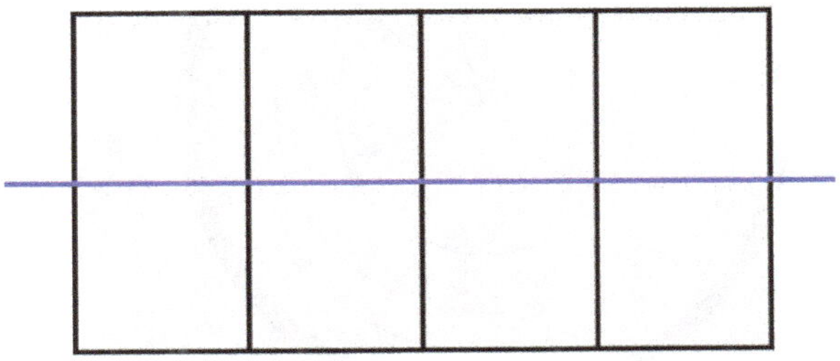

(b)

Figure 1. Proposed multi cell accelerator units (a) vs. those of SLAC (b), assuming the same electron transit time through the cavity.

cavities. Unfortunately, since superconductivity of the walls disappears if the magnetic field on the wall exceeds a critical value, the maximal values of accelerating fields in these highly complicated cavities are not much higher than those reached in SLAC.

The presence of a dielectric in the central part of the resonance cavity shifts the magnetic field maximum from regions close to the metallic wall towards the dielectric surface. This considerably lowers skin effect losses in the wall. Even though additional losses due to dielectric heating are introduced, total losses would nevertheless be lower if one could apply ultra low loss dielectrics (with $\tan\delta \sim 10^{-7}$) as described in [7,8]. In that case, a resonator quality reaching $Q \sim 10^{7}$ can be obtained, as compared to $Q \sim 10^{4}$ for SLAC.

2. A Spherical Conductive Cavity with Dielectric Layers

Our approach to describe electromagnetic oscillations in a resonant cavity assumes that the cavity can be divided into regions in which the fields can be determined analytically. The resonant frequency is defined by the fact that the fields must satisfy boundary conditions at the cavity wall along with continuity conditions at the interfaces. This frequency will be determined by numerically solving the consistency condition for these requirements.

In general, we assume that the cavity is bounded by a conducting spherical surface, and filled concentrically with $N(\geq 1)$ either dielectric or vacuum layers. Each dielectric layer is assumed to be homogeneous. We introduce a spherical coordinate system (r,θ,ϕ) with its origin at the cavity center. The layers are bounded by $r = a_1, a_2, \cdots, r_{N-1}$, up to $r = a_N \equiv b$ for the metallic boundary, see **Figure 2**.

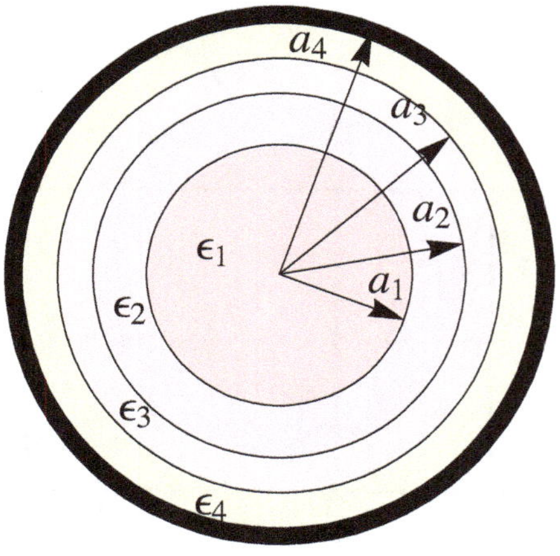

Figure 2. An example of a spherical cavity with dielectric layers (N = 4).

The harmonically oscillating electromagnetic fields in each concentric layer are described by Maxwell's equations (Gaussian units, magnetic permeability $\mu = 1$, and complex fields proportional to $\exp(-i\omega t)$):

$$\nabla \times \boldsymbol{E} = ik\boldsymbol{B},\ \nabla \times \boldsymbol{B} = -ik\epsilon\boldsymbol{E} \tag{1}$$

where

$$k = \omega/c \tag{2}$$

ω is the angular frequency, and ϵ denotes complex dielectric permittivity

$$\epsilon = \epsilon' + i\epsilon'',\ |\epsilon''| \ll \epsilon' \tag{3}$$

These fields split into transverse electric (TE) or transverse magnetic (TM), which have no radial components of either field [9]. In an ideal resonator with perfectly conducting walls and perfect dielectrics, pure TE or TM modes can be excited. They will also be approximately valid in real resonators if their energy losses are not too high.

Using (9.116), and (9.119) in [9], which describe the vacuum TE field in spherical coordinates, and replacing there $k \to \sqrt{\epsilon}k$ we obtain the most general form of the TE field in the uniform dielectric:

$$\begin{aligned} \boldsymbol{E}_{lm}(\boldsymbol{r},t) &= \boldsymbol{E}_{lm}(\boldsymbol{r},\omega)\mathrm{e}^{-i\omega t} \\ \boldsymbol{E}_{lm}(\boldsymbol{r},\omega) &= E_t(kr)\boldsymbol{X}_{lm}(\theta,\phi) \end{aligned} \tag{4}$$

where

$$E_t(r) = A_l^{(1)} j_l\left(\sqrt{\epsilon}kr\right) + A_l^{(2)} y_l\left(\sqrt{\epsilon}kr\right) \tag{5}$$

$\boldsymbol{X}_{lm}(\theta,\phi)$ are vector spherical harmonics as defined by Eqation (9.119) in [9], and

$$j_l(\rho) \equiv \sqrt{\frac{\pi}{2\rho}} J_{l+\frac{1}{2}}(\rho)$$

$$\text{and } y_l(\rho) \equiv \sqrt{\frac{\pi}{2\rho}} Y_{l+\frac{1}{2}}(\rho)$$

are spherical Bessel and Neumann functions.

The corresponding magnetic induction can be determined from the first Maxwell Equation (1). Using also (10.60) in [9] we obtain

$$\boldsymbol{B} \equiv \boldsymbol{B}_{lm}(\boldsymbol{r},t) = \boldsymbol{B}_{lm}(\boldsymbol{r},\omega)\mathrm{e}^{-i\omega t} \tag{6}$$

where $\boldsymbol{B}_{lm}(\boldsymbol{r},\omega)$ involves both the transverse and radial component:

$$\boldsymbol{B}_{lm}(\boldsymbol{r},\omega) = \boldsymbol{B}_{lmt}(\boldsymbol{r},\omega) + \boldsymbol{B}_{lmr}(\boldsymbol{r},\omega) \tag{7}$$

in which

$$\boldsymbol{B}_{lmt}(\boldsymbol{r},\omega) = \mathcal{B}_t(r)\boldsymbol{n} \times \boldsymbol{X}_{lm}(\theta,\phi) \tag{8}$$

$$\boldsymbol{B}_{lmr}(\boldsymbol{r},\omega) = \mathcal{B}_r(r)Y_{lm}(\theta,\phi)\boldsymbol{n} \tag{9}$$

$$\mathcal{B}_t(r) = -\frac{i}{kr}\left[A_l^{(1)} j_l^{\mathrm{D}}\left(\sqrt{\epsilon}kr\right) + A_l^{(2)} y_l^{\mathrm{D}}\left(\sqrt{\epsilon}kr\right)\right] \quad (10)$$

$$\mathcal{B}_r(r) = \frac{\sqrt{l(l+1)}}{kr} E_t(r) \quad (11)$$

Here $Y_{lm}(\theta,\phi)$ are spherical harmonics, $\boldsymbol{n} = \boldsymbol{r}/r$, l is a positive integer related to the integer m by $-l \le m \le l$,

$$j_l^D(\rho) \equiv \frac{\mathrm{d}}{\mathrm{d}\rho}(\rho j_l(\rho))$$

and $$y_l^D(\rho) \equiv \frac{\mathrm{d}}{\mathrm{d}\rho}(\rho y_l(\rho))$$

are derivatives of the Riccati-Bessel and Riccati-Neumann functions.

In a similar way, using (9.118), (9.119) and (10.60) in [9] along with the second Maxwell Equation (1), we obtain for the TM modes in the uniform dielectric:

$$\begin{aligned} \overline{\boldsymbol{B}}_{lm}(\boldsymbol{r},t) &= \overline{\boldsymbol{B}}_{lm}(\boldsymbol{r},\omega)\mathrm{e}^{-i\omega t} \\ \overline{\boldsymbol{B}}_{lm}(\boldsymbol{r},\omega) &= \overline{\mathcal{B}}_t(r)\boldsymbol{X}_{lm}(\theta,\phi) \end{aligned} \quad (12)$$

where

$$\overline{\mathcal{B}}_t(r) = \overline{A}_l^{(1)} j_l\left(\sqrt{\epsilon}kr\right) + \overline{A}_l^{(2)} y_l\left(\sqrt{\epsilon}kr\right) \quad (13)$$

$$\overline{\boldsymbol{E}} \equiv \overline{\boldsymbol{E}}_{lm}(\boldsymbol{r},t) = \overline{\boldsymbol{E}}_{lm}(\boldsymbol{r},\omega)\mathrm{e}^{-i\omega t} \quad (14)$$

where $\overline{\boldsymbol{E}}_{lm}(\boldsymbol{r},\omega)$ involves both the transverse, and radial component:

$$\overline{\boldsymbol{E}}_{lm}(\boldsymbol{r},\omega) = \overline{\boldsymbol{E}}_{lmt}(\boldsymbol{r},\omega) + \overline{\boldsymbol{E}}_{lmr}(\boldsymbol{r},\omega) \quad (15)$$

in which

$$\overline{\boldsymbol{E}}_{lmt}(\boldsymbol{r},\omega) = \overline{E}_t(r)\boldsymbol{n}\times\boldsymbol{X}_{lm}(\theta,\phi) \quad (16)$$

$$\overline{\boldsymbol{E}}_{lmr}(\boldsymbol{r},\omega) = \overline{E}_r(r)Y_{lm}(\theta,\phi)\boldsymbol{n} \quad (17)$$

$$\overline{E}_t(r) = \frac{i}{kr\epsilon}\left[\overline{A}_l^{(1)} j_l^{\mathrm{D}}\left(\sqrt{\epsilon}kr\right) + \overline{A}_l^{(2)} y_l^{\mathrm{D}}\left(\sqrt{\epsilon}kr\right)\right] \quad (18)$$

$$\overline{E}_r(r) = -\frac{\sqrt{l(l+1)}}{kr\epsilon}\overline{\mathcal{B}}_t(r) \quad (19)$$

In our analysis we will admit small energy losses in both the wall and the dielectric layers. However, when calculating the resonant frequency of the cavity $\omega_0 \equiv \omega^l$, these losses will be neglected. Thus the wall is assumed to be perfectly conducting. This means that it carries no electric or magnetic field. Then continuity of the tangential components of the electric field and the normal components of the magnetic induction at each interface require vanishing of these components at the boundary of Nth layer at $r = b$. In view of (4)-(9), and (12)-(19) this will be the case if the following boundary conditions at $r = b$ are fulfilled:

$$\begin{aligned} A_{lN}^{(1)} j_l\left(\sqrt{\epsilon_N}kb\right) + A_{lN}^{(2)} y_l\left(\sqrt{\epsilon_N}kb\right) &= 0 \\ \overline{A}_{lN}^{(1)} j_l^{\mathrm{D}}\left(\sqrt{\epsilon_N}kb\right) + \overline{A}_{lN}^{(2)} y_l^{\mathrm{D}}\left(\sqrt{\epsilon_N}kb\right) &= 0 \end{aligned} \quad (20)$$

where the upper line refers to TE modes and the lower one to TM modes.

In the first layer which contains the origin $r = 0$ we must choose

$$A_{l1}^{(2)} = 0 \text{ or } \overline{A}_{l1}^{(2)} = 0 \quad (21)$$

to avoid singularities of $y_l(\rho)$ or $y_l^D(\rho)$ at $\rho = 0$.

In the simplest case of a spherical cavity filled completely with the dielectric (or vacuum), *i.e.*, $N = 1$, the boundary conditions (20) lead to

$$j_l\left(\sqrt{\epsilon_1}kb\right) = 0 \text{ or } j_l^D\left(\sqrt{\epsilon_1}kb\right) = 0 \quad (22)$$

where $\sqrt{\epsilon_1} \ge 1, k = \omega_0/c$. This defines the resonant frequency ω_0 of either TE or TM modes, depending only on $\epsilon_1 (= \epsilon_N)$ and b.

In the presence of layers $(N > 1)$, ω_0 must also depend on ϵ_{N-1}, a_{N-1} etc. and therefore condition (22) cannot be fulfilled. In fact, for the same reason, we can assume that also the remaining functions in conditions (20) are non-vanishing. Therefore these conditions can be satisfied by choosing

$$A_{lN}^{(1)} = \frac{\mathcal{N}_l}{j_l\left(\sqrt{\epsilon_N}kb\right)},\ A_{lN}^{(2)} = -\frac{\mathcal{N}_l}{y_l\left(\sqrt{\epsilon_N}kb\right)} \quad (23)$$

$$\overline{A}_{lN}^{(1)} = \frac{\overline{\mathcal{N}}_l}{j_l^{\mathrm{D}}\left(\sqrt{\epsilon_N}kb\right)},\ \overline{A}_{lN}^{(2)} = -\frac{\overline{\mathcal{N}}_l}{y_l^{\mathrm{D}}\left(\sqrt{\epsilon_N}kb\right)} \quad (24)$$

for TE modes (upper line) or TM modes, where $\mathcal{N}_l$ and $\overline{\mathcal{N}}_l$ are normalization factors.

At the interfaces between dielectrics, the following quantities must be continuous: the tangential components of the electric field and normal ones of the magnetic induction and furthermore, the tangential components of the magnetic induction, due to vanishing of the surface currents at the dielectric surface. For the TE modes, this leads to the following conditions at $r = a_n, n = 1,\cdots,N-1$:

$$\begin{aligned} A_{ln}^{(1)} j_l(\rho_n) + A_{ln}^{(2)} y_l(\rho_n) &= A_{ln+1}^{(1)} j_l(\rho_n^+) + A_{ln+1}^{(2)} y_l(\rho_n^+) \\ A_{ln}^{(1)} j_l^D(\rho_n) + A_{ln}^{(2)} y_l^D(\rho_n) &= A_{ln+1}^{(1)} j_l^D(\rho_n^+) + A_{ln+1}^{(2)} y_l^D(\rho_n^+) \\ \rho_n = \sqrt{\epsilon_n}ka_n,\ \rho_n^+ &= \sqrt{\epsilon_{n+1}}ka_n. \end{aligned} \quad (25)$$

This can be written in matrix form

$$\boldsymbol{M}_n \cdot \boldsymbol{A}_n = \boldsymbol{M}_n^+ \cdot \boldsymbol{A}_{n+1} \quad (26)$$

where

$$\boldsymbol{A}_n = \begin{bmatrix} A_{ln}^{(1)} \\ A_{ln}^{(2)} \end{bmatrix}, \boldsymbol{M}_n = \begin{bmatrix} j_l(\rho_n) & y_l(\rho_n) \\ j_l^D(\rho_n) & y_l^D(\rho_n) \end{bmatrix},$$

$$\boldsymbol{M}_n^+ = \begin{bmatrix} j_l(\rho_n^+) & y_l(\rho_n^+) \\ j_l^D(\rho_n^+) & y_l^D(\rho_n^+) \end{bmatrix}. \tag{27}$$

The $\boldsymbol{M}$ matrices are non-singular:

$$\begin{vmatrix} j_l(\rho) & y_l(\rho) \\ j_l^D(\rho) & y_l^D(\rho) \end{vmatrix} = \frac{\pi}{2} \begin{vmatrix} J_{l+\frac{1}{2}}(\rho) & Y_{l+\frac{1}{2}}(\rho) \\ J'_{l+\frac{1}{2}}(\rho) & Y'_{l+\frac{1}{2}}(\rho) \end{vmatrix} = \frac{1}{\rho} \neq 0 \tag{28}$$

where last equality follows from the fact that $J_{l+\frac{1}{2}}(\rho)$ and $Y_{l+\frac{1}{2}}(\rho)$ are solutions of the Bessel equation

$$u''(\rho) + \frac{1}{\rho}u'(\rho) + \left[1 - \frac{\left(l+\frac{1}{2}\right)^2}{\rho^2}\right]u(\rho) = 0.$$

Multiplying (26) by

$$\boldsymbol{M}_n^{-1} = \rho_n \begin{bmatrix} y_l^D(\rho_n) & -y_l(\rho_n) \\ -j_l^D(\rho_n) & j_l(\rho_n) \end{bmatrix} \tag{29}$$

we arrive at the recurrence relation

$$\boldsymbol{A}_n = \boldsymbol{M}_n^{-1} \cdot \boldsymbol{M}_n^+ \cdot \boldsymbol{A}_{n+1}, n = 1, \cdots, N-1. \tag{30}$$

For the n th interface between dielectrics, this relation defines the vector $\boldsymbol{A}$ at the lower layer in terms of that at the upper one. Using this relation successively for $n = N-1, N-2, \cdots, 1$, we can express all $\boldsymbol{A}_n$ vectors in terms of

$$\boldsymbol{A}_N = \mathcal{N}_l \begin{bmatrix} \dfrac{1}{j_l(\rho_N)} \\ -\dfrac{1}{y_l(\rho_N)} \end{bmatrix} \equiv \mathcal{N}_l \boldsymbol{a}_N, \rho_N = \sqrt{\epsilon_N} kb \tag{31}$$

i.e.,

$$\boldsymbol{A}_n = \left(\boldsymbol{M}_n^{-1} \cdot \boldsymbol{M}_n^+\right) \cdot \left(\boldsymbol{M}_{n+1}^{-1} \cdot \boldsymbol{M}_{n+1}^+\right) \cdots \left(\boldsymbol{M}_{N-1}^{-1} \cdot \boldsymbol{M}_{N-1}^+\right) \cdot \boldsymbol{A}_N \equiv \mathcal{N}_l \boldsymbol{a}_n. \tag{32}$$

We recall that in the first layer we must satisfy $A_{l1}^{(2)} = 0$, see (21). In view of this requirement, Equation (25) for $n = 1$ can be written as

$$\begin{aligned} A_{l1}^{(1)} j_l(\rho_1) - \mathcal{N}_l\left[a_2^{(1)} j_l(\rho_1^+) + a_2^{(2)} y_l(\rho_1^+)\right] = 0 \\ A_{l1}^{(1)} j_l^D(\rho_1) - \mathcal{N}_l\left[a_2^{(1)} j_l^D(\rho_1^+) + a_2^{(2)} y_l^D(\rho_1^+)\right] = 0 \end{aligned} \tag{33}$$

where $\rho_1 = \sqrt{\epsilon_1} ka_1, \rho_1^+ = \sqrt{\epsilon_2} ka_1$, and $a_2^{(1,2)}$ are components of the vector $\boldsymbol{a}_2 \equiv \boldsymbol{A}_2 / \mathcal{N}_l$. This vector is defined by (32) and (31) if $N > 2\left(\rho_N = \sqrt{\epsilon_N} kb\right)$:

$$\boldsymbol{a}_2 = \left(\boldsymbol{M}_2^{-1} \cdot \boldsymbol{M}_2^+\right) \cdot \left(\boldsymbol{M}_3^{-1} \cdot \boldsymbol{M}_3^+\right) \cdots \left(\boldsymbol{M}_{N-1}^{-1} \cdot \boldsymbol{M}_{N-1}^+\right) \cdot \begin{bmatrix} \dfrac{1}{j_l(\rho_N)} \\ -\dfrac{1}{y_l(\rho_N)} \end{bmatrix}. \tag{34}$$

For $N = 2$, $\boldsymbol{a}_2$ is defined by (31), *i.e.*, is given by the last factor in (34).

The linear and homogeneous set of Equation (33) for $A_{l1}^{(1)}$ and $\mathcal{N}_l$ will have non-zero solutions if and only if its determinant vanishes,

$$\begin{aligned} & j_l(\rho_1)\left[a_2^{(1)} j_l^D(\rho_1^+) + a_2^{(2)} y_l^D(\rho_1^+)\right] \\ & - j_l^D(\rho_1)\left[a_2^{(1)} j_l(\rho_1^+) + a_2^{(2)} y_l(\rho_1^+)\right] = 0. \end{aligned} \tag{35}$$

If this condition is fulfilled, $A_{l1}^{(1)}$ is given by either of Equation (33), which are equivalent. Like all remaining coefficients $A_{ln}^{(1)}$ and $A_{ln}^{(2)}, n = 2, \cdots, N$, also $A_{l1}^{(1)}$ will be proportional to the normalization factor $\mathcal{N}_l$, see (32) and (33).

If there are only two layers $(N = 2)$, $a_2^{(1)}$ and $a_2^{(2)}$ in (33) and (35) are given by (31) and the resonant frequency ω_0 defined by (35) can be found from

$$\begin{aligned} & j_l\left(\sqrt{\epsilon_1} ka_1\right)\left[\frac{j_l^D\left(\sqrt{\epsilon_2} ka_1\right)}{j_l\left(\sqrt{\epsilon_2} kb\right)} - \frac{y_l^D\left(\sqrt{\epsilon_2} ka_1\right)}{y_l\left(\sqrt{\epsilon_2} kb\right)}\right] \\ & - j_l^D\left(\sqrt{\epsilon_1} ka_1\right)\left[\frac{j_l\left(\sqrt{\epsilon_2} ka_1\right)}{j_l\left(\sqrt{\epsilon_2} kb\right)} - \frac{y_l\left(\sqrt{\epsilon_2} ka_1\right)}{y_l\left(\sqrt{\epsilon_2} kb\right)}\right] = 0. \end{aligned} \tag{36}$$

and

$$A_{l1}^{(1)} = \mathcal{N}_l \frac{1}{j_l\left(\sqrt{\epsilon_1} ka_1\right)}\left[\frac{j_l\left(\sqrt{\epsilon_2} ka_1\right)}{j_l\left(\sqrt{\epsilon_2} kb\right)} - \frac{y_l\left(\sqrt{\epsilon_2} ka_1\right)}{y_l\left(\sqrt{\epsilon_2} kb\right)}\right]. \tag{37}$$

By replacing in (25)-(37)

$$\begin{aligned} & A_{lm}^{(1,2)} \to \overline{A}_{lm}^{(1,2)}, \\ & j_l^D\left(\sqrt{\epsilon_m} ka\right) \to j_l^D\left(\sqrt{\epsilon_m} ka\right)/\epsilon_m, \\ & y_l^D\left(\sqrt{\epsilon_m} ka\right) \to y_l^D\left(\sqrt{\epsilon_m} ka\right)/\epsilon_m \end{aligned} \tag{38}$$

for any m and a, we obtain the corresponding equations for the TM modes.

Any standard software like *Mathematica* or *Maple* can be used to solve the non-linear Equations (35) or (36) defining the resonant frequency $\omega_0 \equiv \omega^l$, along with the pertinent linear algebra for $N > 2$. We did it for $N = 2$, see the following section, and also for $N = 3$, by using *Mathematica*.

Note that for the TM modes, where $\overline{\mathcal{B}}_t(r)$ in (19) is continuous at each dielectric interface, $\overline{E}_r(r)$ will have jumps, due to discontinuities in ϵ. However, the radial component of the electric displacement $\overline{\mathcal{D}}_r(r) \equiv \epsilon \overline{E}_r(r)$ will be continuous. This will also be true of the TE modes where the radial displacement is identically zero. These facts imply the vanishing of surface charges at each dielectric interface. And this in turn means that the multi-layer dielectric structure resembles (and can approximate) a smooth dielectric with some permittivity profile $\epsilon(r)$, in spite of jumps in ϵ.

It was pointed out to us by Paul Martin of SIAM, that our matrix Equation (26), which can be used to relate the EM fields of a given mode for two layers of a stratified sphere, is not new. It was probably first used by A. Moroz [10] when calculating forced oscillations in such a sphere but without a conducting wall, induced by an oscillating electric dipole. In this application, the frequency ω is arbitrary.

3. A Spherical Conductive Cavity with a Dielectric Sphere

The general theory given in the previous section will now be illustrated by calculations pertinent to the TE modes in a spherical cavity with a dielectric sphere of radius a and dielectric permittivity ϵ, *i.e.*, for

$$N = 2, a_1 \equiv a, \epsilon_1 \equiv \epsilon, \text{ and } \epsilon_2 = 1.$$

Fields in such a system will be described by Equations (4)-(9) both in the sphere and the surrounding vacuum. In view of (21) and (23) their radial profiles will be given by

$$E_t(r) = \mathcal{N}_l \times \begin{cases} \mathcal{A}_l j_l\left(\sqrt{\epsilon} kr\right), & \text{if } 0 \le r \le a \\ \dfrac{j_l(kr)}{j_l(kb)} - \dfrac{y_l(kr)}{y_l(kb)}, & \text{if } a \le r \le b \end{cases} \quad (39)$$

$$\mathcal{B}_t(r) = -\frac{i\mathcal{N}_l}{kr} \times \begin{cases} \mathcal{A}_l j_l^D\left(\sqrt{\epsilon} kr\right), & \text{if } 0 \le r \le a \\ \dfrac{j_l^D(kr)}{j_l(kb)} - \dfrac{y_l^D(kr)}{y_l(kb)}, & \text{if } a \le r \le b. \end{cases} \quad (40)$$

Replacing $\epsilon_1 \to \epsilon, \epsilon_2 \to 1$ and $A_{l1}^{(1)} \to \mathcal{A}_l$ in (36) and (37), we obtain equations defining the resonant frequency ω^l and the amplitude coefficient $\mathcal{A}_l$.

We verified that for $\omega = \omega^l$, the average energies associated with the electric and the magnetic field in the cavity are equal:

$$\int_V \epsilon' \left|\boldsymbol{E}_{lm}\left(r, \omega^l\right)\right|^2 \mathrm{d}v = \int_V \left(\left|\boldsymbol{B}_{lmt}\left(r, \omega^l\right)\right|^2 + \left|\boldsymbol{B}_{lmr}\left(r, \omega^l\right)\right|^2\right) \mathrm{d}v. \quad (41)$$

(This was a check on the correctness of our formulas and accuracy of calculations.) The normalization constant $\mathcal{N}_l$ was chosen so as to satisfy:

$$\frac{1}{2}\left(\int_0^a \epsilon' \left|E_t(r)\right|^2 r^2 \mathrm{d}r + \int_a^b \left|E_t(r)\right|^2 r^2 \mathrm{d}r + \int_0^b \left(\left|\mathcal{B}_t(r)\right|^2 + \left|\mathcal{B}_r(r)\right|^2\right) r^2 \mathrm{d}r\right) = 1. \quad (42)$$

(The corresponding average energy associated with the electric and the magnetic field over our cavity is $1/(8\pi)$ erg.)

In **Figure 3**, $\nu(l) = \omega^l/(2\pi)$ as a function of l is presented for three spherical cavities with dielectric spheres. Note that ν is m independent (due to degeneracy). This fact is one of the reasons why the spherical resonator shown in **Figure 1** cannot be used in the final project of a real accelerator even though it is very convenient for a general analysis. The degeneration in question can be broken e.g., by replacing the spherical resonator by an ellipsoidal one, or shifting the center of the dielectric sphere. Another possibility could be to use an anisotropic dielectric. In any case, however, a separate numerical analysis would be necessary.

In **Figure 4** we give an example of radial functions in the equations describing fields in our spherical cavity with a dielectric sphere, (4), (8) and (9). Large values of these functions in a vicinity of the dielectric boundary can be observed.

3.1. The Motion of Relativistic Electrons

The trajectory $\boldsymbol{r}(t)$ of a relativistic electron crossing the spherical cavity shown in **Figure 1** can be parametrized by the electron's closest approach $\boldsymbol{r}_0 = \boldsymbol{r}(t_0)$ and electron velocity $c\boldsymbol{\beta}, |\boldsymbol{\beta}| = 1$:

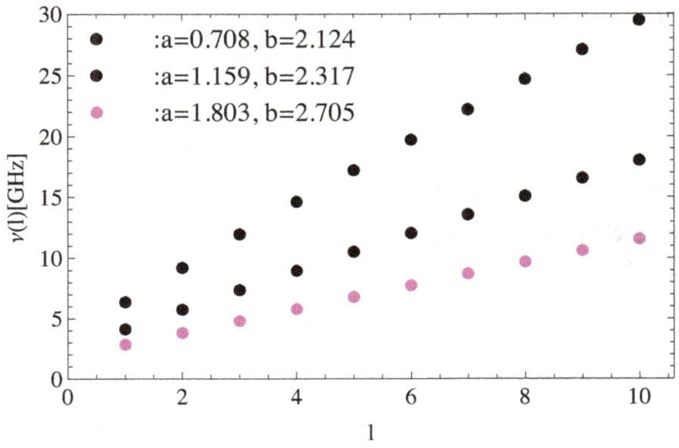

Figure 3. The resonant frequencies $\nu(l) = \omega^l/(2\pi)$ vs. l for three spherical cavities with dielectric spheres (a and b in cm). Note that ν is m independent (due to degeneracy).

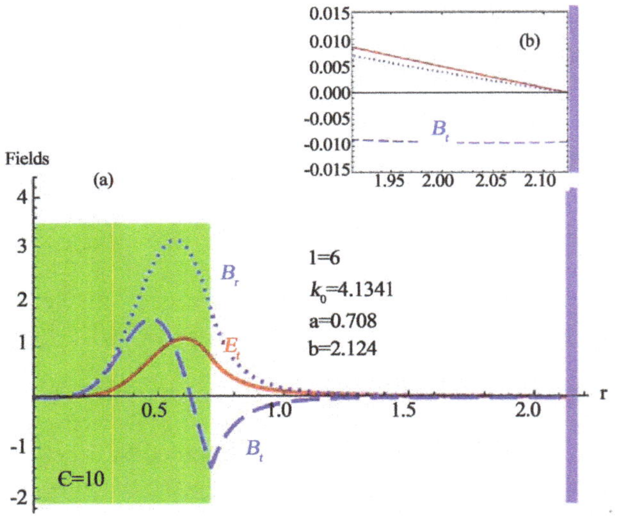

Figure 4. (a) Field radial functions for a spherical cavity with dielectric sphere and perfect metallic wall (Gaussian units, $k_0 = \omega^l/c$, a and b in cm). And the complete fields are given by (4)-(11). Note that E_t and $\mathcal{B}_r$ vanish at the wall, whereas $\mathcal{B}_t$ is non-zero but small, see (b).

$$\boldsymbol{r}(t) = \boldsymbol{r}_0 + c\boldsymbol{\beta}(t - t_0). \quad (43)$$

The origin of the Cartesian coordinate system (x, y, z) was chosen at the center of the dielectric sphere, and the electron moving along the x axis was passing just above the dielectric sphere as shown in **Figure 1**. We chose

$$\boldsymbol{r}_0 = 1.01a(0, \sin\theta_0, \cos\theta_0), \ \boldsymbol{\beta} = (1, 0, 0). \quad (44)$$

The effective accelerating field felt by the electron as it passes through the cavity, E_{eff}, is equal to the real part of $(c\mathrm{d}t = \mathrm{d}x)$

$$\bar{E}(l, m, \theta) \equiv |\bar{E}| \mathrm{e}^{i\varphi} = \frac{c}{d} \int_{t_0 - \frac{d}{2c}}^{t_0 + \frac{d}{2c}} E_x(\boldsymbol{r}(t), \omega^l) \mathrm{e}^{-i\omega^l t} \mathrm{d}t, \quad (45)$$

where $d = 2\sqrt{b^2 - (1.01a)^2}$ is the electron trajectory segment within the cavity, E_x is the x component of the electric field $\boldsymbol{E}_{lm}(\boldsymbol{r}, \omega^l)$ given by (4) and $\boldsymbol{r}(t)$ is given by (43). Thus

$$E_{\text{eff}} = |\bar{E}| \cos\varphi, \ \varphi = f(l, m, \theta) - \omega^l t_0. \quad (46)$$

Maximal acceleration is obtained $(E_{\text{eff}} = |\bar{E}|)$ if t_0 is chosen so that the accelerating phase $\varphi = 0$. With this choice, the relativistic electron is never decelerated within the spherical cavity, see **Figure 5**, where two examples are given. Typical results for $E_{\text{eff}}|_{\varphi=0}$ obtained with our normalization (42) are shown in **Figures 6** and **7**.

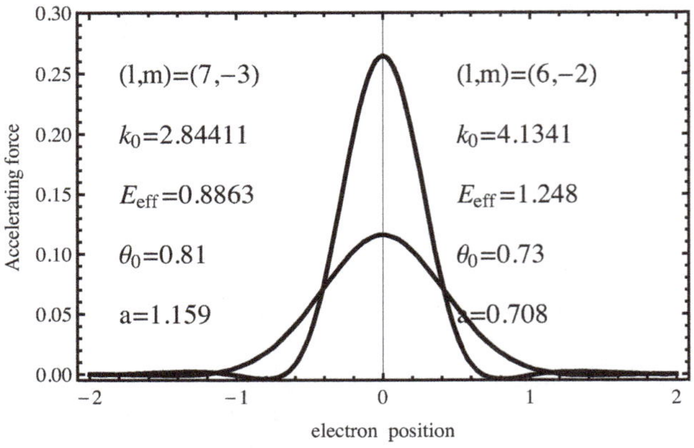

Figure 5. The accelerating force (in units of 4.8032×10^{-10} dyne) acting on an electron as it moves through the cavity shown in Figure 1 ($k_0 = \omega^l/c$, E_{eff} in kV/m, and dimensions in cm).

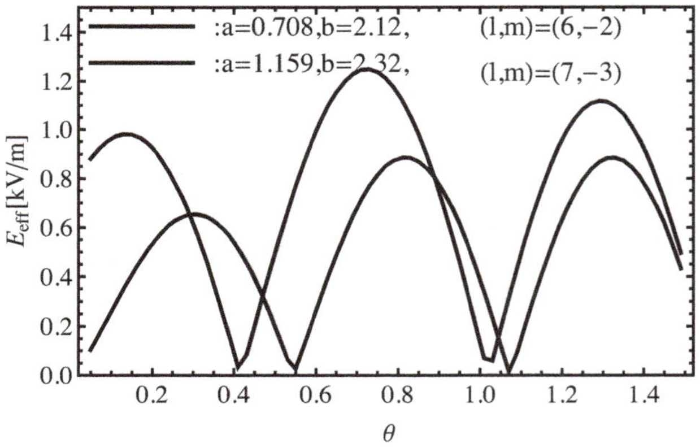

Figure 6. Effective accelerating field as a function of θ_0 for two spherical cavities (a and b in cm).

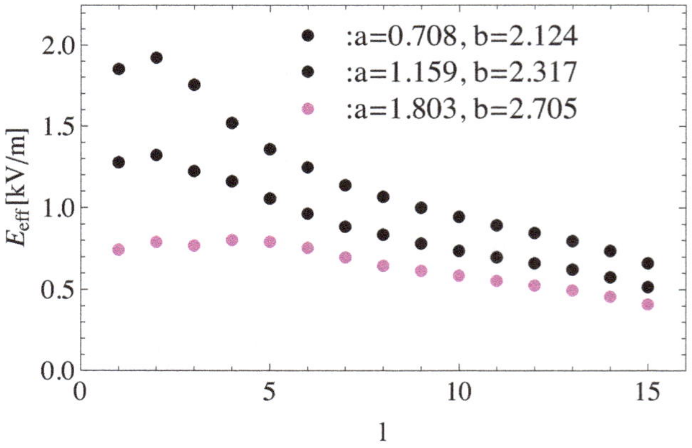

Figure 7. Effective accelerating fields vs. l for three spherical cavities with the assumed normalization of the RF field, see (42) (a and b in cm). For each value of l, these values were optimized with respect to m as well as θ_0, see Figure 6.

The electromagnetic field given by the real parts of (4), (8) and (9) is strongly non-uniform. Therefore one should check how much the relativistic electron will deflect from the assumed trajectory $\boldsymbol{r}(t)$ given by (43), due to interaction with this field. A nice feature of our model is that the field in question is described analytically by (4)-(9) so that the pertinent equations of the transversal motion can easily be integrated numerically.

In a real accelerator, where we are dealing with an electron beam of finite cross section, the electromagnetic fields $\boldsymbol{E}(\boldsymbol{r}(t),t)$ and $\boldsymbol{B}(\boldsymbol{r}(t),t)$ acting on each electron will be superpositions of the external fields and the fields due to the electron charge and current. However, in the lowest approximation (and particularly for not too large beam densities) the latter fields can be neglected. Furthermore, if as in our case, the transversal deflections are small, the deflecting fields can be calculated on the unperturbed trajectory given by (43). It will also be assumed that the electron mass $m_e = m_{e0}\gamma, \gamma \gg 1$, is time independent within the spherical cavity. With these approximations, and within the Cartesian coordinate system $(\overline{x},\overline{y},\overline{z})$ with its center at $\boldsymbol{r}_0$, the $\overline{x}$ axis along the unperturbed trajectory, and the $\overline{y}$ axis along $\boldsymbol{r}_0$, the electron's transversal motion will be described by

$$m_e \frac{\mathrm{d}v_{\overline{y}}}{\mathrm{d}\overline{t}} = -e\left[E_{\overline{y}} - B_{\overline{z}}\right],\ v_{\overline{y}} = \frac{\mathrm{d}\overline{y}}{\mathrm{d}\overline{t}} \qquad (47)$$

$$m_e \frac{\mathrm{d}v_{\overline{z}}}{\mathrm{d}\overline{t}} = -e\left[E_{\overline{z}} + B_{\overline{y}}\right],\ v_{\overline{z}} = \frac{\mathrm{d}\overline{z}}{\mathrm{d}\overline{t}} \qquad (48)$$

where $\overline{t} = t - t_0$, the field components $E_{\overline{y}}(\overline{x},\overline{t}), B_{\overline{z}}(\overline{x},\overline{t})$, etc. are given by the real parts of Equations (4), (8) and (9), taken at $\overline{x} = c\overline{t}$, and $\overline{y} = \overline{z} = 0$. Integrating these equations with zero initial conditions we end up with $(\mathrm{d}\overline{x} = c\mathrm{d}\overline{t})$

$$v_{\overline{y}}(\overline{x}) = -\frac{e}{m_e c}\int_{-d/2}^{\overline{x}} F_{\overline{y}}(\overline{x}')\mathrm{d}\overline{x}' \qquad (49)$$

$$v_{\overline{z}}(\overline{x}) = -\frac{e}{m_e c}\int_{-d/2}^{\overline{x}} F_{\overline{z}}(\overline{x}')\mathrm{d}\overline{x}' \qquad (50)$$

$$\overline{y}(\overline{x}) = \frac{1}{c}\int_{-d/2}^{\overline{x}} v_{\overline{y}}\mathrm{d}\overline{x}',\ \overline{z}(\overline{x}) = \frac{1}{c}\int_{-d/2}^{\overline{x}} v_{\overline{z}}\mathrm{d}\overline{x}' \qquad (51)$$

where $F_{\overline{y}} = E_{\overline{y}} - B_{\overline{z}}$, and $F_{\overline{z}} = E_{\overline{z}} + B_{\overline{y}}$.

In **Figures 8** and **9** we give an example of the coordinates $F_{\overline{y}}(\overline{x})$ and $F_{\overline{z}}(\overline{x})$ of the deflecting force. They correspond to $a = 0.708$ cm, $b = 2.124$ cm, and $(l,m) = (6,-2)$, for which the the accelerating field will be our reference value

$$E_{\text{eff}}\big|_{\varphi=0} \equiv E_{\text{effref}} = 1.2366\ \text{kV/m}. \qquad (52)$$

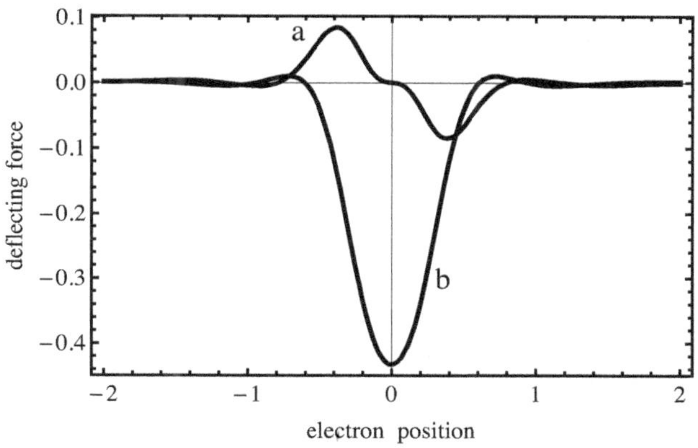

Figure 8. The deflecting force coordinate $F_{\overline{y}}(\overline{x})$ for the accelerating phase $\varphi = 0$ (a) and $\varphi = \pi/2$ (b), see Figure 5 for units and parameters (right column).

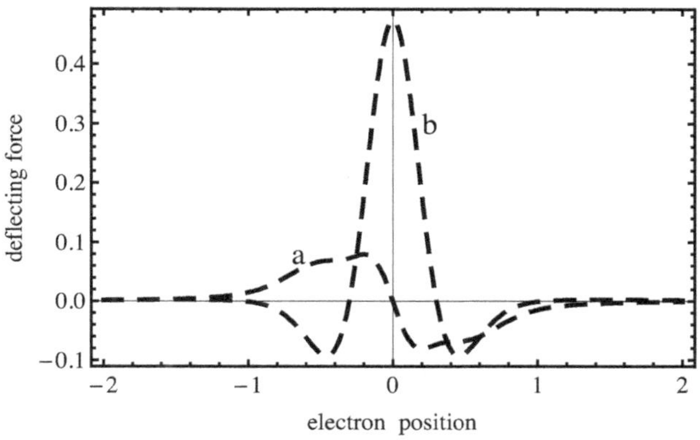

Figure 9. The deflecting force coordinate $F_{\overline{z}}(\overline{x})$ for the accelerating phase $\varphi = 0$ (a) and $\varphi = \pi/2$ (b), see Figure 5 for units and parameters (right column).

It can be seen that if the accelerating phase $\varphi = 0$, both $F_{\overline{y}}(\overline{x})$ and $F_{\overline{z}}(\overline{x})$ are odd functions. Therefore in this case of maximal acceleration, there will be no transversal velocity increments over the cavity, $v_{\overline{y}}(d/2) = v_{\overline{z}}(d/2) = 0$. At the same time the velocity components $v_{\overline{y}}(\overline{x})$ and $v_{\overline{z}}(\overline{x})$ in (51) will be even functions of $\overline{x}$ tending to zero as $\overline{x} \to d/2$. Hence the transversal deflections $\overline{y}(\overline{x})$ and $\overline{z}(\overline{x})$ will be increasing functions tending to constants as $\overline{x} \to d/2$, see **Figure 10(a)**.

For the worst case of accelerating phase $(\varphi = \pi/2)$ for which $E_{\text{eff}} = 0, v_{\overline{y}}(\overline{x})$ and $v_{\overline{z}}(\overline{x})$ will be increasing functions soon reaching their limiting values $v_{\overline{y}}(d/2)$ and $v_{\overline{z}}(d/2)$ for $\overline{x} > 0$. The corresponding transversal motions $\overline{y}(\overline{x})$ and $\overline{z}(\overline{x})$ will soon become uniform for $\overline{x} > 0$, leading to much larger deflections at $\overline{x} = d/2$, see **Figure 10(b)**.

The actual transversal deflections per cavity in an

accelerator involving our spherical cavities can be obtained by multiplying the normalized values given in **Figure 10** by the factor

$$\alpha = \frac{e}{m_e}\frac{E_{\text{eff}}}{E_{\text{effref}}}\text{cm} = 5.9\times10^{-4}\frac{E_{\text{eff}}/E_{\text{effref}}}{\gamma}\text{cm} \tag{53}$$

where E_{effref} is given by (52), E_{eff} is the assumed value of the effective accelerating field, and $\gamma = m_e/m_{e0}$. For large values of E_{eff}, small α requires γ to be sufficiently large. Assuming that $\bar{z}_{\max}\left(=\bar{z}(d/2)\right) > \bar{y}(d/2)$, see **Figure 10(a)** is not larger than $p\%$ of the spacing between the dielectric sphere and the electron trajectory, $0.01a, a = 0.708$ cm, the required minimal electron energy is given by

$$m_e c^2\,[\text{GeV}] = \frac{20.69}{p}\frac{E_{\text{eff}}\,[\text{MV/m}]}{100}. \tag{54}$$

Thus, if we assume that $E_{\text{eff}} = 100$ MV/m, the transversal displacements will be smaller than 1% of the spacing in question, if the electron energy $m_e c^2 \geq 21$ GeV, *i.e.*, for typical output energies from SLAC. Whether the real dielectric can withstand this value of E_{eff} is another question beyond the scope of this paper. More comments will be given later on.

3.2. Quality Factors

An important parameter of any linear accelerator is the quality Q of its resonant cavities:

$$Q = \omega_0\frac{U}{P} \equiv 2\pi\frac{U}{T_0 P} \tag{55}$$

where ω_0 is the resonant angular frequency of the ideal cavity $(\sigma = \infty, \epsilon'' = 0)$, T_0 is the corresponding resonant period, U is the time-averaged energy stored in the cavity $(\mu = 1)$

$$U = \frac{1}{16\pi}\int_V\left(\epsilon'|\boldsymbol{E}|^2 + |\boldsymbol{H}|^2\right)\mathrm{d}v \tag{56}$$

and P is time-averaged cavity power loss.

The power loss caused by the skin current in the metallic wall bounded by the surface S is given by

$$P_{\text{met}} = \alpha\int_S|\boldsymbol{H}|^2\,\mathrm{d}s \tag{57}$$

where

$$\alpha = \frac{c}{8(2\pi)^{3/2}}\sqrt{\frac{\omega_0}{\sigma}} \equiv \frac{c^2}{32\pi^2\sigma\delta}, \delta = \frac{c}{\sqrt{2\pi\sigma\omega_0}} \tag{58}$$

δ is the skin depth, σ is conductivity of the wall, and the magnetic field intensity $\boldsymbol{H}$ refers to the ideal cavity, *i.e.*, its normal component is vanishing $(\boldsymbol{H} = \boldsymbol{H}_t)$.

The quality of the cavity related to losses in the metallic wall is thus given by

$$Q_{\text{met}} = \omega_0\frac{U}{P_{\text{met}}}. \tag{59}$$

Using the fact that at resonance, the averaged energies stored in the electric and magnetic fields are equal, see (41), we end up with $(\boldsymbol{H} = \boldsymbol{B})$:

$$Q_{\text{met}} = \frac{2}{\delta}\frac{\int_V|\boldsymbol{B}|^2\,\mathrm{d}v}{\int_S|\boldsymbol{B}|^2\,\mathrm{d}s}. \tag{60}$$

This formula is quite general, and in particular can also be used for a traditional cylindrical cavity of radius R_c and height h. In that case, the cylindrically symmetric $n = 0$ TM mode used for acceleration is given by

$$E_x(\rho,t) = \mathcal{N}J_0(k_0\rho)\exp(-i\omega_0 t) \tag{61}$$

$$B_\varphi(\rho,t) = i\mathcal{N}J_0'(k_0\rho)\exp(-i\omega_0 t) \tag{62}$$

where $k_0 = \omega_0/c$, J_0 is a Bessel function, and ρ and φ are cylindrical coordinates (cylindrical axis along x).

The vanishing of E_x on an ideally conducting cylindrical wall requires that $(\omega_0/c)R_c = 2.405$ (the smallest zero of J_0) which defines the angular resonant frequency in terms of R_c. Equations (62) and (60) lead to the well known formula for the quality of the cylindrical pill box cavity

$$Q_c = 2.405\sqrt{\frac{2\pi\sigma}{\omega_0}}\frac{1}{1 + R_c/h}. \tag{63}$$

For the SLAC pill box cavity shown in **Figure 1**

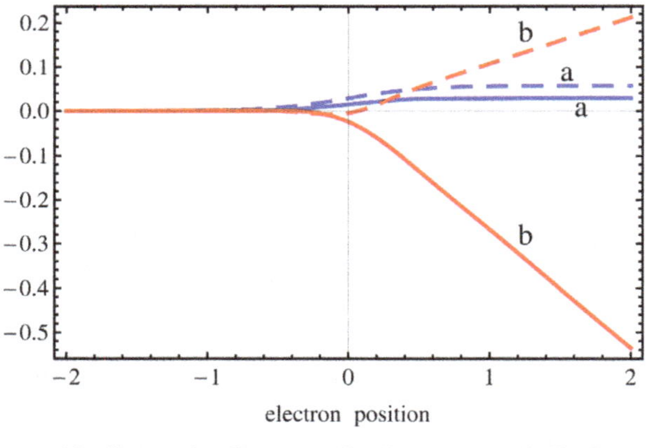

Figure 10. Conveniently normalized transversal displacements $\bar{y}$ (solid curves) and $\bar{z}$ (dashed curves) as functions of $\bar{x}$ for the accelerating phase $\varphi = 0$ (a) and $\varphi = \pi/2$ (b), see Figure 5 for units and parameters (right column).

($R_c = h = d = 4$ cm, and $\sigma = 5.294\times10^{17}\ \mathrm{s}^{-1}$ for copper wall in room-temperature), this formula leads to $Q_c = 1.633\times10^4$. The corresponding values for spherical cavities with ideal dielectric spheres and the same values of d and σ reach much larger values, see **Figure 11**.

In the presence of the dielectric sphere, one is also dealing with losses due to an imperfect dielectric specified by $\epsilon'' = \mathrm{Im}\epsilon$. The non-vanishing value of ϵ'' leads to $\mathrm{Im}k \neq 0$, for k defined by (35). In view of Equation (2) this implies a complex value of $\omega = \omega' + i\omega''$.

For the fields given by (4)-(9), we obtain

$$U(t) = U(t=0)\mathrm{e}^{2\omega'' t}$$

where $\omega'' < 0$ for the energy U being dissipated rather than generated. Using this result we find for the power losses in the dielectric:

$$P_{\mathrm{diel}} = -\frac{\mathrm{d}U}{\mathrm{d}t} = -2\omega'' U.$$

In view of (55), the corresponding quality will thus be given by

$$Q_{\mathrm{diel}} = -\frac{\omega_0}{2\mathrm{Im}\omega}. \tag{64}$$

This value is of the order of $(\tan\delta)^{-1} \equiv \epsilon'/\epsilon''$. It is approximately l independent.

In our calculations we took $\epsilon' = 10$ and $\epsilon'' = 10^{-6}$. Dielectrics with such ultra small losses were investigated in [8].

The total power loss in the spherical cavity encasing the dielectric sphere P_s is due to the power loss in the metallic wall and that in the dielectric sphere:

$$P_s = P_{\mathrm{met}} + P_{\mathrm{diel}}. \tag{65}$$

Dividing both sides of this relation by $\omega_0 U$ and using (55) we obtain

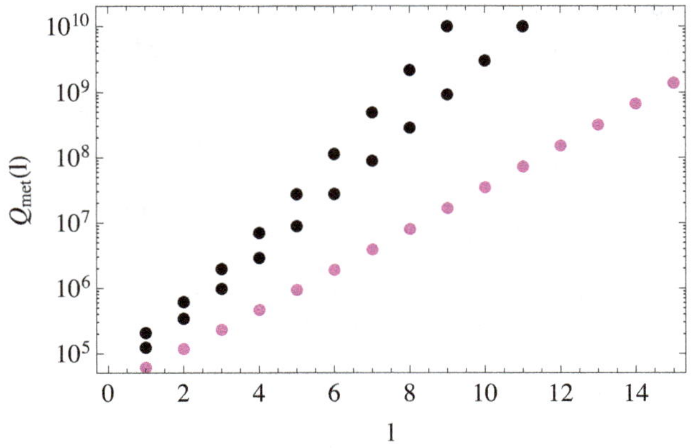

Figure 11. The quality Q_{met} vs. l for three spherical cavities with ideal dielectric spheres (see Figure 7).

$$\frac{1}{Q_s} = \frac{1}{Q_{\mathrm{met}}} + \frac{1}{Q_{\mathrm{diel}}}. \tag{66}$$

where Q_s is the total Q-factor of the spherical cavity. Values of Q_s versus l for three spherical cavities with dielectric spheres and $d = 4$ cm are shown in **Figure 12**. They are about three orders of magnitude larger than $Q_c = 1.633\times10^4$.

In a real accelerator, openings in the metallic wall are necessary for free penetration of the cavity by the electron beam, and to enable coupling between neighboring cavities. This will lower the quality Q_{met} but should have little effect on the total quality of the spherical cavity Q_s. The latter is defined by losses in the dielectric, see (66) where $Q_{\mathrm{diel}} \ll Q_{\mathrm{met}}$. The resonant frequency should not be drastically changed either, as the EM fields at the iris $(r = b)$ are very small fractions of their maxima, see **Figure 4**.

3.3. Discussion and Summary

When comparing the effective accelerating fields in the traditional pill box cavity with that in our spherical cavities with dielectric spheres, we first assume that U, the time-averaged energy stored in the cavity, see (56), is the same in both situations. Therefore the normalization factor $\mathcal{N}$ in (61) and (62) will first be chosen so that

$$2\pi h\frac{1}{2}\int_0^{R_c} \rho\mathrm{d}\rho\left[\left|E_x(\rho,t)\right|^2 + \left|B_\varphi(\rho,t)\right|^2\right] = 1 \tag{67}$$

see (42) ($U = 1/(8\pi)$ erg).

The complex effective accelerator field $\bar{E}$ for the cylindrical resonator shown in **Figure 1** $(R_c = h = d)$ is given by the right hand side of (45) in which E_x is defined by (61) with $\rho = 0$, and $\omega' = \omega_0$. The result is

$$\bar{E}_c = \mathcal{N}\frac{\sin\alpha}{\alpha}\mathrm{e}^{-i\omega_0 t_0},\ \alpha = \frac{\omega_0 d}{2c} \tag{68}$$

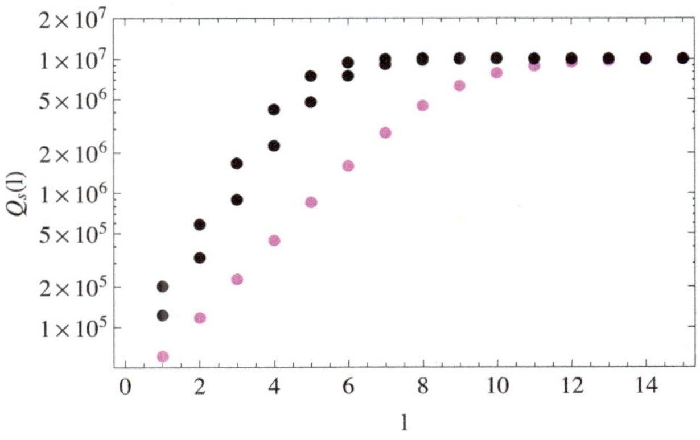

Figure 12. The quality Q_s vs. l for three spherical cavities with dielectric spheres (see Figure 7).

where t_0 is the time at which the electron passes the center of the cavity.

We now denote by E_{eff_s} and E_{eff_c} the maximal effective accelerating fields (equal to $|\bar{E}|$) for the spherical cavity with a dielectric sphere and the traditional cylindrical cavity, for any values of the average energies in the cavities, U_s and U_c. In view of the fact that E_x in (61) is proportional to $\sqrt{U}$ we can write, using the definition (55) of Q,

$$\frac{E_{\text{eff}_s}}{E_{\text{eff}_c}} = \sqrt{\frac{P_s}{P_c}} G(l) \tag{69}$$

where $G(l)$, the "gain factor", is given by

$$G(l) = \left.\frac{E_{\text{eff}_s}}{E_{\text{eff}_c}}\right|_{U_s=U_c} \sqrt{\frac{Q_s}{Q_c}\frac{\nu_c}{\nu_s}}. \tag{70}$$

Here ν_s and ν_c are the resonant frequencies of the spherical and the cylindrical cavities $(\nu = \omega_0/(2\pi))$ and P_s and P_c are the corresponding power losses. They are equal to the powers that must be supplied from external sources to sustain the oscillations. They should be as large as possible to avoid breakdown in the dielectric or at the metallic wall. Further research is necessary to give an estimate of the ratio P_s/P_c. We can only hope that it is not smaller than unity.

For our typical SLAC pill box cavity shown in **Figure 1** $(R_c = h = d = 4\text{ cm})$ we obtain $\nu_c = 2.87$ GHz and

$$E_{\text{eff}_c}\Big|_{U_c=1/(8\pi)\text{erg}} = 3.16 \text{ kV/m}$$

to be used in (70). The resulting values of the gain factor are shown in **Figure 13**.

The results of our calculation are shown in **Figures 3-13** for three reasonable values of

$$b/a = 3, 2, \frac{3}{2}.$$

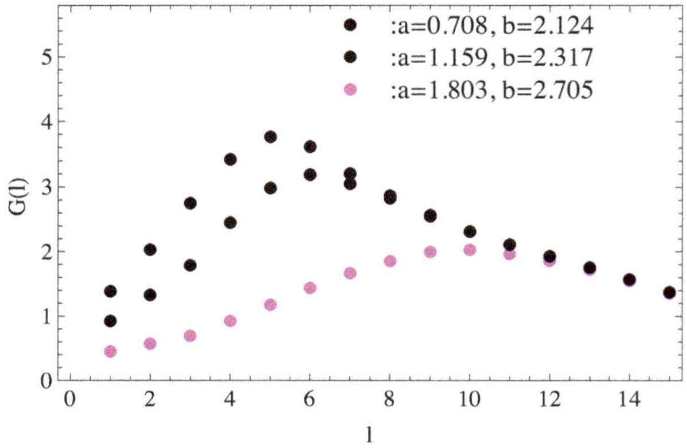

Figure 13. Gain factor $G(l)$ in (69) and (70) for three spherical cavities (a and b in cm).

The electron trajectory segment inside the cavity shown in **Figure 1** was equal to the typical length of the pill-box cavity of SLAC (4 cm). Calculations were performed for various values of l and m. The optimal parameters found were: $a = 0.708$ cm, $b = 2.124$ cm, and $(l,m) = (6,-2)$.

4. Conclusion

An electric field, intensified by structural resonance, can be used to accelerate electrons. This is demonstrated here by placing a dielectric sphere concentrically inside a spherical resonator, in which an appropriate whispering gallery mode is excited. A strong, accelerating field appears next to the surface of the dielectric. At the same time, the tangential component of the magnetic field at the wall of the resonator is minimal. This makes losses at the metallic walls negligible without engaging expensive cryogenic systems ensuring superconductivity of the walls. The Q factor of the resonator only depends on losses in the dielectric. For existing dielectrics, this gives a Q factor three orders of magnitude better than obtained in existing cylindrical cavities. Furthermore, for the proposed spherical cavity, all field components at the metallic wall are either zero or very small, see Figure 4. Therefore, one can expect the proposed spherical cavity to be less prone to electrical breakdowns than the traditional cylindrical cavity.

5. Acknowledgements

The authors would like to thank Professor Stanisław Kuliński for useful discussions.

REFERENCES

[1] W. Zakowicz, "Whispering-Gallery-Mode Resonances: A New Way to Accelerate Charged Particles," *Physical Review Letters*, Vol. 95, No. 11, 2005, Article ID: 114801. doi:10.1103/PhysRevLett.95.114801

[2] W. Zakowicz, "Erratum: Whispering-Gallery-Mode Resonances: A New Way to Accelerate Charged Particles [Phys. Rev. Lett. 95, 114801 (2005)]," *Physical Review Letters*, Vol. 97, No. 10, 2006, Article ID: 109901. doi:10.1103/PhysRevLett.97.109901

[3] W. Zakowicz, "Particle Acceleration by Wave Scattering off Dielectric Spheres at Whispering-Gallery-Mode Resonance," *Physical Review Special Topics—Accelerators and Beams*, Vol. 10, No. 10, 2007, Article ID: 101301. doi:10.1103/PhysRevSTAB.10.101301

[4] M. Ornigotti and A. Aiello, "Theory of Anisotropic Whispering-Gallery-Mode Resonators," *Physical Review A*, Vol. 84, No. 1, 2011, Article ID: 013828. doi:10.1103/PhysRevA.84.013828

[5] T. V. Liseykina, S. Pirner and D. Bauer, "Relativistic Attosecond Electron Bunches from Laser-Illuminated Dro-

plets," *Physical Review Letters*, Vol. 104, No. 9, 2010, Article ID: 095002. doi:10.1103/PhysRevLett.104.095002

[6] V. S. Ilchenko, A. A. Savchenkov, A. B. Matsko and L. Maleki, "Nonlinear Optics and Crystalline Whispering Gallery Mode Cavities," *Physical Review Letters*, Vol. 92, No. 4, 2004, Article ID: 043903. doi:10.1103/PhysRevLett.92.043903

[7] R. C. Taber and C. A. Flory, "Microwave Oscillators Incorporating Cryogenic Sapphire Dielectric Resonators," *IEEE Transactions on Ultrasonics, Ferroelectrics and Frequency Control*, Vol. 42, No. 1, 1995, pp. 111-119. doi:10.1109/58.368306

[8] J. Krupka, K. Derzakowski, M. E. Tobar, J. Hartnett and R. G. Geyer, "Complex Permittivity of Some Ultralow Loss Dielectric Crystals at Cryogenic Temperatures," *Measurement Science and Technology*, Vol. 10, No. 5, 1999, pp. 387-392. doi:10.1088/0957-0233/10/5/308

[9] J. D. Jackson, "Classical Electrodynamics," 3rd Edition, John Wiley, New York, 1998.

[10] A. Moroz, "A Recursive Transfer-Matrix Solution for a Dipole Radiating inside and outside a Stratified Sphere," *Annals of Physics*, Vol. 315, No. 2, 2005, pp. 352-418. doi:10.1016/j.aop.2004.07.002

Dielectric Relaxation in Complex Perovskite Oxide $Sr(Gd_{0.5}Nb_{0.5})O_3$

Pritam Kumar[1], Ajay Kumar Sharma[1], Bhrigunandan Prasad Singh[2], Tripurari Prasad Sinha[3], Narendra Kumar Singh[1*]

[1]Department of Physics, V. K. S. University, Ara, India; [2]Department of Physics, T. M. Bhagalpur University, Bhagalpur, India; [3]Department of Physics, Bose Institute, Kolkata, India.
Email: [*]singh_nk_phy27@yahoo.com

ABSTRACT

The complex perovskite oxide a strontium gadolinium niobate (SGN) synthesized by solid-state reaction technique has single phase with tetragonal structure. X-ray diffraction (XRD) technique and scanning electron microscopy (SEM) were used to study the structural and microstructural properties of the ceramics, respectively. The XRD patterns of SGN at room temperature show a tetragonal structure. Studies of the dielectric constant and dielectric loss of compound as a function of frequency (50 Hz to 1 MHz) at room temperature, and as a function of temperature (60˚C to 420˚C) indicate polydispersive nature of the material. The logarithmic angular frequency dependence of the loss peak is found to obey the Arrhenius law with activation energy ~0.18 eV. The small value of activation energy of the compound (~0.18 eV) can be explained by mixed ionic-polaronic conductivity mechanism. The grain size of the pellet sample was found to be 1.92 μm. The frequency-dependant electrical data are analyzed in the framework of conductivity and electric modulus formalisms. The complex plane impedance plot shows the grain boundary contribution for higher value of dielectric constant in the law frequency region.

Keywords: Dielectrics; Perovskite Oxides; X-Ray Diffraction; Scanning Electron Micrograph

1. Introduction

The family of compounds with general formula ABO_3 is generally called perovskite oxides, as their structural is similar to the naturally obtained $CaTiO_3$. The study on this compounds is important as they find several applications in non-linear optics, memory devices, pyroelectric, piezoelectric sensors etc. apart from the academic point of view due to physical properties they exhibits. The well-known examples are $BaTiO_3$, $PbZrO_3$, $PbTiO_3$ etc. Inherent definition of ferroelectricity is the existence of a polarization state, the direction of which can be reversed by an externally applied electric field [1]. This property makes ferroelectric materials obvious candidates for new device applications [2-4]. First attempt on the synthesis of complex perovskite was reported by Galasso and Pyle [5] and Galasso and Pinto [6] with the modification in the B-site. The structures that result when there exits perfect ordering in B-site with divalent and pentavalent ions in one set of compounds and trivalent and pentavalent ions in the other set of compounds (A = Pb and Ba: B′ = Mg, Zn, Y, Fe, Nd and Gd etc., and B″ = Nb and Ta). Some of the well known complex perovskite are $Ba(Zn_{1/3}Nb_{2/3})O_3$, $Sr(Zn_{1/3}Nb_{2/3})O_3$ [7,8], $(Sr_xLa_{1-x})MnO_3$ [9], etc. Perovskite materials with high dielectric constants have been used for technological applications such as wireless communication system, cellular phones and global positioning systems in the form of capacitors resonators and filters. Since, most of the devices are operated in the ac electrical mode, the investigation of ac electrical conduction of these materials is interesting. Complete information on correlation of sample topology with its electrical properties can be obtained by the complex impedance spectroscopy (CIS) technique [10-13]. This technique is used to investigate the electrical process occurring in the material on applying ac signal at the input. An analysis of the impedance data provides unique relaxation frequency describing the relaxation process occurring within a polycrystalline sample. High dielectric constant permits smaller capacitive components, thus enabling smaller size of electronic devices [14]. Complex perovskites are very promising for electroceramic applications and many researchers have shown considerable interest in the dielectric properties of these compounds. Materials having a diffuse phase transition (relaxors) have attracted the most attention due to their broad maximum in the temperature dependence of their dielec-

[*]Corresponding author.

tric constant [15,16]. COMPLEX Impedance Spectroscopy (CIS) allows measurement of the capacitance and tangent loss (tanδ) and/or conductance over a frequency range at various temperatures. From the measured capacitance and tanδ, four complex dielectric functions can be computed: impedance (Z^*), electric modulus (M^*), permittivity (ε^*), and admittance (Y^*). Studying dielectric data in the different functions allows different features of the materials to be recognized. The dielectric study of lead-free ternary compounds has been the focus of researchers in recent years and some interesting results have already been reported [17-21]. Recently, our group has also studied the dielectric relaxation behavior of some $A(B'B'')O_3$ type perovskite oxides [22-30]. In this paper we investigate the structural and dielectric relaxation properties of the strontium gadolinium niobate, $Sr(Gd_{0.5}Nb_{0.5})O_3$, (SGN) ceramics in the temperature range from 60˚C to 420˚C and in the frequency 50 Hz to 1 MHz by means of dielectric spectroscopy.

2. Experimental Procedures

Polycrystalline samples of SGN have been prepared by conventional solid-state reaction method. High purity reagents of $SrCO_3$ (99.9%), Nb_2O_5 (99.99%) and Gd_2O_3 (99.95%), are used for the preparation of SGN. The mixed raw materials are calcined at 1300˚C. First of all we have sintered the pellets at temperature 1350˚C but pellets were broken quickly. So we have annealed at 1100˚C for 10 h and cooled down to room temperature by adjusting the cooling rate. Pellets of SGN are made after mixing with binder (polyvinyl alcohol) using uniaxial press. The dielectric measurement is carried out on a pellet of 2.32 mm thickness and 8.47 mm diameter with silver paste for electrical contacts. After the application of silver paste, the samples are fired at 500˚C for 30 min. From the measurement, we have obtained capacitance (C) and tangent loss (tanδ) by using an LCR meter (Hioki) in the temperature range from 60˚C to 420˚C and in the frequency 50 Hz to 1 MHz. Using(C) and (tand), we have computed dielectric constant ($\varepsilon' = (C/C_0)$) and conductivity ($\sigma = \omega\varepsilon_0\varepsilon''$), where ε_0 is the dielectric permittivity in air, C/C_0 the ratio of capacitance measured with dielectric and without dielectric, ω the angular frequency and $\varepsilon'' = \tan\delta\varepsilon'$. The complex electric modulus M^* (= $1/\varepsilon^*$) and impedance Z^* (= $M^*/j\omega C_0$) are obtained from the temperature dependence of the real (ε') and imaginary (ε'') components of the dielectric permittivity ε^* (= $\varepsilon' - j\varepsilon''$).

3. Results and Discussion

The room temperature X-ray diffraction (XRD) pattern of $Sr(Gd_{0.5}Nb_{0.5})O_3$, (SGN) ceramics is shown in **Figure 1**. The XRD pattern (*i.e.*, peak position and intensity), which is different from those of ingredients, clearly shows

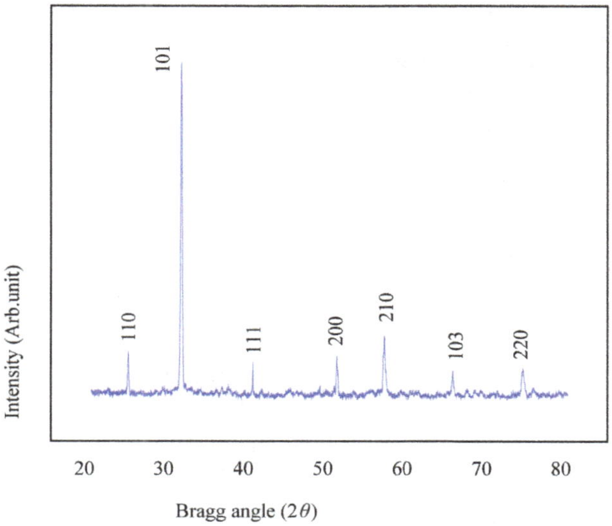

Figure 1. XRD of $Sr(Gd_{0.5}Nb_{0.5})$, (SGN), ceramics at room temperature.

the formation of a new compound. Using 2θ, interplaner spacing (d) of each peak was calculated. All the peaks were indexed in different crystal systems and cell configurations. The best agreement in observed (obs) and calculated (cal) 2θ and d values ($\sum \Delta d = d_{obs} - d_{cal}$ = minimum) was found for the tetragonal system. The least-squares refined unit cell parameters of the compound are: a = 3.6042 Å, c = 4.6481 Å (with estimated standard deviation in parenthesis).

These values suggest that cubic perovskite unit cell of ABO_3 (A = mono/divalent, B = tri-penta valent ions) type has been distorted to the tetragonal structure on the substitution of Gd^{3+}, Nb^{5+} of different radius at the B-site. Comparison of some observed (obs) and calculated (cal) d-values (in Å) of some reflections of SGN ceramics at room temperature with intensity ratio I/I_0 is shown in **Table 1**.

The scanning electron micrograph of the SGN material at magnification of 2 and 3 µm are shown in **Figure 2**. The scanning electron micrograph of the sample was recorded by FEI Quanta 200 equipment to check proper compactness of the sample.

The nature of the micrographs exhibits the polycrystalline texture of the material having highly distinctive and compact rectangular/cubical grain distributions (with less voids). The grain size of the pellet sample was found to be 1.92 µm. Careful examinations (scanning) of the complete surface of the sample exhibits that the grains are homogeneously distributed through out the surface of the sample.

The relation of angular frequency ω (= $2\pi\nu$) with dielectric constant (ε') and tangent loss (tanδ) at various temperatures for SGN is described in **Figure 3** and **Figure 4** respectively. The variation of ε' with frequency

Table 1. Comparison of some observed (obs) and calculated (cal) d-values (in Å) of some reflections of SGN ceramics at room temperature with intensity ratio I/I_0.

(hkl)	SGN d (Å)
(110)	[o] 3.6042 (24) [c] 3.6042
(101)	[o] 2.8482 (100) [c] 2.8482
(111)	[o] 2.2339 (18) [c] 2.2347
(200)	[o] 1.7905 (22) [c] 1.8021
(210)	[o] 1.6169 (28) [c] 1.6118
(103)	[o] 1.4234 (19) [c] 1.4234
(220)	[o] 1.2749 (19) [c] 1.2743

explains relaxation phenomena of the material, which are associated with a frequency-dependant orientation polarization. The value of ε' decreases to a constant value with increase in frequency (**Figure 3**) in the SGN compound, which may be attributed to the fact that, at lower frequency region the permanent dipoles align themselves along the direction of the field and contribute to the total polarization of the dielectric material. On the other hand, at higher frequency the variation in field is too rapid for the dipoles to align themselves in the direction of field, *i.e.*, dipoles can no longer follow the field, so their contribution to the total polarization and hence to the dielectric permittivity be come negligible. Therefore the dielectric constant (ε') decreases with increase in frequency. The high value of dielectric constant (ε') at lower frequencies, increases with decreasing frequency and increasing temperature correspond to bulk effect of the system [31].

Dielectric loss is the electrical energy lost as heat in the polarization process in the presence of an applied ac field. The energy is absorbed from the ac voltage and converted to heat during the polarization of the molecule. The dielectric loss is a function of frequency and temperature and is related to relaxation polarization, in which a dipole cannot follow the field variation without a measurable lag because of the retarding or friction forces of the rotating dipoles. Also, due to the change in polarization of the dielectric, a polarizing current flowing in the dielectric is induced by the relaxation rate. This current induces dielectric loss in the material. It is well known that in high frequency alternating fields there is always a phase difference between polarization and field, which gives the dissipation factor $\tan\delta$ (= $\varepsilon''/\varepsilon'$) and is propor-

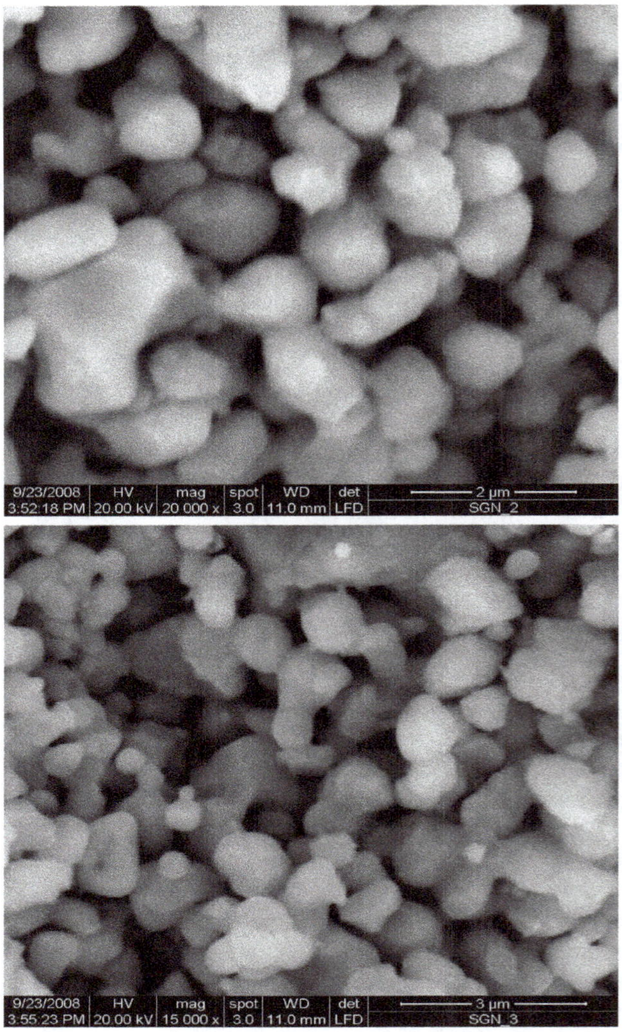

Figure 2. Scanning electron micrographs of $Sr(Gd_{0.5}Nb_{0.5})$, (SGN), ceramics at 2 and 3 µm.

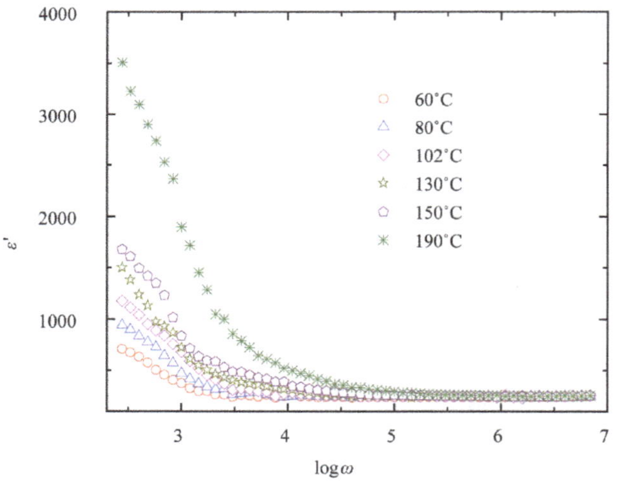

Figure 3. Frequency dependence of dielectric constant (ε') of SGN at various temperatures.

tional to the energy absorbed per cycle by the dielectric from the field. It is observed that from **Figure 4** tangent loss (tanδ) increases with increasing temperature and decreasing frequency indicating that the concentration of conduction electrons increases as temperature due to thermal activation [18].

The variation of dielectric constant (ε') of SGN as a function of temperature at few selected frequencies is shown in **Figure 5**. The variation in ε' with temperature shows that it is strongly frequency and temperature dependent.

At all the frequencies, the relative dielectric constant increases linearly with rise in temperature. The increase in ε' is more prominent at lower frequencies. However, no phase transition was observed in SGN in the experimental temperature range [32].

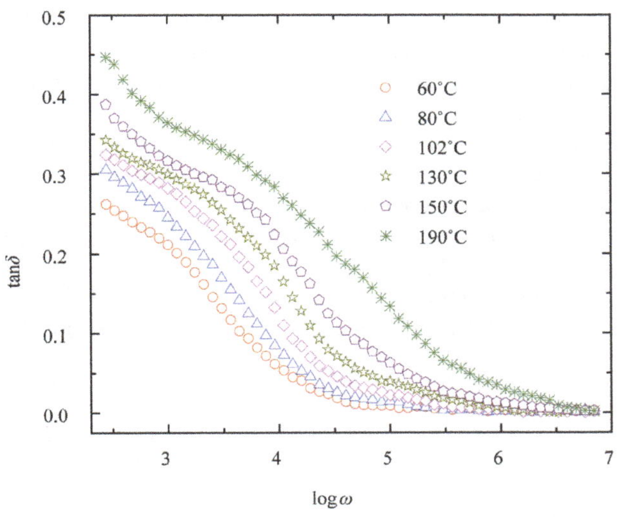

Figure 4. Variation of tanδ with frequency of SGN at various temperatures.

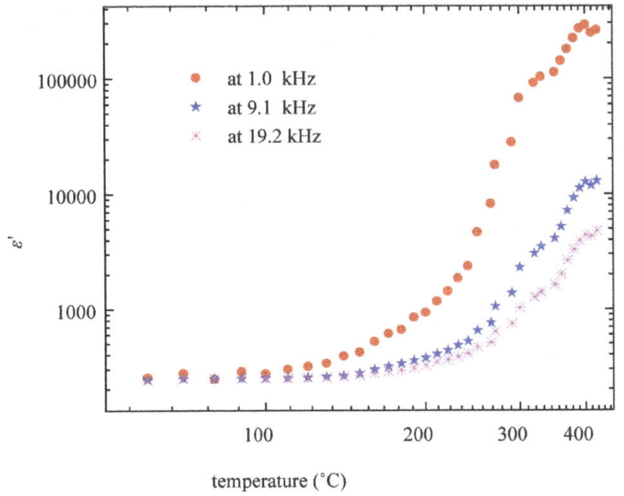

Figure 5. Variation of dielectric constant (ε') of SGN with temperature at selected frequencies.

Figure 6 shows the variation of loss tangent (tanδ) of SGN as a function of temperature at selected frequencies. It revealed 1) a peak in tanδ at a certain temperature for a particular frequency, 2) shift in the peak-temperature towards higher temperature side with rise in frequency, and 3) increase in the broadening of the peak with rise in frequency. It is clear from the figure that a temperature dependent dielectric relaxation is observed at the experimental frequencies (1, 9.1 and 19.2 kHz).

The variation of real (Z') and imaginary (Z'') part of impedance with frequency at different temperatures is shown in **Figure 7**. A monotonous decrease in the real part of impedance with frequency up to near about 15 kHz, followed by a (nearly) constant value, was observed at all the temperatures. Similar behavior was found in the variation of Z'' with frequency. The merger/coincidence of the impedance (Z') at higher frequencies for all the temperatures indicates a possible release of space charge polarization at high temperatures and frequencies.

Complex impedance spectroscopy is a technique that enables us to separate the real and imaginary component of the complex electrical parameters so as to get the true picture of material properties. It is used to characterize microstructural and electrical properties of some electronic and/or ionic materials. Generally, the impedance properties of materials arise due to intragrain, intergrain and electrode processes. The motion of charges could occur in a number of ways: 1) dipole reorientation, 2) space charge formalism and 3) charge displacement. Thus, the complex impedance formalism allows for a direct separation of the bulk, grain boundary and the electrode phenomena.

The plot of the imaginary (Z'') versus real (Z') parts of the complex impedance (Cole-Cole plot) at temperature 150°C is shown in **Figure 8**. For a bulk crystal contain-

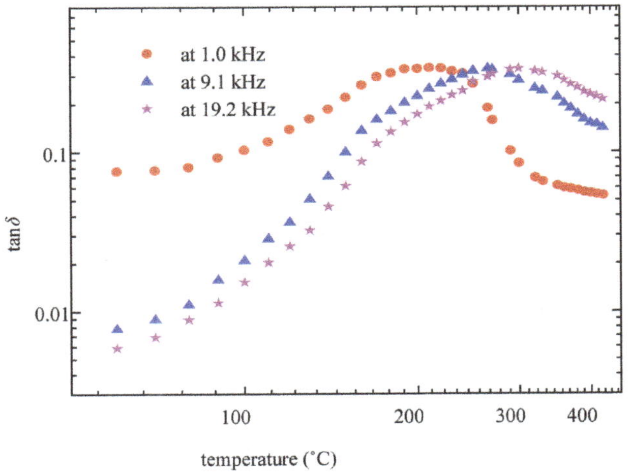

Figure 6. Dielectric loss (tanδ) of SGN with temperature at selected frequencies.

ing interfacial boundary layer (grain-boundary), the equivalent circuit may be considered as two parallel RC elements connected in series (inset of **Figure 8**) and gives rise to two arcs in complex plane, one for bulk crystal (grain) and the other for the interfacial boundary (grain-boundary) response [12]. The real (Z′) and imaginary (Z″) part of total impedance of the equivalent circuit are defined as [33].

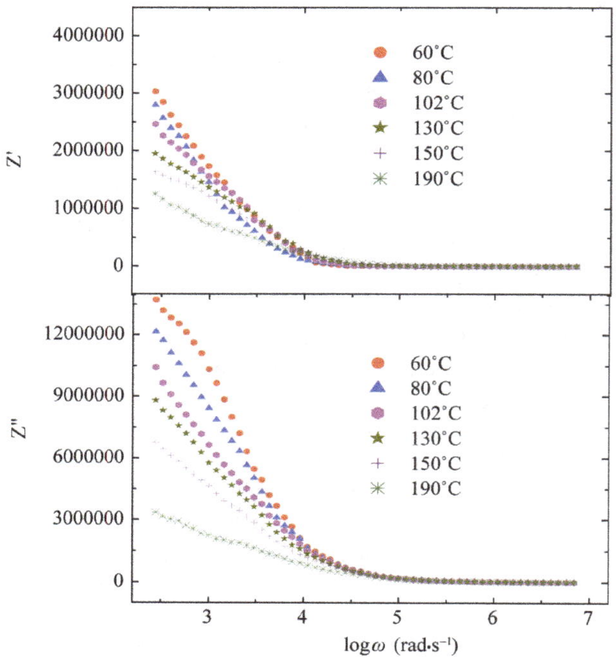

Figure 7. Plots of Z′ and Z″ with frequency at different temperatures of SGN.

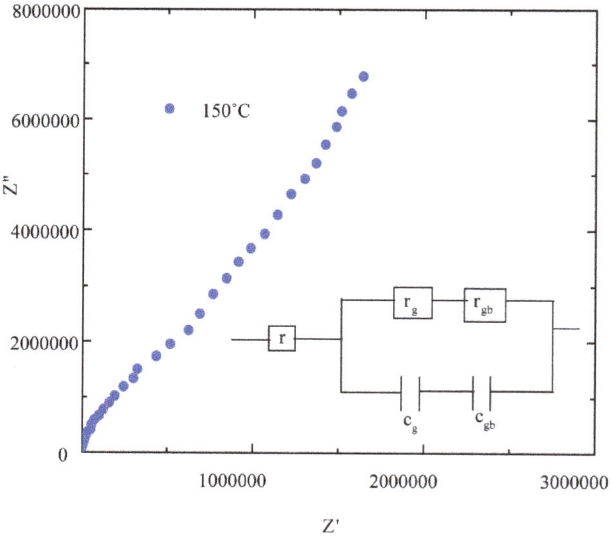

Figure 8. Complex plane impedance plot of SGN at 150°C and parallel RC equivalent circuit (inset).

$$Z' = \frac{r_g}{1+\left(\omega r_g c_g\right)^2} + \frac{r_{gb}}{1+\left(\omega r_{gb} c_{gb}\right)^2} \quad (1)$$

$$Z'' = r_g\left[\frac{\omega r_g c_g}{1+\left(\omega r_g c_g\right)^2}\right] + r_{gb}\left[\frac{\omega r_{gb} c_{gb}}{1+\left(\omega r_{gb} c_{gb}\right)^2}\right] \quad (2)$$

where r_g, c_g are the bulk (grain) resistance and capacitance respectively and r_{gb}, c_{gb} are the corresponding quantities for interfacial boundary (grain boundary). The relative position of the two arcs in a complex plane can be identified by the frequency. The arc of bulk generally lies on the frequency range higher than that of interfacial boundary since the relaxation time $\tau_m = 1/\omega_m$ for the interfacial boundary is much larger than that for the bulk crystal. Hence, when the bulk resistance (r_g) is much lower and the resistance in the equivalent circuit is dominated by the interfacial boundary resistance (r_{gb}), the arc of bulk (grain) may be masked in the limited frequency range.

Advantage of complex electric modulus formalism is that it can easily distinguish electrode polarization effect from grain boundary conduction process. It is also useful in detecting bulk properties as apparent conductivity relaxation times [34,35].

The complex impedance modulus ($M^*(\omega)$) has been calculated from the impedance data using the following relations:

$$M^*(\omega) = \frac{1}{\varepsilon^*} = \frac{1}{\varepsilon' - j\varepsilon''} = \frac{\varepsilon'}{\varepsilon'^2 + \varepsilon''^2} + j\frac{\varepsilon'}{\varepsilon'^2 + \varepsilon''^2} \quad (3)$$
$$= M'(\omega) + jM''(\omega)$$

where M′ is the real and M″ the imaginary electric modulus, and ε' the real and ε'' the imaginary permittivity.

Figure 9 shows the angular frequency dependence of $M'(\omega)$ and $M''(\omega)$ for SGN as a function of temperature. $M'(\omega)$ shows a dispersion tending towards M_∞ (the asymptotic value of $M'(\omega)$ at higher frequencies in **Figure 9(a)**, while $M''(\omega)$ exhibits a maximum (M''_m) centered at the dispersion region of $M'(\omega)$. In **Figure 9(b)** the position of the peak M''_m shifts to higher frequencies as the temperature is increased. The frequency region below peak maximum M″ determines the range in which charge carriers are mobile on long distances. At frequency above peak maximum M″ the carriers are confined to potential wells, being mobile on short distances.

The most probable relaxation time follows the Arrhenius law given by

$$\omega_m = \omega_0 \exp\left[\frac{-E_a}{k_B T}\right] \quad (4)$$

where ω_0 is the pre-exponential factor and E_a is the activation energy. A plot of $\log\omega$ vs. $1/T$ is shown in **Figure 10**, where the symbols are the experimental data points and the solid line is the least squares straight line fit to the data.

Using Equation (4), the calculated value of the activtion energy E_a of the sample was found to be ~0.18 eV.

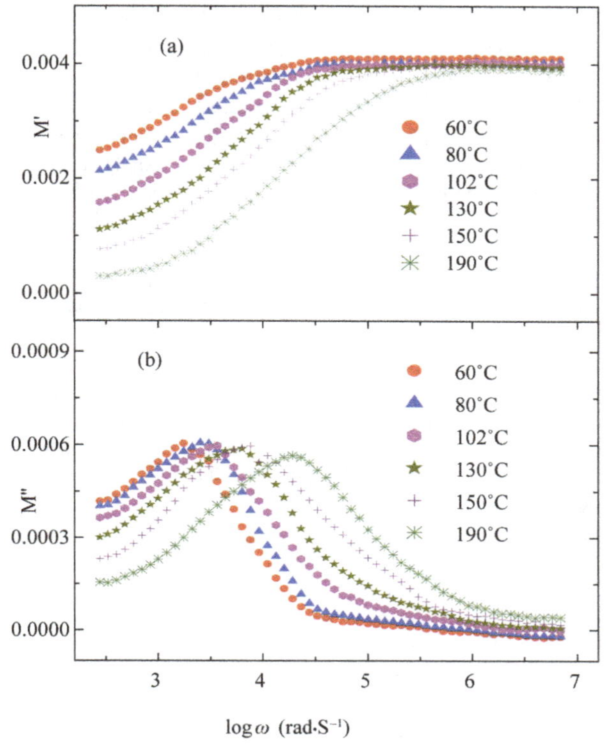

Figure 9. Logarithmic angular frequency dependence of M′ (a) and M″ (b) of $Sr(Gd_{0.5}Nb_{0.5})O_3$, SGN at various temperatures.

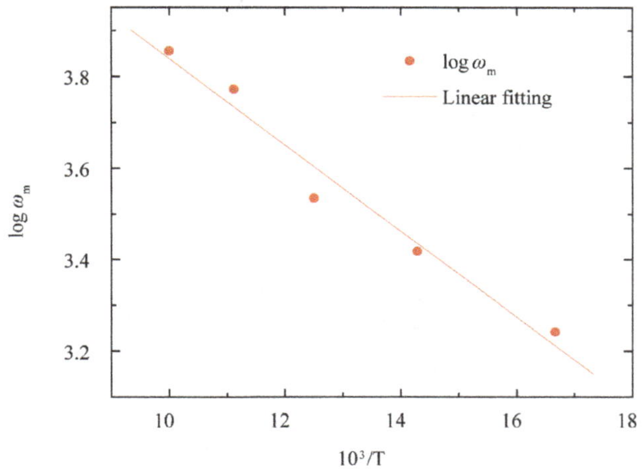

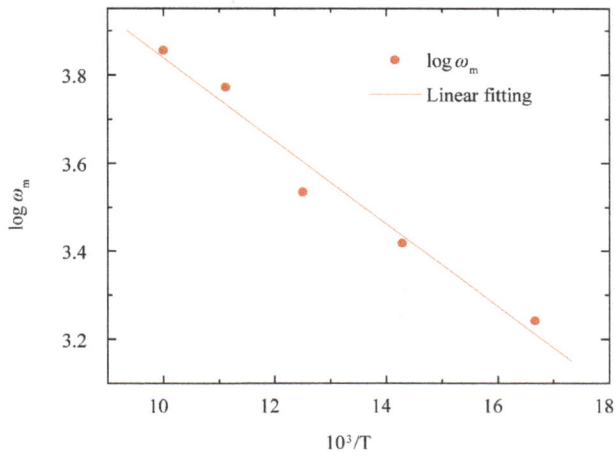

Figure 10. The Arrhenius plot of ω_m corresponding to M″ for SGN, where symbols are the experimental points and the solid line is the least squares fit to the experimental data.

The low value of activation energy (~0.18 eV), may be due to the single ionized oxygen vacancies. Therefore, the role of the mixed ionic-polaronic conductivity is realized, since it includes the charge carriers generated from the vacancies. In addition, a collective motion of oxygen ions and vacancies play important role in the conduction process [36,37].

Conductivity analysis provides significant information related to transport of charge carriers, *i.e.*, electron/holes or cations/anions that predominate the conduction process and their response as a function of temperature and frequency. If we assumes that all dielectric loss in the temperature range studied are due to conductivity, the conductivity can be expressed as $\sigma(\omega) = \omega\varepsilon_0\varepsilon''$, here σ is the real part of the conductivity and ε'' is the imaginary part of complex dielectric permittivity (ε^*).

The frequency spectra of the conductivity for SGN at different measuring temperatures are shown in **Figure 11**. The conductivity shows dispersion which shifts to higher frequency side with the increase of temperature. It is seen from figure that $\log\sigma_{ac}$ decreases with decreasing frequency but at higher temperature (200˚C), it is observed that $\log\sigma_{ac}$ becomes independent of frequency after a certain value. Extrapolation of this part towards lower frequency will give σ_{dc}. The very basic fact about AC conductivity in SGN is that σ is an increasing function of frequency. The real parts of conductivity spectra can be explained by the power law define as [38]

$$\sigma = \sigma_{dc}\left[1+\left(\frac{\omega}{\omega_H}\right)^n\right] \qquad (5)$$

where σ_{dc} is the DC conductivity, ω_H is the hopping frequency of the charge carriers, and n is the dimensionless

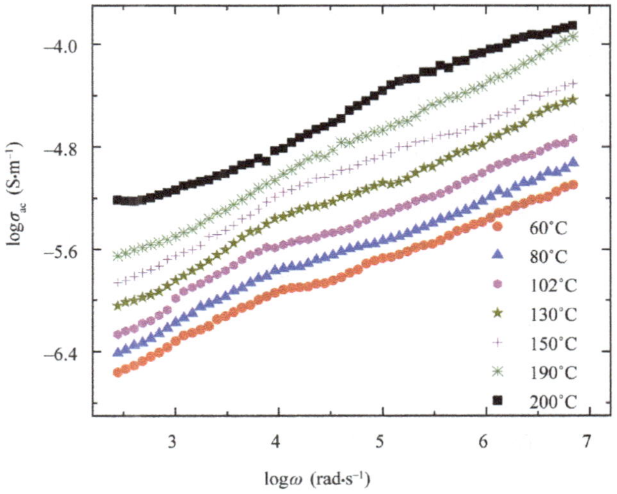

Figure 11. Logarithmic angular frequency dependence of conductivity ($\log\sigma_{ac}$) for $Sr(Gd_{0.5}Nb_{0.5})O_3$ at various temperatures.

frequency exponent.

4. Conclusion

The frequency-dependent dielectric dispersion of polycrystalline $Sr(Gd_{0.5}Nb_{0.5})O_3$, (SGN) ceramic synthesized by a high-temperature solid-state reaction technique is investigated in the temperature range from 60˚C to 420˚C for the first time. The increasing dielectric constant (ε') with increasing temperature is attributed to the conductivity which is directly related to an increase in mobility of localized charge carriers. The logarithmic angular frequency dependence of the electric modulus loss peak ($\log\omega_m$ vs. M'') is found to obey the Arrhenius law with activation energy of ~0.18 eV. The frequency-dependent electrical data are also analyzed in the framework of conductivity and modulus formalisms.

REFERENCES

[1] M. E. Lines and A. M. Glass, "Principles and Applications of Ferroelectrics and Related Materials," Clarendon, Oxford, 2001.

[2] J. F. Scott, "Ferroelectric Memories," Springer, Berlin, 2000.

[3] O. Auciello, J. F. Scott and R. Ramesh, "The Physics of Ferroelectric Memories," *Physics Today*, Vol. 51, No. 7, 1998, p. 22. doi:10.1063/1.882324

[4] A. K. Singh, S. K. Barik, R. N. P. Choudhary and P. K. Mahapatra, "Ac Conductivity and Relaxation Mechanism in $Ba_{0.9}Sr_{0.1}TiO_3$," *Journal of Alloys and Compounds*, Vol. 479, No. 1-2, 2009, pp. 39-42. doi:10.1016/j.jallcom.2008.12.130

[5] F. Galasso and J. Pyle, "Ordering in Compounds of the $A(B'_{0.33}Ta_{0.67})O_3$ Type," *Inorganic Chemistry*, Vol. 2, No. 3, 1963, pp. 482-484. doi:10.1021/ic50007a013

[6] F. Galasso and J. Pinto, "Growth of Single Crystals of $Ba(B'_{0.33}Ta_{0.67})O_3$ Perovskite-Type Compounds," *Nature*, Vol. 207, 1965, pp. 70-72. doi:10.1038/207070b0

[7] E. L. Colla, I. M. Reaney and N. Setter, "Effect of Structural Changes in Complex Perovskite on the Temperature Coefficient of the Relative Permittivity," *Journal of Applied Physics*, Vol. 74, No. 5, 1993, pp. 3414-3425. doi:10.1063/1.354569

[8] M. Onada, J. Kuwata, K. Toyama and S. Nomura, "Ba $(Zn_{1/3}Nb_{2/3}O_3$-$Sr(Zn_{1/3}Nb_{2/3})O_3$, Solid Solution Ceramics with Temperature Stable High Dielectric Constant and Low Microwave Loss," *Japanese Journal of Applied Physics*, Vol. 21, 1982, pp. 1707-1710. doi:10.1143/JJAP.21.1707

[9] E. Graado, O. Moreno, A. Gracia, J. A. Sanjurjo, C. Rettori, I. Torriani, S. Oseroff, J. J. Neumeier, K. J. McClallan, S. W. Cheong and Y. Tokura, "Phonon Raman Scattering in $R_{1-x}A_xMnO_{3+\delta}$ (R = La, Pr; A = Ca, Sr)," *Physical Review B*, Vol. 58, No. 17, 1998, pp. 11435-11440. doi:10.1103/PhysRevB.58.11435

[10] N. Ortega, A. Kumar, P. Bhattacharya, S. B. Majumder and R. S. Katiyar, "Impedance Spectroscopy of Multiferroic $PbZr_xTi_{1-x}O_3/CoFe_2O_4$ Layered Thin Films," *Physical Review B*, Vol. 77, No. 1, pp. 14111-14120. doi:10.1103/PhysRevB.77.014111

[11] D. Perez-Coll, D. Marrero-Lopez, P. Nunez, S. Pinol and J. R. Frade, "Grain Boundary Conductivity of $Ce_{0.8}Ln_{0.2}$ $O_{2-\delta}$ Ceramics (Ln = Y, La, Gd, Sm) with and without Co-Doping," *Electrochimica Acta*, Vol. 51, No. 28, 2006, pp. 6463-6469. doi:10.1016/j.electacta.2006.04.032

[12] S. Saha and T. P. Sinha, "Dielectric Relaxation in $SrFe_{1/2}$ $Nb_{1/2}O_3$," *Journal of Applied Physics*, Vol. 99, No. 1, 2006, Article ID: 014109. doi:10.1063/1.2160712

[13] A. Shukla, R. N. P. Choudhary and A. K. Thakur, "Thermal, Structural and Complex Impedance Analysis of Mn^{4+} Modified $BaTiO_3$ Electroceramic," *Journal of Physics and Chemistry of Solids*, Vol. 70, No. 11, 2009, pp. 1401-1407. doi:10.1016/j.jpcs.2009.08.015

[14] R. J. Cava, "Dielectric Materials for Applications in Microwavecommunications," *Journal of Materials Chemistry*, Vol. 11, No. 1, 2001, pp. 54-62. doi:10.1039/b0036811

[15] U. Intatha, S. Eitssayeam, K. Pengpat, J. D. MacKenzie and T. T. Kenneth, "Dielectric Properties of Low Temperature Sintered LiF Doped $BaFe_{0.5}Nb_{0.5}O_3$," *Materials Letters*, Vol. 61, No. 1, 2007, pp. 196-200. doi:10.1016/j.matlet.2006.04.030

[16] F. Roulland, R. Terras, G. Allainmat, M. Pollet and S. Marinel, "Lowering of $BaB'_{1/3}B''_{2/3}$ Complex Perovskite Sintering Temperature by Lithium Salt Additin," *Journal of the European Ceramic Society*, Vol. 24, No. 6, 2004, pp. 1019-1023. doi:10.1016/S0955-2219(03)00553-3

[17] S. Priya, A. Ando and Y. Sakebe, "Non-Lead Perovskite Materials $Ba(Li_{1/4}Nb_{3/4})O_3$ and $Ba(Cu_{1/3}Nb_{2/3})O_3$," *Journal of Applied Physics*, Vol. 94, No. 2, 2003, pp. 1171-1177. doi:10.1063/1.1585121

[18] I. Levin, J. Y. Chan, J. E. Maslar and T. A. Vanderah, "Phase Transitions and Microwave Dielectric Properties in the Perovskite-Like $Ca(Al_{0.5}Nb_{0.5})O_3$-$CaTiO_3$ System," *Journal of Applied Physics*, Vol. 90, No. 2, 2001, pp. 904-914. doi:10.1063/1.1373705

[19] R. Zurmuhlen, J. Petzelt, S. Kamba, V. V. Voitsekhovskii, E. Colla and N. Setter, "Dielectric Materials for Wireless Communication," *Journal of Applied Physics*, Vol. 77, 1995, p. 5341.

[20] R. Zurmuhlen, J. Petzelt, S. Kamba, G. Kozlov, A. Volkov, B. Gorshunov, D. Dube, A. Tagantsev and N. Setter, "Dielectric Spectroscopy of $Ba(B_{1/2}'B_{1/2}'')O_3$ complex Perovskite Ceramics: Correlations between Ionic Parameters and Microwave Dielectric Properties. II. Studies Below the Phonon Eigenfrequencies (10^2 - 10^{12} Hz)," *Journal of Applied Physics*, Vol. 77, No. 10, 1995, pp. 5351-5364. doi:10.1063/1.359290

[21] S. Saha and T. P. Sinha, "Structural and Dielectric Studies of $Ba(Fe_{0.5}Nb_{0.5})O_3$," *Journal of Physics: Condensed Matter*, Vol. 14, No. 2, 2002, pp. 249-258. doi:10.1088/0953-8984/14/2/311

[22] N. K. Singh, P. Kumar, H. Kumar and R. Rai, "Structural and Dielectric Properties of $Dy_2(Ba_{0.5}R_{0.5})_2O_7$ (R = W, Mo) Ceramics," *Advanced Materials Letters*, Vol. 1, 2010,

pp. 79-82. doi:10.5185/amlett.2010.3102

[23] N. K. Singh, P. Kumar, O. P. Roy and R. Rai, "Structural, and Dielectric Properties of $Eu_2(B'_{0.5}\ B''_{0.5})_2O_7$ (B' = Ba; B'' = Mo, W) Ceramics," *Journal of Alloys and Compounds*, Vol. 507, No. 2, 2010, pp. 542-546. doi:10.1016/j.jallcom.2010.08.015

[24] N. K. Singh, R. N. P. Choudhary and B. Banarji, "Dielectric and Electrical Characteristics of $Nd_2(Ba_{0.5}R_{0.5})_2O_7$ (R = W, Mo) Ceramics," *Physica B: Condensed Matter*, Vol. 403, No. 10-11, 2008, pp. 1673-1677. doi:10.1016/j.physb.2007.09.083

[25] P. Kumar, B. P. Singh, T. P. Sinha and N. K. Singh, "Ac Conductivity and Dielectric Relaxation in $Ba(Sm_{1/2}Nb_{1/2})O_3$, Ceramic," *Physica B: Condensed Matter*, Vol. 406, No. 2, 2011, pp. 139-143. doi:10.1016/j.physb.2010.09.019

[26] N. K. Singh, P. Kumar and R. Rai, "Study of Structural, Dielectric and Electrical Behavior of $(1-x)Ba(Fe_{0.5}Nb_{0.5})O_{3-x}SrTiO_3$ Ceramics," *Journal of Alloys and Compounds*, Vol. 509, No. 6, 2011. pp. 2957-2963. doi:10.1016/j.jallcom.2010.11.168

[27] P. Kumar, B. P. Singh, T. P. Sinha and N. K. Singh, "X-Ray and Electrical Properties of $Ba(Gd_{0.5}Nb_{0.5})O_3$ Ceramic," *Advanced Materials Letters*, Vol. 2, No. 1, 2011, p. 76. doi:10.5185/amlett.2010.11176

[28] N. K. Singh, P. Kumar and R. Rai, "Comparative Study of Structure, Dielectric and Electrical Behavior of $Ba(Fe_{0.5}Nb_{0.5})O_3$ Ceramics and Their Solid Solutions with $BaTiO_3$," *Advanced Materials Letters*, Vol. 2, No. 3, 2011, pp. 200-205. doi:10.5185/amlett.2010.11178

[29] N. K. Singh and P. Kumar, "Studies of Structural and Electrical Behavior of Samarium Barium Tungstate Ceramics," *Journal of Advanced Dielectrics*, Vol. 1, No. 4, 2011, pp. 465-470. doi:10.1142/S2010135X11000495

[30] N. K. Singh, P. Kumar, A. K. Sharma and R. N. P. Choudhary, "Structural and Impedance Spectroscopy Studies of $Ba(Fe_{0.5}Nb_{0.5})O_3$-$SrTiO_3$ Ceramic System," *Materials Science and Applications*, Vol. 2, 2011, pp. 1593-1600. doi:10.4236/msa.2011.211213

[31] D. K. Mahato, A. Dutta and T. P. Sinha, "Dielectric Relaxation and Ac Conductivity of Double Perovskite Oxide Ho_2ZnZrO_6," *Physica B: Condensed Matter*, Vol. 406, No. 13, 2011, pp. 2703-2708. doi:10.1016/j.physb.2011.04.012

[32] R. N. P. Choudhary, D. K. Pradhan, C. M. Tirado, G. E. Bonilla and R. S. Katiyar, "Structural, Dielectric and Impedance Properties of $Ca(Fe_{2/3}W_{1/3})O_3$ Nanoceramics," *Physica B*, Vol. 393, No. 1-2, 2007, pp. 24-31. doi:10.1016/j.physb.2006.12.006

[33] A. Dutta, C. Bharti and T. P. Sinha, "AC Conductivity and Dielectric Relaxation in $CaMg_{1/3}Nb_{2/3}O_3$," *Materials Research Bulletin*, Vol. 43, No. 5, 2008, pp. 1246-1254. doi:10.1016/j.materresbull.2007.05.023

[34] D. C. Sinclair and A. R. West, "Impedance and modulus Spectroscopy of Semiconducting $BaTiO_3$ Showing Positive Temperature Coefficient of Resistance," *Journal of Applied Physics*, Vol. 66, No. 8, 1989, pp. 3850-3856. doi:10.1063/1.344049

[35] I. M. Hodge, M. D. Ingram and A. R.West, "A New Method for Analysing the a.c. Behaviour of Polycrystalline Solid Electrolytes," *Journal of Electroanalytical Chemistry and Interfacial Electrochemistry*, Vol. 58, No. 2, 1975, pp. 429-432. doi:10.1016/S0022-0728(75)80102-1

[36] G. Deng, G. Li, A. Ding and Q. Yin, "Evidence for Oxygen Vacancy Including Spontaneous Normal-Relaxor Transition in Complex Perovskite Ferroelectrics," *Applied Physics Letters*, Vol. 87, No. 192, 2005, p. 905.

[37] A. Molak, E. Talik, M. Kruczek, M. Plauch, A. Ratuszna and Z. Ujma, "Characterisation of $Pb(Mn_{1/3}Nb_{2/3})O_3$ Ceramics by SEM, XRD, XPS and Dielectric Permittivity Tests," *Materials Science and Engineering: B*, Vol. 128, No. 1-3, 2006, pp. 16-24. doi:10.1016/j.mseb.2005.11.011

[38] E. F. Hairetdinov, N. F. Uvarov, H. K. Patel and S. W. Martin, " Estimation of the Free Charge Carrier Concentration in Fast Ion Conducting $Na_2Sb_2S_3$ Glasses from an Analysis of the Frequency Dependent Conductivity," *Physical Review B*, Vol. 50, No. 18, 1994, pp. 13259-13266. doi:10.1103/PhysRevB.50.13259

Permissions

All chapters in this book were first published by Scientific Research Publishing; hereby published with permission under the Creative Commons Attribution License or equivalent. Every chapter published in this book has been scrutinized by our experts. Their significance has been extensively debated. The topics covered herein carry significant findings which will fuel the growth of the discipline. They may even be implemented as practical applications or may be referred to as a beginning point for another development.

The contributors of this book come from diverse backgrounds, making this book a truly international effort. This book will bring forth new frontiers with its revolutionizing research information and detailed analysis of the nascent developments around the world.

We would like to thank all the contributing authors for lending their expertise to make the book truly unique. They have played a crucial role in the development of this book. Without their invaluable contributions this book wouldn't have been possible. They have made vital efforts to compile up to date information on the varied aspects of this subject to make this book a valuable addition to the collection of many professionals and students.

This book was conceptualized with the vision of imparting up-to-date information and advanced data in this field. To ensure the same, a matchless editorial board was set up. Every individual on the board went through rigorous rounds of assessment to prove their worth. After which they invested a large part of their time researching and compiling the most relevant data for our readers.

The editorial board has been involved in producing this book since its inception. They have spent rigorous hours researching and exploring the diverse topics which have resulted in the successful publishing of this book. They have passed on their knowledge of decades through this book. To expedite this challenging task, the publisher supported the team at every step. A small team of assistant editors was also appointed to further simplify the editing procedure and attain best results for the readers.

Apart from the editorial board, the designing team has also invested a significant amount of their time in understanding the subject and creating the most relevant covers. They scrutinized every image to scout for the most suitable representation of the subject and create an appropriate cover for the book.

The publishing team has been an ardent support to the editorial, designing and production team. Their endless efforts to recruit the best for this project, has resulted in the accomplishment of this book. They are a veteran in the field of academics and their pool of knowledge is as vast as their experience in printing. Their expertise and guidance has proved useful at every step. Their uncompromising quality standards have made this book an exceptional effort. Their encouragement from time to time has been an inspiration for everyone.

The publisher and the editorial board hope that this book will prove to be a valuable piece of knowledge for researchers, students, practitioners and scholars across the globe.

List of Contributors

Taha A. Hanafy
Department of Physics, Faculty of Science, Tabuk University, Tabuk, KSA
Physics Department, Faculty of Science, Fayoum University, El Fayoum, Egypt

Natt Makul
Faculty of Industrial Technology, Phranakhon Rajabhat University, Bangkok, Thailand

Fang Chen, Sheng Mao and Xiaohui Wang
Electrical and Computer Engineering Department, Michigan Technological University, Houghton, USA

Elena Semouchkina
Electrical and Computer Engineering Department, Michigan Technological University, Houghton, USA
Materials Research Insti- tute, The Pennsylvania State University, University Park, USA

Michael Lanagan
Materials Research Insti- tute, The Pennsylvania State University, University Park, USA

Daniel Tan, Yang Cao, Enis Tuncera and Patricia Irwin
Dielectrics & Electrophysics Lab, GE Global Research Center, Niskayuna, Schenectady, USA

Ryuhei Kinjo, Iwao Kawayama, Hironaru Murakami and Masayoshi Tonouchi
Institute of Laser Engineering, Osaka University, Osaka, Japan

P. Koteeswari and P. Mani
Department of Physics, Hindustan Institute of Technology, Padur, India

S. Suresh
Department of Physics, Loyola College, Chennai, India

Mohd. Nasir and M. Zulfequar
Department of Physics, Jamia Millia Islamia, New Delhi, India

Moganti Venkata Someswara Rao
Department of Physics, S.R.K.R. Engineering College, Bhimavaram, India

Kocharlakota Venkata Ramesh, Majeti Naga Venkata Ramesh and Bonthula Srinivasa Rao
Department of Physics, GIT, GITAM University, Visakhapatnam, India

Xiaogang Yao, Wei Chen and Lan Luo
Shanghai Institute of Ceramics, Chinese Academy of Science, Shanghai, China

Safenaz M. Reda
Chemistry Department, Faculty of Science, Benha University, Benha, Egypt

Sheikha M. Al-Ghannam
Chemistry Department, College of Girls for Science, University of Dammam, Dammam, KSA

Viktoria Evgenèvna Buravtsova, Elena Alexandrovna Ganshina and Sergey Alexandrovich Kirov
Physics Faculty, Lomonosov Moscow State University, Moscow, Russia

Yuriy Egorovich Kalinin and Alexandr Viktorovich Sitnikov
Voronezh State Technical University, Voronezh, Russia

Juan Manuel Navarrete
Inorganic and Nuclear Chemistry Department, Faculty of Chemistry, National University of Mexico, Mexico City, Mexico

Gustavo Leonardo Martínez
National Coordination to Restore the Cultural Inheritance, National Institute of Anthropology and History, Mexico City, Mexico

Hany M. Zamel and Essam El Diwany
Electronics Research Institute (ERI), Microwave Engineering Department, Cairo, Egypt

Hadia El Hennawy
Faculty of Engineering, Ain Shams Uni-versity, Cairo, Egypt

Hayet Menasra, Zelikha Necira, Karima Bouneb, Abdelhak Maklid and Ahmed Boutarfaia
Applied Chemistry Laboratory, Exact and Natural and Life Sciences Faculty, Materials Science Department, Mohamed Kheider Uni- versity of Biskra, Biskra, Algeria

M. Raghasudha
Department of Chemistry, Jayaprakash Narayan College of Engineering, Mahabubnagar, India

D. Ravinder
Department of Physics, Nizam College, Basheerbagh Osmania University, Hyderabad, India

P. Veerasomaiah
Department of Chemistry, Osmania University, Hyderabad, India

M. P. Binitha and P. P. Pradyumnan
Department of Physics, University of Calicut, Kerala, India

Yongdong Li, Qing-da Meng, Po, Yang Zheyuan Zhao, Wei Zhang and Zhuo Pan
Jibei Electric Power Maintenance Company, Beijing, China

Sagadevan Suresh
Crystal Growth Centre, Anna University, Chennai, India

K. Rama Krishna
Department of Physics, Malla Reddy College of Engineering & Technology, Secunderabad, India

Dachepalli Ravinder
Department of Physics, P.G. College of Science, Osmania University, Hyderabad, India

K. Vijaya Kumar
Department of Physics, Jawaharlal Nehru Technological University, Hyderabad, India

Utpal S. Joshi and V. A. Rana
Department of Physics, Gujarat University, Ahmedabad, India;

Abrham Lincon
Department of Chemsitrys, P.G. College of Science, Osmania University, Hyderabad, India

Sakri Adel and Boutarfaia Ahmed
Laboratory of Applied Chemistry, University of Biskra, Biskra, Algeria

Delci Zion and Shyamala Devarajan
Department of Physics, D.G.Vaishnav College, Chennai, India

Thayumanavan Arunachalam
Department of Physics, A.V.V.M. Sri Pushpam College, Thanjavur, India

Khalid Siraj and Rafiuddin
Physical Chemistry Division, Department of Chemistry, Faculty of Sciences, Aligarh Muslim University, Aligarh, India

Soheil Sharifi
Department of Physics, University of Sistan and Baluchestan, Zahedan, Iran

Mahdi Ghasemifard and Misagh Ghamari
Nanotechnology Laboratory, Esfarayen University, Esfarayen, Iran

Meisam Daneshvar
Department of Physisc (Electroceramics Laboratory), Shahid Beheshti University, Tehran, Iran

P. Koteeswari and P. Mani
Department of Physics, Hindustan Institute of Technology, Padur, India

S. Suresh
Department of Physics, Loyola College, Chennai, India

Wladyslaw Zakowicz
Institute of Physics, Polish Academy of Sciences, Warsaw, Poland

Andrzej A. Skorupski and Eryk Infeld
Department of Theoretical Physics, National Centre for Nuclear Research, Warsaw, Poland

Pritam Kumar, Ajay Kumar Sharma and Narendra Kumar Singh
Department of Physics, V. K. S. University, Ara, India

Bhrigunandan Prasad Singh
Department of Physics, T. M. Bhagalpur University, Bhagalpur, India

Tripurari Prasad Sinha
Department of Physics, Bose Institute, Kolkata, India